Heavy Minerals

Heavy Minerals

Methods & Case Histories

Special Issue Editor

Sergio Andò

MDPI • Basel • Beijing • Wuhan • Barcelona • Belgrade • Manchester • Tokyo • Cluj • Tianjin

Special Issue Editor
Sergio Andò
University of Milano-Bicocca
Italy

Editorial Office
MDPI
St. Alban-Anlage 66
4052 Basel, Switzerland

This is a reprint of articles from the Special Issue published online in the open access journal *Minerals* (ISSN 2075-163X) (available at: https://www.mdpi.com/journal/minerals/special_issues/heavy_minerals).

For citation purposes, cite each article independently as indicated on the article page online and as indicated below:

LastName, A.A.; LastName, B.B.; LastName, C.C. Article Title. *Journal Name* **Year**, *Article Number*, Page Range.

ISBN 978-3-03936-130-4 (Hbk)
ISBN 978-3-03936-131-1 (PDF)

Cover image courtesy of Sergio Andò.
Heavy-mineral placer deposit with ephemeral sawtooth structures (width 15 cm).
Location: Beach of Punta Penna Grossa, Torre Guaceto Nature Reserve, Brindisi, Italy, coordinates 40°43′08″ N, 17°46′44″E.

Contents

About the Special Issue Editor

Sergio Andò graduated in Geology with a Ph.D. in Sedimentology and he is teaching as an Associate Professor of Physical Geography and Sedimentary Petrography at the University of Milano-Bicocca, Italy. He is a researcher with a strong and genuine passion for sharing and promoting heavy mineral studies. He is the host and organizer of an International School for the study and recognition of heavy minerals, with a style built on creating a dynamic interaction with his attendees. He has been a Guest Editor for two Special Issues in preparation by Earth Science Review and Minerals. He has been a co-supervisor for different Ph.D. Students and with formal assignment of teaching at qualified universities in Europe, USA, and China. In the last 20 years, he has studied heavy mineral assemblages in different geological settings and depositional environments, applying this knowledge to the field of provenance studies, paleoclimate reconstructions, and source-to-sink investigations, from passive margins in Africa to orogenic settings in the Alps and Himalaya. He is interested in sharing his recent studies dealing with the application of classical and innovative techniques in silt and sand to identify the mineralogy of sediments. He has grown an interest in the exploration of sedimentary records stored in turbiditic deposits in the Indus Fan and Bengal Fan and in coastal sediments of the South Africa and Mozambique margin. With deep curiosity, he loves to explore the differentiation of heavy mineral assemblages during transport and burial, developing new protocols for the laboratory, applying classical optical microscopy and innovative Raman spectroscopy. He has devoted his research merging different techniques to allow the proper identification of each single mineral encoding a piece of our geological history in the sedimentary archive. Raman spectroscopy combined with polarizing microscope represent a powerful tool in provenance studies to reconstruct the possible source rocks in modern and ancient sediments and give us a chance to link the mineralogy of sediment with tectonic and climatic overprinting through time. Innovative protocols created to analyze such fine sediments will be of interest for future studies of the sedimentary record and for petroleum and gas exploration.

An Interview with John Dewey: Maria Mange, the Mother of Heavy Mineral Analysis

Maria Mange was an internationally distinguished geologist and the foremost expert and leader on the use of detrital heavy minerals in sandstones for the study of provenance and unroofing of mountain belts, fine-scale correlation of rock sequences, the differentiation of heavy mineral assemblages during transport and burial, and associated chemical reactions in the deep subsurface and in forensics. She invented and developed the technique of high-resolution-heavy-mineral-analysis; her research was the most precise in the world and her advice and collaboration were widely sought by academia and industry.

Her work rate was enormous and labor-intensive, involving huge amount of laboratory work apart from all the basic preparatory fieldwork and the subsequent massive amount of microscopy, computer data analysis, synthesis, and the writing of papers. Her work was characterized by attention to detail, clever experimental design, seeking out important problems, global networking with the best people in her field, and a massive amount of dedication and sheer hard work in the field, laboratory and office, typically 14-hour days, seven days a week. She spent most of her time, including weekends, in her office and laboratory. She unselfishly worked on a wide variety of research projects including with me on the Caledonides of Scotland and Ireland, and Paleozoic terranes of the U.S. Cordillera, Sarah Roeske on Argentina and Alaska, with a University of Bern group on unroofing of the Alps, with a Louisiana group on the evolution of the Lower Mississippi River, and with a University of Vienna group on provenance of Roman amphorae in the Mediterranean region.

Throughout her career in Budapest, Berne, and Oxford, she gave excellent mini-courses on heavy minerals to academia and industry, and spent a large amount of time training graduate students in the techniques of heavy mineral analysis. She was in constant demand as a reviewer many articles for the top journals in her field and gave many invited and contributed lectures at national and international meetings.

She was a fine geologist who constantly advanced the frontiers of sedimentology, tectonics, palaeogeography, and stratigraphy. She was and will remain unique as a brilliant practitioner of the use of heavy minerals in geology, and a humble, kind and generous woman loved by all who knew her.

John Dewey
February 2020

minerals

Editorial

Editorial for Special Issue "Heavy Minerals"

Sergio Andò

Laboratory for Provenance Studies, Department of Earth and Environmental Sciences, University of Milano-Bicocca, Piazza della Scienza 4, 20126 Milan, Italy; sergio.ando@unimib.it

Received: 9 April 2020; Accepted: 14 April 2020; Published: 16 April 2020

This special volume, published 13 years after the monumental volume *"Heavy Minerals in Use"* edited by Maria Mange and David Wright, demonstrates that the use of heavy minerals as provenance tracers is alive and in full health. This collection of scientific contributions is born as an ideal continuation and update of that milestone of provenance studies given to us by Maria as a gift a few years before her untimely death. For this reason, this volume is opened by a testimony by John Dewey, one of the greatest protagonists of the plate tectonics revolution, who worked on many geological problems with Maria through some 25 years.

The methodological articles and case histories collected in this book aim at documenting the main developments of the discipline achieved during the last decade, which has seen the birth of new concepts, new technologies, and new applications in heavy-mineral studies.

The first part of the volume includes papers embracing general themes. To emphasize the importance of artisanal laboratory processes, which are the starting point of any mineralogical research, the opening article is a technical handbook by Sergio Andò (Laboratory for Provenance Studies, University of Milano-Bicocca) [1], conceived as a state-of-the-art simple practical guide for students of any level willing to tackle the challenges posed by mineral separation from rocks and loose sediments of any grain size from silt to sand. The second article by Eduardo Garzanti and Sergio Andò [2] illustrates the general criteria that should guide provenance studies based on heavy-mineral suites, from the problems posed by sampling in the field and the choice of the size window for analysis to the complex issues associated with data interpretation. These two articles condense experience gained through the last decades and aim at providing simple useful advice using a direct, and at places colloquial and entertaining language including anecdotes, which are difficult to find in traditional scientific literature forcibly characterized by a stern, more official attitude.

The articles that follow include a guide to data analysis using *provenance*, a software package within the statistical programming environment R, by Pieter Vermeesch (University College London) [3], one of the leading experts in the field of statistics applied to geological data. The aid of statistical techniques has become indispensable at a time when the flourishing of new technologies allows the collection of bigger and bigger datasets that make objective interpretation by visual inspection impossible. Next, the article by Keno Lünsdorf and colleagues of the University of Goettingen [4] presents the most updated report on the semi-automated identification of detrital minerals by Raman Spectroscopy, a user-friendly tool that allows targeting detrital minerals as small as a few microns in diameter, thus opening up a new frontier for provenance studies. The validity and robustness of Raman spectroscopy is demonstrated in the following study by Danilo Bersani and colleagues of the Universities of Parma and Milano-Bicocca [5] focusing on calcic amphiboles of the tremolite-actinolite series, and articles that show how this innovative technique allows a rapid and accurate identification and subtle discrimination within mineral series. Last but far from least, the first part is concluded by the article by Andrew Morton and Paula MacGill (HM Research and CASP Cambridge) [6], who summarize their pluri-decal experience on the application of heavy-mineral analysis for the correlation of hydrocarbon reservoirs and in general of strata barren of age-diagnostic fossils cored in oil boreholes.

The second part is dedicated to case histories and begins with the regional study by Wendong Liang and co-workers of the University of Milano-Bicocca [7], dealing with the geochemical characterization at the scanning electron microprobe of the four most common mineral groups in orogenic sediments (amphiboles, epidotes, garnets, and pyroxenes). The article focuses on the provenance of Thal Desert sand, an eolian dune field in central-northern Pakistan formed by Indus River sediments derived from the western Himalayan syntaxis at a time of dry climate and weak monsoon following the Last Glacial Maximum. The two companion papers that follow, by Laura Borromeo, Eduardo Garzanti, and co-workers including Christian France Lanord (CRPG, Nancy) and Annette Hahn (Marum Center, Bremen) [8,9], are dedicated to provenance analysis of silt and sand deposited on the Bengal shelf. These sediments represent the link between Ganga and Brahmaputra bedload and suspended load and Bengal Fan turbidites in the largest sedimentary system on Earth. Another two studies are dedicated to the two largest sedimentary systems of China, represented by the Changjiang River (Yang Tze) e Huang He (Yellow River). The first article, by Wei Yue and co-authors of the Jiangsu Normal University (Xuzhou), Tongji University (Shanghai), Binzhou University (Binzhou) and Ludong University (Yantai) [10], is a quantitative analysis of textural and compositional modifications of detrital sediments used to unravel the sediment provenance and environmental changes in the Changjiang basin. The second article, by Zhao Wang and co-workers of diverse universities in China, United States, and Italy [11], investigates the early evolution of the upper reaches of the Yellow River based on heavy-mineral and REE geochemical signatures recorded in the sequence of nine terraces formed during the progressive incision of the Yellow River in the last 1.7 Ma. The volume is closed by the article by João Cascalho [12], a regional study of heavy minerals from the Portuguese continental margin, that points to the existence of contrasting sources, namely felsic igneous and metamorphic rocks and igneous basic rocks next to dolomitic limestones affected by thermal metamorphism. This article also integrates issues presented originally in the *Heavy Mineral in Use* book edited by Maria Mange and David Wright.

Conflicts of Interest: The author declares no conflict of interest.

References

1. Andò, S. Gravimetric separation of heavy-minerals in sediments. *Minerals* **2020**, *10*, 273. [CrossRef]
2. Garzanti, S.; Andò, S. Heavy minerals for junior woodchucks. *Minerals* **2019**, *9*, 148. [CrossRef]
3. Vermeesch, P. Exploratory Analysis of Provenance Data Using R and the Provenance Package. *Minerals* **2019**, *9*, 193. [CrossRef]
4. Lünsdorf, N.K.; Kalies, J.; Ahlers, P.; Dunkl, I.; von Eynatten, H. Semi-automated heavy-mineral analysis by Raman spectroscopy. *Minerals* **2019**, *9*, 385. [CrossRef]
5. Bersani, D.; Andò, S.; Scrocco, L.; Gentile, P.; Salvioli-Mariani, E.; Fornasini, L.; Lottici, P.P. Composition of amphiboles in the tremolite–ferro–actinolite series by Raman spectroscopy. *Minerals* **2019**, *9*, 491. [CrossRef]
6. Morton, A.; McGill, P. Correlation of hydrocarbon reservoir sandstones using heavy mineral provenance signatures: Examples from the North Sea and adjacent areas. *Minerals* **2019**, *9*, 564. [CrossRef]
7. Liang, W.; Garzanti, E.; Andò, S.; Gentile, P.; Resentini, A. Multimineral Fingerprinting of Transhimalayan and Himalayan Sources of Indus-Derived Thal Desert Sand (Central Pakistan). *Minerals* **2019**, *9*, 457. [CrossRef]
8. Borromeo, L.; Andò, S.; France-Lanord, C.; Hahn, A.; Coletti, G.; Garzanti, E. Provenance of Bengal Shelf Sediments: 1. Mineralogy of Silt. *Minerals* **2019**, *9*, 640. [CrossRef]
9. Garzanti, E.; Andò, S.; France-Lanord, C.; Limonta, M.; Borromeo, L.; Vezzoli, G. Provenance of Bengal Shelf Sediments. 2. Petrology and geochemistry of sand. *Minerals* **2019**, *9*, 642. [CrossRef]
10. Yue, W.; Yue, X.; Panwar, S.; Zhang, L.; Jin, B. The Chemical Composition and Surface Texture of Transparent Heavy Minerals from Core LQ24 in the Changjiang Delta. *Minerals* **2019**, *9*, 454. [CrossRef]

11. Wang, Z.; Zhang, H.; Garzanti, E.; Nie, J.; Peng, W.; Andò, S.; Hu, X.; Pan, B.; Pfaff, K. Evolution of the Upper Yellow River as revealed by changes in heavy-mineral and geochemical (REE) signatures of fluvial terraces (Lanzhou, China). *Minerals* **2019**, *9*, 603. [CrossRef]
12. Cascalho, J. Provenance of Heavy Minerals: A Case Study from the WNW Portuguese Continental Margin. *Minerals* **2019**, *9*, 355. [CrossRef]

minerals

MDPI

Article

Gravimetric Separation of Heavy Minerals in Sediments and Rocks

Sergio Andò

Laboratory for Provenance Studies, Department of Earth and Environmental Sciences, University of Milano-Bicocca, Piazza della Scienza 4, 20126 Milan, Italy; sergio.ando@unimib.it

Received: 15 February 2020; Accepted: 17 March 2020; Published: 18 March 2020

Abstract: The potential of heavy minerals studies in provenance analysis can be enhanced conspicuously by using a state-of-the-art protocol for sample preparation in the laboratory, which represents the first fundamental step of any geological research. The classical method of gravimetric separation is based on the properties of detrital minerals, principally their grain size and density, and its efficiency depends on the procedure followed and on the technical skills of the operator. Heavy-mineral studies in the past have been traditionally focused on the sand fraction, generally choosing a narrow grain-size window for analysis, an approach that is bound to introduce a serious bias by neglecting a large, and sometimes very large, part of the heavy-mineral spectrum present in the sample. In order to minimize bias, not only the largest possible size range in each sample should be considered, but also, the same quantitative analytical methods should be applied to the largest possible grain-size range occurring in the sediment system down to 5 µm or less, thus including suspended load in rivers, loess deposits, and shallow to deep-marine muds. Wherever the bulk sample cannot be used for practical reasons, we need to routinely analyze the medium silt to medium sand range (15–500 µm) for sand and the fine silt to sand range (5–63 or > 63 µm) for silt. This article is conceived as a practical handbook dedicated specifically to Master and PhD students at the beginning of their heavy-mineral apprenticeship, as to more expert operators from the industry and academy to help improving the quality of heavy-mineral separation for any possible field of application.

Keywords: handbook for laboratory procedures; nontoxic heavy liquids; wet sieving of silt; size-window for analysis; zircon separation; heavy-mineral mounts; provenance analysis

---

"Day by day, what you choose, what you think, and what you do is who you become"

Heraclitus

1. Introduction

A detailed and efficient protocol for heavy-mineral separation from sediments and rocks represents a handy practical support tool designed to be consulted routinely in the laboratory. Continuing a tradition started in Italy more than a century ago by De Filippi (1839) [1], who performed a pioneering quantitative provenance study of heavy minerals in sediments of the Ticino River and Artini (1891) [2], who investigated mineralogy of the Po plain sediments, this little handbook will enable Master and PhD students to use it as a substantial help to carry out autonomously a quick and effective heavy-mineral separation, but also, more expert researchers will discover simple practical solutions to speed the separation procedure.

As in Garzanti and Andò (2019) [3], the definition of heavy minerals used in this article includes only minerals of certain extrabasinal terrigenous origin (i.e., ultimately eroded from bedrock exposed in source areas), denser than 2.90 g/cm^3, and occurring either as single detrital grains or in rock fragments. Grains of suspect intrabasinal (e.g., carbonates, bioclasts, and glaucony); pedogenic or diagenetic (e.g., aggregates of iron or titanium oxides); and anthropogenic origin (e.g., barite in core samples and moissanite) are thus neglected. Phyllosilicates are neglected as well. Transparent heavy minerals identified under the microscope are considered separately from opaque and altered heavy minerals.

The study of heavy minerals, which was quite popular in the past—for a historical overview, the reader is referred to Carver, 1971 [4] and to the monumental book edited by Mange and Wright, 2007 [5])—has recently seen a renovated and increasing interest also in source-to-sink studies aimed at the search for hydrocarbon reservoirs (e.g., Morton and McGill, 2018 [6]). Laboratory practice for sample treatment and heavy-mineral separation was thoroughly investigated and illustrated a half century ago (e.g., [7–10]), but, later on, the methodology has seen a sort of standardization, and not only for the best, and the stimulus for the betterment of laboratory procedures has seen little progress since then. An updated recent publication in which the state-of-the art techniques for heavy-mineral separation are thoroughly explained through the entire process, from sampling to mineralogical analysis, thus does not exist to the best of my knowledge. The purpose of this article is to fill this gap, especially as the study of recent unconsolidated silts and sands is concerned. Only a limited attention would be dedicated instead to lithified sandstones, and chemical methods for disaggregation will not be illustrated here. We strongly suggest to minimize the use of chemicals in the lab to a very minimum and especially to avoid the use of carcinogenic organic compounds such as bromoform, which has represented for a century the standard dense liquid used in mineral separation and which is still unfortunately in use in several laboratories worldwide. Other standard procedures include chemical attack with oxalic, acetic, or even chloridric acids, but such attacks are in most cases not only scarcely helpful but also counterproductive. Besides incremental time and cost, every additional passage involves material loss during cleaning, precipitation of insoluble salts, and even loss of key provenance indicators such as olivine or apatite. Another critical aspect that we emphasize here is first and foremost the necessary care taken in the field to collect pristine samples following criteria apt to guarantee both consistency and representativeness.

2. Sampling

No study can be better than the samples collected in the field. Great attention should thus be dedicated to the sampling plan, concerning locations, sampling spacing, and representativity of the targeted sedimentary system [3]. In the case of modern sediments, it is vital to avoid the detrimental practice of panning, which concentrates the densest species and consequently modifies irreparably the original proportions among different detrital minerals.

Once the sample is collected it is important to adopt a good systematic practice in sample labelling, in order to simplify laboratory procedures and avoid bad mistakes. For instance, in the case of modern sediment samples, we simply label them in progressive numerical order specifying the name of a nearby site and, in case of fluvial sediments, the name of the river.

During the successive steps of laboratory protocol, it is wise to add a series of simple coded symbols to readily identify the separated grain-size or density fraction (e.g., 15–500 μm, L = light, <2.90 g/cm^3 and H = heavy, >2.90 g/cm^3). Color-coded dots can be used to designate sediment fractions separated for specific petrographic, mineralogical, or geochemical analyses.

3. Safety Rules and Pre-treatments

This section illustrates the practical duties that need to be carried out before starting to process a series of sediment samples. Specific care should be dedicated to health and safety issues following good practices in the lab and especially avoiding wherever possible the use of toxic chemicals. The laboratory for mineral separation should be clean, safe, and well-organized. The use of blotting paper before starting any procedure is advisable. Reading carefully safety data sheets for each chemical product used (e.g., acids, dense liquids, etc.) is essential.

NB*: Never ever use bromoform. This organic liquid is toxic and carcinogenic!*

3.1. Preparation of Na-Polytungstate

Sodium-polytungstate (SPT; salt formula Na_6 $(H_2W_{12}O_{40})\cdot H_2O$) is a more suitable heavy liquid, perfectly soluble in water and widely used for heavy-mineral separations. The use of SPT [11] combined with the centrifuge [12,13] has replaced the traditional but very dangerous use of bromoform, also using a much smaller amount of dense liquid and thus saving both costs and time during the procedure. Solutions with density up to 3.15 g/cm^3 can be prepared with SPT, which may be helpful to concentrate mineral of interest for geochronological and thermochronological analysis (see Section 6.2 below). Lower densities can be simply achieved by adding deionized or distilled water and higher densities by evaporation in a fume hood. This versatility allows us to separate isodensimetric fractions to concentrate specific minerals (e.g., quartz or K-feldspar for cosmonuclide or optically stimulated luminescence analysis; [14]).

One liter of solution with density 2.90 g/cm^3 is obtained from 2420 g of SPT and 478 g of deionized or distilled water poured in a 5-L beaker. Add the SPT in progressive steps, a few grams at a time. Put the beaker on a magnetic stirrer with an anchor for an hour until a homogeneous solution, transparent and pale yellow in color, is obtained. Stop the stirrer and check the density with a lead densimeter with density range 2.50–3.00 g/cm^3, in a cylinder with a 4-cm diameter, adding 250 mL of SPT solution.

3.2. Preparation of Na–Dithionite–Citrate–Bicarbonate

The presence of iron oxides and hydroxides (e.g., hematite, lepidocrocite, and goethite) as coatings on single minerals or rock fragments may modify their density and hamper their proper identification under the microscope. This problem needs to be faced while studying surficial textures of weathered minerals in deeply weathered tropical and equatorial soils and paleosols. As a most effective way for iron-oxide removal from clays, the use of *Na–Dithionite–Citrate–Bicarbonate* (DCB) was proposed by Mehra and Jackson (1958) [15]. This mildly acid solution does not corrode apatite, monazite, or olivine and can thus be used for treatment of sedimentary samples.

The following protocol, derived from procedures used at the Natural History Museum of Milano, is recommended. Take a 5-L beaker with 2 L of deionized water and put the beaker on a magnetic stirrer at maximum speed. Add 120 g of Na-citrate, 40 g of Na-ditionite, and 16 g of Na-bicarbonate. After ca 15 min, when salts are all in solution, the DCB is ready. The DCB is poured in a 500-mL labeled beaker containing the sediment until the sediment is submerged; the rest can be stored for future use. Place the beaker under a fume hood and let the reaction to go on for 12 h at least at room temperature; after which, the sampled is cleaned with abundant (1–2 L) tap water to eliminate acid residues and finally wash with deionized water. This procedure can be used also on rock samples and large crystals and is most effective if followed by energic cleaning with universal degreaser and a brush.

3.3. Preparation of Nylon Sieves with 5 μm and 15 μm Mesh

Steel sieves, which can be cleaned in ultrasonic bath, are commercially available in the market down to 32 (or 20) μm. Since in sediments deposited by tractive currents the finest tail of the size distribution is markedly enriched in ultra-dense minerals (e.g., monazite, magnetite, and zircon; [16]), in very-coarse silt to very-fine sand samples, it is crucial to consider the finest size classes of the sample as well in order not to obtain a biased heavy-mineral suite [17,18]. This holds true also for poorly sorted sediments and especially for cohesive muds and mudrocks, for which including even classes as fine as 5–15 μm is compulsory. For this purpose, specific tools must be prepared.

Currently, in our lab, we use handmade tissue sieves with 15 μm, 10 μm, and 5 μm mesh. Nylon mesh rolls that are commercially available are cut in 10-cm-wide strips and then cut in turn in 10 × 10-cm-square pieces that are stored in a clean plastic bag on which the mesh size is clearly indicated. A one-square-piece is mounted on a PVC or plexiglass ring with a diameter of 8 cm obtained by cutting a PVC or plexiglass gutter pipe in 10-cm-high pieces. Both basal and top surfaces of the ring must be made flat and smooth by using sandpaper and then carefully washed.

The tissue sieve is glued well-centered on the surface of the ring using a nontoxic and nonrapid glue spread thinly and evenly all around the rim of the ring. Next, to seal the tissue sieve, a few drops of glue are spread all around the rim with a toothpick, and the handmade tool is dried in a clean oven at ca. 40 °C for several hours. Finally, the tissue in excess is cut with scissors (Figure 1).

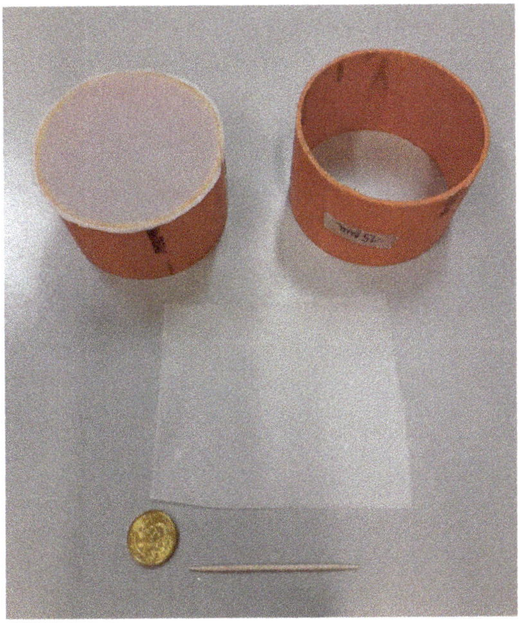

Figure 1. Tissue sieve made with nylon mesh. From bottom to top: toothpick and 100-cm^2-piece of nylon mesh; PVC ring and tissue sieve ready for use.

NB: Carefully wash the PVC ring with tap water to eliminate residues of abrasive grains used in sandpaper (e.g., moissanite but sometimes also garnet) that may contaminate the heavy-mineral fraction.

3.4. Sample Drying

The sample may reach the lab in diverse conditions and packages. In case of loose sediment, in order to reduce loss or contamination risks the sample is best laid on a large sheet of clean paper. If the sample is wet, then it may be placed in an aluminum tray and left overnight in the oven at a temperature not exceeding 40–60 °C.

4. Let Us Start!

4.1. Rock and Consolidated Samples

The initial laboratory procedures depend on sample type. In the case of hard rocks, the first step is to mechanically disaggregate the sample minimizing pulverization. Rocks are split with a manual or hydraulic press in small centimetric chips (Figure 2). Between 20 and 40 g are generally sufficient to obtain the desired amount of heavy minerals, and the rest of the chips can be archived. The part of the sample collected in a tray is weighed and placed in an agate mortar with 10–20 mL of deionized water to prevent loss of material and the production of powder, which is harmful to breathe. A pressure is exerted repetitively with the agate pestle, avoiding rotational movements that may lead to grain grinding. Gentle percussions have been experimentally demonstrated to cause negligible breakage of heavy minerals (Henningsen, 1967 [19] and Mange and Maurer, 1992 [20] (p. 11)).

In the case of indurated silt or sand, it may suffice to place the weighed sample in a glass beaker, to add 100 mL of deionized water and to stir for several minutes with a metal spatula until complete disaggregation. The obtained sediment suspension can be poured onto the 500-μm mesh sieve as illustrated in Section 5.1 below.

Figure 2. Manual press designed to split sedimentary rocks using protective eye goggles.

4.2. Micro-Sampling of Loose Sediment

The representativity of the sample is granted not only by careful collection in the field but in the laboratory as well, by following suitable procedures that largely depend on sample grain size. In the simplest case of clean sand, the sample can be split repeatedly in two parts by means of a riffle box (Figure 3). The operation is repeated until the desired amount is obtained, weighed, and placed in a suitable labeled plastic vial. The discarded quantity is archived in the original container. More complex is the case of muddy samples, which are dried, placed on a clean sheet of paper, and well-homogenized horizontally by rolling the paper until a cone of sediment is obtained. Following Parfenoff, 1970 [21] (pp. 45–47), the cone is divided into four parts; a quarter of it is taken, and the rest is stored. The procedure is repeated until an appropriate quantity is obtained and transferred with a suitable tool (Even a bus ticket would do, but do not use electrostatic plastic!) on another clean sheet of paper and weighed. The same Parfenoff method is used in the case of gravelly sand.

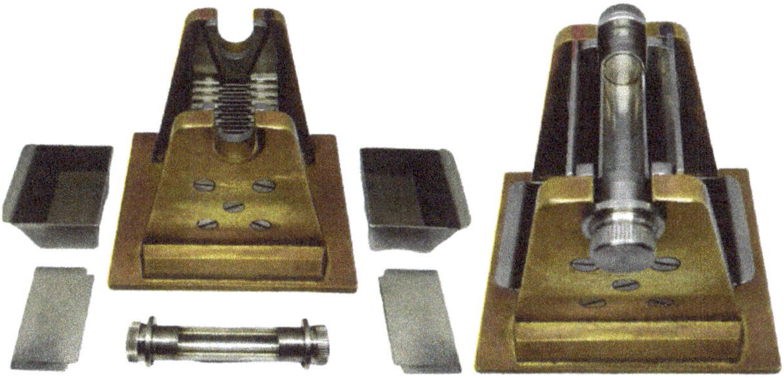

Figure 3. Riffle box designed to split sediments (length 10 cm, width 8 cm, and height 8 cm).

4.3. *How Much Do We Need?*

The amount of sample needed for heavy-mineral separation depends on several factors, including expected heavy-mineral concentration, grain size, sorting, and type of study. Heavy-mineral concentration ranges widely from typically << 1% in case of ancient sandstone or quartzose modern sand heavily weathered in equatorial environments or recycled from ancient quartzarenites (e.g., [22]) to 5%–10% in modern orogenic sediments, reaching up to >50% in placer sands (e.g., [23]).

***NB**: For modern sands, heavy-mineral concentration can be expeditiously assessed by weighing a well-filled plastic vial of known volume.*

As a general standard, we use up to 50–60 g of very coarse sand and 30–40 g of coarse sand, whereas 5–15 g are sufficient for medium sand, 2–3 g for fine to very fine sand, and even only 1 g for silt. For clayey silt and silty clay, it is advisable to take about 5 g.

A semi-quantitative operational approach can be assisted by a simple formula that takes into account not only grain size but also the age of the sample (a rough proxy of the intensity of selective diagenetic dissolution), the average density of expected source rocks, and the degree of heavy-mineral enrichment (as judged empirically by sample color and weight). The required amount of sample *X* is thus given by: $\mathbf{X = t \cdot \rho R \cdot S \cdot H}$, where **t** ranges between 1 for Holocene samples to 10 for Mesozoic or older samples; ρ**R** ranges between 1 for medium/high-grade source rocks to 10 for sedimentary source rocks; **S** is 0.5 for silt; 1 is for fine sand, 2 for medium sand, 3 for coarse sand, 10 for very coarse sand, and 30 for pebbly sand; and **H** ranges between 0.5 for blackish-reddish placer lags to 5 for white antiplacer sand.

***NB**: Playing with examples:*

Silt from a collision orogen unaffected by hydraulic sorting: $t = 1$, $\rho R = 2$, $S = 0.5$, and $H = 1 \rightarrow \mathbf{X = 1g}$

Modern sand from granitoid rocks: $t = 1$, $\rho R = 6$, $S = 2$, and $H = 4 \rightarrow \mathbf{X = 48g}$

Gravelly sand from a magmatic arc: $t = 1$, $\rho R = 2$, $S = 30$, and $H = 1 \rightarrow \mathbf{X = 60g}$

Mesozoic fine-grained sandstone from a cratonic basement: $t = 10$, $\rho R = 10$, $S = 1$, and $H = 1 \rightarrow \mathbf{X = 100g}$

4.4. *Do We Really Need Large Samples for U-Pb Zircon Dating?*

In the last decade, detrital-zircon geochronology has become one of the most popular approach to provenance analysis. In order to be sure to extract a sufficient amount of zircon grains and to minimize logistic problems and costs, researchers may think of preconcentrating heavy minerals in the field with expeditious techniques such as panning. Panning, however, concentrates denser grains at the expense of less dense grains and therefore introduces bias, for instance because metamictic zircon is less dense than crystalline zircon (e.g., [24]). Is the only correct alternative to collect very big samples (many kilograms; e.g., [25])?

Considering the hydraulic-equivalence principle that tells us that zircon grains in water-laid fine to medium sand are a 0.6–0.7ϕ class finer than the bulk sample [16,26], the appropriate grain size of the sample to be collected is assessed readily and even precisely by taking advantage of ad hoc software [27]. The volume of sand that needs to be collected in the field may thus be reduced by orders of magnitude. By adopting a correct protocol in the lab, that may be applied even for coarse silt, even small sediment samples (2–10 g) can yield a sufficient amount of datable zircon grains. In this way, a series of problems are circumvented, not only as regards field logistics and shipping costs, but also, mineral separation in the laboratory becomes more rapid and much less cumbersome. This is particularly valuable when only a few grams of material are available (e.g., deep-sea cores).

After following the procedure illustrated here below, without doing any further preconcentration by toxic liquids such as methylene iodide, or by Wilfley table, Franz magnetometer, or hand-picking that inevitably introduce a grain-selection bias, the heavy-mineral separate is mounted in araldite, and each grain in the mount is located and identified by Cartesian coordinates under either the Raman spectroscope or the scanning electron microscope [28]. Fixing scanning electron microscope grids or other easily identified locator grids to the mount is useful and offers a simple solution for identifying the exact position of points of interest within a sample, where grains can then be located by reference to these.

5. Sieving Procedures

NB*: The worst mistakes that may occur in the lab are accidental sample switch or sample contamination. To stay on the safe side, samples are best processed in numerical order. In case of a modern fluvial system, you may proceed from upstream to downstream.*

5.1. Wet Sieving with Steel Sieves

Sediment samples may contain a mixture of clay, silt, sand, and gravel and thus need to be sieved before heavy-mineral separation. In poorly sorted sediments, the presence of detrital grains with great size differences within a single concentrate makes mounting and identification difficult [20]. For these reasons, we routinely consider a size window as large as possible, including as many as 5ϕ classes (15–500 μm; see [3] for a detailed discussion of this crucial issue).

Wet sieving is particularly recommended for poorly sorted sediments including mud. The weighted aliquot of the original sample is placed in a glass beaker with 100 mL of deionized water to facilitate disaggregation of particles. For recent sediments, a gentle pressure with a spatula or spoon will be enough. In order to extract heavy minerals from the 15–500-μm-size window, then a 500-μm sieve is cleaned and mounted onto a sieve bottom pan with drainage outlet. This is useful to collect in another beaker the passing < 500-μm sediment suspension.

NB*: During this process, dense minerals may concentrate and remain in the sieve bottom pan. The bias introduced by an inaccurate recovery from the beaker may be serious, especially for heavy-mineral poor samples and for the densest heavy minerals (Figure 4).*

In order to recover the entire amount of the <500-μm fraction, the sieve bottom pan must be cleaned several times with tap water until no grains are detected on the sieve bottom pan.

NB*: Since we choose to avoid chemical attack with hydrogen peroxide, samples very rich in plant debris pose a problem; their mechanical comminution during sieving produces a flux of particles that pass through the steel meshes. The weight of the passing fraction is thus somewhat overestimated and the heavy-mineral concentration slightly underestimated. Repeated washing beyond 3–5 times is therefore not recommended.*

The <500-μm sediment suspension is left to settle in the beaker overnight. After sand and silt have settled, most of the water is siphoned off with a little tube, taking advantage of the principle of communicating vessels. Some water is left in the beaker for the subsequent sieving step.

> ***NB***: *Sediment in suspension can be recovered from the beaker for further investigations of clay mineralogy and chemistry or organic matter.*

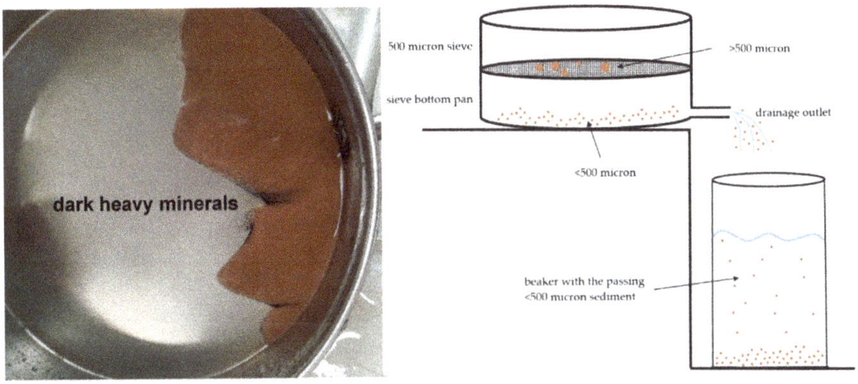

Figure 4. Dense minerals concentrate in the sieve bottom pan during wet sieving with a sketch of the apparatus described in Section 5.1.

5.2. Wet Sieving with Nylon Sieves

A handmade sieve with 15-μm, 10-μm or 5-μm mesh (prepared as in Section 3.3) is labeled by writing the sample name and grain-size range (e.g., 15–500 μm) on adhesive paper. The sieve is placed in a funnel sustained by a metal rack, centered above a 2l beaker labeled by writing the sample name and grain-size range (e.g., <15 μm) on adhesive paper. A supplementary beaker is prepared in the same way. A little water and sediment are poured in the sieve, exerting a gentle pressure with fingers using gloves.

> ***NB***: *A light pressure and continuous stirring are necessary to speed up the sieving process and avoid that the clay is deposited, thus clogging the meshes. Excess pressure, however, can break the nylon sieve.*

The process needs to be repeated over and over, adding a small quantity of sediment each time to avoid clogging of the meshes. After each step, the sieve is placed in the empty supplementary beaker and moved up and down to let the water pass through and clean the net.

> ***NB***: *The use of ultrasonic bath is not recommended, as it might lead to the deformation of nylon meshes, thus jeopardizing the quality of the sieving process.*

At the end of this multistep procedure, which may take between tens of minutes to several hours, the cleaned 15–500-μm-class thus obtained is transferred into a labeled paper filter and dried in an oven overnight at 40–60 °C. The tissue is then removed with a knife from the handmade sieve and disposed.

The passing <15-μm fraction collected in the beaker is left to settle for up to a full day. Once settled, the excess water is siphoned off. The sediment is transferred in a smaller beaker (e.g., 500 mL) by pipetting deionized water and left again to settle. The day after, the excess water is siphoned off, and the small beaker with sediment is placed in the oven. Once dried, the <15-μm fraction is weighed and archived.

NB: While recovering sediment from the paper filter, be careful and avoid exerting an excessive pressure with your fingers that may generate paper fibers in the sample. In this event, the fibers can be easily removed with an electrostatically charged tip. Successively, whenever the sediment is poured on paper for weighing or splitting, the use of transparent paper is recommended to avoid undesired electrostatic effects and material loss which may be significant for silt samples and platy minerals.

Since this procedure takes three days, it is convenient to process batches of 4–5 samples and possibly to work in pairs to optimize efficiency.

5.3. Clay-rich Samples: Settling Tubes and Wet Sieving at 5 µm

Clay-rich samples cannot be processed by wet-sieving alone in a reasonable amount of time, because the nylon meshes are quickly clogged. In these cases, a hybrid procedure is recommended to extract the silt fraction, which may be essential in heavy-mineral studies of soils, suspended load, eolian dust, lagoonal or offshore muds, or deep-sea turbidites [29–34].

The weighted sample is passed through a 63-µm steel sieve mounted onto a sieve bottom pan with drainage outlet. Sand-size grains, commonly including micas, bioclasts or plant debris, are thus separated and weighed. The passing <63-µm fraction, collected in a glass beaker, is poured in a 40-cm-high graduated cylinder with a 6-cm diameter and 1000-mL volume filled up with filtered tap water. After adding 1 g of sodium hexaphosphate, the suspension is homogenized for 1 min with a 55-cm-high PVC or steel stirrer with a 5-cm diameter, thus preventing flocculation. After 20 min, silt particles are settled, and the dirty water in excess of 100 mL containing clay (<2 µm fraction) is siphoned off and left in a 5-L beaker for days. It is recommended to siphon water off up to the 100 mL line of the cylinder in order to avoid siphoning off any silt. The procedure is repeated up to five times, adding filtered tap water and Na-hexaphosphate each time.

The silt fraction thus obtained is carefully pipetted out of the cylinder in a glass beaker and next wet-sieved with handmade sieves at 5 µm following the procedure illustrated in Section 5.2. The cleaned silt fraction (5–63 µm) thus obtained is dried in an oven at 40–60 °C and weighed. The passing (<5 µm) very fine silt class is left to settle in the beaker. Heavy minerals can be separated from very fine silt, but they are too small to be recovered with paper filter. The clay fraction can be also recovered and weighed.

6. Gravimetric Separation by Centrifugation

The use of the centrifuge has become a most practical method of heavy-mineral separation since the development of the partial freezing method [4,20]. Nevertheless, the use of funnels is still largely adopted, even though it involves larger volumes of the dense liquid, longer times for recovery, low replicability of results because of operator bias, and lower quality of the heavy-mineral separation. Moreover, the funnel method is strongly discouraged for silt-sized sediments for a variety of reasons. The use of the centrifuge with 50-mL conic test tubes combined with partial freezing in liquid nitrogen represents a better procedure by far, allowing us to obtain a very sharp separation of heavy minerals from tectosilicates and other low-density particles that are common in sediments.

NB: The conic tubes must have externally printed graduation; otherwise, grains may be trapped by the internal graduation steps.

6.1. Ordinary Separation with SPT at 2.90 g/cm^3

Once dried, the sieved grain-size fraction is ready for gravimetric separation with sodium politungstate (SPT) at density 2.90 g/cm^3. If a common centrifuge with four positions is used, then it is convenient to treat four samples at a time. Each conic tube is labeled with the sample name written both on the lid and on the sides. One by one, the sediment samples are poured in the appropriate test tube, which is inserted in a holder placed on a large piece of a thoroughly cleaned paper. The tubes are progressively stored in a plastic tube rack possibly orderly in increasing numerical order. Next, a tube holder is placed on a precision balance and, shifting the tare to zero, each test tube is weighed, adding SPT to the sediment until the total weight reaches 100 ± 0.1 g. It is advisable to add drop by drop the last grams of SPT using a pipette.

The operative conditions of the centrifuge are set according to the chosen grain-size window. If a wide-size window was chosen (e.g., 5–500 μm or 15–500 μm), then 2500–3000 revolutions per minute (rpm) and 2–3 min are appropriate.

NB: *Do note ever press the conic tubes in the holder of the centrifuge. Always use a balanced setting with tubes having the same weight to avoid undesired vibration leading to automatic blockage.*

The test tubes are firmly closed and shaken until the all sediment is in suspension. The test tubes are placed in the centrifuge either filling all positions or in a balanced configuration, the cover of the centrifuge is locked, and the engine started. Once the centrifuge automatically stops and the cover is unlocked, the test tubes are collected and placed vertically in a plastic rank where they are left for 15–30 min. This time interval is necessary to complete in static conditions the sharp separation of slow-settling platy phyllosilicates and other grains with density close to 2.90 g/cm^3.

NB: *Instead of liquid nitrogen, you may use a freezing mixture consisting of ice and salt at temperatures −11/−14 °C, although the procedure results to be laborious and time-consuming.*

Dry ice is also effective, cheaper, and easier to handle than liquid nitrogen. Freezing of SPT is achieved within ca. *5 min where a centrifuge tube is half-embedded in dry ice within an insulated, lidded container. A further advantage is the slow sublimation of dry ice; provided the container is insulated, 1 liter of dry ice will persist for at least 12 h at 18 °C and can be used repeatedly during a separation session.*

To retrieve the separated dense fractions, the test tubes are partially frozen by using liquid nitrogen. Safety reasons advise to wear a lab coat and protective goggles and gloves. The test tubes are placed in a plastic container insulated by polystyrene that can be obtained by cutting the top of a 1l plastic bottle. Liquid nitrogen at −196 °C is poured in the container to a height of 2 cm. The liquid nitrogen rapidly evaporates and thus needs to be continuously refilled until the container is frozen. The test tubes are placed in the container and fixed in vertical position with a piece of paper. After a couple of minutes, the lower part of the test tube containing the dense fraction is frozen, whereas the upper part containing the low-density fraction remains liquid (Figure 5).

Such a partial freezing allows to recover the liquid suspension containing the low-density fraction in a folded and properly labeled paper filter by pipetting deionized water with care. Washing with 0.5–1 L deionized water dilutes the liquid, speeding up the filtering process, and avoids the clogging of grains with SPT salt after drying. The diluted SPT is recovered in glass beakers.

NB: *Do not touch the interior of the conic tube with the tip of the wash bottle to avoid picking up grains and preventing the contamination of successive samples.*

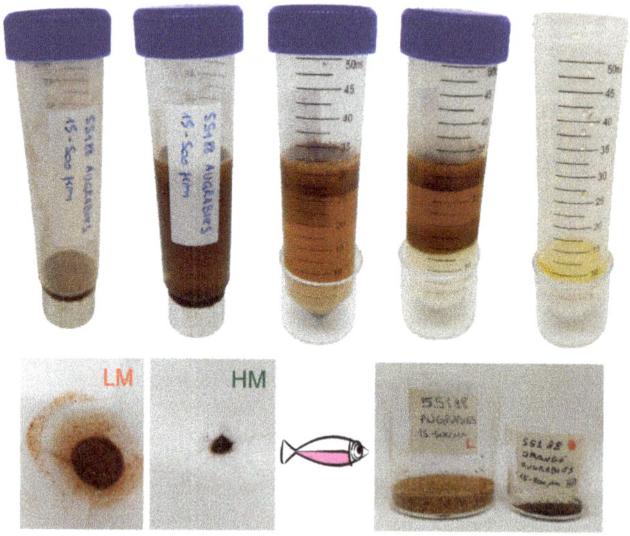

Figure 5. Gravimetric separation sequence with SPT and conic tubes, to separate the dense and light fractions in a centrifuge. From left to right in the first line: initial dry sample, sample mixed with SPT after shaking, the light minerals (LM) and the heavy minerals (HM) separated by SPT after the centrifugation, sample with frost-heavy fraction in the bottom and unfrozen light fraction in the top of the conic tube, and empty conic tube after recovering of the light fraction with the heavy minerals still in the ice. Second line: washed light minerals in the paper filter, washed dark heavy minerals in the paper filter, and light and heavy minerals dried and stored and labeled in plastic boxes after weighing.

The lower part of the test tube is defrosted by adding deionized water, and the dense fraction is recovered in another folded, properly labelled paper filter, washed with 0.3–0.5 L deionized water (more where dense grains are particularly abundant). The diluted SPT is recovered in glass beakers.

For each sample, the two paper filters containing the low-density and dense fractions are folded, closed with a peg, and placed to dry in oven at 40–60 °C overnight. The dried fractions are both weighed with a precision balance with four decimals, and weights are noted on an appropriate data sheet (Table 1).

Table 1. Laboratory data sheet for sieving medium silt to medium sand range (15–500 μm). Percentage of dense (>2.90 g/cm^3) fraction separated in the lab (%HM); grams (g).

Sample Name		g				
	Initial dry weight					
	>500 μm	g used	g Light Minerals	g Heavy Minerals	%HM	Mount ready
	15–500 μm					
	<15 μm					
	Total weight					
Splitter with Parfenoff	○	Dry sieving	○	Settling tubes	○	
Micro-Splitter	○	Wet sieving	○			

6.2. *Separation with Nontoxic SPT at 3.15 g/cm³*

In the separation of minerals commonly used in geochronological and thermochronological analysis such as zircon, apatite, monazite, rutile, or titanite, a valid alternative to using a toxic carcinogenic organic liquid such as methylene iodide is represented by SPT liquid with density reaching up to 3.15 g/cm^3 using goggles/visor when handling SPT, as it can precipitate on the retina if it gets into the eyes. This is obtained by slowly evaporating under a fume hood the SPT liquid, starting from 2.90 g/cm^3, on a magnetic stirrer at room temperature 20–25 °C (not on the hot plate to avoid rapid evaporation and crystallization of SPT). Density is checked periodically with a lead densimeter including the density range 3.00–4.00 g/cm^3. Good results are achieved up to 3.15 g/cm^3, whereas obtaining an even denser liquid at room temperature is difficult because of incipient crystallization of SPT turning the solution turbid.

6.3. *Recovery of Diluted SPT*

The diluted SPT collected in glass beakers is evaporated on the hot plate under a fume hood at 100 °C. To enhance aspiration and speed up the evaporation process, it is advisable to nearly close the frontal screen of the hood, leaving a space of ca. 1 cm. To obtain a clean heavy liquid, while evaporating the SPT solution on the hot plate, it is necessary to filter the solution with double-piled paper filters when a density of ≤2.5 g/cm^3 is reached in order to remove impurities, tiny SPT crystals formed during evaporation, or mineral grains left over by inaccurate previous operators.

When the SPT solution's density approaches 2.90 g/cm^3, the density is checked with a lead densimeter in a cylinder with a 4-cm diameter adding 250 mL of SPT solution. If the density is too low, then the SPT solution is placed again in a beaker under the fume hood to slowly evaporate at room temperature. If the density is too high, then the SPT is placed again in the 2-L beaker under the fume hood and on a magnetic stirrer, and a few drops of deionized water are added with a 5-mL dropper.

If the procedure is followed carefully, the used SPT can be recycled and used for years, thus reducing costs and environmental pollution.

7. Preparation of Grain Mounts

Canada Balsam

The recovered amount of the heavy fraction is generally much greater than what is strictly needed to prepare a grain mount. In order to obtain a representative aliquot of the heavy fraction, the latter needs to be repeatedly split into parts by using a micro-riffle box with 2-mm-wide slots and a 4-cm-long holder (Figure 3). The use of the Parfenoff cone or, worse, of a spoon or spatula is not recommended, especially when particles with different grain sizes and shapes occur in the sample.

> *NB*: *The minimum quantity needed for the grain mount is 3–5 mg. It is wise to keep in the archive the remaining part of the heavy-mineral fraction for replicate analyses and future investigations.*

The split aliquot is temporarily poured out on a small (e.g., 7 × 5 cm) piece of clean white paper folded and refolded to prevent contamination and appropriately labeled. In the meanwhile, a digital hot plate covered by a clean aluminum foil is warmed up to ~130 °C. I advise to place between three and five glass slides (76 × 24 mm), carefully cleaned with paper and denatured alcohol to remove any trace of haloes and dust, on the upper left part of the hot plate. In the same way, an equal number of 22 × 22-mm cover slips are cleaned and warmed up. A glass slide and a cover slip are gently pushed with a toothpick to the lower right corner of the hot plate. Next, a drop of Canada balsam ($n = 1.54$) with the appropriate viscosity (not all brands perform in the best way) is picked up with a spatula and placed on the cover slip.

> *NB*: *The heavy-mineral mount and the cover slip should be positioned in an asymmetric position, approximately in the middle of the left half of the slide. This enables using the entire area of the mount during point counting.*

After a couple of minutes, the Canada balsam is warm enough. The refolded paper containing the split heavy fraction is opened and used to pour out carefully all grains within the Canada balsam. To prepare a single layer of grains avoiding overlaps, we suggest spreading them around evenly with a clean toothpick conveniently held inclined by ca. 45° and rotating it quickly, avoiding to concentrate the grains close to the edges of the glass slide. With another toothpick, the cover slip is moved off the edge of the hot plate and picked up between fingers holding it by the sides to avoid burns. To avoid formation of air bubbles, the cover slip is overturned, placed tilted by ca. 45° by the edge of the glass slide, and let lying well-centered on the melted Canada balsam.

NB: The hot plate should be very slightly inclined toward the back of the fume hood to facilitate spreading the balsam and expelling all air bubbles.

After a couple of minutes necessary to polymerize the hot Canada balsam and let the air out, a light pressure is applied with a toothpick to minimize the thickness of the grain mount. This precaution is necessary, because thick grain mounts may not be observed with objectives with large magnification (e.g., 63×). If the amount of Canada balsam is insufficient to fill the space among grains —which may happen with bimodal mounts, including medium-sand particles—an additional drop of bounding resin picked up with a spatula can be added with a clean toothpick from the edge of the cover slip. The drop will melt soon and, thanks to the gentle tilt of the hot plate, will gradually flow and fill up the voids among the grains. The mount is now ready. You can rotate it with the toothpick to the edge of the hot plate, pick it up with a tong, and immerge in a small beaker filled with tap water to quench it. Finally, the mount is cleaned with alcohol and paper to eliminate all balsam residues, labeled with the proper sample name, and archived in slide holders.

8. Conclusions

This article, born from the experience gained in twenty years of heavy-mineral studies, illustrates in detail practical solutions apt to optimize workflow and avoid common mistakes. Our goal is to fill a gap in technical knowledge opened in the last half-century—a period during which both provenance-analysis techniques and sensitivity to environmental safety issues have rapidly increased—thus responding to the growing needs of students, researchers, and professionals from the industry. We indicate how, with handmade tools and a clever protocol design, mineral separations can be performed, minimizing time and cost on one side and improving quality of the results on the other. The illustrated procedure allows a skilled operator to obtain clean fractions of any grain size and to separate minerals not only from sand but also from finer and finest sediments, thus enabling grain by a grain analysis of particles down to a size of few microns. High-resolution quantitative provenance analysis of the entire grain-size spectrum thus becomes possible, targeting soils, fluvial suspended loads, eolian dust, and muds deposited in any environment, from the land to the deep sea. Such a performance, however, requires to make the correct decisions, from the first essential step represented by sampling in the field to the choice of the appropriate size-window for analysis, and to follow the correct procedures aimed at a complete recovery of the targeted minerals in the laboratory. The proposed practical handbook provides the beginner, as well as the experienced researcher, with a friendly protocol, standard but also adaptable to diverse needs, as an essential requirement for the comparison and consistency of data produced in different laboratories.

Funding: Funding provided by Projects MIUR-PRIN 2015EC9PJ5 "The subduction and exhumation of the continental lithosphere: Their effects on the structure and evolution of the orogens" and MIUR—Dipartimenti di Eccellenza 2018–2022, Department of Earth and Environmental Sciences, University of Milano-Bicocca.

Acknowledgments: This little handbook is dedicated to my wife, Giovanna Castiglioni and to all my students and colleagues who have helped me to apply and continuously refine the separation techniques illustrated here, including Pietro Vignola, Paolo Gentile, Danilo Bersani, Nicoletta Fusi, Marta Padoan, Mara Limonta, Laura Borromeo, Marta Barbarano, Guido Pastore, Eleonora Botti, and Christian France-Lanord. A special thought is for Maria Mange for her fundamental contribution to heavy-mineral studies and to Eduardo Garzanti for years of encouragements and continuing challenges to push forward the limits of heavy-mineral research.

Conflicts of Interest: The author declares no conflicts of interest.

References

1. De Filippi, F. Sulla costituzione geologica della pianura e delle colline della Lombardia. *Annu. Univers. Stat.* **1839**, *59*, 225–248.
2. Artini, E. Intorno alla composizione mineralogica delle sabbie del Ticino. *Giornale di Mineralogia Cristallografia e Petrologia Pavia* **1891**, *2*, 177–195.
3. Garzanti, E.; Andò, S. Heavy Minerals for Junior Woodchucks. *Minerals* **2019**, *9*, 148. [CrossRef]
4. Carver, R.E. *Heavy-Mineral Separation. Chap. 18 in Procedures in Sedimentary Petrology*; Carver, R.E., Ed.; John Wiley: Hoboken, NJ, USA, 1971.
5. Mange, M.A.; Wright, D.T. *Heavy Minerals in Use*; Developments in Sedimentology Series; Elsevier: Amsterdam, The Netherlands, 2007; Volume 58, 1283p.
6. Morton, A.; McGill, P. Correlation of Hydrocarbon Reservoir Sandstones Using Heavy Mineral Provenance Signatures: Examples from the North Sea and Adjacent Areas. *Minerals* **2018**, *8*, 564. [CrossRef]
7. Krumbein, W.C.; Pettijohn, F.J. *Manual of Sedimentary Petrography*; D. Appleton-Century: New York, NY, USA, 1938; 549p.
8. Hutton, C.O. Studies of heavy detrital minerals. *Bull. Geol. Soc. Am.* **1950**, *61*, 635–710. [CrossRef]
9. Pettijohn, F.J.; Potter, P.E.; Siever, R. *Sand and Sandstone*; Springer: New York, NY, USA, 1972; 618p.
10. Hutchison, C.S. *Laboratory Handbook of Petrographic Techniques*; John Wiley & Sons: Hoboken, NJ, USA, 1974.
11. Callahan, J. A nontoxic heavy liquid and inexpensive filters for separation of mineral grains. *J. Sediment. Petrol.* **1987**, *57*, 765–766. [CrossRef]
12. Fessenden, F.W. Removal of heavy liquid separates from glass centrifuge tubes. *J. Sediment. Petrol.* **1959**, *29*, 621. [CrossRef]
13. Scull, B.J. Removal of heavy liquid separates from glass centrifuge tubes—Alternate method. *J. Sediment. Petrol.* **1960**, *30*, 626. [CrossRef]
14. Galli, A.; Panzeri, L.; Martini, M.; Sibilia, E.; Vignola, P.; Andò, S.; Pini, R.; Rossi, P.M. Optically stimulated luminescence dating of a stratigraphic Late Glacial–Holocene sequence in the Po plain (Bubano quarry, Bologna, Italy). *Quat. Int.* **2009**, *199*, 45–55. [CrossRef]
15. Mehra, O.P.; Jackson, M.L. Iron Oxide Removal from Soils and Clays by a Dithionite-Citrate System Buffered with Sodium Bicarbonate. *Clays Clay Miner.* **1958**, *7*, 317–327. [CrossRef]
16. Garzanti, E.; Andò, S.; Vezzoli, G. Settling equivalence of detrital minerals and grain-size dependence of sediment composition. *Earth Planet. Sci. Lett.* **2008**, *273*, 138–151. [CrossRef]
17. Garzanti, E.; Andò, S.; Vezzoli, G. Grain-size dependence of sediment composition and environmental bias in provenance studies. *Earth Planet. Sci. Lett.* **2009**, *277*, 422–432. [CrossRef]
18. Andò, S.; Vignola, P.; Garzanti, E. Raman counting: A new method to determine provenance of silt. *Rend. Lincei* **2011**, *22*, 327–347. [CrossRef]
19. Henningsen, D. Crushing of sedimentary rock samples and its effect on shape and number of heavy minerals. *Sedimentology* **1967**, *8*, 537–546. [CrossRef]
20. Mange, M.A.; Maurer, H.F.W. *Heavy Minerals in Colour*; Chapman and Hall: London, UK, 1992; 147p.
21. Parfenoff, A.; Pomerol, C.; Tourenq, J. *Les Minéraux en Grains—Méthodes d' Étude et Détermination*; Masson: Paris, France, 1970; 578p.
22. Garzanti, E.; Vezzoli, G.; Andò, S.; Limonta, M.; Borromeo, L.; France-Lanord, C. Provenance of Bengal Shelf Sediments: 2. Petrology and Geochemistry of Sand. *Minerals* **2019**, *9*, 642.
23. Garzanti, E.; Andò, S.; France-Lanord, C.; Vezzoli, G.; Censi, P.; Galy, V.; Najman, Y. Mineralogical and chemical variability of fluvial sediments 1. Bedload sand (Ganga–Brahmaputra, Bangladesh). *Earth Planet. Sci. Lett.* **2010**, *299*, 368–381. [CrossRef]
24. Chakoumakos, B.C.; Murakami, T.; Lumpkin, G.R.; Ewing, R.C. Alpha-decay—Induced fracturing in zircon: The transition from the crystalline to the metamict state. *Science* **1987**, *236*, 1556–1559. [CrossRef]
25. Chisholm, E.K.; Sircombe, K.; DiBugnara, D. *Handbook of Geochronology Mineral Separation Laboratory Techniques*; Chisholm, E.K., Sircombe, K., DiBugnara, D., Eds.; Geoscience Australia: Canberra, Australia, 2014; Volume 46, pp. 1–45.

26. Malusà, M.; Garzanti, E. The Sedimentology of Detrital Thermochronology. In *Fission-Track Thermochronology and its Application to Geology*; Springer: Cham, Switzerland, 2019; pp. 123–143.
27. Resentini, A.; Malusà, M.G.; Garzanti, E. MinSORTING: An Excel® worksheet for modelling mineral grain-size distribution in sediments, with application to detrital geochronology and provenance studies. *Comput. Geosci.* **2013**, *59*, 90–97. [CrossRef]
28. Vermeesch, P.; Rittner, M.; Petrou, E.; Omma, J.; Mattinson, C.; Garzanti, E. High Throughput Petrochronology and Sedimentary Provenance Analysis by Automated Phase Mapping and LAICPMS. *Geochem. Geophys. Geosyst.* **2017**, *18*, 4096–4109. [CrossRef]
29. Yokoyama, K.; Amano, K.; Taira, A.; Saito, Y. Mineralogy of Silts from the Bengal Fan. *Proc. ODP Sci. Results* **1990**, *116*, 59–73.
30. Garzanti, E.; Andó, S.; France-Lanord, C.; Censi, P.; Vignola, P.; Galy, V.; Lupker, M. Mineralogical and chemical variability of fluvial sediments 2. Suspended-load silt (Ganga–Brahmaputra, Bangladesh). *Earth Planet. Sci. Lett.* **2011**, *302*, 107–120. [CrossRef]
31. Delmonte, B.; Paleari, C.I.; Andò, S.; Garzanti, E.; Andersson, P.S.; Petit, J.R.; Crosta, X.; Narcisi, B.; Baroni, C.; Salvatore, M.C.; et al. Causes of dust size variability in central East Antarctica (Dome B): Atmospheric transport from expanded South American sources during Marine Isotope Stage 2. *Quat. Sci. Rev.* **2017**, *168*, 55–68. [CrossRef]
32. Caracciolo, L.; Andò, S.; Vermeesch, P.; Garzanti, E.; McCabe, R.; Barbarano, M.; Paleari, C.; Rittner, M.; Pearce, T. A multidisciplinary approach for the quantitative provenance analysis of siltstone. Mesozoic Mandawa Basin, southeastern Tanzania. *Geol. Soc. Lond. Spec. Publ.* **2019**, *484*, SP484-2018. [CrossRef]
33. Andò, S.; Aharonovich, S.; Hahn, A.; George, S.C.; Clift, P.D.; Garzanti, E. Integrating heavy-mineral, geochemical and biomarker analyses of Plio-Pleistocene sandy and silty turbidites: A novel approach for provenance studies (Indus Fan, IODP Expedition 355). *Geol. Mag.* **2019**. [CrossRef]
34. Borromeo, L.; Andò, S.; France-Lanord, C.; Coletti, G.; Hahn, A.; Garzanti, E. Provenance of Bengal Shelf Sediments: 1. Mineralogy and Geochemistry of Silt. *Minerals* **2019**, *9*, 640. [CrossRef]

minerals

MDPI

Review

Heavy Minerals for Junior Woodchucks

Eduardo Garzanti and Sergio Andò *

Laboratory for Provenance Studies, Department of Earth and Environmental Sciences, Università di Milano-Bicocca, 20126 Milano, Italy; eduardo.garzanti@unimib.it

* Correspondence: sergio.ando@unimib.it; Tel.: +39-64482088

Received: 29 January 2019; Accepted: 23 February 2019; Published: 28 February 2019

Abstract: In the last two centuries, since the dawn of modern geology, heavy minerals have been used to investigate sediment provenance and for many other scientific or practical applications. Not always, however, with the correct approach. Difficulties are diverse, not just technical and related to the identification of tiny grains, but also procedural and conceptual. Even the definition of "heavy minerals" is elusive, and possibly impossible. Sampling is critical. In many environments (e.g., beaches), both absolute and relative heavy mineral abundances invariably increase or decrease locally to different degrees owing to hydraulic-sorting processes, so that samples close to "neutral composition" are hard to obtain. Several widely shared opinions are misleading. Choosing a narrow size-window for analysis leads to increased bias, not to increased accuracy or precision. Only point-counting provides real volume percentages, whereas grain-counting distorts results in favor of smaller minerals. This paper also briefly reviews the heavy mineral associations typically found in diverse plate-tectonic settings. A mineralogical assemblage, however, only reproduces the mineralogy of source rocks, which does not correlate univocally with the geodynamic setting in which those source rocks were formed and assembled. Moreover, it is affected by environmental bias, and by diagenetic bias on top in the case of ancient sandstones. One fruitful way to extract information on both provenance and sedimentological processes is to look for anomalies in mineralogical–textural relationships (e.g., denser minerals bigger than lower-density minerals; harder minerals better rounded than softer minerals; less durable minerals increasing with stratal age and stratigraphic depth). To minimize mistakes, it is necessary to invariably combine heavy mineral investigations with the petrographic analysis of bulk sand. Analysis of thin sections allows us to see also those source rocks that do not shed significant amounts of heavy minerals, such as limestone or granite, and helps us to assess heavy mineral concentration, the "outer" message carrying the key to decipher the "inner message" contained in the heavy mineral suite. The task becomes thorny indeed when dealing with samples with strong diagenetic overprint, which is, unfortunately, the case of most ancient sandstones. Diagenesis is the Moloch that devours all grains that are not chemically resistant, leaving a meager residue difficult or even impossible to interpret when diagenetic effects accumulate through multiple sedimentary cycles. We have conceived this friendly little handbook to help the student facing these problems, hoping that it may serve the purpose.

Keywords: relative and absolute abundances; sampling strategy; size-window for analysis; heavy mineral point-counting; provenance and plate-tectonic setting; chemical weathering; hydraulic sorting; recycling; diagenesis

---

Well, according to the Junior Woodchuck's Guide Book, to get there we've have to take about 537 million steps straight up, till we reach the moon (Huey *in* Duck Tales).

1. Introduction

The aim of this script is to suggest a few "rules of thumb", challenge some widely held beliefs, pose some unsolvable problems, tell stories, and highlight some key aspects concerning heavy minerals, especially when they are used to infer provenance and reconstruct modern or ancient "source-to-sink" sediment-dispersal systems. Its content largely reflects the experience obtained by the writers mostly on modern sediments, with no ambition to reach any final truth. The reader can complete the panorama of heavy mineral research by consulting the monumental book by Mange and Wright [1] and the reviews by Morton [2] and von Eynatten and Dunkl [3] that focus on single-mineral techniques.

This paper deals first with only apparently trivial operational choices, and summarizes next the information gathered from modern sedimentary systems where everything is knowable in principle. Finally, it discusses the complex problems posed by ancient sedimentary systems, where little is known and most is lost by erosion from the sources and by chemical dissolution from the sink.

What are Heavy Minerals?

The simplest questions are often the hardest. A theoretically simple operational definition of heavy minerals includes all detrital components with density exceeding a given threshold, traditionally that of the once universally used bromoform (i.e., 2.89 g/cm^3). Because bromoform was discovered to be carcinogenic, other dense liquids came into use (generally polytungstates) and the standard density value is now generally rounded off to 2.90 g/cm^3.

In most continental or marine sediments, however, the 2.90 g/cm^3 fraction contains particles of diverse origin, including micas (e.g., biotite, chlorite), composite grains and rock fragments, chunks of ferruginous soils, altered grains of uncertain origin, iron oxides, titanium oxides, sulfides, sulfates (e.g., barite, celestite), ferriferous carbonates (e.g., ankerite, siderite), bioclasts with pyrite-filled chambers, glaucony, or phosphates (e.g., vertebrate teeth and bones). In some cases, even most of the dense fraction is formed by chemical or biochemical precipitation in the sedimentary basin rather than derived from erosion of source rocks. In ancient sandstones, dense grains may be of diagenetic or anchimetamorphic origin (e.g., anatase, brookite, barite, fluorite, epidote, iron oxides, sulfides, or even titanite and tourmaline) [4]. Anthropogenic grains such as moissanite or corundum may be present in modern sediments [5], and core samples and cuttings may contain barite or other heavy mineral contaminants associated with drilling muds [6]. Moreover, separation in the laboratory may not be perfect, and the recovered dense fraction occasionally includes tectosilicate grains or precipitated polytungstate crystals. Finally, among those that we should consider as heavy minerals proper, many are opaque to transmitted light and thus not readily identified under the microscope (e.g., Fe–Ti–Cr oxides), others are too altered to be safely identified or may belong to rare unknown species. These issues are rarely handled systematically or even exposed in scientific articles, with the result that researchers tend to use different and generally unexplicit criteria to decide what is included and what excluded from the heavy mineral string. Raman spectroscopy is a handy tool to solve many problems [7], but some remain elusive nonetheless.

2. Heavy Mineral Concentration

The definition of heavy minerals used in this article includes only minerals of certain extrabasinal terrigenous origin (i.e., ultimately eroded from bedrock exposed in source areas), denser than 2.90 g/cm^3, and occurring either as single detrital grains or in rock fragments. Grains of suspect intrabasinal (e.g., carbonates, bioclasts, glaucony), pedogenic or diagenetic (e.g., aggregates of iron or titanium oxides), and anthropogenic origin (e.g., barite in core samples) are thus neglected. Phyllosilicates are neglected as well. Transparent heavy minerals (tHM) identified under the microscope are considered separately from opaque and altered heavy minerals. Two indices are thus calculated, in either weight or volume, expressing the concentration of transparent heavy minerals (tHMC) and of total heavy minerals (HMC) relative to the bulk sediment. To avoid uncertainties involved in the identification of opaque grains and make datasets consistent and comparable, we choose here to refer to transparent heavy minerals only. Transparent-heavy mineral suites are considered conventionally as extremely poor (tHMC < 0.1), very poor ($0.1 \leq$ tHMC < 0.5), poor ($0.5 \leq$ tHMC < 1), moderately poor ($1 \leq$ tHMC < 2), moderately rich

(2 ≤ tHMC < 5), rich (5 ≤ tHMC < 10), very-rich (10 ≤ tHMC < 20), or extremely rich (tHMC > 20). Heavy mineral-dominated sediments are called "placers" (HMC ≥ 50).

The Inner and Outer Messages

Heavy mineral studies take the relative percentages of heavy mineral species in the sample as the starting point for provenance interpretation. Relative percentages do contain the information we are looking for, but to read such an "inner" message correctly we need an "outer" message, a key that helps us to get it right [8] (p. 162ff). That key is heavy mineral concentration (Figure 1A). Otherwise, we would be tempted to call amphibole-rich a sample with 90% tHM amphibole and tHMC 0.1, and instead amphibole-poor another sample with 9% tHM amphibole but tHMC 10, which is wrong by a full order of magnitude. A particularly instructive case is presented by chromian spinel (Figure 1B), a mineral widely valued as a robust indicator of provenance from ultramafic rocks in ophiolitic complexes. Cr-spinel does occur in ultramafic rocks, where it is however rarer than olivine and pyroxenes by almost two orders of magnitude (Figure 3 in Reference [9]) [9]. Unless olivine and pyroxenes are massively destroyed by either serpentinization in the parent rock or by diagenesis during burial of the daughter sediment, Cr-spinel thus can represent only a low percentage of the very rich or extremely rich transparent-heavy mineral suite contained in ophioliticlastic sediments [10,11]. Conversely, because Cr-spinel resists diagenesis well [12], a high percentage of Cr-spinel in a very poor transparent-heavy mineral suite points to recycling of older sandstones ultimately derived in part from mafic or ultramafic rocks [5,13–15] (Figure 12 in Reference [14]; Figure 9 in Reference [15]).

Therefore, to avoid gross mistakes and misleading generalizations we need to estimate not only the relative abundance of heavy mineral species but also and always their absolute abundance (i.e., their concentration in the bulk sample). This is done simply by weighing the separated dense fraction and by making due corrections for spurious dense grains [9] (pp. 521–523). The concentration (and not just the spectrum) of heavy minerals in a sedimentary deposit depends on the composition of parent rocks, and increases by one order of magnitude or more during progressive unroofing of denser rocks found at deeper-seated crustal levels [16]. Drastic modifications of their concentration (as well as of their spectrum) may also occur by selective entrainment of low-density grains in the depositional environment [17], or by selective leaching of unstable species during diagenesis [18,19]. The concentration of heavy minerals in the sample (as well as their spectrum) is therefore *per se* crucial in provenance diagnoses and in the correct assessment of recycling and hydraulic or diagenetic processes.

The distortive fertility effect related to the different potential of different rock types to generate heavy minerals must always be taken into full account in the interpretation of heavy mineral suites, which tend to document aberrantly a limited number of sources (e.g., mafic igneous and medium/high-grade metamorphic rocks) whereas several others are barely recorded (limestone, chert, shale, granite). In the absence of significant chemical weathering and diagenetic dissolution, igneous and metamorphic rocks may impose their mark on the heavy mineral spectrum even where their outcrops are sparse (Figure 1A). As an extreme case, the heavy mineral assemblage of granite sand may be dominated by heavy minerals from medium/high-grade metamorphic country rocks contained as xenoliths within the granite body [16] (p. 541).

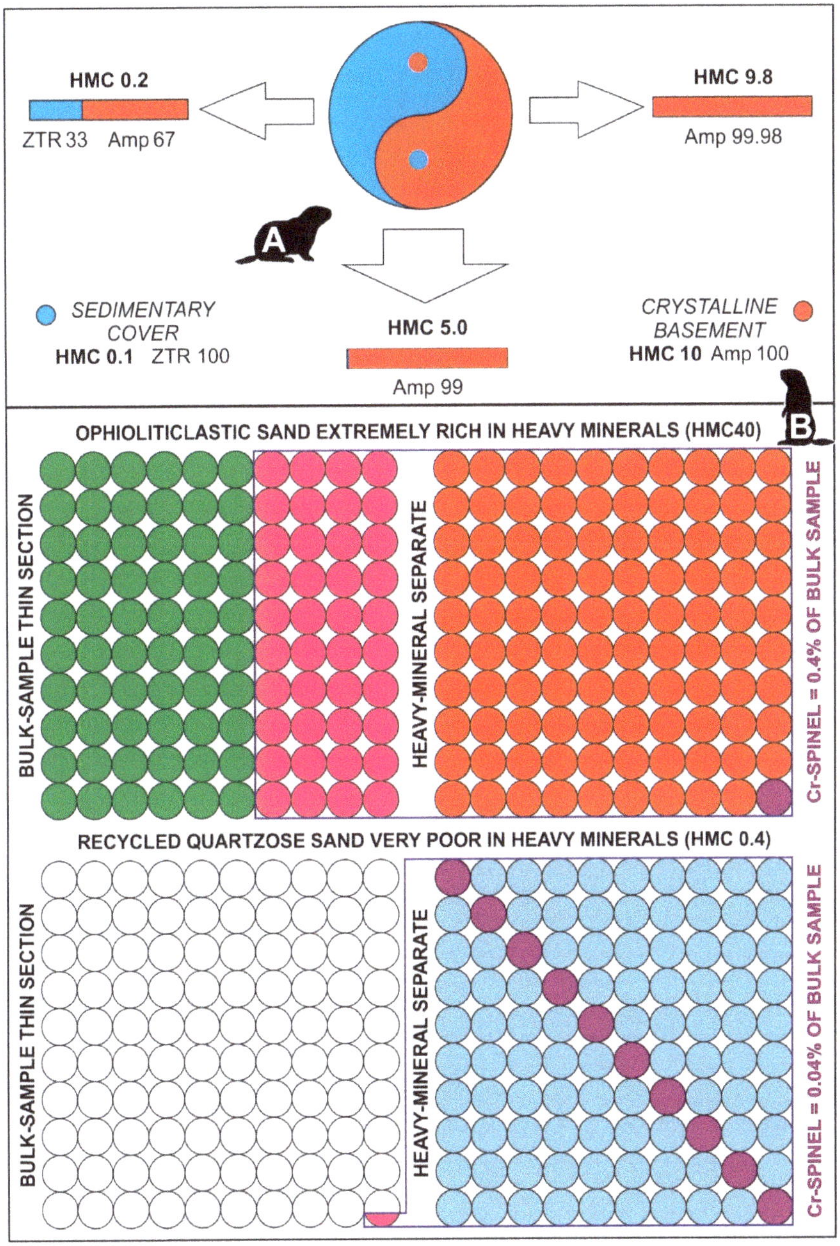

Figure 1. Heavy mineral concentration is the key to decipher provenance of heavy mineral suites. (**A**) The fertility effect (granitoid basement is supposed to supply 100% amphibole and cover strata 100% ZTR minerals; redrawn from Figure 1 in [9]). Amphibole shed from sparse basement outcrops dominates the heavy mineral assemblage in sand derived 98% from sedimentary covers (Amp 67% HM). Amphibole is $\geq$99% HM no matter whether crystalline basement represents 50% or 100% of total outcrops, although HMC is half in the former case. (**B**) Relative and absolute concentrations (stylistic representation

of the real case illustrated in Figure 9 in [15]). The sediment sample depicted schematically above is entirely derived first-cycle from an ophiolite complex: it contains 60% framework grains (plagioclase, mafic and ultramafic rock fragments; green circles) and 40% heavy minerals (magenta) including mostly pyroxene and olivine (red) and only 1% Cr-spinel (purple). The sample below is entirely recycled from older sandstones: it contains mostly quartz (white) with only 0.4% heavy minerals, all durable (greyish blue) and 10% of which is Cr-spinel. In the sample below, the absolute abundance of Cr-spinel is lower by one order of magnitude, but its relative abundance is higher by one order of magnitude.

3. Passing through Scylla and Charybdis

The results of heavy mineral analysis are strongly dependent on methodological choices, the very first of which is made in the field. Sampling and analytical biases are present in every heavy mineral study, although generally undeclared, uninvestigated, and consequently unquantified.

3.1. *What are We Sampling?*

What is apparently easier than collecting a sand sample in a dune field or along a sandy beach? Sand being everywhere, however, does not make things easier. Grain size may change from place to place, and colour may change from place to place indicating local enrichment or depletion in heavy minerals (Figure 2A). A commonly held prejudice is that, while sampling for heavy minerals, we should go for heavy minerals. Consequently, we may be attracted by beautifully coloured sand patches where heavy minerals are found in abundance (Figure 2B). Researchers may even choose to concentrate heavy minerals further by panning in the field, to reduce sample size and make the tedious job of separating them in the lab simpler and quicker. Although apparently reasonable and thus commonly adopted when aiming at zircon, apatite, or monazite for thermochronological or geochronological analysis, such an approach ends up to maximise bias [20]. Not only our artificially concentrated samples will be enriched anomalously in heavy minerals, hence leading to a gross overestimate of the natural heavy mineral concentration, but also all proportions among heavy minerals will be markedly altered. Denser and denser grains will be enriched more and more, hence leading to overestimate higher-density minerals such as zircon and to underestimate lower-density minerals such as amphibole.

The best sample is the one that most closely approximates "neutral" composition, i.e., the composition that sand would have everywhere in the absence of local hydraulic-sorting effects. Both coloured to black heavy mineral-enriched lags and lighter sand patches conversely enriched in slow-settling low-density tectosilicates or platy phyllosilicates should be carefully avoided. As grain size is concerned, it is advisable to collect samples as close as possible to the dominant grain size in the site for representativeness, and in the lower-fine to lower-medium sand range for practical reasons. Sediment finer than ~3ϕ is less suitable for petrographic analysis, and sand coarser than ~1.5ϕ is less suitable for heavy mineral analysis. Sample volume does not need to exceed 50–100 g and, provided the collected grain-size window is appropriate and separation procedures designed to maximise recovery, a few hundred grams are generally more than enough for the geochemical or geochronological study of any accessory mineral [21]. If automated-phase-mapping procedures by Scanning Electron Microscopy (SEM/EDS), Qemscan or Raman spectroscopy are used [22], then even a few grams may be enough for detrital-zircon geochronology. Single-mineral techniques can thus be routinely employed also on very small core samples.

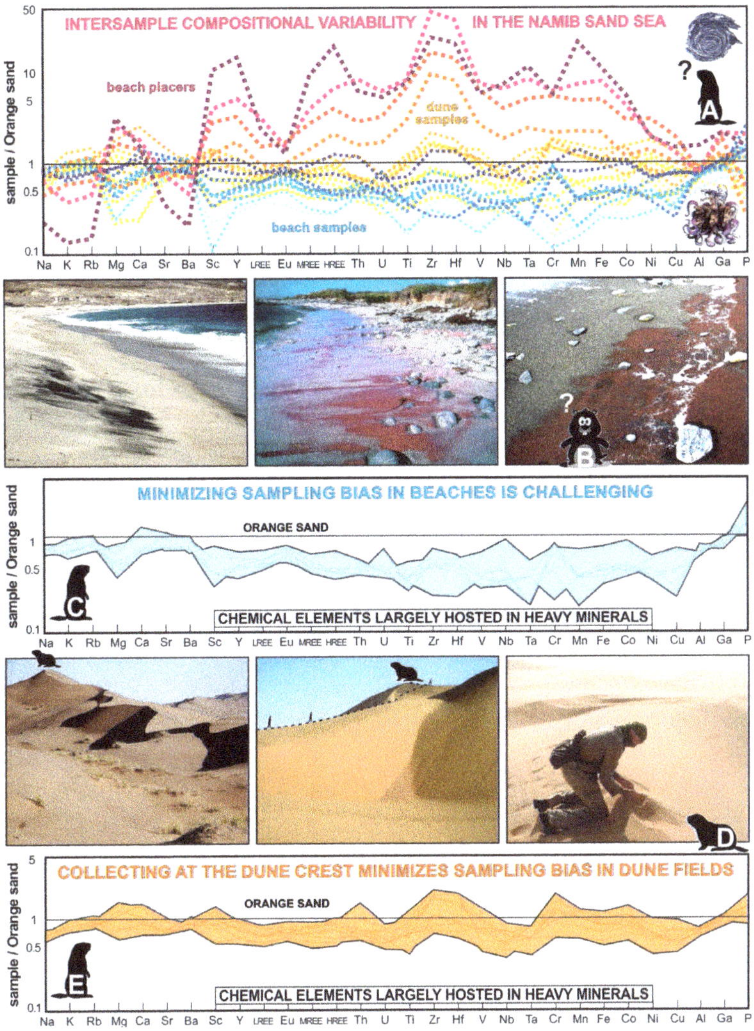

Figure 2. Sampling bias. Sampling, the most critical first step in any provenance study, should fulfill two basic requirements: (1) representativeness (sample composition as close as possible to "neutral" composition); (2) consistency (same sampling criterion applied for all samples). Geochemical data, particularly sensitive to anomalous concentrations of heavy minerals caused by selective entrainment at the local scale and made visible by marked colour differences, are most useful to check for sampling bias, as done for the Namib Sand Sea (**A**). In (**A**,**C**,**E**), elements are arranged following the periodic table group by group, and concentrations are normalized to average composition of Orange River sand. Colours: dunes in yellow-orange, beaches in blue, beach placers in purple. Data after [23]. For beaches (**B**), we could not identify a criterion that guarantees for both consistency and representativeness, and the geochemical test indicates that we ended up sampling heavy mineral depleted sites (Scylla's antiplacers) to avoid heavy mineral-enriched sites (Charybdis' placers) (**C**). For dune fields (**D**), the geochemical test indicates that collecting systematically at the crest of the highest dune is reproducible and reduces sampling bias (**E**). Unless we have specific aims (e.g., investigating grain-size dependent intersample variability or hydraulic-sorting effects), we routinely collect one sample in each locality.

Besides representativeness, the sampling plan must follow criteria that maximise consistency and comparability among samples. The criterion we currently use in dune fields is to collect sand systematically from the crest at the top of the largest dune in the area (Figure 2D). Geochemical analysis proves that such samples contain an amount of heavy minerals acceptably close to the "neutral" average heavy mineral concentration in the dune field (Figure 2E). On the contrary, beach sampling proved to be far less consistent. The geochemical test failed, indicating that in order to avoid being engulfed into the whirlpool of Charybdis (samples anomalously enriched in heavy minerals called "semiplacers"), we fell instead into the mouth of Scylla (samples anomalously depleted in heavy minerals called "antiplacers"; Figure 2C). The attention paid to design an appropriate sampling plan is never enough.

3.2. The Size-Window Problem

Contrary to what recommended by several authors, heavy mineral analysis of a narrow size class (e.g., 125–250 μm [24]; 63–125 μm [25]; 90–125 μm [26]) leads to biased results. The densest grains, markedly concentrated in the fine sediment tail [27,28], may be either notably overestimated or even completely cut off. The narrower is the class chosen for analysis, the larger the bias (Figure 3). A faithful characterization of detrital assemblages is obtained only if samples are analysed in bulk, which is feasible and fully advisable for well-sorted beach and eolian-dune sands.

In poorly sorted sediments, however, the presence of detrital grains with great size differences within a single concentrate makes mounting and identification difficult [29], and practical reasons force us to set lower and upper size limits for heavy mineral analysis in many cases. Heavy minerals in the finest silt fraction are difficult to identify with confidence under the petrographic microscope, whereas in coarse sand they are rare and commonly contained in rock fragments. For these reasons, we routinely consider a size window including 4ϕ (32–500 μm) or 5ϕ classes (15–500 μm) for heavy mineral analysis of moderately to poorly sorted sands. Not losing the finest tail of the size distribution is crucial, because the densest minerals such as zircon or monazite concentrate there. For a correct presentation of heavy mineral data, it is necessary to weigh the excluded finest and coarsest fractions, and to indicate what percentage of the total sediment the analysed size window represents and how much of the tails were cut off.

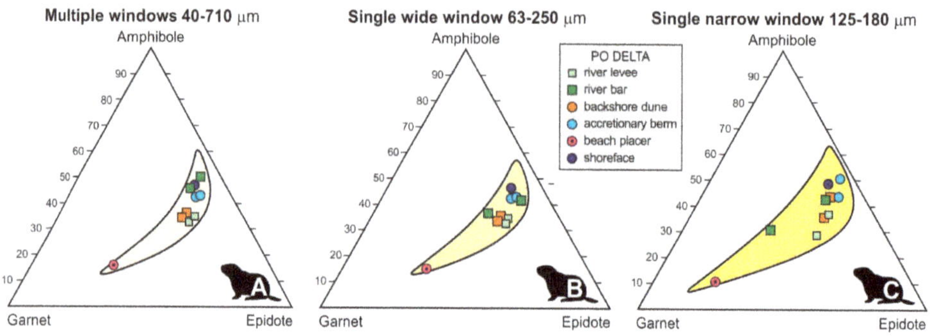

Figure 3. Analytical bias. Multiple-window analyses at 0.25ϕ or 0.5ϕ sieve intervals faithfully assess natural intersample mineralogical variability in Po Delta sand (**A**). Instead, single-window analyses introduce bias which decreases with the width of the analysed size class (**B**,**C**). All pale yellow fields are 90% predictive regions for data points. Data after [30].

3.3. Beyond Grain Counting

None of the three different methods generally used in grain counting (i.e., "ribbon/area", "line", and "Fleet"; [29,31] provides data which can be converted to correct volume or weight percentages [32]. Accurate determination of volume percentages by grain counting requires either the analysis of numerous closely-spaced sieve fractions for each sample, so that mineral grains on each slide can be considered to be of the same size [33] ("multiple-window" strategy of Garzanti et al. [34]), or measuring the size of all counted grains in an unsieved fraction [35]. Being these procedures very time-consuming, researchers seldom attempt to determine volume or weight percentages, and number percentages or number frequencies are used instead [36]. In such a way, however, abundances in volume or weight are systematically overestimated for higher-density heavy minerals and underestimated for lower-density heavy minerals, because in sorted sediments deposited by traction currents denser grains are smaller than settling-equivalent less dense grains. Errors may exceed 100% for high-density zircon, ilmenite, monazite and magnetite, and thus lead to significant imprecision in quantitative provenance analysis, comparison of mineralogical and geochemical data, calculation of sediment budgets, and assessment of mineral resources in placer deposits.

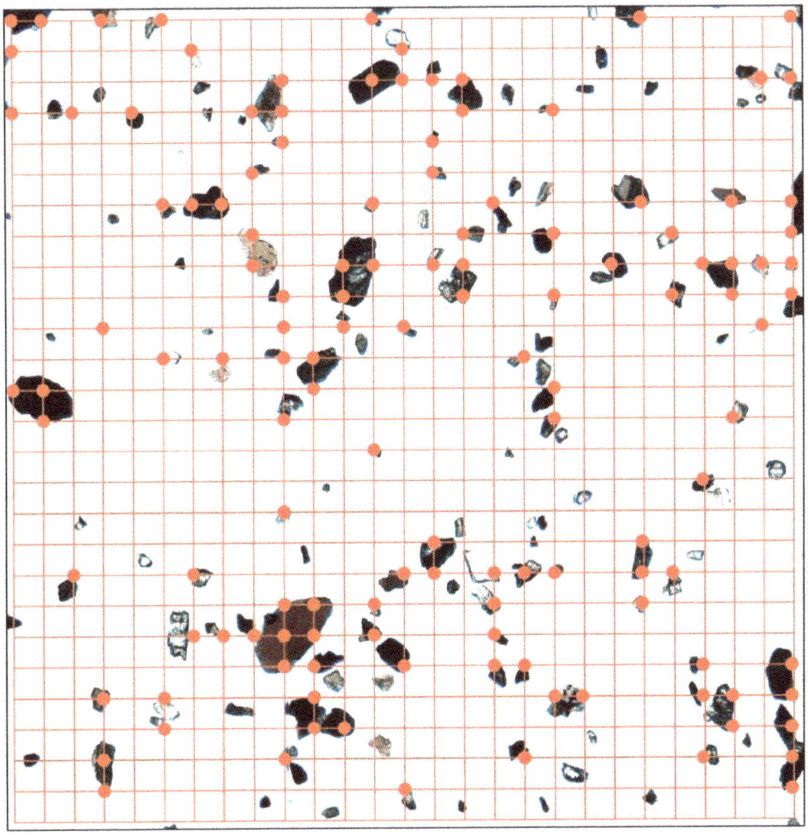

Figure 4. Point-counting allows obtaining real areal and therefore volumetric percentages of heavy minerals in grain mounts [32,37]. Choosing an appropriate grid is critical. In the case represented here—heavy mineral mount from the size window 15–500 m of a turbiditic silty sand from the Indus Fan (IODP1456A), square grid 125 μm—several grains are counted more than once, but with a larger spacing more than a single slide would be needed to count a representative number of transparent heavy minerals (usually $\geq$ 200).

In petrographic analysis, this problem was long solved by point-counting of impregnated sand or lithified sandstone in thin section following a grid of equally-spaced points along equally-spaced linear traverses (Figure 4; Glagolev-Chayes method of quantitative mineralogical analysis) [32]. Because the probability of a grain being hit by the cross-hair along this grid is proportional to its visible area, the areal percentages of various detrital components are obtained, which are equivalent to volume percentages [37]. Surprisingly, the Glagolev-Chayes method was hardly ever used in heavy mineral analysis. Nonetheless, it can be done, it is not significantly more time-consuming than grain-counting, and should be used routinely. As for petrographic analysis of thin sections, grid spacing needs to be large enough to ensure that grains are not counted twice. Heavy mineral mounts must be properly prepared to avoid that grains are either too numerous, otherwise they overlap and cannot be identified properly, or too sparse, otherwise a large number of voids will be counted before a sufficient number of heavy minerals is encountered and identified.

4. Heavy Minerals as Provenance Tracers

The investigation of multi-mineral suites provides crucial provenance information for paleotectonic reconstructions [38], especially if coupled with petrographic observations under the microscope. Four decades have passed since Dickinson and Suczek [39] have attempted to establish a link between sediment composition and plate-tectonic setting. Detrital modes, however, can only reflect—in many cases unfaithfully—the mineralogy of source rocks, which is not necessarily a good proxy for geodynamic environment [40,41]. Volcanic and plutonic rocks with very similar mineralogy are found in magmatic arcs, orogenic belts, rift shoulders, and continental interiors. The relationship between detrital modes and geodynamic setting is even looser for heavy mineral suites for a range of reasons, first of all because of the fertility bias: some sources lack heavy minerals (carbonates, chert), whereas others are extremely prolific (mafic and ultramafic igneous rocks). Hydraulic effects may alter both absolute and relative abundances of heavy minerals by an order of magnitude or more. In ancient sandstones, most heavy minerals may be leached out by diagenetic dissolution. And how can we assess the extent of sedimentary recycling? These are the main reasons why heavy mineral and petrographic analyses should always be coupled. Petrographic investigations prevent us from losing precious information conveyed especially by rock fragments and enable us to assess heavy mineral concentration more accurately. Only with this knowledge provenance interpretations can be based on firm ground. A brief summary of typical transparent-heavy mineral assemblages found in different anorogenic and orogenic geodynamic settings is provided in the following paragraphs as an update of Garzanti and Andò [42]. The expected unroofing trends produced by the progressive erosion of deeper-seated rocks are also envisaged.

4.1. Anorogenic Provenances

Geodynamic settings unrelated to oceanic or continental subduction include continental rifts, passive margins, and continental interiors. Three main sources of detritus may be distinguished in these settings: (a) siliciclastic to carbonate sedimentary covers; (b) generally polymetamorphic old basement rocks; (c) intraplate volcanic rocks including continental flood basalts. Three are the corresponding provenances: (A) undissected continental block and/or recycled clastic; (B) dissected continental block; (C) anorogenic volcanic [40].

Siliciclastic cover strata are commonly distinguished by notably poor heavy mineral suites containing zircon, tourmaline, and rutile (ZTR). Carbonate rocks and chert do not supply heavy minerals. Crystalline basement including granitoids and gneisses typically sheds rich assemblages including amphibole and epidote derived from mafic and intermediate rocks of respectively medium and low metamorphic grade, with subordinate metamorphic minerals including garnet and generally minor staurolite, kyanite, andalusite, or sillimanite. Anorogenic volcanic sources are revealed by very rich transparent-heavy mineral suites dominated by clinopyroxene, locally associated with olivine, apatite, zircon, pigeonite, Cr-spinel, or hypersthene. The expected unroofing trend in non-volcanic settings will see a progressive increase in heavy mineral concentration, with a progressive increase in amphibole relative to ZTR minerals as wider areas of basement rocks are erosionally exhumed through time. In volcanic settings, a decrease in pyroxene will be compensated by an increase in amphibole and metamorphic minerals derived from the underlying crystalline rocks [43–45].

4.2. Magmatic Arcs

Magmatic arcs formed above an oceanic subduction zone include two main sources of detritus: (a) volcanic rocks of the "calc-alkaline" series; (b) granitoid batholiths including granodiorite and tonalite. The two corresponding provenances are: (A) undissected magmatic arc; (B) dissected magmatic arc [46–48].

Volcaniclastic heavy mineral assemblages are commonly rich to extremely rich and dominated by clinopyroxene. Arc basalts also commonly shed olivine, whereas hypersthene and kaersutitic hornblende or oxy-hornblende occur in andesites and more felsic rocks. The orthopyroxene/clinopyroxene ratio may thus be used as an indicator of increasing silica contents [49]. Pigeonite may occur in tholeiitic andesites lacking olivine, and clinoenstatite in boninites. Relatively Ti-poor titanomagnetite is the dominant opaque mineral. Garnet (e.g., melanite) and Cr-spinel are rare. Such arc-related suites are unfortunately not readily discriminated from those derived from anorogenic lavas.

Granitoid batholiths supply rich, hornblende-dominated suites. Epidote and actinolite shed from metavolcanic complexes, kaersutitic hornblende and hypersthene shed from gabbro-norites, as well as titanite, allanite, zircon, and monazite may occur. The predicted unroofing trend would see a progressive increase of the hornblende/pyroxene ratio as granitoid batholiths are progressively exhumed through time.

4.3. Axial Belts and Obducted Ophiolites

Orogenic belts consist of stacked and juxtaposed tectonic domains that may include any kind of rock assemblages and thus cover all previously described provenances. Two additional domains typical of orogenic settings can, however, be identified: (a) fossil continental subduction zones, characterized by high-pressure neometamorphic rocks; (b) obducted ophiolitic complexes, characterized by mafic and ultramafic rocks. The two corresponding provenances are: (A) axial belt; (B) ophiolite [50].

The neometamorphic axial backbone of orogenic belts associated with continental subduction consists of exhumed high-pressure to ultra-high-pressure crustal and mantle rocks shedding rich to extremely rich transparent-heavy mineral suites dominated by garnet, amphibole or epidote depending on protoliths (continental vs. oceanic), depth reached during subduction, and pressure/temperature paths followed during exhumation [51]. Diverse metamorphic minerals, including glaucophane, chloritoid, staurolite, kyanite, sillimanite, and diopsidic clinopyroxene may also occur. Ultramafic mantle rocks formerly belonging to the lower plate and transformed into antigorite schists during subduction at eclogitic depths shed mainly magnetite clusters formed during release of iron from the olivine lattice during serpentinization.

Tectonic deformation is much less extensive in obducted ophiolites belonging to the upper plate, which supply rich to extremely rich suites typically dominated by enstatite derived from mantle harzburgites and including olivine from mantle rocks and cumulates, diopside and diallage from gabbros, hypersthene from gabbro-norites, hornblende from high-level gabbros, epidote and actinolite from the sheeted-dyke complex, and clinopyroxene from pillow basalts. Cr-spinel is derived from ultramafic rocks, zircon from plagiogranites, orthopyroxene and clinoenstatite from boninites [10,11].

4.4. Mixed Orogenic Provenances

Orogenic provenance is a complex issue, difficult to summarize in brief. At least five radically different types of orogenic belts can be identified on our rotating planet, depending on subduction polarity and on the continental or oceanic nature of the lower and upper plates involved in orogeny. Himalayan-type collision orogens are generated by continent-beneath-continent eastward subduction, Andean-type cordilleras by oceanic-beneath-continent eastward subduction, Oman-type obduction orogens by continent-beneath-ocean eastward subduction, Burman-Andaman and Barbados-type subduction complexes by ocean-beneath-ocean subduction, and finally Apennine-type orogens by retreating westward subduction [50,52]. Each of these archetypal orogenic belts consists of peculiar rock assemblages, and thus sheds different quantities and types of heavy minerals. Alpine-Himalayan collision orogens are characterized by neometamorphic axial-belt detrital signatures of trunk-river sands [17,51,53], Andean cordilleras by magmatic-arc signatures on the pro-side and by largely recycled detrital suites on the retro-side [54], Oman-type ophiolitic allochthons by enstatite and other pyroxenes, emerged subduction complexes and Apennine-type belts by largely recycled heavy mineral suites [5,11,15,55,56].

5. Environmental Bias: Turning Problems into Opportunities

Provenance signals may become progressively distorted by physical and chemical processes during transfer from source to sink along the sediment-routing system, producing what we are used to considering as noise that limits our capacity to understand. Particles with different size, density, and shape are segregated in different sediment fractions according to physical laws in all three stages of the sedimentary cycle: erosion, transport, and deposition. Most environmental "noise" is therefore coherent, and closer inspection allows us to discover that it can be converted profitably into both environmental and clearer provenance information.

5.1. Heavy Minerals as Tracers of Hydraulic Processes

There are two distinct ways to consider mineralogical variability, either among the different size classes of the same sediment sample (intrasample variability) or among different samples (intersample variability). Each is controlled by a different hydraulic process of size-density sorting: intrasample variability is fundamentally a consequence of settling equivalence (Figure 5A), intersample variability of selective entrainment (Figure 5B).

The settling-equivalence principle states that grains found together in a sediment layer deposited by a traction current under a single set of hydraulic conditions must have the same settling velocity [27]. A temporary deposit, such as a fluvial bar or a beach, therefore, consists of coarser low-density grains such as quartz or feldspar associated with a range of heavy mineral species the size of which decreases progressively as their density increases. Low-density tectosilicates will thus protrude above the sediment layer higher than heavy minerals. When the deposit is impacted by a high-energy current, during a storm or a flood, the larger grains that have smaller pivoting angles and experience greater flow velocities and drag forces will be selectively removed (Figure 5B) [57]. As a consequence of such a selective-entrainment process, lag deposits are progressively enriched in heavy minerals in proportion to their density, until placers are formed [58–60].

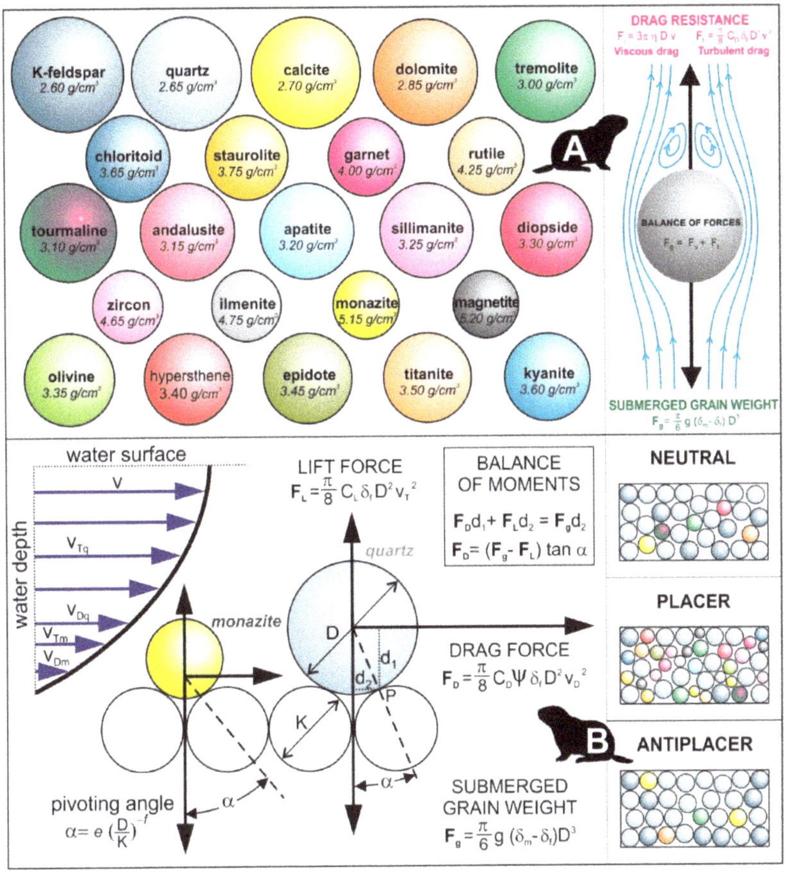

Figure 5. Hydraulic sorting. Grains are sorted by their size, density, and shape during erosion, transport, and deposition by traction currents. Complex shape effects are not taken into account here. (**A**) The settling-equivalence principle [27] controls size-density relationships among deposited grains and is responsible for intrasample compositional variability [34]. Spherical minerals in the picture have all the same settling velocity of 2.67 cm/s (calculated for a quartz sphere 250 μm in diameter according to the empirical formula of Cheng [61]). (**B**) Selective entrainment of coarser low-density grains while smaller settling-equivalent heavy minerals are left behind [62] is responsible for intersample compositional variability [30]. Sediments of theoretical "neutral" composition are thus partitioned into "placer" and "antiplacer" deposits readily distinguished by colour contrast even at the very local scale (Figure 2B).

5.2. Extracting Environmental and Provenance Information

Understanding how detrital minerals of different size, density, and shape behave under the action of traction currents allows us to add details to sedimentological interpretations and to choose correctly among provenance options. By means of settling-equivalence analysis, we can reconstruct hydraulic conditions at the instant of deposition, and calculate the settling velocity of specific laminae and laminasets. We can discriminate between sorting processes occurring in water or air, and thus distinguish between eolian backshore dunes and shoreface bars in deltaic settings (Figure 6 in Reference [34]) [34].

Violations of the settling-equivalence principle offer additional clues to provenance diagnoses. If one mineral or group of minerals shows anomalous textural relationships with another mineral or mineral group, then we may suspect mixing of detrital populations derived from different sources or transported in different modes or by different media (e.g., wind versus ephemeral streams in deserts, wind versus waves in deltas, local rivers versus longshore drift along high-energy coasts (Figure 8 in Reference [34]; Figures 22 and 26 in Reference [63]; Figure 6 in Reference [64]; Figures 9 and 10 in Reference [65]) [34,63–65].

Anomalous roundness relationships may convey useful information on both provenance and depositional environment. As a first approximation, we may consider the roundability of heavy minerals as inversely related to their hardness measured by the empirical Mohs' scale [66] (p. 299), [67], [68] (p. 13). Grain roundness has long been held to be suggestive of eolian sand transport [69,70], because rounding is far more effectively achieved by strong grain-to-grain impacts in air than during transport under water [23,71,72]. Association of heavy mineral species with similar hardness but markedly different roundness is another clue revealing mixing of detrital populations from different sources (e.g., angular local-river sand and rounded longshore-drifting grains) (Figure 10 in Reference [65]) [65].

6. Illusions, Shortcuts, and Logical Traps

Geological problems are intricate. Nature's products typically result from the interplay of several controlling factors, and we seldom have adequate tools to detangle the relative contribution of each. Laboratory experiments can hardly reproduce natural complexities at the prohibitively huge spatial and temporal scales of geological processes. Our equations thus generally remain with far too many unknowns and poorly constrained knowns to solve. In face of difficulty and with the desire to hit ground, it may be tempting to recur to instinct, which often merely amounts to prejudice, and fix ourselves on one preferred plausible solution. But plausibility, and worse fashion, seldom paves the way of scientific progress [73].

For instance, considering grain-size classes as transport-invariant sub-populations [74] may lead to the widely shared but wrong assumption that narrowing the size-window in heavy mineral analysis leads to increased consistency, whereas in fact, it maximizes bias (Figure 3). As seen above, high-density minerals settle at the same velocity of, and hence are deposited together with, coarser low-density or platy minerals [27]. Therefore, the different grain-size classes of sorted sediment deposited by a traction current have invariably a notably different heavy mineral suite, and bulk-sample point-counting or multiple-window grain-counting or point-counting represent the only options to estimate the volume percentages of detrital minerals in the sample correctly [30].

6.1. *How Could Zircon Be Enough?*

Owing to its great durability, detrital zircon is preserved widely in polycyclic sands and in ancient sandstones, where it commonly represents one of the few minerals that survived chemical dissolution during diagenesis. Moreover, zircon grains can be dated robustly and routinely at a reasonable cost. These unique features have made zircon the preferred target of provenance research in the last decade. Zircon-age spectra, especially if coupled with Hf isotopic fingerprinting, represent indeed a powerful tool to identify sources characterized by specific age-windows of magmatism and crustal growth. Using detrital-zircon data alone to extract provenance information, and even worse to calculate sediment budgets, is however a risky business [75,76].

First, it requires handling the thorny fertility issue [77–79], which is hard to do precisely and robustly [80]. Zircon-free sources, including mafic and ultramafic rocks, limestone and chert, will remain unseen, and worse, not looked for [81] (p. 85). Second, durable zircon is likely to undergo recycling even several times [82,83]. The information it carries, therefore, may not relate to igneous or metamorphic rocks exposed during the sedimentary cycle in question, but to the igneous or metamorphic sources of the parent sandstones, or even to the igneous or metamorphic sources of the sandstones that sourced the parent sandstones, and so on in a recursive iteration that may climb backwards an unknown number of steps.

Last but not least, the opportunities offered by zircon grains should not lead us to disregard the wide spectrum of other detrital minerals. The expected average zircon content in a siliciclastic sediment is about 1 grain out of 5000, considering that the average zirconium concentration in the upper continental crust is estimated as 190–193 ppm [84,85], that zircon in sediments does not occur exclusively as sand-sized grains but also commonly as tiny inclusions within micas and other minerals [22], and that not all zirconium is hosted in zircon. If the fascination exerted by zircon grains leads us to forget everything else, then we are bound to miss all of the information potentially offered by the other 99.98% of framework grains in the sample (Figure 6).

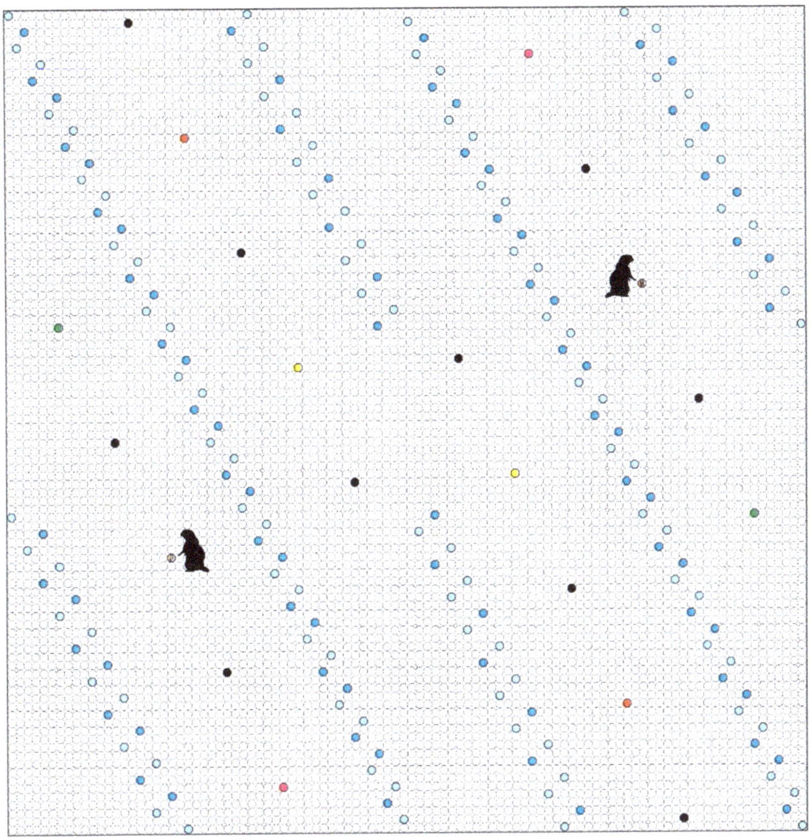

Figure 6. Wonder zircon is hardly enough. This schematic picture represents the quartzose composition typical of an ancient quartzarenite or recycled sand as that carried by the Congo River. Quartz is 98% (white), K-feldspar 1% (light blue) and plagioclase 0.8% (darker blue); opaque (black) and transparent heavy minerals (colourful) are 0.1% each. In such cases, zircon is undoubtedly the most precious vehicle of provenance information, although it represents only 2 grains (both spotted by brave woodchucks) out of 10,000.

6.2. *Jumping to Conclusions: the Plausibility Trap*

To handle complexity, we have to make assumptions about those controlling factors that we believe to be commanding. This is legitimate, provided that our starting assumptions are not rapidly forgotten, thus increasing the risk of moving in a circle to make discoveries that are merely tautological.

The development of sequence stratigraphy [86,87] provides an archetypal example illustrating the pitfalls related to such a faulty conceptual procedure. Although we are well aware that stratigraphic architecture is the product of complexly interacting local autocyclical and regional to global allocyclical factors including tectonics, climate, and sediment supply [88], at a certain point most of the scientific community found advantageous to believe that the stratigraphic record could be almost mechanically interpreted as the univocal response to eustatic change [89]. Eustasy, however, has no plausible cause at the required amplitude and frequency through geological time [90,91], and thus ended up being used as a convenient ruling factor reinforced by circular reasoning rather than by observational evidence [92,93]. A similar illusory way to cut the Gordian knot is to assume that *"detrital modes of sandstone suites primarily reflect the different tectonic settings of provenance terranes"* [46] (p. 333), although they rather reflect the mineralogy of source rocks, which is not univocally related to a geodynamic setting, and are variously affected by environmental and diagenetic bias.

Because the controlling processes and modifying factors are far too many and too poorly constrained to be evaluated with fair accuracy, paleogeographic reconstructions based on detrital modes of ancient sandstones are inevitably as simplistic as "spherical cows" [94]. If on the one hand, our cartoons appear too often as hopeless exercises when observed with a critical mind, on the other hand exposing ourselves to failure may represent the only path to progress in the long run [95]. Rather, the game is really lost when we choose to defend a comfortable theory by dismissing observations. Examples include the widely held belief that mafic detrital minerals such as olivine and pyroxene, or even amphibole and feldspar, can be rapidly eliminated during transport, a fact long demonstrated untrue by studies of natural river systems and laboratory experiments [23,63,73,96,97]. Another idea that *"owes its popularity to plausible reasoning rather than to observational evidence"* is that sand-sized detrital minerals get effectively rounded during fluvial transport, thus enabling us to guess whether the source was near or far [73] (p. 1348).

6.3. The Maturity Misconcept

When confronted with natural phenomena that we do not understand, we humans are prone to believe in mythical narrative. Myths are nice to hear, credible, fearsome, and appeasing. Geological literature is full of mythical scenarios plunged in deep time. Some are skilfully reconstructed (e.g., the dinosaur-inhabited Earth) [98], but others are simply born in our mind rather than inferred by induction. An example is the maturity concept, extrapolated from biology to numerous other disciplines including psychology, sociology and economics, which conveys a positive feeling of natural progress and betterment. Under such a teleologic spell [99], sediments are expected to become cleaner and purer with time and destined to reach a final stage of perfection in both texture (well sorted and rounded "supermature" sandstone) [100] and mineralogy (quartzarenite with "ultrastable" heavy mineral suite) [101]. Quartzose sand containing only zircon, tourmaline and rutile may indeed be the result of polycyclicity [102], but recycling may even produce lithic sands poorer in quartz and less "mature" than their parent sedimentary rocks [13,103,104]. Pure quartzose sand was apparently more common on Earth during the Proterozoic and Lower Paleozoic [70,105] than today [106].

7. The Chemical Moloch

Chemical processes are far more efficient than mechanical processes in removing unstable grains such as ferromagnesian silicates. Dissolution occurs while fluids circulate through soil profiles and more extensively when acting through geological time at the progressively higher temperatures reached during burial diagenesis.

7.1. Pre-Depositional and Post-Depositional Dissolution

Minerals found in igneous and metamorphic rocks grew at temperature and pressure conditions very different from those existing at the surface of the Earth. In sediments, therefore, they find themselves "out of place" by different degrees, as empirically indicated by the Bowen series:

olivine > pyroxene > amphibole, biotite > muscovite, plagioclase > K-feldspar > quartz [107]. Chemical reactions, however, proceed slowly in the lack of circulating fluids and/or at low temperatures, and dissolution rates tend therefore to be lower in arid and cold climates [108,109]. In humid and warm climates, instead, prolonged feldspar hydrolysis can transform granitoid and gneissic source rocks into residual sand consisting almost entirely of quartz [110]. Tourmaline [111], zircon [112,113], and even quartz may be extensively leached [114–116], although generally slower than other minerals. No detrital species is chemically stable in all conditions.

Two important aspects are emphasized in van Loon and Mange [111] (p. 189). First: "*the effects of extreme chemical weathering on a heavy mineral assemblage differ fundamentally from those of burial diagenesis*". Apatite and garnet are widely considered among the most durable species, which is generally true only during diagenesis [12]. Apatite can be highly unstable even in mildly acidic conditions [26], and garnet is rapidly leached out in lateritic soils of equatorial regions, where it may prove to be far more unstable than amphibole (Figure 9 in Reference [44]) [44], thus reversing the stability sequence observed in most ancient siliciclastic successions.

Second: "*the joint occurrence of fresh and strongly weathered grains (with the same chemical composition) of one heavy mineral species indicates that the degree of chemical weathering is a statistical rather than a fixed parameter*". In ancient sandstones, deeply etched or skeletal grains of pyroxene, amphibole, or staurolite may coexist with surviving grains of the same mineral that are only weakly corroded. This "statistical" aspect must be kept in mind while trying to assess the degree of pre-depositional or post-depositional chemical dissolution from surface textures of heavy minerals [19].

7.2. Diagenetic Bias: What You See is not All There Was

The conviction that what is observed is all that matters is typical of naive "fast" thinking [81]. This easygoing attitude proves to be often misleading in ordinary life as in psychological or economic issues, and the potential pitfalls are not less insidious in geological reconstructions. For instance, although we are aware that most was lost from the sedimentary record, we prefer to believe "*that the stratigraphical column in any one place is a long record of sedimentation with occasional gaps*", rather than "*a long gap with only very occasional sedimentation … a lot of holes tied together with sediment*" [88] (p. 35). Since Steensen [117], paleontologists have long realised that most fossils represent a part of the hard skeleton and only a fraction of the living body, but it is much harder to realize that the very same is true in heavy mineral research. We know that most detrital minerals do not resist prolonged diagenesis in most cases (Figure 7), and yet, because we can work only with what we have and cannot retrieve what has been lost, the "*what you see is all there is*" pitfall [81] is difficult to avoid.

Many sedimentary geologists, asked about the most common heavy minerals, will name zircon, tourmaline, and rutile. Probably a minority only will mention amphibole, epidote, garnet, or pyroxene, which are by far the most abundant species in igneous and metamorphic rocks and consequently in detritus derived from them. Researchers working with ancient sandstones indeed find zircon, tourmaline, and rutile far more commonly than other minerals in their heavy mineral mounts, but this is just because zircon, tourmaline, and rutile are the durable ones that stand the best chance to survive chemical attack through multiple sedimentary cycles.

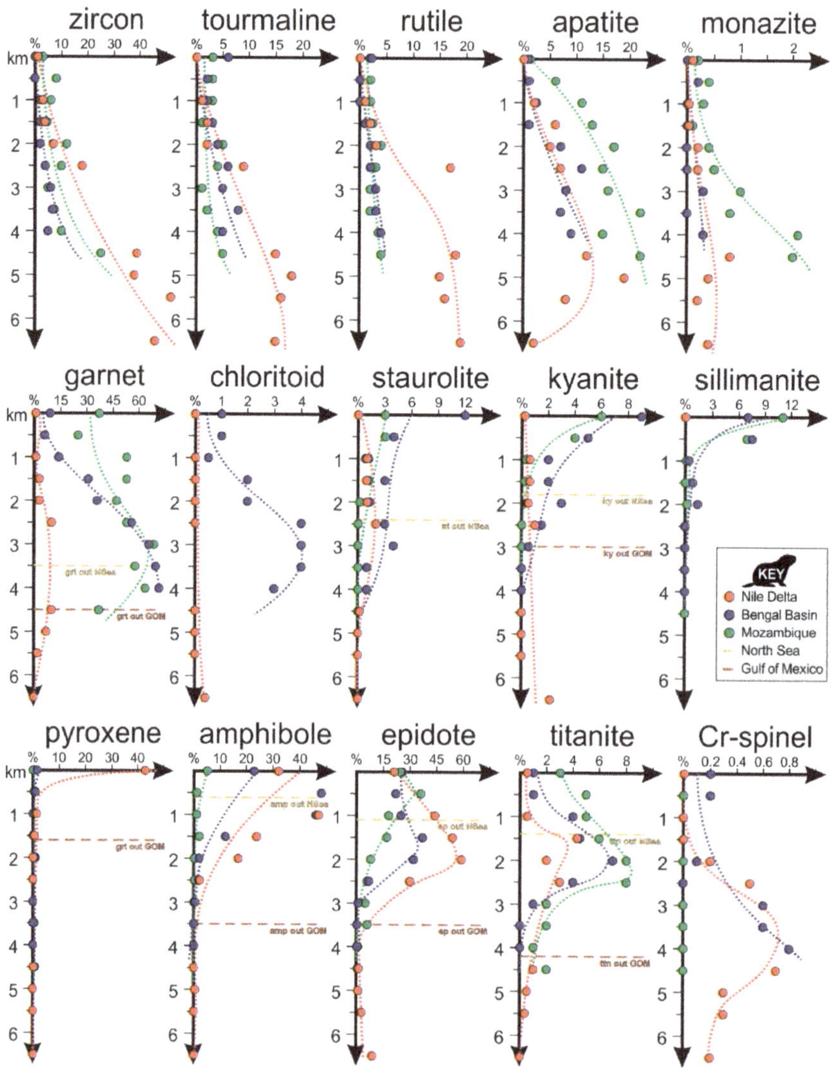

Figure 7. Diagenetic bias. Comparison between the burial-depth distribution of transparent heavy minerals in subsurface sedimentary successions of the Gulf of Mexico, North Sea, Bay of Bengal, Nile Delta, and coastal Mozambique. Data after Figure 2 in [4], Figure 2 in [12], [19], and [118,119]. Heavy minerals are progressively dissolved during burial diagenesis and finally leached out at depths varying notably from basin to basin depending on original mineral abundance, pore-fluid composition, and geothermal gradient (18–26.5 °C/Km for the Nile Delta, 20–30 °C/Km for the Gulf of Mexico, 30–40 °C/Km for the North Sea) [120–122]. The indicative order of relative mineral durability would be zircon ≥ rutile ≥ tourmaline ≥ Cr-spinel ≥ apatite ≥ monazite > chloritoid ≥ garnet > staurolite > kyanite ≥ titanite > epidote > sillimanite > amphibole > pyroxene.

7.3. How to Deal with Ancient Sandstones?

Heavy mineral investigations are important for the industry [123] and not of academic interest only. In unfossiliferous strata, heavy minerals may represent one of the most dependable tools for correlation [6]. In the common unlucky case that only durable minerals have been preserved, one good way to extract information on the ultimate metamorphic or igneous sources is to consider the ratio between couples of minerals having similar density and thus hydraulic behaviour. To this goal, Morton and Hallsworth [124] formally defined the ATi [100 × %apatite/(%apatite + %tourmaline], GZi [100 × %garnet/(%garnet + %zircon)], RZi [100 × %rutile/(%rutile + %zircon)], and CZi [100 × %Cr-spinel/(%Cr-spinel + %zircon)] indices. Because garnet and apatite are particularly sensitive to dissolution in soils [26,44,111,125], the ATi and GZi indices are affected by weathering during the sedimentary cycle, and thus may provide information on paleoclimatic conditions. Care should be taken in handling these indices, which generally reflect selective diagenetic dissolution as well, not all durable minerals being equally durable (e.g., garnet generally far less than zircon) [12] (Figure 3 and p. 239 in reference [12]).

The classical ZTR index (sum of zircon, tourmaline, and rutile over total transparent heavy minerals) [101], widely used to evaluate the durability of a heavy mineral assemblage, integrates the effects of both pre-depositional weathering and post-depositional dissolution accumulated through an undetermined number of sedimentary cycles, and thus it is fundamentally a rough estimator of the extent of recycling. Other indicators such as the Hornblende Colour Index, the Metasedimentary Mineral Index, or the Sillimanite Index (Table 1) can seldom be used, because of the insufficient occurrence of these minerals in ancient sandstones. Varietal studies on durable (e.g., tourmaline, rutile, Cr-spinel, apatite) [126–129] or semi-durable detrital minerals (e.g., garnet) [130] represent another good viable option [131].

Table 1. Heavy mineral indices to be used in modern sediments or sedimentary rocks poorly affected by diagenetic dissolution (based on data from modern Alpine sands) [16,132].

Mineral Index	Greenschist Facies	Amphibolite Facies			Granulite Facies	
		Lower	Middle	Upper	Metasediments	Metagabbro
Hornblende Colour Index	-	$\leq$10	10–30	30–60	>60	$\geq$90
Metasedimentary Mineral Index	0	50	75	100	100	-
Sillimanite Index	-	-	0	$\leq$30	>80	-

Hornblende Colour Index: HCI = (1/3 green + 2/3 green/brown + brown hornblende)/total hornblende × 100; Metasedimentary Mineral Index: MMI = (St/2 + Ky/2 + And/2 + Sil)/(chloritoid + staurolite + kyanite + andalusite + sillimanite) × 100; Sillimanite Index = prismatic sillimanite/total (fibrolitic + prismatic) sillimanite × 100.

7.4. The Recycling Nightmare

In J.L.Borges's novel *"The writing of God"* [133], the Aztec priest Tzinacan is imprisoned in a stone-walled cell in the threatening company of a jaguar. While striving to decipher the divine formula inscribed on the animal's skin (Figure 8), he falls into a noteworthy sedimentological nightmare *"I dreamt there was a grain of sand on the floor of the prison. Indifferent, I slept again. I dreamt I awoke and that on the floor there were two grains of sand. I slept again. I dreamt that the grains of sand were three. They went on multiplying in this way until they filled the prison and I lay dying beneath that hemisphere of sand. I realized that I was dreaming; with a vast effort I roused myself and awoke. It was useless to awake; the innumerable sand was suffocating me. Someone said to me: You have not awakened to wakefulness, but to a previous dream. This dream is enclosed within another, and so on to infinity, which is the number of grains of sand. The path you must retrace is interminable and you will die before you ever really awake"*. This poetical prose describes the sense of doom felt when trying to retrace the ultimate provenance of recycled sand grains. How could we know whether the information carried by a zircon grain does refer to the present cycle, or instead to the previous one, or perhaps to the one before, and so on, and so on? How could we tie such information with any of the many panoramas created and destroyed during the successive evolutionary episodes that shaped the face of our planet? The path is fraught with difficulties, but the junior woodchuck will not be discouraged to take the few hundred million steps needed to reach the moon!

Figure 8. Messages encrypted in animal's skin as bar codes. A few years after Jorge Luis Borges wrote *"The writing of God"* fascinated by the jaguar's fur at Buenos Aires zoo, Alan Turing [134] envisaged a reaction-diffusion model of colour pigments to explain patterns on animal skins [135,136].

8. Conclusions

In this playful little handbook, we have tried to condense the experience acquired in twenty years of heavy mineral research and to provide practical and conceptual advice pointing at those mistakes most commonly made in heavy mineral analysis. The most important piece of advice is not to use heavy minerals alone (or worse a single heavy mineral species only), but to combine heavy mineral and petrographic analysis in thin section routinely. This also helps to visualize the heavy mineral concentration in our sample, the outer message that provides the key to understand the inner message carried by the heavy mineral suite. Our suggestions are based largely on modern case studies, which allow combining information on sources, processes, and products with theoretically complete control of all factors potentially affecting detrital mineralogy. In the study of modern sediments, we can evaluate, and in optimal cases quantify, the relative importance of each process. Moreover, we do not have to tackle the thorny problem of diagenesis, which becomes a nightmare to solve if combined with recycling, because the effect of recycling basically equates to multiple events of chemical dissolution cumulated through successive sedimentary cycles. Sooner or later, the junior woodchuck will find the answer, as Tzinacan eventually deciphered the divine design encrypted in the spots of the jaguar skin.

Author Contributions: Methodology, S.A.; Conceptualizations and writing, E.G.

Funding: Funding provided by Projects MIUR-PRIN 2015EC9PJ5 "The subduction and exhumation of the continental lithosphere: their effects on the structure and evolution of the orogens" and MIUR—Dipartimenti di Eccellenza 2018–2022, Department of Earth and Environmental Sciences, University of Milano-Bicocca.

Acknowledgments: The heavy mineral team at Milano-Bicocca could count through the last decade on the skillful dedication of Mara Limonta, Marta Padoan, and numerous graduate and undergraduate students.

Conflicts of Interest: The authors declare no conflict of interest.

References

1. Mange, M.A.; Wright, D.T. *Heavy Minerals in Use*; Developments in Sedimentology Series; Elsevier: Amsterdam, The Netherlands, 2007; Volume 58, 1283p.
2. Morton, A.C. Value of heavy minerals in sediments and sedimentary rocks for provenance, transport history and stratigraphic correlation. In *Quantitative Mineralogy and Microanalysis of Sediments and Sedimentary Rocks*; Sylvester, P., Ed.; Mineralogical Association of Canada Short Course Series; Mineralogical Association of Canada: Quebec City, QC, Canada, 2012; Volume 42, pp. 133–165.
3. von Eynatten, H.; Dunkl, I. Assessing the sediment factory: The role of single grain analysis. *Earth-Sci. Rev.* **2012**, *115*, 97–120. [CrossRef]
4. Milliken, K.L. Provenance and diagenesis of heavy minerals, Cenozoic units of the northwestern Gulf of Mexico sedimentary basin. In *Heavy Minerals in Use*; Mange, M.A., Wright, D.T., Eds.; Developments in Sedimentology Series; Elsevier: Amsterdam, The Netherlands, 2007; Volume 58, pp. 247–261.
5. Garzanti, E.; Canclini, S.; Moretti Foggia, F.; Petrella, N. Unraveling magmatic and orogenic provenances in modern sands: The back-arc side of the Apennine thrust-belt (Italy). *J. Sediment. Res.* **2002**, *72*, 2–17. [CrossRef]
6. Morton, A.C.; McGill, P. Correlation of hydrocarbon reservoir sandstones using heavy mineral provenance signatures: Examples from the North Sea and adjacent areas. *Minerals* **2019**, *8*, 564. [CrossRef]
7. Andò, S.; Garzanti, E. Raman spectroscopy in heavy-mineral studies. In *Sediment Provenance Studies in Hydrocarbon Exploration and Production*; Scott, R.A., Smyth, H.R., Morton, A.C., Richardson, N., Eds.; Geological Society: London, UK, 2014; Special Publication 386; pp. 395–412.
8. Hofstadter, D.R. The location of meaning. In *Gödel, Escher, Bach: An Eternal Golden Braid; A Metaphorical Fugue on Minds and Machines in the Spirit of Lewis Carroll*; Penguin Books: London, UK, 1979; Chapter VI; pp. 166–188.
9. Garzanti, E.; Andò, S. Heavy-mineral concentration in modern sands: Implications for provenance interpretation. In *Heavy Minerals in Use*; Mange, M.A., Wright, D.T., Eds.; Developments in Sedimentology; Elsevier: Amsterdam, The Netherlands, 2007; Volume 58, pp. 517–545.
10. Garzanti, E.; Andò, S.; Scutellà, M. Actualistic ophiolite provenance: The Cyprus Case. *J. Geol.* **2000**, *108*, 199–218. [CrossRef] [PubMed]
11. Garzanti, E.; Vezzoli, G.; Andò, S. Modern sand from obducted ophiolite belts (Oman, U.A.E.). *J. Geol.* **2002**, *110*, 371–391. [CrossRef]
12. Morton, A.C.; Hallsworth, C. Stability of detrital heavy minerals during burial diagenesis. In *Heavy Minerals in Use*; Mange, M.A., Wright, D.T., Eds.; Developments in Sedimentology Series; Elsevier: Amsterdam, The Netherlands, 2007; Volume 58, pp. 215–245.
13. Garzanti, E.; Al-Juboury, A.I.; Zoleikhaei, Y.; Vermeesch, P.; Jotheri, J.; Akkoca, D.B.; Allen, M.; Andò, S.; Limonta, M.; Padoan, M.; et al. The Euphrates-Tigris-Karun river system: Provenance, recycling and dispersal of quartz-poor foreland-basin sediments in arid climate. *Earth-Sci. Rev.* **2016**, *162*, 107–128. [CrossRef]
14. von Eynatten, H. Petrography and chemistry of sandstones from the Swiss Molasse Basin: An archive of the Oligocene to Miocene evolution of the Central Alps. *Sedimentology* **2003**, *50*, 703–724. [CrossRef]
15. Garzanti, E.; Limonta, M.; Resentini, A.; Bandopadhyay, P.C.; Najman, Y.; Andò, S.; Vezzoli, G. Sediment recycling at convergent plate margins (Indo-Burman Ranges and Andaman-Nicobar Ridge). *Earth-Sci. Rev.* **2013**, *123*, 113–132. [CrossRef]
16. Garzanti, E.; Andò, S.; Vezzoli, G. The Continental Crust as a Source of Sand (Southern Alps cross-section, Northern Italy). *J. Geol.* **2006**, *114*, 533–554. [CrossRef]
17. Garzanti, E.; Andò, S.; France-Lanord, C.; Vezzoli, G.; Najman, Y. Mineralogical and chemical variability of fluvial sediments. 1. Bedload sand (Ganga-Brahmaputra, Bangladesh). *Earth Planet. Sci. Lett.* **2010**, *299*, 368–381. [CrossRef]
18. Gazzi, P. On the heavy mineral zones in the geosyncline series, recent studies in the Northern Appennines, Italy. *J. Sediment. Petrogr.* **1965**, *35*, 109–115.
19. Andò, S.; Garzanti, E.; Padoan, M.; Limonta, M. Corrosion of heavy minerals during weathering and diagenesis: A catalog for optical analysis. *Sediment. Geol.* **2012**, *280*, 165–178. [CrossRef]

20. Malusà, M.G.; Garzanti, E. The sedimentology of detrital thermochronology. In *Fission-track Thermochronology and Its Application to Geology*; Malusà, M.G., Fitzgerald, P.G., Eds.; Springer: Berlin, Germany, 2019; pp. 123–143.
21. Resentini, A.; Malusà, M.G.; Garzanti, E. MinSORTING: An Excel$^{(R)}$ worksheet for modelling mineral grain-size distribution in sediments, with application to detrital geochronology and provenance studies. *Comput. Geosci.* **2013**, *59*, 90–97. [CrossRef]
22. Vermeesch, P.; Rittner, M.; Petrou, E.; Omma, J.; Mattinson, C.; Garzanti, E. High throughput petrochronology and sedimentary provenance analysis by automated phase mapping and LAICPMS. *Geochem. Geophys. Geosystems* **2017**, *18*, 4096–4109. [CrossRef]
23. Garzanti, E.; Resentini, A.; Andò, S.; Vezzoli, G.; Vermeesch, P. Physical controls on sand composition and relative durability of detrital minerals during long-distance littoral and eolian transport (coastal Namibia). *Sedimentology* **2015**, *62*, 971–996. [CrossRef]
24. Carver, R.E. Heavy-mineral separation. In *Procedures in Sedimentary Petrology*; Carver, R.E., Ed.; Wiley: New York, NY, USA, 1971; pp. 427–452.
25. Morton, A.C. Heavy minerals in provenance studies. In *Provenance of Arenites*; Zuffa, G.G., Ed.; NATO-ASI Series; Springer: Dordrecht, The Netherlands, 1985; Volume 148, pp. 249–277.
26. Bateman, R.M.; Catt, J.A. Provenance and palaeoenvironmental interpretation of superficial deposits, with particular reference to post-depositional modification of heavy-mineral assemblages. In *Heavy Minerals in Use*; Mange, M.A., Wright, D.T., Eds.; Developments in Sedimentology Series; Elsevier: Amsterdam, The Netherlands, 2007; Volume 58, pp. 151–188.
27. Rubey, W.W. The size-distribution of heavy minerals within a water-laid sandstone. *J. Sediment. Petrol.* **1933**, *3*, 3–29.
28. Rittenhouse, G. Transportation and deposition of heavy minerals. *Geol. Soc. Am. Bull.* **1943**, *54*, 1725–1780. [CrossRef]
29. Mange, M.A.; Maurer, H.F.W. *Heavy Minerals in Colour*; Chapman and Hall: London, UK, 1992; 147p.
30. Garzanti, E.; Andò, S.; Vezzoli, G. Grain-size dependence of sediment composition and environmental bias in provenance studies. *Earth Planet. Sci. Lett.* **2009**, *277*, 422–432. [CrossRef]
31. Parfenoff, A.; Pomerol, C.; Tourenq, J. *Les minéraux en grains—méthodes d' étude et détermination*; Masson: Paris, France, 1970; 578p.
32. Galehouse, J.S. Point counting. In *Procedures in Sedimentary Petrology*; Carver, R.E., Ed.; Wiley: New York, NY, USA, 1971; pp. 385–407.
33. Berman, R. A nomogram for obtaining per cent composition by weight from mineral-grain counts. *J. Sediment. Petrol.* **1953**, *23*, 120–123.
34. Garzanti, E.; Andò, S.; Vezzoli, G. Settling-equivalence of detrital minerals and grain-size dependence of sediment composition. *Earth Planet. Sci. Lett.* **2008**, *273*, 138–151. [CrossRef]
35. Hunter, R. E A rapid method for determining weight percentages of unsieved heavy minerals. *J. Sediment. Petrol.* **1967**, *37*, 521–529.
36. Galehouse, J.S. Counting grain mounts; number percentage vs. number frequency. *J. Sediment. Petrol.* **1969**, *39*, 812–815. [CrossRef]
37. Chayes, F. The theory of thin-section analysis. *J. Geol.* **1954**, *62*, 92–101. [CrossRef]
38. Dewey, J.F. Orogeny can be very short. *Proc. Natl. Acad. Sci. USA* **2005**, *102*, 15286–15293. [CrossRef] [PubMed]
39. Dickinson, W.R.; Suczek, C.A. Plate tectonics and sandstone composition. *Am. Assoc. Pet. Geol. Bull.* **1979**, *63*, 2164–2172.
40. Garzanti, E. From static to dynamic provenance analysis—Sedimentary petrology upgraded. *Sediment. Geol.* **2016**, *336*, 3–13. [CrossRef]
41. Garzanti, E. Petrographic classification of sand and sandstone. *Earth-Sci. Rev.* **2019**. [CrossRef]
42. Garzanti, E.; Andò, S. Plate tectonics and heavy-mineral suites of modern sands. In *Heavy Minerals in Use*; Mange, M.A., Wright, D.T., Eds.; Developments in Sedimentology Series; Elsevier: Amsterdam, The Netherlands, 2007; Volume 58, pp. 741–763.
43. Garzanti, E.; Vezzoli, G.; Andò, S.; Castiglioni, G. Petrology of rifted-margin sand (Red Sea and Gulf of Aden, Yemen). *J. Geol.* **2001**, *109*, 277–297. [CrossRef]

44. Garzanti, E.; Padoan, M.; Andò, S.; Resentini, A.; Vezzoli, G.; Lustrino, M. Weathering and relative durability of detrital minerals in equatorial climate: Sand petrology and geochemistry in the East African Rift. *J. Geol.* **2013**, *121*, 547–580. [CrossRef]
45. Garzanti, E.; Vermeesch, P.; Padoan, M.; Resentini, A.; Vezzoli, G.; Andò, S. Provenance of passive-margin sand (Southern Africa). *J. Geol.* **2014**, *122*, 17–42. [CrossRef]
46. Dickinson, W.R. Interpreting provenance relations from detrital modes of sandstones. In *Provenance of Arenites*; Zuffa, G.G., Ed.; NATO-ASI Series; Springer: Dordrecht, The Netherlands, 1985; Series 148; pp. 333–361.
47. Marsaglia, K.M.; Ingersoll, R.V. Compositional trends in arc-related, deep-marine sand and sandstone: A reassessment of magmatic-arc provenance. *Geol. Soc. Am. Bull.* **1992**, *104*, 1637–1649. [CrossRef]
48. Garzanti, E.; Limonta, M.; Vezzoli, G.; An, W.; Wang, J.; Hu, X. Petrology and multimineral fingerprinting of modern sand generated from a dissected magmatic arc (Lhasa River, Tibet). In *Tectonics, Sedimentary Basins, and Provenance: A Celebration of William R. Dickinson's Career*; Ingersoll, R.V., Lawton, T.F., Graham, S.A., Eds.; Geological Society of America: Boulder, CO, USA, 2019; Special Paper 540; pp. 197–221.
49. Gill, J. *Orogenic Andesites and Plate Tectonics*; Springer: Berlin, Germany, 1981; 390p.
50. Garzanti, E.; Doglioni, C.; Vezzoli, G.; Andò, S. Orogenic Belts and Orogenic Sediment Provenances. *J. Geol.* **2007**, *115*, 315–334. [CrossRef]
51. Garzanti, E.; Resentini, A.; Vezzoli, G.; Andò, S.; Malusà, M.G.; Padoan, M.; Paparella, P. Detrital fingerprints of fossil continental-subduction zones (Axial Belt Provenance, European Alps). *J. Geol.* **2010**, *118*, 341–362. [CrossRef]
52. Doglioni, C.; Harabaglia, P.; Merlini, S.; Mongelli, F.; Peccerillo, A.; Piromallo, C. Orogens and slabs vs. their direction of subduction. *Earth Sci. Rev.* **1999**, *45*, 167–208. [CrossRef]
53. Garzanti, E.; Vezzoli, G.; Lombardo, B.; Andò, S.; Mauri, E.; Monguzzi, S.; Russo, M. Collision-orogen Provenance (Western and Central Alps): Detrital signatures and unroofing trends. *J. Geol.* **2004**, *112*, 145–164. [CrossRef]
54. Morton, A.C.; Johnsson, M.J. Factors influencing the composition of detrital heavy mineral suites in Holocene sands of the Apure River drainage basin, Venezuela. *Geol. Soc. Am. Spec. Pap.* **1993**, *284*, 171–185.
55. Limonta, M.; Garzanti, E.; Resentini, A.; Andò, S.; Boni, M.; Bechstädt, T. Multicyclic sediment transfer along and across convergent plate boundaries (Barbados, Lesser Antilles). *Basin Res.* **2015**, *27*, 696–713. [CrossRef]
56. Gazzi, P.; Zuffa, G.G.; Gandolfi, G.; Paganelli, L. Provenienza e dispersione litoranea delle sabbie delle spiagge adriatiche fra le foci dell' Isonzo e del Foglia: Inquadramento regionale. *Mem. Soc. Geol. Ital.* **1973**, *12*, 1–37.
57. Komar, P.D. The entrainment, transport and sorting of heavy minerals by waves and currents. In *Heavy Minerals in Use*; Mange, M.A., Wright, D.T., Eds.; Developments in Sedimentology Series; Elsevier: Amsterdam, The Netherlands, 2007; Volume 58, pp. 3–48.
58. Trask, C.B.; Hand, B.M. Differential transport of fall-equivalent sand grains, Lake Ontario, New York. *J. Sedim. Petrol.* **1985**, *55*, 226–234.
59. Slingerland, R.; Smith, N.D. Occurrence and formation of water-laid placers. *Ann. Rev. Earth Planet. Sci.* **1986**, *14*, 113–147. [CrossRef]
60. Garzanti, E.; Dinis, P.; Vermeesch, P.; Andò, S.; Hahn, A.; Huvi, J.; Limonta, M.; Padoan, M.; Resentini, A.; Rittner, M.; et al. Sedimentary processes controlling ultralong cells of littoral transport: Placer formation and termination of the Orange sand highway in southern Angola. *Sedimentology* **2018**, *65*, 431–460. [CrossRef]
61. Cheng, N.S. Simplified settling velocity formula for sediment particle. *J. Hydraul. Eng.* **1997**, *123*, 149–152. [CrossRef]
62. Komar, P.D.; Li, Z. Application of grain-pivoting and sliding analyses to selective entrainment of gravel and to flow-competence evaluations. *Sedimentology* **1988**, *35*, 681–695. [CrossRef]
63. Garzanti, E.; Andò, S.; Padoan, M.; Vezzoli, G.; El Kammar, A. The modern Nile sediment system: Processes and products. *Quat. Sci. Rev.* **2015**, *130*, 9–56. [CrossRef]
64. Garzanti, E.; Vermeesch, P.; Al-Ramadan, K.A.; Andò, S.; Limonta, M.; Rittner, M.; Vezzoli, G. Tracing transcontinental sand transport: From Anatolia-Zagros to the Rub' al Khali Sand Sea. *J. Sediment. Res.* **2017**, *87*, 1196–1213. [CrossRef]
65. Resentini, A.; Andò, S.; Garzanti, E. Quantifying roundness of detrital minerals by image analysis: Sediment transport, shape effects, and provenance implications. *J. Sediment. Res.* **2018**, *88*, 276–289. [CrossRef]

66. Mackie, W. On the laws that govern the rounding of particles of sand. *Trans. Edinb. Geol. Soc.* **1897**, *7*, 298–311. [CrossRef]
67. Marsland, P.S.; Woodruff, J.G. A study of the effects of wind transportation on grains of several minerals. *J. Sediment. Petrol.* **1937**, *7*, 18–30.
68. Folk, R.L. *Petrology of Sedimentary Rocks*; Hemphill Publishing Co.: Austin, TX, USA, 1980; 182p.
69. Berkey, C.P. Paleogeography of Saint Peter time. *Geol. Soc. Am. Bull.* **1906**, *17*, 229–250. [CrossRef]
70. Dott, R.H. The importance of eolian abrasion in supermature quartz sandstones and the paradox of weathering on vegetation-free landscapes. *J. Geol.* **2003**, *111*, 387–405. [CrossRef]
71. Twenhofel, W.H. The rounding of sand grains. *J. Sediment. Petrol.* **1945**, *15*, 59–71. [CrossRef]
72. Kuenen, P.H. Experimental abrasion: 4, Eolian action. *J. Geol.* **1960**, *68*, 427–449. [CrossRef]
73. Russell, R.D. Mineral composition of Mississippi River sands. *Geol. Soc. Am. Bull.* **1937**, *48*, 1307–1348. [CrossRef]
74. Weltje, G.J. A quantitative approach to capturing the compositional variability of modern sands. *Sediment. Geol.* **2004**, *171*, 59–77. [CrossRef]
75. Malusà, M.G.; Carter, A.; Limoncelli, M.; Villa, I.M.; Garzanti, E. Bias in detrital zircon geochronology and thermochronometry. *Chem. Geol.* **2013**, *359*, 90–107. [CrossRef]
76. Vezzoli, G.; Garzanti, E.; Limonta, M.; Andó, S.; Yang, S. Erosion patterns in the Changjiang (Yangtze River) catchment revealed by bulk-sample versus single-mineral provenance budgets. *Geomorphology* **2016**, *261*, 177–192. [CrossRef]
77. Moecher, D.P.; Samson, S.D. Differential zircon fertility of source terranes and natural bias in the detrital zircon record: Implications for sedimentary provenance analysis. *Earth Planet. Sci. Lett.* **2006**, *247*, 252–266. [CrossRef]
78. Dickinson, W.R. Impact of differential zircon fertility of granitoid basement rocks in North America on age populations of detrital zircons and implications for granite petrogenesis. *Earth Planet. Sci. Lett.* **2008**, *275*, 80–92. [CrossRef]
79. Sláma, J.; Košler, J. Effects of sampling and mineral separation on accuracy of detrital zircon studies. *Geochem. Geophys. Geosystems* **2012**, *13*, Q05007. [CrossRef]
80. Malusà, M.G.; Resentini, A.; Garzanti, E. Hydraulic sorting and mineral fertility bias in detrital geochronology. *Gondwana Res.* **2016**, *31*, 1–19. [CrossRef]
81. Kahneman, D. *Thinking, Fast and Slow*; Penguin: London, UK, 2011; 499p.
82. Garzanti, E.; Vermeesch, P.; Andó, S.; Vezzoli, G.; Valagussa, M.; Allen, K.; Khadi, K.A.; Al-Juboury, I.A. Provenance and recycling of Arabian desert sand. *Earth-Sci. Rev.* **2013**, *120*, 1–19. [CrossRef]
83. Andersen, T.; Kristoffersen, M.; Elburg, M.A. How far can we trust provenance and crustal evolution information from detrital zircons? A South African case study. *Gondwana Res.* **2016**, *34*, 129–148. [CrossRef]
84. Taylor, S.R.; McLennan, S.M. The geochemical evolution of the continental crust. *Rev. Geophys.* **1995**, *33*, 241–265. [CrossRef]
85. Rudnick, R.L.; Gao, S. Composition of the continental crust. In *Treatise on Geochemistry, The Crust*; Rudnick, R.L., Holland, H.D., Turekian, K.K., Eds.; Elsevier Pergamon: Oxford, UK, 2003; Volume 3, pp. 1–64.
86. Vail, P.R.; Mitchum, R.M., Jr.; Thompson, S., III. Seismic stratigraphy and global changes of sea level, part 3: Relative changes of sea level from coastal onlap. In *Seismic Stratigraphy—Applications to Hydrocarbon Exploration*; Payton, C.E., Ed.; American Association of Petroleum Geologists: San Antonio, TX, USA, 1977; Memoir 26; pp. 63–81.
87. Haq, B.; Hardenbol, J.; Vail, P.R. Chronology of fluctuating sea levels since the Triassic (250 million years ago to present). *Science* **1987**, *235*, 1156–1167. [CrossRef] [PubMed]
88. Ager, D. More gaps than record. In *The Nature of the Stratigraphical Record*; Halsted Press, Wiley: New York, NY, USA, 1971; Chapter 3; pp. 28–35.
89. Miall, C.E.; Miall, A.D. The Exxon factor: The roles of corporate and academic science in the emergence and legitimation of a new global model of sequence stratigraphy. *Sociol. Q.* **2002**, *43*, 307–334. [CrossRef]
90. Pitman, W.C.; Golovchenko, X. *The Effect of Sealevel Change on the Shelfedge and Slope of Passive Margins*; Society of Economic Paleontologists and Mineralogists: Broken Arrow, OK, USA, 1983; Special Publication 33; pp. 41–58.

91. Dewey, J.F.; Pitman, W.C. Sea-level changes: Mechanisms, magnitudes and rates. In *Paleogeographic Evolution and Non-Glacial Eustasy, Northern South America*; Pindell, J.L., Drake, C., Eds.; Society of Economic Paleontologists and Mineralogists: Broken Arrow, OK, USA, 1998; Special Publication 58; pp. 1–16.
92. Miall, A.D. Eustatic sea level changes interpreted from seismic stratigraphy: A critique of the methodology with particular reference to the North Sea Jurassic record. *Am. Assoc. Pet. Geol. Bull.* **1986**, *70*, 131–137.
93. Miall, A.D. Exxon global cycle chart: An event for every occasion? *Geology* **1992**, *20*, 787–790. [CrossRef]
94. Paola, C.; Leeder, M. Environmental dynamics: Simplicity versus complexity. *Nature* **2011**, *469*, 38–39. [CrossRef] [PubMed]
95. Arreguin-Toft, I. How the weak win wars: A theory of asymmetric conflict. *Int. Secur.* **2001**, *26*, 93–128. [CrossRef]
96. Shukri, N.M. The mineralogy of some Nile sediments: Geological Society of London. *Q. J.* **1950**, *105*, 511–534.
97. Kuenen, P.H. Experimental abrasion: 3. Fluviatile action on sand. *Am. J. Sci.* **1959**, *257*, 172–190. [CrossRef]
98. Gould, S.J. Dinomania. *N. Y. Rev. Books* **1993**, *40*, 52–53.
99. Popper, K. Historicism and the myth of destiny. In *The Open Society and Its Enemies*; Princeton University Press: Princeton, NJ, USA, 1994; pp. 7–9.
100. Folk, R.L. Stages of textural maturity in sedimentary rocks. *J. Sediment. Petrol.* **1951**, *21*, 127–130. [CrossRef]
101. Hubert, J.F. A zircon–tourmaline–rutile maturity index and the interdependence of the composition of heavy minerals assemblages with the gross composition and texture of sandstones. *J. Sediment. Petrol.* **1962**, *32*, 440–450.
102. Muhs, D.R. Mineralogical maturity in dunefields of North America, Africa and Australia. *Geomorphology* **2004**, *59*, 247–269. [CrossRef]
103. Cavazza, W.; Zuffa, G.G.; Camporesi, C.; Ferretti, C. Sedimentary recycling in a temperate climate drainage basin (Senio River, north-central Italy): Composition of source rock, soil profiles, and fluvial deposits. In *Processes Controlling the Composition of Clastic Sediments*; Johnsson, M.J., Basu, A., Eds.; Geological Society of America: Boulder, CO, USA, 1993; Special Paper 284; pp. 247–262.
104. Fontana, D.; Parea, G.C.; Bertacchini, M.; Bessi, P. Sand production by chemical and mechanical weathering of well lithified siliciclastic turbidites of the Northern Apennines (Italy). *Memorie Descrittive della Carta Geologica d'Italia* **2003**, *61*, 51–60.
105. Avigad, D.; Sandler, A.; Kolodner, K.; Stern, R.J.; McWilliams, M.O.; Miller, N.; Beyth, M. Mass-production of Cambro-Ordovician quartz-rich sandstone as a consequence of chemical weathering of Pan-African terranes. Environmental implications. *Earth Planet. Sci. Lett.* **2005**, *240*, 818–826. [CrossRef]
106. Garzanti, E. The maturity myth in sedimentology and provenance analysis. *J. Sediment. Res.* **2017**, *87*, 353–365. [CrossRef]
107. Goldich, S.S. A study in rock-weathering. *J. Geol.* **1938**, *46*, 17–58. [CrossRef]
108. Nesbitt, H.W.; Young, G.M. Petrogenesis of sediments in the absence of chemical weathering: Effects of abrasion and sorting on bulk composition and mineralogy. *Sedimentology* **1996**, *43*, 341–358. [CrossRef]
109. Potter, P.E.; Huh, Y.; Edmond, J.M. Deep-freeze petrology of Lena River sand, Siberia. *Geology* **2001**, *29*, 999–1002. [CrossRef]
110. Johnsson, M.J.; Stallard, R.F.; Meade, R.H. First-cycle quartz arenites in the Orinoco River basin: Venezuela and Colombia. *J. Geol.* **1988**, *96*, 263–277. [CrossRef]
111. van Loon, A.J.; Mange, A.M. "In situ" dissolution of heavy minerals through extreme weathering, and the application of the surviving assemblages and their dissolution characteristics to correlation of Dutch and German silver sands. In *Heavy Minerals in Use*; Mange, M.A., Wright, D.T., Eds.; Developments in Sedimentology Series; Elsevier: Amsterdam, The Netherlands, 2007; Volume 58, pp. 189–213.
112. Carroll, D. Weatherability of zircon. *J. Sediment. Res.* **1953**, *23*, 106–116.
113. Colin, F.; Alarcon, C.; Vieillard, P. Zircon: An immobile index in soils? *Chem. Geol.* **1993**, *107*, 273–276. [CrossRef]
114. Crook, K.A.W. Weathering and roundness of quartz sand grains. *Sedimentology* **1968**, *11*, 171–182. [CrossRef]
115. Cleary, W.J.; Conolly, J.R. Distribution and genesis of quartz in a piedmont-coastal plain environment. *Geol. Soc. Am. Bull.* **1971**, *82*, 2755–2766. [CrossRef]
116. Schulz, M.S.; White, A.F. Chemical weathering in a tropical watershed, Luquillo Mountains, Puerto Rico III: Quartz dissolution rates. *Geochim. Et Cosmochim. Acta* **1999**, *63*, 337–350. [CrossRef]

117. Steensen, N. *De Solido Intra Solidum Naturaliter Contento Dissertationis Prodromus*; ex Typographia sub signo Stellae: Florentiae, Italy, 1669; 78p.
118. Chanvry, E.; Andò, S.; Garzanti, E.; Guillocheau, F.; Dall'Asta, M.; Beaufort, D.; Mas, P.P. Impact of hinterland evolution in mineralogy of clastic sediments: First results from mineralogical analysis focused on the Zambezi system during Meso-Cenozoic times. In Proceedings of the EGU General Assembly 2018, Vienna, Austria, 8–13 April 2018. Geophysical Research Abstracts, 20, EGU2018-18077.
119. Garzanti, E.; Andò, S.; Limonta, M.; Fielding, L.; Najman, Y. Diagenetic control on mineralogical suites in sand, silt, and mud (Cenozoic Nile Delta): Implications for provenance reconstructions. *Earth-Sci. Rev.* **2018**, *185*, 122–139. [CrossRef]
120. Riad, S.; Abdelrahman, E.M.; Refai, E.; El-Ghalban, H.M. Geothermal studies in the Nile Delta, Egypt. *J. Afr. Earth Sci.* **1989**, *9*, 637–649. [CrossRef]
121. Sharp, J.M.; Galloway, W.E.; Land, L.S.; McBride, E.F.; Blanchard, P.E.; Bodner, D.P.; Dutton, S.P.; Farr, M.R.; Gold, P.B.; Jackson, T.J.; et al. Diagenetic processes in Northwest Gulf of Mexico sediments. In *Diagenesis II*; Chilingarian, G.V., Wolf, K.H., Eds.; Developments in Sedimentology Series; Elsevier: Amsterdam, The Netherlands, 1988; Volume 43, pp. 43–133.
122. Evans, T.R.; Coleman, N.C. North Sea geothermal gradients. *Nature* **1974**, *247*, 28–30. [CrossRef]
123. Smyth, H.R.; Morton, A.; Richardson, N.; Scott, R.A. Sediment provenance studies in hydrocarbon exploration and production: An introduction. In *Sediment Provenance Studies in Hydrocarbon Exploration and Production*; Scott, R.A., Smyth, H.R., Morton, A.C., Richardson, N., Eds.; Geological Society: London, UK, 2014; Special Publication 386; pp. 1–6.
124. Morton, A.C.; Hallsworth, C.R. Identifying provenance specific features of detrital heavy mineral assemblages in sandstones. *Sediment. Geol.* **1994**, *90*, 241–256. [CrossRef]
125. Velbel, M.A. Formation of protective surface layers during silicate-mineral weathering under well-leached, oxidizing conditions. *Am. Mineral.* **1993**, *78*, 405–414.
126. Henry, D.J.; Guidotti, C.V. Tourmaline as a petrogenetic indicator mineral: An example from the staurolite-grade metapelites of NW Maine. *Am. Mineral.* **1985**, *70*, 1–15.
127. Meinhold, G. Rutile and its applications in Earth sciences. *Earth-Sci. Rev.* **2010**, *102*, 2–28. [CrossRef]
128. Hu, X.; An, W.; Wang, J.; Garzanti, E.; Guo, R. Himalayan detrital chromian spinels and timing of Indus-Yarlung ophiolite erosion. *Tectonophysics* **2014**, *621*, 60–68. [CrossRef]
129. Malusà, M.G.; Wang, J.; Garzanti, E.; Liu, Z.C.; Villa, I.M.; Wittmann, H. Trace-element and Nd-isotope systematics in detrital apatite of the Po river catchment: Implications for provenance discrimination and the lag-time approach to detrital thermochronology. *Lithos* **2017**, *290–291*, 48–59.
130. Krippner, A.; Meinhold, G.; Morton, A.C.; von Eynatten, H. Evaluation of garnet discrimination diagrams using geochemical data of garnets derived from various host rocks. *Sediment. Geol.* **2014**, *306*, 36–52. [CrossRef]
131. Mange, M.A.; Morton, A.C. Geochemistry of heavy minerals. In *Heavy Minerals in Use*; Mange, M.A., Wright, D.T., Eds.; Developments in Sedimentology; Elsevier: Amsterdam, The Netherlands, 2007; Volume 58, pp. 345–391.
132. Andò, S.; Morton, A.; Garzanti, E. Metamorphic grade of source rocks revealed by chemical fingerprints of detrital amphibole and garnet. In *Sediment Provenance Studies in Hydrocarbon Exploration and Production*; Scott, R.A., Smyth, H.R., Morton, A.C., Richardson, N., Eds.; Geological Society: London, UK, 2017; Special Publication 386; pp. 351–371.
133. Borges, J.L. La escritura del dios. *Sur* **1949**, *172*, 7–12.
134. Turing, A. The chemical basis of morphogenesis. *Philos. Trans. R. Soc. Lond.* **1952**, *237*, 37–72.
135. Liu, R.T.; Liaw, S.S.; Maini, P.K. Two-stage Turing model for generating pigment patterns on the leopard and the jaguar. *Phys. Rev. E* **2006**, *74*, 011914. [CrossRef] [PubMed]
136. Dougoud, M.; Mazza, C.; Schwaller, B.; Pecze, L. The phenomenon of growing surface interference explains the rosette pattern of jaguar. *arXiv*, 2017; arXiv:1711.05574.

minerals

MDPI

Article

Exploratory Analysis of Provenance Data Using R and the Provenance Package

Pieter Vermeesch

University College London, Gower Street, London WC1E 6BT, UK; p.vermeesch@ucl.ac.uk

Received: 17 January 2019; Accepted: 15 March 2019; Published: 22 March 2019

Abstract: The provenance of siliclastic sediment may be traced using a wide variety of chemical, mineralogical and isotopic proxies. These define three distinct data types: (1) compositional data such as chemical concentrations; (2) point-counting data such as heavy mineral compositions; and (3) distributional data such as zircon U-Pb age spectra. Each of these three data types requires separate statistical treatment. Central to any such treatment is the ability to quantify the 'dissimilarity' between two samples. For compositional data, this is best done using a logratio distance. Point-counting data may be compared using the chi-square distance, which deals better with missing components (zero values) than the logratio distance does. Finally, distributional data can be compared using the Kolmogorov–Smirnov and related statistics. For small datasets using a single provenance proxy, data interpretation can sometimes be done by visual inspection of ternary diagrams or age spectra. However, this no longer works for larger and more complex datasets. This paper reviews a number of multivariate ordination techniques to aid the interpretation of such studies. Multidimensional Scaling (MDS) is a generally applicable method that displays the salient dissimilarities and differences between multiple samples as a configuration of points in which similar samples plot close together and dissimilar samples plot far apart. For compositional data, classical MDS analysis of logratio data is shown to be equivalent to Principal Component Analysis (PCA). The resulting MDS configurations can be augmented with compositional information as biplots. For point-counting data, classical MDS analysis of chi-square distances is shown to be equivalent to Correspondence Analysis (CA). This technique also produces biplots. Thus, MDS provides a common platform to visualise and interpret all types of provenance data. Generalising the method to three-way dissimilarity tables provides an opportunity to combine several datasets together and thereby facilitate the interpretation of 'Big Data'. This paper presents a set of tutorials using the statistical programming language R. It illustrates the theoretical underpinnings of compositional data analysis, PCA, MDS and other concepts using toy examples, before applying these methods to real datasets with the `provenance` package.

Keywords: sediment; provenance; statistics; zircon; heavy minerals; point counting; petrography

1. Introduction

At its most basic level, sedimentary provenance analysis identifies the mineralogical, chemical or isotopic composition of individual grains, or assemblages of multiple grains in siliclastic sediment. These properties can then be used to group samples of similar affinity, and thereby trace the flow of sediment through a sediment routing system, e.g., [1–5]. Different levels of statistical complexity arise when multiple samples are compared to each other, or when multiple provenance proxies are applied to multiple samples.

Using a number of short tutorials, this paper will introduce several simple but effective exploratory data analysis techniques that can help to make geological sense of 'Big Data' in a sedimentary provenance context. The term 'exploratory' means that these techniques allow the user to explore the data independent of any prior knowledge about the geological setting [6–9]. It groups a number of graphical methods to visualise the data and reveal patterns of similarity and differences between samples and variables. This paper will not introduce methods such as discriminant analysis that formally assign samples to pre-defined provenance areas or petrotectonic settings [10,11].

These notes do by no means claim to give a comprehensive overview of exploratory data analysis. The selection of methods presented herein is heavily biased towards techniques that are implemented in a software package for (sedimentary) geology that was created by Vermeesch et al. [12].

`provenance` is available free of charge at the Comprehensive R Archive Network (CRAN, https://cran.r-project.org/package=provenance), on GitHub (http://github.com/pvermees/provenance), or via http://provenance.london-geochron.com. The package is written in the statistical programming language `R`, which is available for Windows, Mac OS-X and Linux/Unix. The easiest way to install the latest stable version of the package is to first install `R` from http://r-project.org and then type the following code at the command prompt (i.e., the '>'):

```
> install.packages("provenance")
```

Once installed, the package can be loaded by typing:

```
> library(provenance)
```

There are two ways to use `provenance`. The first of these is through a query-based user interface. To access this interface, type:

```
> provenance()
```

The main advantage of the query-based user interface is that it does not require any knowledge of `R`. Its main disadvantage is the relative lack of flexibility and the difficulty to automate complex and/or repetitive tasks. The second way to use `provenance` is via the `R` language itself. This is the quicker and more flexible option, whose only downside is a steeper learning curve compared to the query-based interface. This tutorial will help the reader to climb this learning curve whilst explaining the theoretical underpinnings of the methods that are implemented in the package.

This text assumes that the reader has a basic understanding of the `R` programming language, although a short tutorial is provided in the Appendix A for readers who lack such prior knowledge. The paper also assumes that the reader has some basic statistical knowledge. More specifically, (s)he is expected to be familiar with the normal distribution, and understand the meaning of the arithmetic mean, standard deviation and confidence intervals. The normal distribution underpins much of 'conventional' statistics, but we will see that it rarely applies to provenance data. This, in fact, is the main take-home message of this paper.

There exist three fundamental types of provenance data:

1. Chemical data such as major and trace element concentrations are known as *compositional* data. Sections 2 and 3 show that the statistical analysis of this class of data is fraught with difficulties. Fortunately, these are easily overcome by means of 'Aitchison's logratio transformation'. This transformation is a prerequisite to further statistical treatment, including Principal Component Analysis and compositional biplots of multi-sample datasets (Sections 2 and 3).
2. Categorical data such as bulk petrography and heavy mineral compositions are known as *point-counting* data. These are closely related to, but are fundamentally different from, compositional data. Compositional data consist of strictly positive real numbers that are subject to a constant-sum constraint and whose analytical precision can generally be ignored. In contrast, point-counting data contain integer values that may be greater than or equal to zero, and whose multinomial uncertainty is significant compared to the underlying compositional dispersion. Section 4 shows that both of these differences can be captured by a combination of logistic normal and multinomial statistics.

3. Detrital age spectra form a third class of data that will be referred to as *distributional* data. Sections 5 and 7 introduce kernel density estimation, the Kolmogorov–Smirnov statistic, and multidimensional scaling as ways to visualise, compare, and interpret distributional data.

Finally, Section 11 will consider the case where multiple compositional, point-counting and/or distributional datasets are combined. Procrustes analysis and 3-way multidimensional scaling are statistical techniques that aim to extract geologically meaningful trends from such 'Big Data' [13].

2. Ratio Data

Summary: *This tutorial investigates the ratios of two sets of random numbers. It shows that the arithmetic mean and confidence intervals of these synthetic data yield nonsensical results. These problems are solved by a logarithmic transformation. This simple example has important implications because ratio data are common in sedimentary provenance analysis, and are closely related to compositional data, which are introduced in Section 3.*

Many statistical operations assume normality. This includes averaging, the construction of confidence intervals, regression, etc. Although Gaussian distributions are common, it would be unwise to assume normality for all datasets. This paper makes the point that, more often than not, the normality assumption is invalid in the context of sedimentary provenance analysis. Ignoring this non-normality can lead to counter-intuitive and plainly wrong results.

To illustrate this point, we will now consider the simple case of *ratio data*, which are quite common in the Earth Sciences. Take, for example, the ratio of apatite to tourmaline in heavy mineral analysis, which has been used to indicate the duration of transport and storage prior to deposition [14]. In this part of the tutorial, we will investigate the statistics of ratio data using a synthetic example.

1. Create two vectors A and B, each containing 100 random numbers between 0 and 1:

```
ns <- 100
A <- runif(ns)
B <- runif(ns)
```

Intuitively, given that $A/B = 1/(B/A)$ and $B/A = 1/(A/B)$, we would expect the same to be true for their means $\overline{(A/B)}$ and $\overline{(B/A)}$. However, when we define two new variables for the (inverse) of the (reciprocal) mean ratios:

```
AB.mean <- mean(A/B)
inv.BA.mean <- 1/mean(B/A)
```

then we find that `AB.mean`≠`inv.BA.mean`. So $\overline{(A/B)} \neq 1/\overline{(B/A)}$ and $\overline{(B/A)} \neq 1/\overline{(A/B)}$! This is a counterintuitive and clearly wrong result.

2. Calculate the standard deviation of A/B and multiply this by two to obtain a '2-sigma' confidence interval for the data:

```
AB.sd <- sd(A/B)
LL <- AB.mean - 2*AB.sd
UL <- AB.mean + 2*AB.sd
```

then we find that $LL < 0$, which is nonsensical since A and B are both strictly positive numbers and their ratio is therefore not allowed to take negative values either. Herein lies the root of the problem. The sampling distribution of A/B is positively skewed, whereas the normal distribution is symmetric with tails ranging from $-\infty$ to $+\infty$. Geologists frequently encounter strictly positive numbers. *Time*, for example, is a strictly positive quantity, expressed by geochronologists as 'years before present', where 'present' is equivalent to zero.

3. The problems caused by applying normal theory to strictly positive data can often be solved by simply taking logarithms [15]. The transformed data are then free to take on any value, including negative values, and this often allows normal theory to be applied with no problems. For example, when we calculate the (geometric) mean after taking the logarithm of the ratio data:

```
logAB <- log(A/B)
logBA <- log(B/A)
AB.gmean <- exp(mean(logAB))
inv.BA.gmean <- 1/exp(mean(logBA))
```

then we find that `AB.gmean` = `inv.BA.gmean`, which is a far more sensible result.

4. Calculating the 2-sigma interval for the log-transformed data:

```
LL <- exp( mean(logAB) - 2*sd(logAB) )
UL <- exp( mean(logAB) + 2*sd(logAB) )
```

also produces strictly positive values, as expected.

3. Compositional Data

Summary: *Compositional data such as chemical concentrations suffer from the same problems as the ratio data of Section 2. The tutorial uses a geochemical dataset of Al_2O_3 – (CaO+Na_2O) – K_2O data to demonstrate that the 'conventional' arithmetic mean and confidence intervals are inappropriate for data that can be constrained to a constant sum. A logratio transformation solves these problems.*

Like the ratios of the previous Section, the chemical compositions of rocks and minerals are also expressed as strictly positive numbers. They, however, do not span the entire range of positive values, but are restricted to a narrow subset of that space, ranging from 0 to 1 (if fractions are used) or from 0 to 100% (using percentage notation). The compositions are further restricted by a constant sum constraint:

$$\sum_{i=1}^{n} C_i = 1$$

for an n-component system. Consider, for example, a three-component system $\{x, y, z\}$, where $x + y + z = 1$. Such compositions can be plotted on ternary diagrams, which are very popular in geology. Well-known examples are the Q-F-L diagram of sedimentary petrography [16], the A-CN-K diagram in weathering studies [17], and the A-F-M, Q-A-P and Q-P-F diagrams of igneous petrology [18]. The very fact that it is possible to plot a ternary diagram on a two-dimensional sheet of paper already tells us that it really displays only two and not three dimensions worth of information. Treating the ternary data space as a regular Euclidean space with Gaussian statistics leads to incorrect results, as illustrated by the following example.

1. Read a compositional dataset containing the Al_2O_3 – (CaO+Na_2O) – K_2O composition of a number of synthetic samples:

```
ACNK <- read.csv("ACNK.csv",row.names=1,header=TRUE,check.names=FALSE)
```

where `row.names=1` indicates that the sample names are contained in the first column; and the `header=TRUE` and `check.names=FALSE` arguments indicate that the first column of the input table contains the column headers, one of which contains a special character ('+').

2. Calculate the arithmetic mean composition and 95% confidence limits for each column of the dataset:

```
mu <- colMeans(ACNK)
sig <- apply(ACNK,MARGIN=2,FUN="sd")
```

and construct the 2-sigma confidence confidence bounds:

```
LL <- mu - 2*sig
UL <- mu + 2*sig
```

3. In order to plot the compositional data on a ternary diagram, we will need to first load the `provenance` package into memory:

```
library(provenance)
```

Now plot the Al_2O_3, (CaO + Na_2O) and K_2O compositions on a ternary diagram alongside the arithmetic mean composition:

```
plot(ternary(ACNK),pch=20,labels=NA)
points(ternary(mu),pch=22,bg="blue")
```

where `ternary(x)` creates a ternary data 'object' from a variable `x`, and `pch = 20` and `pch = 22` produce filled circles and squares, respectively. Notice how the arithmetic mean plots outside the data cloud, and therefore fails to represent the compositional dataset (Figure 1).

4. Add a 2-sigma confidence polygon to this figure using the `ternary.polygon()` function that is provided in the auxiliary `helper.R` script (see Online Supplement):

```
source("helper.R")
ternary.polygon(LL,UL,col="blue")
```

Note that the polygon partly plots outside the ternary diagram, into physically impossible negative data space. This nonsensical result is diagnostic of the dangers of applying 'normal' statistics to compositional data. It is similar to the negative limits for the ratio data in Section 2.

A comprehensive solution to the compositional data conundrum was only found in the 1980s, by Scottish statistician John Aitchison [19]. It is closely related to the solution of the ratio averaging problem discussed in the previous section. The trick is to map the n-dimensional composition to an (n-1)-dimensional Euclidean space by means of a logratio transformation. For example, in the ternary case, we can map the compositional variables x, y and z to two transformed variables v and w:

$$v = \ln\left(\frac{x}{z}\right), w = \ln\left(\frac{y}{z}\right) \tag{1}$$

After performing the statistical analysis of interest (e.g., calculating the mean or constructing a 95% confidence region) on the transformed data, the results can then be mapped back to compositional space with the inverse logratio transformation. For the ternary case:

$$x = \frac{e^v}{e^v + e^w + 1}, y = \frac{e^w}{e^v + e^w + 1}, z = \frac{1}{e^v + e^w + 1} \tag{2}$$

This transformation is implemented in the provenance package. Let us use this feature to revisit the K-CN-A dataset, and add the geometric mean and 95% confidence region to the ternary diagram for comparison with the arithmetic mean and confidence polygon obtained before.

5. Compute the geometric mean composition and add it to the existing ternary diagram as a red square:

```
mug <- exp(colMeans(log(ACNK)))
points(ternary(mug),pch=22,bg="red")
```

This red square falls right inside the data cloud, an altogether more satisfying result than the arithmetic mean shown in blue (Figure 1).

6. To add a compositional confidence contour, we must re-read `ACNK.csv` into memory using the `read.compositional()` function. This will tell the provenance package to treat the resulting variable as compositional data in subsequent operations:

```
ACNK2 <- read.compositional("ACNK.csv",check.names=FALSE)
```

Adding the 95% confidence contour using provenance's `ternary.ellipse()` function:

```
ternary.ellipse(ACNK,alpha=0.05)
```

creates a 95% confidence ellipse in logratio space, and maps this back to the ternary diagram. This results in a 'boomerang'-shaped contour that tightly hugs the compositional data whilst staying inside the boundaries of the ternary diagram (Figure 1).

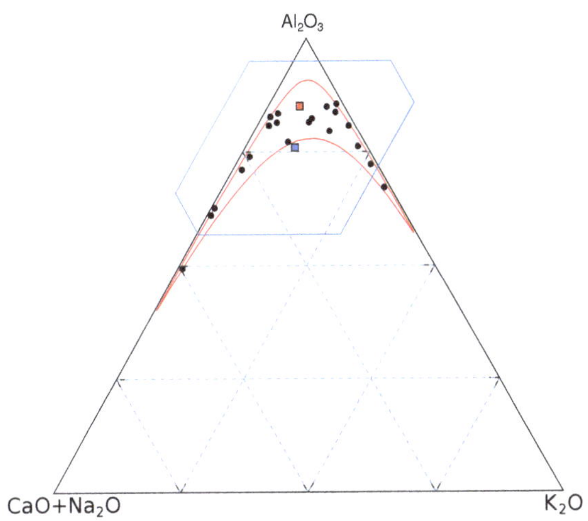

Figure 1. Graphical output of Section 3. Black circles mark 20 synthetic Al_2O_3, (CaO + Na_2O) and K_2O compositions, drawn from a logistic normal distribution. The blue square marks the arithmetic mean, which falls outside the data cloud. The blue polygon marks a 2-σ confidence polygon, which plots outside the ternary diagram, in physically impossible negative space. The red square represents the geometric mean, which firmly plots inside the data cloud. The red confidence envelope marks a 95% confidence region calculated using Aitchison's logratio approach. This confidence envelope neatly fits inside the ternary diagram and tightly hugs the data.

This Section (and Section 8) only touched the bare essentials of compositional data analysis. Further information about this active field of research can be found in Pawlowsky-Glahn et al. [20]. For additional R-recipes for compositional data analysis using the `compositions` package, the reader is referred to Van den Boogaart and Tolosana-Delgado [21,22].

4. Point-Counting Data

Summary: *Point-counting data such as heavy mineral counts are underlain by compositional distributions. However, they are not amenable to the logratio transformations introduced in Section 3 because they commonly contain zero values. Averages and confidence intervals for this type of data require hybrid statistical models combining compositional and multinomial aspects.*

The mineralogical composition of silicilastic sediment can be determined by tallying the occurrence of various minerals in a representative sample of (200–400, say) grains [23,24]. Such *point-counting* data are closely related to the compositional data that were discussed in the previous section. However, there are some crucial differences between these two data classes [25].

Point-counting data are associated with significant (counting) uncertainties, which are ignored by classical compositional data analysis. As a consequence, point-counting data often contain zero values, which are incompatible with the log-ratio transformation defined in Equation (1). Although 'rounding zeros' also occur in compositional data, where they can be removed by 'imputation' methods [26,27], these methods are ill-suited for point-counting datasets in which zeros are the rule rather than the exception.

1. Download the auxiliary data file `HM.csv` from the Online Supplement. This file contains a heavy mineral dataset from the Namib Sand Sea [13]. It consists of 16 rows (one for each sample) and 15 columns (one for each mineral). Read these data into memory and tell `provenance` to treat it as point-counting data in all future operations:

```
HM <- read.counts("HM.csv")
```

   Galbraith [28]'s *radial plot* is an effective way to visually assess the degree to which the random counting uncertainties account for the observed scatter of binary point-counting data. Applying this to the epidote/garnet-ratio of the heavy mineral data (Figure 2):

```
radialplot(HM,num="ep",den="gt")
```

   Each circle on the resulting scatter plot represents a single sample in the `HM` dataset. Its epidote/garnet-ratio can be obtained by projecting the circle onto the circular scale. Thus, low and high ratios are found at negative and positive angles to the origin, respectively. The horizontal distance of each point from the origin is proportional to the total number of counts in each sample and, hence, to its precision. An (asymmetric) 95% confidence interval for the ep/gt-ratio of each sample can be obtained by projecting both ends of a 2-sigma confidence bar onto the circular scale.

Suppose that the data are underlain by a single true population and random counting uncertainties are the sole source of scatter. Let θ be the true but unknown proportion of the binary subpopulation that consists of the first mineral (epidote, say). Then $(1-\theta)$ is the fraction of grains that belong to the second mineral (garnet). Further suppose that we have counted a representative sample of N grains from this population. Then the probability that this sample contains n grains of the first mineral and $m = N - n$ grains of the second mineral follows a binomial distribution:

$$p(n) = \binom{n+m}{n}\theta^n(1-\theta)^m \tag{3}$$

If multiple samples in a dataset are indeed underlain by the same fraction θ, then approximately 95% of the samples should fit within a horizontal band of two standard errors drawn on either side of the origin. In this case, θ can be estimated by *pooling* all the counts together and computing the proportion of the first mineral as a fraction of the total number of grains counted [25].

However, the ep/gt-ratios in `HM` scatter significantly beyond the 2-sigma band (Figure 2i). The data are therefore said to be *overdispersed* with respect to the counting uncertainties. This indicates the presence of true geological dispersion in the compositions that underly the point-counting data. The dispersion can be estimated by a *random effects model* with two parameters:

$$\beta \equiv \ln\left(\frac{\theta}{1-\theta}\right) \approx \mathcal{N}(\mu, \sigma^2) \tag{4}$$

where β is a new variable that follows a normal distribution with mean μ and standard deviation σ, both of which have geological significance.

The 'central ratio' is given by $\exp[\hat{\mu}]$ where $\hat{\mu}$ is the maximum likelihood estimate for μ. This estimates the geometric mean (`ep/gt-`) ratio of the true underlying composition. The 'dispersion' ($\hat{\sigma}$) estimates the geological scatter [25,29]. In the case of our heavy mineral dataset, the epidote-garnet subcomposition is 75% dispersed. This means that the coefficient of variation (geometric standard deviation divided by geometric mean) of the true epidote/garnet-ratios is approximately 0.75.

2. The continuous mixtures from the previous section can be generalised from two to three or more dimensions. The following code snippet uses it to construct a 95% confidence contour for the ternary subcomposition of garnet, epidote and zircon (Figure 2ii). Note that this dataset contains four zero values, which would have rendered the logratio approach of Figure 1 unusable.

```
tern <- ternary(HM,x="gt",y="ep",z="zr")
plot(tern,pch=1,labels=NA)
ternary.ellipse(tern,alpha=0.05)
```

3. For datasets comprising more than three variables, the central composition can be simply obtained as follows:

```
> central(HM)
```

This produces a matrix with the proportions of each component; its standard error; the dispersion of the binary subcomposition formed by the component and the amalgamation of all remaining components; and the outcome of a chi-square test for homogeneity.

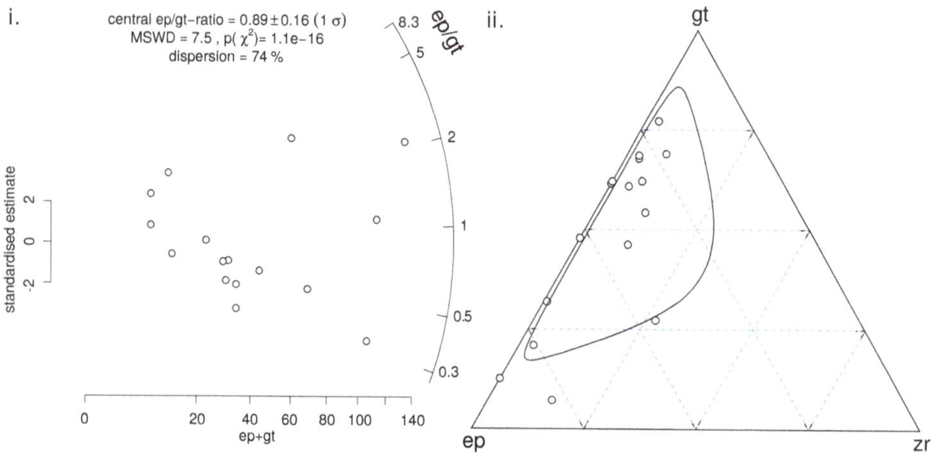

Figure 2. (**i**) Radial plot of the epidote/garnet-ratios of 16 samples of Namibian desert sand. These data are overdispersed with respect to the point-counting uncertainties, indicating 74% of geological scatter in the underlying compositional data. (**ii**) Ternary diagram of garnet, epidote and zircon, with a 95% confidence envelope for the underlying population, using a ternary generalisation of the random effects model. Note that four of the samples contain zero zircon counts. However, this does not pose a problem for the random effects model, unlike the logratio-procedure used for Figure 1.

5. Distributional Data

Summary: *This Section investigates a 16-sample, 1547-grain dataset of detrital zircon U-Pb ages from Namibia. It uses Kernel Density Estimation and Cumulative Age Distributions to visualise this dataset, and introduces the Kolmogorov–Smirnov statistic as a means of quantifying the dissimilarity between samples.*

Compositional data such as the chemical concentrations of Sections 3 and 8 are characterised by the relative proportions of a number of *discrete categories*. A second class of provenance proxies is based on the sampling distribution of *continuous* variables such as zircon U-Pb ages [30,31]. These *distributional* data do not fit in the statistical framework of the (logistic) normal distribution.

1. Download auxiliary data file `DZ.csv` from the Online Supplement. This file contains a detrital zircon U-Pb dataset from Namibia. It consists of 16 columns—one for each sample—each containing the single grain U-Pb ages of their respective sample. Let us load this file into memory using `provenance`'s `read.distributional()` function:

```
DZ <- read.distributional("DZ.csv")
```

`DZ` now contains an object of class `distributional` containing the zircon U-Pb ages of 16 Namibian sand samples. To view the names of these samples:

```
> names(DZ)
```

2. One way to visualise the U-Pb age distributions is as Kernel Density Estimates. A KDE is defined as:

$$KDE_x(t) = \frac{1}{n}\sum_{i=1}^{n}\mathcal{K}(t|x_i, bw) \quad (5)$$

where $\mathcal{K}$ is the 'kernel' and bw is the 'bandwidth' [32,33]. The kernel can be any unimodal and symmetric shape (such as a box or a triangle), but is most often taken to be Gaussian (where x_i is the mean and bw the standard deviation). The bandwidth can either be set manually, or selected automatically based on the number of data points and the distance between them. `provenance` implements the automatic bandwidth selection algorithm of Botev et al. [34] but a plethora of alternatives are available in the statistics literature. To plot all the samples as KDEs:

```
kdes <- KDEs(DZ)
plot(kdes,ncol=2)
```

where `ncol` specifies the number of columns over which the KDEs are divided.

3. Alternatively, the Cumulative Age Distribution (CAD) is a second way to show the data [35]. A CAD is a step function that sets out the rank order of the dates against their numerical value:

$$\mathrm{CAD}(t) = \sum_{i=1}^{n} 1(t < t_i)/n \tag{6}$$

where $1(*) = 1$ if $*$ is true and $1(*) = 0$ if $*$ is false. The main advantages of CADs over KDEs are that (i) they do not require any smoothing (i.e., there is no 'bandwidth' to choose), and (ii) they can superimpose multiple samples on the same plot. Plotting samples N1, N2 and N4 of the Namib dataset:

```
plot(DZ,snames=c("N1","N2","N4"))
```

we can see that (1) the CADs of samples N1 and N2 plot close together with steepest sections at 500 Ma and 1000 Ma, reflecting the prominence of those age components; (2) sample N4 is quite different from N1 and N2.

4. We can quantify this difference using the *Kolmogorov–Smirnov* (KS) statistic [36–38], which represents the maximum vertical difference between two CADs:

```
> N124 <- subset(DZ,select=c("N1","N2","N4"))
> diss(N124)
```

This shows that the KS-statistic between N1 and N2 is KS(N1,N2) = 0.18, whereas KS(N1,N4) = 0.44, and KS(N2,N4) = 0.35 (Figure 3). The KS statistic is a *non-negative* value that takes on values between zero (perfect overlap between two distributions) and one (no overlap between two distributions). It is *symmetric* because the KS statistic between any sample x and another sample y equals that between y and x. For example, KS(N1,N2) = 0.18 = KS(N2,N1). Finally, the KS-statistic obeys the *triangle equality*, which means that the dissimilarity between any two samples is always smaller than or equal to the sum of the dissimilarities between those two samples and a third. For example, KS(N1,N2) = 0.18 $<$ KS(N1,N4) + KS(N2,N4) = 0.44 + 0.35 = 0.79. These three characteristics qualify the KS statistics as a *metric*, which makes it particularly suitable for Multidimensional Scaling (MDS) analysis (see Section 7). The KS statistic is just one of many dissimilarity measures for distributional data. However, not all these alternatives to the KS statistic fulfil the triangle inequality [38].

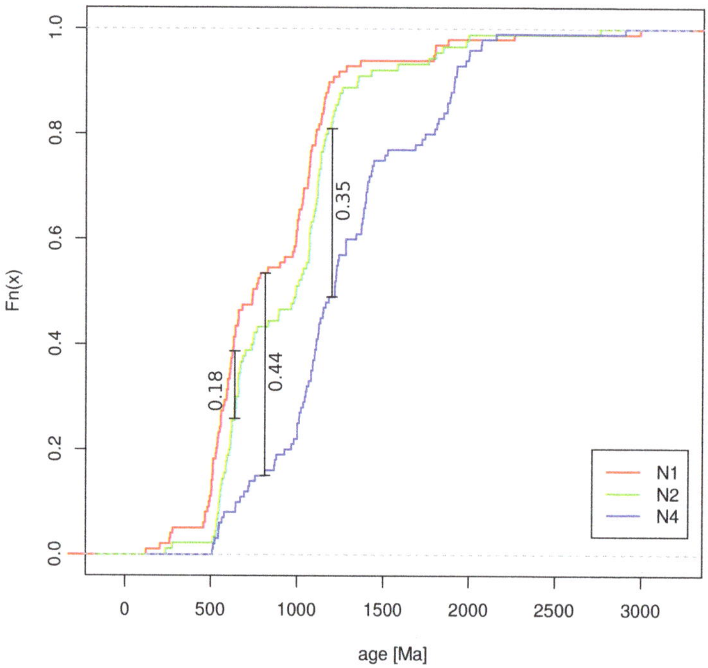

Figure 3. Cumulative Age Distributions (CADs) of Namib desert sand samples N1, N2 and N4 with indication of the Kolmogorov–Smirnov distances between them.

6. Principal Component Analysis (PCA)

Summary: *Principal Component Analysis is an exploratory data analysis method that takes a high dimensional dataset as input and produces a lower (typically two-) dimensional 'projection' as output. PCA is closely related to Multidimensional Scaling (MDS), compositional PCA, and Correspondence Analysis (CA), which are introduced in Sections 7–9. This tutorial introduces PCA using the simplest working example of three two-dimensional points. Nearly identical examples will be used in Sections 7–9.*

1. Consider the following bivariate (a and b) dataset of three (1, 2 and 3) samples:

$$X = \begin{array}{c} \\ 1 \\ 2 \\ 3 \end{array} \begin{array}{c} \begin{array}{cc} a & b \end{array} \\ \begin{bmatrix} -1 & 7 \\ 3 & 2 \\ 4 & 3 \end{bmatrix} \end{array} \quad (7)$$

Generating and plotting X in R:

```
X <- matrix(c(-1,3,4,7,2,3),nrow=3,ncol=2)
colnames(X) <- c("a","b")
plot(X)
```

yields a diagram in which two of the three data points plot close together while the third one plots further away.

2. Imagine that you live in a one-dimensional world and cannot see the spatial distribution of the three points represented by X. Principal Component Analysis (PCA) is a statistical technique

invented by Pearson [39] to represent multi- (e.g., two-) dimensional data in a lower- (e.g., one-) dimensional space whilst preserving the maximum amount of information (i.e., variance). This can be achieved by decomposing X into four matrices (C, S, V and D):

$$\begin{aligned} X &= 1_{3,1}\, C + S\, V\, D \\ &= \begin{bmatrix} 1 \\ 1 \\ 1 \end{bmatrix} \begin{bmatrix} 2 & 4 \end{bmatrix} + \begin{bmatrix} -1.15 & 0 \\ 0.58 & -1 \\ 0.58 & 1 \end{bmatrix} \begin{bmatrix} 3.67 & 0 \\ 0 & 0.71 \end{bmatrix} \begin{bmatrix} 0.71 & -0.71 \\ 0.71 & 0.71 \end{bmatrix} \end{aligned} \tag{8}$$

where C is the centre (arithmetic mean) of the two data columns; S are the *normalised scores*; the diagonals of V correspond to the standard deviations of the two principal components; and D is a rotation matrix (the *principal directions*). S, V and D can be recombined to define two more matrices:

$$P = S\, V = \begin{bmatrix} -4.24 & 0 \\ 2.12 & -0.71 \\ 2.12 & 0.71 \end{bmatrix}, \tag{9}$$

$$\text{and } L = V\, D = \begin{bmatrix} 2.6 & -2.6 \\ 0.5 & 0.5 \end{bmatrix} \tag{10}$$

where P is a matrix of transformed coordinates (the *principal components* or *scores*) and L contains the scaled eigenvectors or *loadings*. Figure 4i shows X as numbers on a scatterplot, C as a yellow square, and $1_{2,1}C \pm L$ as a cross. Thus, the first principal direction (running from the upper left to the lower right) has been stretched by a factor of $(3.67/0.71) = 5.2$ with respect to the second principal component, which runs perpendicular to it. The two principal components are shown separately as Figure 4ii, and their relative contribution to the total variance of the data as Figure 4iv. Figure 4 can be reproduced with the following `R`-code:

```
source("helper.R")
PCA2D(X)
```

3. Although the two-dimensional example is useful for illustrative purposes, the true value of PCA obviously lies in higher dimensional situations. As a second example, let us consider one of `R`'s built-in datasets. `USArrests` contains statistics (in arrests per 100,000 residents) for assault, murder, and rape in each of the 50 US states in 1973. Also given is the percentage of the population living in urban areas. Thus, `USArrests` is a four-column table that cannot readily be visualised on a two-dimensional surface. Applying PCA yields four principal components, the first two of which represent 62% and 25% of the total variance, respectively. Because the four columns of the input data are expressed in different units (arrests per 100,000 or percentage), it is necessary to scale the data to have unit variance before the analysis takes place:

```
pc <- prcomp(USArrests, scale=TRUE)
biplot(pc)
```

The resulting biplot shows that the loading vectors for `Murder`, `Assault` and `Rape` are all pointing in approximately the same direction (dominating the first principal component), perpendicular to `UrbanPop` (which dominates the second principal component). This tells us that crime and degree of urbanisation are not correlated in the United States.

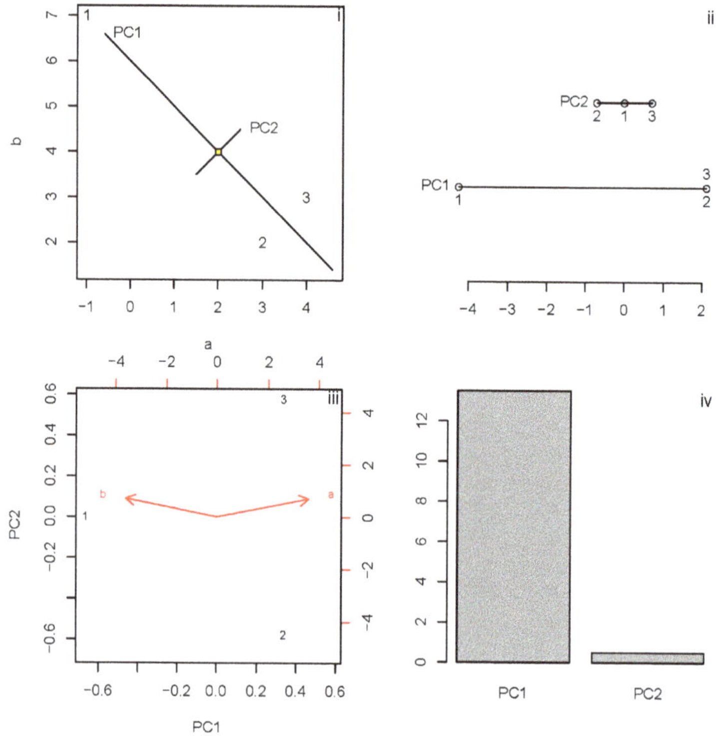

Figure 4. **(i)**—Three samples (1, 2 and 3) of bivariate (a and b) data (X in Equation (7)). The yellow square marks the arithmetic mean (C in Equation (8)), the cross marks the two principal directions (D in Equation (8)) stretched by the diagonal elements (i.e., the standard deviations) of V (Equation (8)); **(ii)**—The projection of the data points on these two directions yields two principal components (P in Equation (9)), representing a one-dimensional representation of the two-dimensional data; **(iii)**—A biplot of both principal components along with the loadings of the two variables shown as arrows; **(iv)**—The squared diagonal values of V (Equation (8)) indicate the relative amounts of variance encoded by the two principal components.

7. Multidimensional Scaling

Summary: *Multidimensional Scaling (MDS) is a less restrictive superset of PCA. This tutorial uses a geographical example to demonstrate how MDS re-creates a map of Europe from a table of pairwise distances between European cities. Applying the same algorithm to the synthetic toy-example of Section 6 yields exactly the same output as PCA.*

1. Multidimensional Scaling (MDS [40–43]) is a dimension-reducing technique that aims to extract two- (or higher) dimensional 'maps' from tables of pairwise distances between objects. This method is most easily illustrated with a geographical example. Consider, for example, the `eurodist` dataset that is built into `R`, and which gives the road distances (in km) between 21 cities in Europe (see `?eurodist` for further details):

```
> eurodist
```

2. The MDS configuration can be obtained by `R`'s built-in `cmdscale()` function

```
conf <- cmdscale(eurodist)
```

Set up an empty plot with a 1:1 aspect ratio, and then label the MDS configuration with the city names:

```
plot(conf,type="n",asp=1)
text(conf,labels=labels(eurodist))
```

Note that the map may be turned 'upside down'. This reflects the rotation invariance of MDS configurations.

3. R's `cmdscale()` function implements so-called 'classical' MDS, which aims to fit the actual distances [42,43]. If these distances are Euclidean, then it can be shown that MDS is equivalent to PCA [44–46]. To demonstrate this equivalence, let us apply MDS to the data in Equation (7). First, run the first two lines of code from part 1 in Section 6. Calculating the Euclidean distances between the three samples produces a dissimilarity matrix d. For example, the distance between points 1 and 2 is $\sqrt{(-1-3)^2+(7-2)^2}=6.4$. This value is stored in `d[1,2]`. In R:

```
d <- dist(X)
```

which produces:

$$d = \begin{array}{c} \\ 1 \\ 2 \\ 3 \end{array}\begin{array}{c} \begin{array}{ccc} 1 & 2 & 3 \end{array} \\ \begin{bmatrix} 0 & 6.4 & 6.4 \\ 6.4 & 0 & 1.4 \\ 6.4 & 1.4 & 0 \end{bmatrix} \end{array} \quad (11)$$

4. Next, calculate the MDS configuration:

```
conf2 <- cmdscale(d)
```

Finally, plot the MDS configuration as a scatterplot of text labels:

```
plot(conf2,type="n")
text(conf2,labels=1:3)
```

which is identical to the PCA configuration of Figure 4iii apart from an arbitrary rotation or reflection.

5. An alternative implementation of MDS loosens the Euclidean distance assumption by fitting the *relative* distances between objects [40,41]. Let us apply this to the dataset of European city distances using the `isoMDS` function of the 'Modern Applied Statistics with S' (`MASS`) package [47]:

```
library(MASS)
```

To compute and plot the non-metric MDS configuration:

```
conf3 <- isoMDS(eurodist)$points
plot(conf3,type="n",asp=1)
text(conf3,labels=labels(eurodist))
```

where `conf3` is a list with two items: `stress`, which expresses the goodness-of-fit of the MDS configuration; and `points`, which contains the configuration. The '$' operator is used to access any of these items.

Non-metric MDS is a less-restrictive superset of classical MDS and, hence, PCA, which opens this methodology up to non-Euclidean dissimilarity measures, such as the KS-distance introduced in Section 5 [48].

8. PCA of Compositional Data

Summary: *PCA can be applied to compositional data after logratio transformation. This tutorial first applies such compositional PCA to a three sample, three variable dataset that is mathematically equivalent to the three sample two variable toy example of Section 6. Then, it applies the same method to a real dataset of major element compositions from Namibia. This is first done using basic* `R` *and then again (and more succinctly) using the* `provenance` *package.*

Consider the following trivariate (a, b and c) dataset of three (1, 2 and 3) compositions:

$$X = \begin{array}{c} \\ 1 \\ 2 \\ 3 \end{array} \begin{array}{c} \begin{array}{ccc} a & b & c \end{array} \\ \begin{bmatrix} 0.03 & 99.88 & 0.09 \\ 70.54 & 25.95 & 3.51 \\ 72.14 & 26.54 & 1.32 \end{bmatrix} \end{array} \quad (12)$$

It would be wrong to apply conventional PCA to this dataset, because this would ignore the constant sum constraint. As was discussed in Section 6, PCA begins by 'centering' the data via the arithmetic mean. Section 3 showed that this yields incorrect results for compositional data. Subjecting the data to a logratio transformation produces:

$$X_a = \begin{array}{c} \\ 1 \\ 2 \\ 3 \end{array} \begin{array}{c} \begin{array}{cc} \ln(a/c) & \ln(b/c) \end{array} \\ \begin{bmatrix} -1 & 7 \\ 3 & 2 \\ 4 & 3 \end{bmatrix} \end{array} \quad (13)$$

which, the observant reader will note, is identical to the example of Equation (7) (Figure 5i). Applying conventional PCA to the log-transformed data of Equation (13) will yield two principal components that are expressed in terms of the logratios $\ln(a/c)$ and $\ln(b/c)$ (Figure 5ii–iii).

Alternatively, the data of Equation (12) can also be subjected to a different type of logratio transformation [44]. The so-called *centred* logratio transformation (as opposed to the *additive* logratio transformation of Equation (1)) maps any set of (for example, two) compositional data vectors $x = \{x_1, \ldots, x_i, \ldots, x_n\}$, $y = \{y_1, \ldots, y_i, \ldots, y_n\}$ to the same number of (centred) logratios $u = \{u_1, \ldots, u_i, \ldots, u_n\}$, $v = \{v_1, \ldots, v_i, \ldots, v_n\}$, where:

$$u_i = \ln(x_i) - [\ln(x_i) + \ln(y_i)]/2 \text{ and } v_i = \ln(y_i) - [\ln(x_i) + \ln(y_i)]/2 \quad (14)$$

Applying this transformation to the data of Equation (12) yields a new trivariate dataset:

$$X_c = \begin{array}{c} \\ 1 \\ 2 \\ 3 \end{array} \begin{array}{c} \begin{array}{ccc} \ln(a/g) & \ln(b/g) & \ln(c/g) \end{array} \\ \begin{bmatrix} -3 & 5 & -2 \\ 1.33 & 0.33 & -1.67 \\ 1.67 & 0.67 & -2.33 \end{bmatrix} \end{array} \quad (15)$$

where g stands for the geometric mean of each row. Subjecting Equation (15) to the same matrix decomposition as Equation (8) yields:

$$X_c = 1_{3,1}\,C + S\,V\,D = \begin{bmatrix} 1 \\ 1 \\ 1 \end{bmatrix} \begin{bmatrix} 0 & 2 & -2 \end{bmatrix} + \begin{bmatrix} -1.15 & 0 & 0.82 \\ 0.58 & -1 & 0.82 \\ 0.58 & 1 & 0.82 \end{bmatrix} \begin{bmatrix} 3.67 & 0 & 0 \\ 0 & 0.41 & 0 \\ 0 & 0 & 0 \end{bmatrix} \begin{bmatrix} 0.71 & -0.71 & 0 \\ 0.41 & 0.41 & -0.82 \\ 0.58 & 0.58 & 0.58 \end{bmatrix} \quad (16)$$

Note that, even though this yields three principal components instead of two, the standard deviation of the third component is zero. Therefore, all the information is contained in the first two components. The PCA map using the centred logratio transformation looks identical to that using the additive logratio transformation. The only difference is that the loadings are expressed in terms of the three centred logratio variables, rather than the two additive logratio variables (Figure 5iv). The centred logratios are easier to interpret than the additive logratios, which is why the centred logratio transformation is preferred in this context.

1. The following script applies compositional PCA to a dataset of major element compositions from Namibia (see Online Supplement) using base R:

```
# load the major element composition of Namib sand:
Major <- read.csv(file="Major.csv",
                  header=TRUE,row.names=1)
# apply the centred logratio transformation:
cMajor <- log(Major) -
       rowMeans(log(Major)) %*% matrix(1,1,ncol(Major))
# perform PCA of the logratio transformed data:
pc <- prcomp(cMajor)
# plot the results of the PCA analysis:
biplot(pc)
```

2. Alternatively, we can also do this more easily in provenance:

```
library(provenance)
# tell R that Major.csv contains compositional data:
Major.comp <- read.compositional("Major.csv")
# perform the principal component analysis:
pc.comp <- PCA(Major.comp)
# create the biplot:
plot(pc.comp)
```

where the `read.compositional` function reads the `.csv` file into an object of class `compositional`, thus ensuring that logratio statistics are used in all `provenance` functions (such as `PCA`) that accept compositional data as input. Also note that the `provenance` package *overloads* the `plot` function to generate a compositional biplot when applied to the output of the `PCA` function.

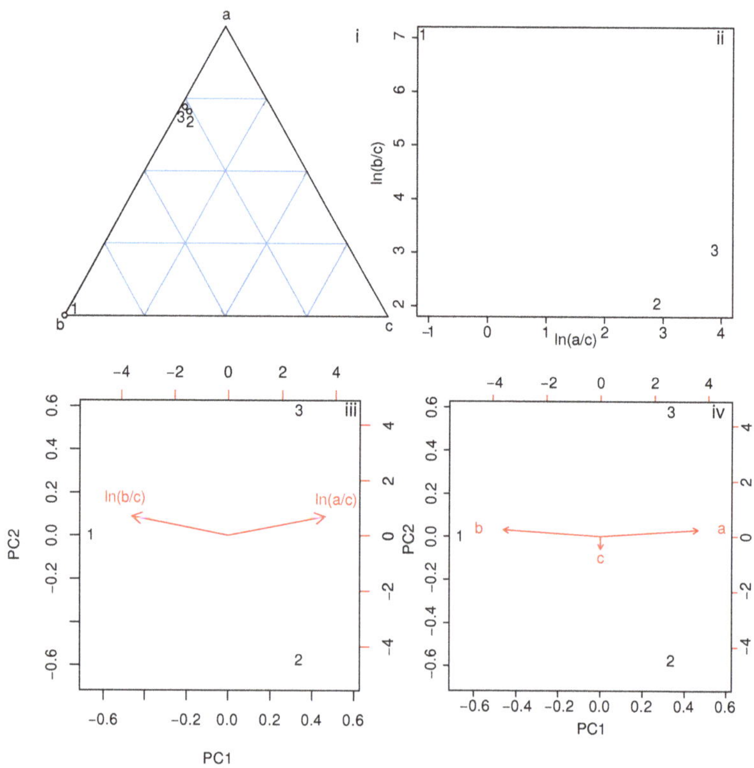

Figure 5. **(i)** —the compositional dataset of Equation (15) shown on a ternary diagram; **(ii)**—subjecting the same dataset to an additive logratio transformation (alr) produces a configuration of points that is identical to Figure 4i; **(iii)**—as a consequence, the PCA biplot of the logratio transformed data looks identical to Figure 4iii; **(iv)**—using a centred logratio transformation (clr) yields the same configuration as panel iii but with more easily interpretable vector loadings.

9. Correspondence Analysis

Summary: *Point-counting data can be analysed by MDS using the Chi-square distance. Correspondence Analysis (CA) yields identical results whilst producing biplots akin to those obtained by PCA. This tutorial first uses a simple three sample, three variable toy example that is (almost) identical to those used in Sections 6–8, before applying CA to a real dataset of heavy mineral counts from Namibia.*

Consider the following three sets of trivariate point-counting data:

$$X = \begin{array}{c} \\ 1 \\ 2 \\ 3 \end{array} \begin{array}{c} \begin{array}{ccc} a & b & c \end{array} \\ \left[\begin{array}{ccc} 0 & 100 & 0 \\ 38 & 13 & 1 \\ 108 & 38 & 0 \end{array}\right] \end{array} \tag{17}$$

This dataset intentionally looks similar on a ternary diagram to the compositional dataset of Section 3. The only difference is the presence of zeros, which preclude the use of logratio statistics. This problem can be solved by replacing the zero values with small numbers, but this only works when their number is small [26,27]. Correspondence Analysis (CA) is an alternative approach that does not require such 'imputation'.

CA is a dimension reduction technique that is similar in many ways to PCA [25,49]. CA, like PCA, is a special case of MDS. Whereas ordinary PCA uses the Euclidean distance, and compositional data can be compared using the Aitchison distance, point-counting data can be compared by means of a chi-square distance:

$$d_{ij} = \sqrt{\sum_{k=1}^{K} \frac{X_{..}}{X_{.k}} \left(\frac{X_{ik}}{X_{i.}} - \frac{X_{jk}}{X_{j.}} \right)^2} \tag{18}$$

where $X_{.k} = \sum_{i=1}^{m} X_{ik}$, $X_{i.} = \sum_{k=1}^{K} X_{ik}$ and $X_{..} = \sum_{i=1}^{m} \sum_{k=1}^{K} X_{ik}$. Applying this formula to the data of Equation (17) produces the following dissimilarity matrix:

$$\begin{array}{c} \\ 1 \\ 2 \\ 3 \end{array} \begin{array}{c} \begin{array}{ccc} 1 & 2 & 3 \end{array} \\ \left[\begin{array}{ccc} 0 & 1.5 & 1.5 \\ 1.5 & 0 & 0.33 \\ 1.5 & 0.33 & 0 \end{array} \right] \end{array} \tag{19}$$

Note that, although these values are different than those in Equation (11), the ratios between them are (approximately) the same. Specifically, $d_{1,2}/d_{1,3} = 1.5/1.5 = 1$ for Equation (19) and $d_{1,2}/d_{1,3} = 6.4/6.4 = 1$ for Equation (11); or $d_{1,2}/d_{2,3} = 1.5/0.33 = 4.5$ for Equation 19) and $d_{1,2}/d_{2,3} = 6.4/1.4 = 4.5$ for Equation (11). Therefore, when we subject our point-counting data to an MDS analysis using the chi-square distance, the resulting configuration appears nearly identical to the example of Section 7.

The following script applies CA to the heavy mineral composition of Namib desert sand. It loads a table called `HM.csv` that contains point counts for 16 samples and 15 minerals. To reduce the dominance of the least abundant components, the code extracts the most abundant minerals (epidote, garnet, amphibole and clinopyroxene) from the datasets and amalgamates the ultra-stable minerals (zircon, tourmaline and ru- tile), which have similar petrological significance.

```
library(provenance)
# tell R that HM.csv contains point-counting data:
dat <- read.counts("HM.csv")
# select and amalgamate the components of interest:
HM <- amalgamate(dat,ztr=c("zr","tm","rt"),ep="ep",
                     gt="gt",amp="amp",cpx="cpx")
# perform the correspondence analysis:
config <- CA(HM)
# plot the results as a biplot:
plot(config)
```

10. MDS Analysis of Distributional Data

Summary: *This brief tutorial applies MDS to the detrital zircon U-Pb dataset from Namibia, using the Kolmogorov–Smirnov statistic as a dissimilarity measure.*

Part 5 in Section 7 introduced non-metric MDS as a less-restrictive superset of classical MDS and, hence, PCA. This opens this methodology up to non-Euclidean dissimilarity measures, such as the KS-distance introduced in part 4 in Section 5 [38,48].

```
library(provenance)
# read the detrital zircon dataset:
DZ <- read.distributional("DZ.csv")
# calculate and plot the (non-metric)
# MDS configuration using the KS distance:
DZ.X <- MDS(DZ)
plot(DZ.X)
```

In this case, the overloaded `plot` function produces not one but two graphics windows. The first of these shows the MDS configuration, whereas the second shows the *Shepard plot* [40,41]. This is a scatterplot that sets out the Euclidean distances between the samples measured on the MDS configuration against the *disparities*, which are defined as:

$$\delta[i,j] = f(KS[i,j]) \tag{20}$$

where $KS[i,j]$ is the KS-distance between the ith and j^{th} sample and f is a monotonic transformation, which is shown as a step-function. The Shepard plot allows the user to visually assess the goodness-of-fit of the MDS configuration. This can be further quantified using the 'stress' parameter:

$$S = \sum_i \sum_j (d[i,j] - \delta[i,j])^2 \Big/ \sum_i \sum_j (d[i,j])^2 \tag{21}$$

The lower the stress, the better the fit. For moderately sized datasets, stress values should be less than 10% [40]. For larger datasets, a higher dimensional solution may be necessary, using the optional parameter `k` of `provenance`'s `MDS` function [50].

11. 'Big' Data

Summary: *The tutorial jointly analyses 16 Namibian samples using five different provenance proxies, including all three data classes introduced in Sections 3–5. It introduces Procrustes Analysis and 3-way MDS as two alternative ways to extract geologically meaningful information from these multivariate 'big' dataset.*

It is increasingly common for provenance studies to combine compositional, point-counting or distributional datasets together [4,13]. Linking together bulk sediment data, heavy mineral data and single mineral data requires not only a sensible statistical approach, but also a full appraisal of the impact of mineral fertility and heavy mineral concentration in eroded bedrock and derived clast sediment [51–53]. Assuming that such an appraisal has been made, this Section introduces some exploratory data analysis tools that can reveal meaningful structure in complex datasets.

1. The full Namib Sand Sea study that we have used as a test case for this tutorial comprises five datasets (see Online Supplement):

    (a) Major element concentrations (`Major.csv`, compositional data)
    (b) Trace element concentrations (`Trace.csv`, compositional data)
    (c) Bulk petrography (`PT.csv`, point-counting data)
    (d) Heavy mineral compositions (`HM.csv`, point-counting data)
    (e) Detrital zircon U-Pb data (`DZ.csv`, distributional data)

    All these datasets can be visualised together in a single summary plot:

```
library(provenance)
# major elements:
Major <- read.compositional("Major.csv")
# trace elements:
Trace <- read.compositional("Trace.csv")
# petrography:
QFL <- read.counts("PT.csv",colmap=cm.colors)
# heavy minerals:
HM <- read.counts("HM.csv",colmap=cm.colors)
# zircon U-Pb dates:
DZ <- read.distributional("DZ.csv")
# generate the plot:
summaryplot(Major,Trace,QFL,HM,KDEs(DZ),ncol=2)
```

where `Major`, `Trace`, `QFL` and `HM` are shown as pie charts (the latter two with a different colour map than the former), and `DZ` as KDEs. Adding `DZ` instead of `KDEs(DZ)` would plot the U-Pb age distributions as histograms.

2. The entire Namib dataset comprises 16,125 measurements spanning five dimensions worth of compositional, distributional and point-counting information. This complex dataset, which may be rightfully described by the internet-era term of 'Big Data', is extremely difficult to interpret by mere visual inspection of the pie charts and KDEs. Applying MDS/PCA to each of the five individual datasets helps but presents the analyst with a multi-plot comparison problem. `provenance` implements two methods to address this issue [13]. The first of these is called 'Procrustes Analysis' [54]. Given a number of MDS configurations, this technique uses a combination of transformations (translation, rotation, scaling and reflection) to extract a 'consensus view' for all the data considered together:

```
proc <- procrustes(Major,Trace,QFL,HM,DZ)
plot(proc)
```

3. Alternatively, '3-way MDS' is an extension of 'ordinary' (2-way) MDS that accepts 3-dimensional dissimilarity matrices as input. `provenance` includes the most common implementation of this class of algorithms, which is known as 'INdividual Difference SCALing' or INDSCAL [55,56]:

```
scal <- indscal(Major,Trace,QFL,HM,DZ)
plot(scal)
```

This code produces two pieces of graphical output (Figure 6). The 'group configuration' represents the consensus view of all provenance proxies considered together. This looks very similar to the Procrustes configuration created by the previous code snippet. The second piece of graphical information displays not the samples but the provenance proxies. It shows the weights that each of the proxies attach to the horizontal and vertical axis of the group configuration.

For example, the heavy mineral compositions of the Namib desert sands can be (approximately) described by stretching the group configuration vertically by a factor of 1.9, whilst shrinking it horizontally by a factor of 0.4. In contrast, the configurations of the major and trace element compositions for the same samples are obtained by shrinking the group configuration vertically by a factor 0.8, and stretching it horizontally by a factor of 1.3. Thus, by combining these weights with the group configuration yields five 'private spaces' that aim to fit each of the individual datasets.

INDSCAL group configurations are not rotation-invariant, in contrast with the 2-way MDS configurations of Section 7. This gives geological meaning to the horizontal and vertical axes of the plot. For example, samples N1 and N10 plot along a vertical line on the group configuration, indicating that they have different heavy mineral compositions, but similar major and trace element compositions. On the other hand, samples N4 and N8 plot along a horizontal line, indicating that they have similar major and trace element compositions but contrasting heavy mineral compositions.

Closer inspection of the weights reveals that the datasets obtained from fractions of specific densities (`HM`, `PT` and `DZ`) attach stronger weights to the vertical axis, whereas those that are determined on bulk sediment (`Major` and `Trace`) dominate the horizontal direction. Provenance proxies that use bulk sediment are more sensitive to winnowing effects than those that are based on density separates. This leads to the interpretation that the horizontal axis separates samples that have been affected by different degrees of hydraulic sorting, whereas the vertical direction separates samples that have different provenance.

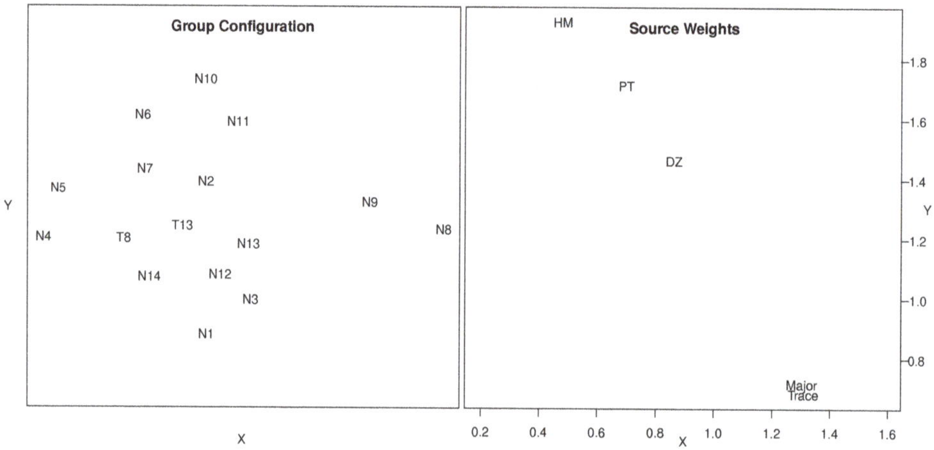

Figure 6. Output of the 3-way MDS analysis of Namib desert sand. Left: the group configurations show the salient similarities and differences between samples as a 'map' in which similar samples plot close together and dissimilar samples plot far apart. Right: the weights for each of the five data sources show that provenance proxies that are performed on the bulk sediment (e.g., the major and trace element compositions) attach a stronger weight to the X- than the Y-axis. In contrast, proxies that are determined on specific density fractions (e.g., zircons, heavy minerals, or quartz—feldspar—lithics), attach stronger weight to the Y-axis. One geological interpretation of these dimensions is that samples that horizontally separated from each other on the group configuration (e.g., N4 and N8) have experienced hydraulic sorting, whereas samples that are vertically separated (e.g., N1 and N10) have a different provenance.

12. Summary, Conclusions and Outlook

The statistical toolbox implemented by the `provenance` package is neither comprehensive nor at the cutting edge of exploratory data analysis. PCA, MDS, CA, and KDEs are tried and tested methods that have been around for many decades. Nothing new is presented here and that is intentional. This paper makes the point that even the most basic statistical parameters like the arithmetic mean and standard deviation cannot be blindly applied to geological data [24,57,58]. Great care must be taken when applying established techniques to sedimentary provenance data such as chemical compositions, point-counts or U-Pb age distributions. Given the difficulty of using even the simplest of methods correctly, geologists may want to think twice before exploring more complicated methods, or inventing entirely new ones.

The set of tutorials presented in this paper did not cover all aspects of statistical provenance analysis. Doing so would fill a book rather than a paper. Some additional topics for such a book could include (1) supervised and unsupervised learning algorithms such as cluster analysis and discriminant analysis, which can group samples into formal groups [10,11,59,60]; (2) the physical and chemical processes that affect the composition of sediment from 'source to sink' [5,61–63]; and (3) quality checks and corrections that must be made to ensure that the data reveal meaningful provenance trends rather than sampling effects [51,52,64–66].

The paper introduced three distinct classes of provenance data. Compositional, point-counting and distributional data each require different statistical treatment. Multi-sample collections of these data can be visualised by Multidimensional Scaling, using different dissimilarity measures (Table 1). Distributional data can be compared using the Kolmogorov–Smirnov statistic or related dissimilarity measures, and plugged straight into an MDS algorithm for further inspection. Compositional data such as chemical concentrations can be visualised by conventional 'normal' statistics after logratio transformation. The Euclidean distance in logratio space is called the Aitchison distance in compositional data space. Classical MDS using this distance is equivalent to Principal Component Analysis. Finally, point-counting data combine aspects of compositional data analysis with multinomial sampling statistics. The Chi-square distance is the natural way to quantify the dissimilarity between multiple point-counting samples. MDS analysis using the Chi-square distance is equivalent to Correspondence Analysis, which is akin to PCA for categorical data.

However, there are some provenance proxies that do not easily fit into these three categories. *Varietal studies* using the chemical composition of single grains of heavy minerals combine aspects of compositional and distributional data [3,60]. Similarly, paired U-Pb ages and Hf-isotope compositions in zircon [1] do not easily fit inside the distributional data class described above. Using the tools provided by the `provenance` package, such data can be processed by procustes analysis or 3-way MDS (Section 11). Thus, U-Pb and ϵ(Hf)-distributions, say, could be entered into the `indscal` function as separate entities. However, by doing so, the single-grain link between the two datasets would be lost. Alternative approaches may be pursued to address this issue, and new dissimilarity measures could be developed for this hybrid data type. Novel approaches to matrix decomposition may be a way forward in this direction [8,67,68].

Table 1. A summary of the three types of provenance data introduced in this paper along with a suitable dissimilarity measure and its corresponding ordination technique.

Data Type	Dissimilarity Measure	Ordination Technique
compositional	Aitchison	Principal Component Analysis
point-counting	Chi-square	Correspondence Analysis
distributional	Kolmogorov–Smirnov	Multidimensional Scaling

Supplementary Materials: The following are available online at http://www.mdpi.com/2075-163X/9/3/193/s1, (a) Major element concentrations (Major.csv, compositional data). (b) Trace element concentrations (Trace.csv, compositional data). (c) Bulk petrography (PT.csv, point-counting data). (d) Heavy mineral compositions (HM.csv, point-counting data). (e) Detrital zircon U-Pb data (DZ.csv, distributional data). (f). ACNK.csv. (g). helper.R.

Funding: This research received no external funding.

Acknowledgments: This paper evolved from a set of lecture notes for an iCRAG workshop in sedimentary provenance analysis at NUI Galway. The author would like to thank Sergio Andò for inviting him to contribute to this Special Issue. The manuscript greatly benefited from four critical but constructive reviews. The ratio averaging example of Section 2 was first suggested by Noah McLean.

Conflicts of Interest: The author declares no conflict of interest.

Appendix A. An Introduction to R

R is an increasingly popular programming language for scientific data processing. It is similar in scope and purpose to Matlab but is available free of charge on any operating system at http://r-project.org. A number of different graphical user interfaces (GUIs) are available for R, the most popular of which are RGui, RStudio, RCommander and Tinn-R. For this tutorial, however, the simple command line console suffices.

1. First, do some arithmetic:

```
> 1 + 1
> sqrt(2)
> exp(log(10))
> 13%%5
```

   where the '>' symbol marks the command prompt.
2. You can use the arrow to assign a value to a variable. Note that the arrow can point both ways:

```
> foo <- 2
> 4 -> bar
> foo <- foo*bar
```

3. Create a sequence of values:

```
> myvec <- c(2,4,6,8)
> myvec*2
```

   Query the third value of the vector:

```
> myvec[3]
```

   Change the third value of the vector:

```
> myvec[3] <- 100
```

   Change the second and the third value of the vector:

```
> myvec[c(2,3)] <- c(100,101)
```

   Create a vector of 1, 2, 3, ..., 10:

```
> seq(from=1,to=10,by=1)
```

   Equivalently:

```
> seq(1,10,1)
> seq(1,10)
> seq(to=10,by=1,from=1)
> seq(to=10)
> 1:10
```

Create a 10-element vector of twos:

```
> rep(2,10)
```

4. Create a 2×4 matrix of ones:

```
> mymat <- matrix(1,nrow=2,ncol=4)
```

Change the third value in the first column of `mymat` to 3:

```
> mymat[1,3] <- 3
```

Change the entire second column of `mymat` to 2:

```
> mymat[,2] <- 2
```

The transpose of `mymat`:

```
> t(mymat)
```

Element-wise multiplication (`*`) vs. matrix multiplication (`%*%`):

```
> mymat * mymat
> mymat %*% t(mymat)
```

5. Lists are used to store more complex data objects:

```
> mylist <- list(v=myvec, m=mymat, nine=9)
> mylist$v
```

6. Plot the first against the second row of `mymat`:

```
> plot(mymat[1,],mymat[2,],type="p")
```

Draw lines between the points shown on the existing plot:

```
> lines(mymat[1,],mymat[2,])
```

Create a new plot with red lines but no points:

```
> plot(mymat[1,],mymat[2,],type="l",col="red")
```

Use a 1:1 aspect ratio for the X- and Y-axis:

```
> plot(mymat[1,],mymat[2,],type="l",col="red",asp=1)
```

7. Save the currently active plot as a vector-editable `.pdf` file:

```
> dev.copy2pdf(file="trigonometry.pdf")
```

8. To learn more about a function, type 'help' or '?':

```
> help(c)
> ?plot
```

9. It is also possible to define one's own functions:

```
> cube <- function(n){
>     return(n^3)
> }
```

Using the newly created function:

```
> cube(2)
> result <- cube(3)
```

10. Create some random (uniform) numbers:

```
> rand.num <- runif(100)
> hist(rand.num)
```

11. List all the variables in the current workspace:

```
> ls()
```

Remove all the variables in the current workspace:

```
> rm(list=ls())
```

To get and set the working directory:

```
> getwd()
> setwd("/path/to/a/valid/directory")
```

12. Collect the following commands in a file called '`myscript.R`'. Note that this text does not contain any '>'-symbols because it is not entered at the command prompt but in a separate text editor:

```
# the "print" function is needed to show intermediate
# results when running commands from an .R file
print(pi)
```

This code can be run by going back to the command prompt (hence the '>' in the next box) and typing:

```
> source("myscript.R")
```

This should result in the number π being printed to the console. Note that everything that follows the '#'-symbol was ignored by R.

13. Conditional statements. Add the following function to `myscript.R`:

```
toss <- function(){
    if (runif(1)>0.5){
        print("head")
    } else {
        print("tail")
    }
}
```

Save and run at the command prompt:

```
> source("myscript.R")
> toss()
```

14. Loops. Add the following function to `myscript.R`:

```
fibonnaci <- function(n){
    if (n < 3) { stop("n must be at least 3") }
    # seed the output vector with 0 and 1:
    s <- c(0,1)
    # loop through all numbers from 3 to n:
    for (i in 3:n){
        s[i] <- s[i-1] + s[i-2]
    }
    return(s)
}
```

Save and run at the command prompt to calculate the first 20 numbers in the Fibonnaci series:

```
> source("myscript.R")
> fibonnaci(20)
```

15. Arguably the greatest power of R is the availability of 10,000 *packages* that provide additional functionality. For example, the `compositions` package implements a number of statistical tools for compositional data analysis [21,22]. To install this package:

```
> install.packages("compositions")
```

Use the newly installed package to plot the built-in `SkyeAFM` dataset, which contains the Al_2O_3—FeO—MgO compositions of 23 aphyric lavas from the isle of Skye.

```
library(compositions) # load the package into memory
dat <- data(SkyeAFM)  # load the Skye lava dataset
AFMcomp <- acomp(dat) # enforce the constant sum constraint
plot(AFMcomp)         # plot as a ternary diagram
```

Note that the `plot()` function has been *overloaded* for compositional data.

References

1. Gerdes, A.; Zeh, A. Combined U–Pb and Hf isotope LA-(MC-) ICP-MS analyses of detrital zircons: Comparison with SHRIMP and new constraints for the provenance and age of an Armorican metasediment in Central Germany. *Earth Planet. Sci. Lett.* **2006**, *249*, 47–61. [CrossRef]
2. Mazumder, R. Sediment provenance. In *Sediment Provenance: Influence on Compositional Change From Source to Sink*; Mazumder, R., Ed.; Elsevier: Amsterdam, The Netherlands, 2017; pp. 1–4.
3. Morton, A.C. Geochemical studies of detrital heavy minerals and their application to provenance research. In *Developments in Sedimentary Provenance Studies*; Morton, A., Todd, S., Haughton, P. D.W., Eds.; Geological Society of London: London, UK, 1991; Volume 57, pp. 31–45.
4. Rittner, M.; Vermeesch, P.; Carter, A.; Bird, A.; Stevens, T.; Garzanti, E.; Andò, S.; Vezzoli, G.; Dutt, R.; Xu, Z.; et al. The provenance of Taklamakan desert sand. *Earth Planet. Sci. Lett.* **2016**, *437*, 127–137. [CrossRef]
5. Weltje, G.J.; von Eynatten, H. Quantitative provenance analysis of sediments: Review and outlook. *Sediment. Geol.* **2004**, *171*, 1–11. [CrossRef]
6. DuToit, S.H.; Steyn, A.G.W.; Stumpf, R.H. *Graphical Exploratory Data Analysis*; Springer Science & Business Media: Berlin, Germany, 1986.
7. Kenkel, N. On selecting an appropriate multivariate analysis. *Can. J. Plant Sci.* **2006**, *86*, 663–676. [CrossRef]
8. Martinez, W.L.; Martinez, A.R.; Solka, J. *Exploratory Data Analysis with MATLAB*; Chapman and Hall/CRC: Boca Raton, FL, USA, 2017; ISBN 9781498776066.
9. Tukey, J.W. *Exploratory Data Analysis*; Addison-Wesley: Boston, MA, USA, 1977; Volume 2, ISBN 978-0-201-07616-5.
10. Bhatia, M.R. Plate tectonics and geochemical composition of sandstones. *J. Geol.* **1983**, *91*, 611–627. [CrossRef]
11. Bhatia, M.R.; Crook, K.A. Trace element characteristics of graywackes and tectonic setting discrimination of sedimentary basins. *Contrib. Mineral. Petrol.* **1986**, *92*, 181–193. [CrossRef]
12. Vermeesch, P.; Resentini, A.; Garzanti, E. An R package for statistical provenance analysis. *Sediment. Geol.* **2016**, *336*, 14–25. [CrossRef]
13. Vermeesch, P.; Garzanti, E. Making geological sense of 'Big Data' in sedimentary provenance analysis. *Chem. Geol.* **2015**, *409*, 20–27. [CrossRef]
14. Morton, A.C.; Hallsworth, C.R. Processes controlling the composition of heavy mineral assemblages in sandstones. *Sediment. Geol.* **1999**, *124*, 3–29. [CrossRef]
15. Aitchison, J.; Brown, J.A. *The Lognormal Distribution*; Cambridge University Press: Cambridge, MA, USA, 1957; ISBN 0521040116.
16. Garzanti, E. Petrographic classification of sand and sandstone. *Earth-Sci. Rev.* **2019**. [CrossRef]
17. Nesbitt, H.; Young, G.M. Formation and diagenesis of weathering profiles. *J. Geol.* **1989**, *97*, 129–147. [CrossRef]
18. LeMaitre, R.W.; Streckeisen, A.; Zanettin, B.; LeBas, M.; Bonin, B.; Bateman, P. *Igneous Rocks: A Classification and Glossary of Terms: Recommendations of the International Union of Geological Sciences Subcommission on the Systematics of Igneous Rocks*; Cambridge University Press: Cambridge, MA, USA, 2002; ISBN 9780511535581.
19. Aitchison, J. *The Statistical Analysis of Compositional Data*; Chapman and Hall: London, UK, 1986.
20. Pawlowsky-Glahn, V.; Egozcue, J.J.; Tolosana-Delgado, R. *Modeling and Analysis of Compositional Data*; John Wiley & Sons: Hoboken, NJ, USA, 2015.
21. Van den Boogaart, K.G.; Tolosana-Delgado, R. "Compositions": A unified R package to analyze compositional data. *Comput. Geosci.* **2008**, *34*, 320–338. [CrossRef]
22. Van den Boogaart, K.G.; Tolosana-Delgado, R. *Analyzing Compositional Data with R*; Springer: Berlin, Germany, 2013; Volume 122.
23. Van der Plas, L.; Tobi, A. A chart for judging the reliability of point counting results. *Am. J. Sci.* **1965**, *263*, 87–90. [CrossRef]
24. Weltje, G. Quantitative analysis of detrital modes: Statistically rigorous confidence regions in ternary diagrams and their use in sedimentary petrology. *Earth-Sci. Rev.* **2002**, *57*, 211–253. [CrossRef]
25. Vermeesch, P. Statistical models for point-counting data. *Earth Planet. Sci. Lett.* **2018**, *501*, 1–7. [CrossRef]

26. Bloemsma, M.R.; Weltje, G.J. Reduced-rank approximations to spectroscopic and compositional data: A universal framework based on log-ratios and counting statistics. *Chemom. Intell. Lab. Syst.* **2015**, *142*, 206–218. [CrossRef]
27. Martín-Fernández, J.A.; Barceló-Vidal, C.; Pawlowsky-Glahn, V. Dealing with zeros and missing values in compositional data sets using nonparametric imputation. *Math. Geol.* **2003**, *35*, 253–278. [CrossRef]
28. Galbraith, R. Graphical display of estimates having differing standard errors. *Technometrics* **1988**, *30*, 271–281. [CrossRef]
29. Galbraith, R.F. The radial plot: Graphical assessment of spread in ages. *Nuclear Tracks Radiat. Meas.* **1990**, *17*, 207–214. [CrossRef]
30. Fedo, C.; Sircombe, K.; Rainbird, R. Detrital zircon analysis of the sedimentary record. *Rev. Mineral. Geochem.* **2003**, *53*, 277–303. [CrossRef]
31. Gehrels, G. Detrital zircon U-Pb geochronology: Current methods and new opportunities. In *Tectonics of Sedimentary Basins: Recent Advances*; Busby, C., Azor, A., Eds.; Wiley Online Library: Hoboken, NJ, USA, 2011; Chapter 2, pp. 45–62.
32. Silverman, B. *Density Estimation for Statistics and Data Analysis*; Chapman and Hall: London, UK, 1986.
33. Vermeesch, P. On the visualisation of detrital age distributions. *Chem. Geol.* **2012**, *312–313*, 190–194. [CrossRef]
34. Botev, Z.I.; Grotowski, J.F.; Kroese, D.P. Kernel density estimation via diffusion. *Ann. Stat.* **2010**, *38*, 2916–2957. [CrossRef]
35. Vermeesch, P. Quantitative geomorphology of the White Mountains (California) using detrital apatite fission track thermochronology. *J. Geophys. Res. (Earth Surf.)* **2007**, *112*, 3004, doi:10.1029/2006JF000671. [CrossRef]
36. DeGraaff-Surpless, K.; Mahoney, J.; Wooden, J.; McWilliams, M. Lithofacies control in detrital zircon provenance studies: Insights from the Cretaceous Methow basin, southern Canadian Cordillera. *Geol. Soc. Am. Bull.* **2003**, *115*, 899–915. [CrossRef]
37. Feller, W. On the Kolmogorov-Smirnov limit theorems for empirical distributions. *Ann. Math. Stat.* **1948**, *19*, 177–189. [CrossRef]
38. Vermeesch, P. Dissimilarity measures in detrital geochronology. *Earth-Sci. Rev.* **2018**, *178*, 310–321. [CrossRef]
39. Pearson, K. On lines and planes of closest fit to systems of points in space. *Lond. Edinburgh Dublin Philos. Mag. J. Sci.* **1901**, *2*, 559–572. [CrossRef]
40. Kruskal, J.B.; Wish, M. *Multidimensional Scaling*; Sage University Paper series on Quantitative Application in the Social Sciences; Sage Publications: Beverly Hills, CA, USA; London, UK, 1978; Volume 7–11.
41. Shepard, R.N. The analysis of proximities: multidimensional scaling with an unknown distance function. i. *Psychometrika* **1962**, *27*, 125–140. [CrossRef]
42. Torgerson, W.S. Multidimensional scaling: I. Theory and method. *Psychometrika* **1952**, *17*, 401–419. [CrossRef]
43. Young, G.; Householder, A.S. Discussion of a set of points in terms of their mutual distances. *Psychometrika* **1938**, *3*, 19–22. [CrossRef]
44. Aitchison, J. Principal component analysis of compositional data. *Biometrika* **1983**, *70*, 57–65, doi:10.1093/biomet/70.1.57. [CrossRef]
45. Cox, T.F.; Cox, M.A. *Multidimensional Scaling*; CRC Press: Boca Raton, FL, USA, 2000.
46. Kenkel, N.C.; Orlóci, L. Applying metric and nonmetric multidimensional scaling to ecological studies: Some new results. *Ecology* **1986**, *67*, 919–928. [CrossRef]
47. Ripley, B. Modern applied statistics with S. In *Statistics and Computing*, 4th ed.; Springer: New York, NY, USA, 2002.
48. Vermeesch, P. Multi-sample comparison of detrital age distributions. *Chem. Geol.* **2013**, *341*, 140–146. [CrossRef]
49. Greenacre, M.J. *Theory and Applications of Correspondence Analysis*; Academic Press: Cambridge, MA, USA, 1984.
50. Stephan, T.; Kroner, U.; Romer, R.L. The pre-orogenic detrital zircon record of the peri-gondwanan crust. *Geol. Mag.* **2018**, *156*, 1–27. [CrossRef]
51. Garzanti, E.; Andò, S. Heavy-mineral concentration in modern sands: implications for provenance interpretation. In *Heavy Minerals in Use, Developments in Sedimentology Series 58*; Mange, M., Wright, D., Eds.; Elsevier: Amsterdam, The Netherlands, 2007; pp. 517–545.

52. Malusà, M.G.; Garzanti, E. The sedimentology of detrital thermochronology. In *Fission-Track Thermochronology and its Application to Geology*; Springer: Berlin, Germany, 2019; pp. 123–143.
53. Malusà, M.G.; Resentini, A.; Garzanti, E. Hydraulic sorting and mineral fertility bias in detrital geochronology. *Gondwana Res.* **2016**, *31*, 1–19. [CrossRef]
54. Gower, J.C. Generalized procrustes analysis. *Psychometrika* **1975**, *40*, 33–51. [CrossRef]
55. Carroll, J.D.; Chang, J.-J. Analysis of individual differences in multidimensional scaling via an N-way generalization of 'Eckart-Young' decomposition. *Psychometrika* **1970**, *35*, 283–319. [CrossRef]
56. DeLeeuw, J. and Mair, P. Multidimensional scaling using majorization: The R package smacof. *J. Stat. Softw.* **2009**, *31*, 1–30.
57. Chayes, F. On ratio correlation in petrography. *J. Geol.* **1949**, *57*, 239–254. [CrossRef]
58. Chayes, F. On correlation between variables of constant sum. *J. Geophys. Res.* **1960**, *65*, 4185–4193. [CrossRef]
59. Armstrong-Altrin, J.; Verma, S.P. Critical evaluation of six tectonic setting discrimination diagrams using geochemical data of Neogene sediments from known tectonic settings. *Sediment. Geol.* **2005**, *177*, 115–129. [CrossRef]
60. Tolosana-Delgado, R.; von Eynatten, H.; Krippner, A.; Meinhold, G. A multivariate discrimination scheme of detrital garnet chemistry for use in sedimentary provenance analysis. *Sediment. Geol.* **2018**, *375*, 14–26. [CrossRef]
61. Weltje, G.J. Quantitative models of sediment generation and provenance: State of the art and future developments. *Sediment. Geol.* **2012**, *280*, 4–20. [CrossRef]
62. Allen, P.A. From landscapes into geological history. *Nature* **2008**, *451*, 274. [CrossRef]
63. Garzanti, E.; Dinis, P.; Vermeesch, P.; Andò, S.; Hahn, A.; Huvi, J.; Limonta, M.; Padoan, M.; Resentini, A.; Rittner, M.; et al. Sedimentary processes controlling ultralong cells of littoral transport: Placer formation and termination of the Orange sand highway in southern Angola. *Sedimentology* **2018**, *65*, 431–460. [CrossRef]
64. Garzanti, E.; Andò, S.; Vezzoli, G. Grain-size dependence of sediment composition and environmental bias in provenance studies. *Earth Planet. Sci. Lett.* **2009**, *277*, 422–432, doi:10.1016/j.epsl.2008.11.007. [CrossRef]
65. Malusà, M.G.; Carter, A.; Limoncelli, M.; Villa, I.M.; Garzanti, E. Bias in detrital zircon geochronology and thermochronometry. *Chem. Geol.* **2013**, *359*, 90–107. [CrossRef]
66. Resentini, A.; Malusà, M.G.; Garzanti, E. MinSORTING: An Excel® worksheet for modelling mineral grain-size distribution in sediments, with application to detrital geochronology and provenance studies. *Comput. Geosci.* **2013**, *59*, 90–97. [CrossRef]
67. Bloemsma, M.; Zabel, M.; Stuut, J.; Tjallingii, R.; Collins, J.; Weltje, G.J. Modelling the joint variability of grain size and chemical composition in sediments. *Sediment. Geol.* **2012**, *280*, 135–148. [CrossRef]
68. Paatero, P.; Tapper, U. Positive matrix factorization: A non-negative factor model with optimal utilization of error estimates of data values. *Environmetrics* **1994**, *5*, 111–126. [CrossRef]

minerals

MDPI

Article

Semi-Automated Heavy-Mineral Analysis by Raman Spectroscopy

Nils Keno Lünsdorf *, Jannick Kalies, Patrick Ahlers, István Dunkl and Hilmar von Eynatten

Department of Sedimentology and Environmental Geology, Geoscience Center, Georg-August University Göttingen, Goldschmidtstraße 3, D-37077 Göttingen, Germany
* Correspondence: kluensd@gwdg.de

Received: 30 May 2019; Accepted: 21 June 2019; Published: 26 June 2019

Abstract: A significant amount of information on sedimentary provenance is encoded in the heavy minerals of a sediment or sedimentary rock. This information is commonly assessed by optically determining the heavy-mineral assemblage, potentially followed by geochemical and/or geochronological analysis of specific heavy minerals. The proposed method of semi-automated heavy-mineral analysis by Raman spectroscopy (Raman-HMA) aims to combine the objective mineral identification capabilities of Raman spectroscopy with high-resolution geochemical techniques applied to single grains. The Raman-HMA method is an efficient and precise tool that significantly improves the comparability of heavy-mineral data with respect to both overall assemblages and individual compositions within solid solution series. Furthermore, the efficiency of subsequent analysis is increased due to identification and spatial referencing of the heavy minerals in the sample slide. The method is tested on modern sediments of the Fulda river (central Germany) draining two Miocene volcanic sources (Vogelsberg, Rhön) resting on top of Lower Triassic siliciclastic sediments. The downstream evolution of the volcanic detritus is documented and the capability to analyze silt-sized grains has revealed an additional eolian source. This capability also poses the possibility of systematically assessing the heavy-mineral assemblages of shales, which are often disregarded in sedimentary provenance studies.

Keywords: Raman spectroscopy; heavy minerals; sedimentary provenance; automatization

1. Introduction

The ultimate goal of sedimentary provenance analysis is to determine source rocks and region and to infer tectonoclimatic forcing of sediment generation (e.g., [1]). Since the object of analysis has already undergone different stages and intensities of alteration during its passage from source to sink (e.g., erosion, transport and burial), the direct link between sediment grains and their source area and lithology is blurred to various degrees. The degree of modification mainly depends on processes such as weathering, hydrodynamic sorting and alteration during burial (e.g., [2–5]), and its verification and especially quantification has remained difficult. Consequently, the toolbox of sedimentary provenance analysis contains a multitude of diverse methods to extract as much information as possible from the sediment. To reduce the impact of differential fractionation, single-grain geochemical and/or isotopic techniques applied to specific mineral phases are often used which provide a tremendous toolbox in sedimentary provenance analysis and usually require electron beam or laser ablation methods. An overview of the different minerals and techniques and their discrimination potential is provided by [6,7].

Considering the variety of different minerals used for dating or chemical/isotopic analysis (14 of the most prominent ones are listed in [7]), it would be advantageous if the most informative minerals, detected during optical mineral characterization, could be directly probed by the above mentioned methods. Most of the relevant minerals used for single-grain techniques are heavy minerals. Heavy-mineral assemblages provide essential information to constrain sediment provenance (e.g., [8]) and the method has been recently revitalized using modern Raman spectroscopic techniques [9]. Thus, the main rationale of the methodology presented in this paper is to efficiently combine an improved, semi-automated Raman spectroscopic identification of heavy minerals with single-grain techniques on the same sample slide. Therefore, the sample has to be precisely characterized, i.e., the exact location and type of the heavy mineral, has to be identified, in order to (i) obtain a robust quantitative measurement of heavy-mineral proportions and (ii) pre-select minerals that are suitable for the subsequent application of single-grain techniques.

The identification of heavy minerals is conventionally done by manually using the polarizing optical microscope exploiting the optical properties of minerals. The refractive index is a key optical property and in order to estimate the refractive index of heavy minerals they are dispersed into a medium of known refractive index, usually Melt Mount^{TM} (n = 1.6, 1.7, ...) or Canada balm (n = 1.54). Thus, heavy minerals separates are most often prepared as strew mounts covered by a glass cover slip (see [10]). Although it is possible to extract the grains for further analyses by liquefying or dissolving the embedding medium, a more efficient approach is to prepare the sample slide in a way that it is accessible to multiple instruments. Most frequently used geochemical techniques such as electron microprobe analyses (EMPA) and laser ablation induced coupled plasma mass spectrometry (LA-ICPMS) interact with the analyte's surface. Therefore, the grains are fixed in a competent polymer on a standard thin section glass slide and polished.

Raman spectroscopy is sensitive to molecular vibration and therefore provides information on the structure and composition of the analyte, which is why Raman spectroscopy is often used as a "finger-printing" technique to identify phases (e.g., [11]). Due to this capability Raman spectroscopy is an ideal method to automate the identification process in heavy-mineral analysis. Compared to optical microscopy Raman micro-spectroscopy has several advantages. The high lateral resolution and confocality allows the measuring of grains less than 10 µm in size. This enables quantification of the composition of silt-sized heavy minerals [12] or to characterize the inclusion inventory of different host phases (e.g., [13–15]). Opaque phases, which are usually neglected by optical heavy-mineral analysis, are readily identified using Raman spectroscopy. Varieties of minerals are confidently recognized [9], which eliminates the operator bias that affects the mineral identification process during optical examination. The automation potential of modern Raman microspectrometers enables the drastic increase of the number of observations (N > 1000) and, thus, decreases the counting-statistical error [16,17] relative to the usually counted 200 to 300 grains by optical mineral identification procedure. The chemical composition can be assessed semi-quantitatively. This is achieved by quantifying the shift of Raman bands with respect to mineral composition [18]. Especially cation substitution, for instance Fe, Mg, or Ca in ferromagnesian systems, give rise to significant band shifts. Such correlations have been quantified for instance for olivine [19], pyroxene [20], garnet [18,21] or amphibole [22]. Furthermore, laser-induced photo-luminescence allows the visualization of the distribution of several tri-valent rare earth elements [23] and the degree of metamictization or structural integrity can be estimated (e.g., [24–26]). The latter is important, for instance, to exclude metamict or structurally defected grains from geochronometric methods such as U–Pb dating (e.g., [27]). Single-grain geothermometers such as the "Zr in Rutile" thermometer [28] require knowledge about the TiO_2-polymorph under investigation, as it has been shown by [29] that detrital Rutiles are frequently mixtures or composite grains of anatase and rutile. These compositional differences are not resolved in back-scattered electron imaging, but are readily observed by Raman spectroscopy [29].

After Raman analysis it is, thus, known where which minerals are situated in the sample slide and their chemical and physical properties can be approximated to pre-select grains of interest for further geochemical and/or isotopic analysis. This approach further improves efficiency because no additional preparation of grain mounts is needed. Only the Raman mount together with the coordinates of selected minerals need to be transferred to the next in situ analytical instrument.

The proposed method was tested on series of modern sediments of the river Fulda (central Germany) and source rocks in the Fulda drainage area.

2. Methodology/Experimental

The workflow of the proposed method is as follows. (1) heavy-mineral samples are prepared as polished strew mounts as described in Section 2.1 (2) the strew mounts are photographed at high magnification in reflected and transmitted light and the topography of the strew mount is approximated as outlined in Section 2.2 (3) Adhering to the "Fleet Method" [30], measurement positions are selected by the user on the created mosaic image and the coordinate list is passed to the Raman spectrometer. Minerals are then analyzed according to their optical properties (Section 2.2). The throughput depends largely on laser power, Raman activity of the mineral and set maximum exposure time (Table 1) and averages at ca. 400 to 500 grains per hour, but can range from ca. 1000 to 120 grains per hour. (4) The Raman spectra are automatically processed and compared to a reference database to infer the heavy-mineral assemblage, as proposed in Section 2.3. (5) Once the heavy-mineral assemblage is characterized, further insight from varieties can be gained by automated curve-fitting of mineral groups (e.g., olivine, pyroxene, garnet, etc.) to estimate chemical composition. In the following sub-sections details on the above outlined steps are presented.

Table 1. A list of measurement parameters for different grain size fractions (gsf) with respect to opaque and transparent minerals. cts: counts, tTL: threshold transmitted light, tRL: threshold reflected light, fS: fine sand, cSi: coarse silt, mSi: medium silt.

Parameter	gsf:	fS (63–125 μm)	cSi (30–63 μm)	mSi (10–30 μm)
Laser wavelength (nm)	*transparent*		532	
	opaque			
Laser power (% of 100 mW)	*transparent*	25	25	10
	opaque	1	1	1
Laser polarization	*transparent*		circular (lambda/4 retarder plate)	
	opaque			
Spectral grating (gr/mm)	*transparent*		1800	
	opaque			
Spectrometer position (cm^{-1})	*transparent*		1000	
	opaque			
Spectral range (cm^{-1})	*transparent*		110–780	
	opaque			
Slit size (μm)	*transparent*		100	
	opaque			
Hole size (μm)	*transparent*		100	
	opaque			
Objective	*transparent*	50x, 0.5 NA, LWD		100x, 0.8 NA, LWD
	opaque			
Maximum exposure time (s)	*transparent*	30		120
	opaque		300	
Number of accumulations	*transparent*		1	
	opaque		2	
Maximum intensity cts	*transparent*		5000	
	opaque		10,000	
tTL (gray value 0–255)			10	
tRL (gray value 0–255)			130	

2.1. Instrumental Setup

2.1.1. Raman Spectrometry

All Raman measurements were performed using the Horiba XploRA Plus system. The configuration for different measurement series is listed in Table 1.

2.1.2. Electron Microprobe Analysis (EMPA)

All EMPA measurements were performed using a Jeol JXA 8900 RL electron microprobe with a tungsten cathode as electron source. Silicates were analyzed for SiO_2, FeO, Na_2O, CaO, MgO, MnO, Cr_2O_3, NiO, Al_2O_3, and TiO_2 with an acceleration voltage of 15 kV, a probe current of 15 nA and a set beam diameter of 20 μm. Counting times were set to 15 s at each peak and 5 s for the upper and lower background, which was linearly corrected.

2.2. Sample Preparation

Using a Retsch BB 50 jaw crusher, hard rock samples (Table 2) were iteratively crushed by setting the gap width to 4, 1.5 and 0.2 to 0.3 mm. The >500 μm fraction of the disintegrated sample was removed by manual dry sieving. The <500 μm fraction was dispersed for 15 min in an ultrasonic bath and then transferred to a wet sieving tower (Fritsch Analysette 3 Pro, 10 min run duration, amplitude 2.3 and 5 s interval time) using 250, 125, 63 and 30 μm mesh sizes, the latter being a sieve cloth and not a metal sieve. For sediment samples, the >2 mm fraction, if present, (Table 2) was removed by manual wet sieving. Afterwards, the amount of sediment was reduced to 500–600 g per sample using a riffle divider. Aggregates of silt and clay were disintegrated by 15 min of ultrasonic treatment in deionized water. In both cases (hard rock, sediment) the rinse water was collected to gain the <30 μm size fraction.

Table 2. A list of sampling locations. VVC: Vogelsberg volcanic complex, Fsp: Feldspar.

Sample Code	N	E	Type	Note
EY55-01	50.69526	9.59377	recent sediment	Fulda river, north of VVC
EY55-02	50.50903	9.25902	Trachyte	Hard rock with weathering crust, cm-sized Fsp phenocrysts
EY55-03	50.52040	9.30481	Basalt	Hard rock with weathering crust, mm-sized olivine phenocrysts
EY55-04	50.51437	9.29809	recent sediment	Fulda tributary on VVC
EY55-05	50.53819	9.40401	recent sediment	Fulda tributary on VVC
EY55-06	50.56057	9.43914	recent sediment	Fulda tributary on VVC
EY55-07	50.45762	9.76565	recent sediment	Fulda river, south of VVC
EY55-08	50.48646	9.94867	recent sediment	near Fulda river spring, limestone gravel, not used for analysis
EY55-09	50.49809	9.93273	Basalt	Hard rock, peridotite xenoliths
EY55-10	50.49091	9.92062	Trachyte	Hard rock with weathering crust, fine grained
EY55-11	50.49251	9.92645	Basalt	Hard rock with weathering crust, fine grained, few phenocrysts
EY55-12	50.52239	9.86492	Phonolite	Hard rock, fresh
EY55-13	51.13599	9.58469	recent sediment	Fulda tributary draining only Bunter sandstone formations
EY55-14	51.21082	9.46757	recent sediment	Fulda river, most distal sample

The rinse water was centrifuged to remove the <10 μm fraction. To this end the particle settling time for a grain size of 10 μm was calculated after [31] under consideration of viscosity, temperature, acceleration, tube height and tube radius [32]. The samples were centrifuged for 91 s in 800 mL tubes at 20 °C and 300 rpm (10 s acceleration and 45 s deceleration phase). After each run the supernatant was decanted and the tube refilled with water. Prior to the next run the sample was dispersed by ultrasonic treatment. This procedure was repeated until the supernatant became clear. Finally, the grain size fractions 63–125 μm (fS: fine sand), 30–63 μm (cSi: coarse silt) and 10–30 μm (mSi: medium silt) were selected for density separation.

Density separation of the fS fraction was achieved by the established funnel technique [10] using sodium polytungstate with density adjusted to 2.86–2.89 g/cm^3. The samples were split by quartering to an amount of 10 to 15 g which was then loaded into the funnels. This approach, however, works poorly for the finer fractions cSi and mSi, e.g., about 70% light minerals remained in the heavy-mineral fraction after density separation. Thus, density separation of these fractions was carried out in the centrifuge. Although up to 3 g per 10 mL heavy liquid are proposed in [10], we conservatively used

1 g of sample and 40 mL of heavy liquid per centrifuge tube. Before centrifuging, the sediment was dispersed in the heavy liquid by ultrasonic treatment. Run duration was set to 5 min and fan speed to 3000 rpm. These settings ensure that a grain of 10 μm diameter will settle to the bottom of the tube. To recover the heavy minerals the lower part of the tubes was frozen by liquid nitrogen. The light minerals are then removed and after thawing of the frozen sodium polytungstate the heavy minerals could be extracted.

To analyze the heavy minerals by Raman spectrometer, EMPA, or LA-ICPMS a polished surface is crucial for the results. Therefore, the heavy minerals need to be fixed and polished. For optimal portability between the different devices the standard thin section format (48 × 28 mm) was chosen as sample carrier. The glass slide was abraded, cleaned with ethanol and placed in the oven at 50 °C to evaporate the remaining ethanol. Before mounting, the sample was split via quartering until an appropriate amount of heavy minerals remained. A flat and flexible silicon plate of 1 mm thickness is covered by double-sided adhesive foil and heavy minerals of up to 4 samples are dispersed on separated fields across the adhesive foil. A drop of ultra-violet (UV) curing glue (Uvrapid 702 by Best Klebstoffe™) is placed in the center of grains and the glass slide is put, with its ground surface facing towards the glue, on top of the drop. The glue is spread out between the grains by gently pressing down the glass slide. Once dispersed, the UV glue is cured by irradiating it with a UV-lamp for five minutes. Due to its flexibility the silicon layer can be easily peeled off from the glass slide.

Because the grains are already exposed to the surface, only little material needs to be removed by abrasion. To this end SiC-emery papers with 1200 and 2500 grit (for fS fraction) and 4000 grit (for mSi and cSi fractions) is used. During gentle, manual abrasion the surface is visually checked on regular basis, using a reflective microscope with 500x magnification, to ensure that a large fraction of grains is exposed to the surface. Polishing is performed iteratively on a Buehler Metaserv 2000 unit at 300 rpm, decreasing the grain size of the polishing agent after each step. Between all abrasion and polishing steps, the samples are cleaned in an ultrasonic bath. If abrasion stopped after 2500 grit, diamond polishing agents (Buehler MetaDi Monocrystalline) of grain size 9, 3, 1 μm were used. If abrasion stopped after 4000 grit, diamond polishing started at 3 μm grain size. In both cases final polishing was carried out using a water-based 0.05 μm Al_2O_3-slurry on a soft polishing cloth (Buehler MicroCloth).

During the time of preparing the manuscript we further improved the preparation routine and currently up to 20 samples can be fixed on a single glass slide which drastically increases the efficiency of this method (Figure S5). Following this approach, it takes about 1 hour to prepare a "ready-to-measure" sample slide containing up to 20 samples.

2.3. Raman Measurement Setup

Ultimately, no distinction between opaque and transparent phases is necessary for Raman analysis, as minerals of both groups are Raman active. However, to avoid photo-oxidation the opaque phases must be measured with less laser power [33,34] than the transparent phases. This requires at least two sets of measurement parameters (Table 1) and a differentiation between opaque and transparent phases before the Raman experiment, which also allows comparison of Raman-HMA data with classic optical HMA data with more confidence. Usually, a 40x or 50x objective together with a raised condenser and an inserted condenser lens is used during optical HMA to determine if a grain is opaque. Therefore, selecting measurement sites for Raman analysis should be performed under similar conditions.

Depending on grain size and grains per unit area a HM sample can cover several square centimeters on a standard thin section slide. For example, at a scaling factor of 0.22 μm/px a single image with 1292 by 968 pixels covers an area of 284.24 by 212.96 μm, which means that ca. 1645 images are needed to photograph an area of 1 cm^2 at high resolution (50x objective). The resulting, uncompressed mosaic image consists of 2063698560 pixels (ca. 2 Gigapixels). Depending on number of channels (red, green, blue, alpha), this is likely going to be a too large file to fit into the RAM of a typical computer and the software packages of the Raman systems often rescale the image at the expense of resolution to reduce its file size. Even if a large enough mosaic image was created within the Raman system's software package the user would waste a significant amount of measuring time by manually selecting measurement sites because automated particle detection by means of image analysis is often not realized by the manufacturers. To satisfy these needs the Coordsetter software has been developed, which relies on the hierarchical data format (HDF). An HDF file is a versatile data container that is stored directly on the hard drive. This allows for large file sizes (>10 GB) and circumvents RAM limitations.

Before selecting measuring sites, the image data needs to be created and saved into an HDF file. The single images (tiles) of the mosaic are photographed using a Zeiss AxioImager M2m microscope with a fully motorized stage using a 50x 0.75 NA objective. The extent of the mosaic is currently defined by selecting the upper left and lower right corner of the area of interest and a focal surface is computed by interpolating through an equally distributed array of support points. At the moment 64 support points are used and at each support point the correct focus value is manually selected. Once defined, the tiles are photographed in reflected (RL) and transmitted light (TL) as gray scale images. All microscope and acquisition related information is stored in a xml file (e.g., scale factor, support point coordinates, tile extent, etc.) and after data acquisition the tiles are stitched to a mosaic that is stored as a large array in the HDF file. Together with the mosaic, the scale factor, aspect ratio, and the support point coordinates are also saved into the HDF file. As the tiles are photographed in RL and TL, two HDF files are created for one sample. The Coordsetter software allows the viewing of the mosaic at full resolution by accessing and rescaling (by default 768x768) a subset of the mosaic array. The subset is defined by the current zoom state, e.g., if totally zoomed out the whole mosaic will be rescaled to 768x768 pixels and when zooming in the subset will get smaller and smaller until full resolution is reached (Figure 1). To be able to relocate the measurement sites a framework of reference points is needed. It is good practice to set the reference points in the corners and center of the mosaic. Measurement sites are then selected by pointing and clicking. With each clicked pixel, the x-, y-, and z-coordinates, and gray values in RL and TL are registered in the output file (JavaScript Object Notation file). The z-coordinates at a given location are calculated by passing the x- and y-coordinates to a two-dimensional spline function which linearly interpolates through the support points. After storing 1000 to 2000 measurements sites, the output file, compressed mosaic image, and sample are transferred to the Raman system.

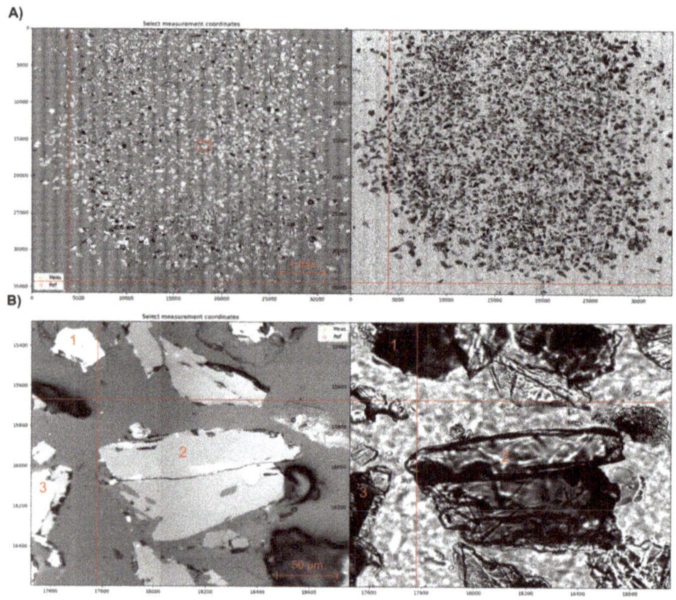

Figure 1. (**A**) Overview of sample EY55-11 (grain size fraction fS) as imaged in the Coordsetter software. The left panel shows the reflected light mosaic image and the right panel the transmitted light mosaic image. The red rectangle marks the position of subfigure B. (**B**) Zoom-in on the mosaic image. At this zoom-level precise selection of measurement sites is possible because fine details such as inclusions at or below the surface are readily visible. The numbered grains show different stages of opacity and are used as an example on the usage of tRL and tTL. Grain 1 will be classified as an opaque grain because the TL gray value is less than tTL. With medium gray values in TL and RL, grain 2 will be classified as a transparent grain. Although the RL gray value is high, grain 3 will be classified as a transparent grain because the TL gray value is greater than tTL.

When mounted, the reference points must be relocated to be able to transfer the selected measurement coordinates into the coordinate system of the Raman-microscope stage. This is achieved by applying a rigid (Euclidean) transformation to the dataset (e.g., [35]). After transformation, the coordinate list is sorted according to the gray values in RL and TL. It is determined, based on threshold gray values (Table 1) estimated during measurement sites selection, if a grain is transparent or opaque. If the lighting during image acquisition is constant, general threshold values (tTL, tRL) could be defined. Nevertheless, an opaque grain will always show low gray values in TL and transparent grains will have relatively high gray values. Thus, tTL would be enough for discrimination; however, relatively thick and dark-colored transparent grains (rutile, Cr-spinel, etc.) will also have low gray values in TL. Compared to "real" opaques, such as Fe- or Mn-oxides, dark transparent grains will have lower gray values in RL. Using tTL in combination with tRL allows a better discrimination between opaque and transparent grains based on following relationship:

- If TL gray value >= tTL: ° transparent grain
- If TL gray value < tTL and RL gray value < tRL: ° transparent grain
- If TL gray value < tTL and RL gray value > tRL: ° opaque grain
- If TL gray value < tTL: ° opaque grain

Sorting the coordinates list by the TL gray value ensures that the measurement parameter set only must be changed once. Because the total measuring time per sample can take several hours, the temporal drift of the Raman system needs to be monitored. As proposed in [36,37] 4-Acetamidophenol was chosen as an internal standard. The average Raman band positions of this substance are reported by [38]. Before mounting the slide onto the stage, a few 4-Acetamidophenol crystals are placed on the sample surface, and after the measurement sites coordinates list has been transformed, the coordinates of the internal standard are inserted, in a user-specified interval (e.g., every 100 measurements), into the transformed coordinate list. Depending on spectral resolution, calibration routine, and temperature variation, the offset in measured and real wavenumbers can be significant and must be corrected for (see Section 2.3).

2.4. *(Semi-) Automated Evaluation Routine*

A recorded spectrum must be pre-processed before it can be compared to a reference material database. The pre-processing and evaluation pipeline used here thus consists of the following steps:

1. Correcting for the temporal drift
2. Background removal
3. Estimation of noise and exclusion of spectra below signal-to-noise (SNR) threshold
4. Smoothing and scaling
5. Correcting for embedding medium spectrum
6. Phase identification and compilation of HM assemblage
7. Curve-fitting of interesting mineral groups (e.g., cpx, ol, grt) to semi-quantitatively assess chemical composition

The temporal drift during a measurement session is documented by recording the spectrum of a reference material (4-Acetamidophenol) in a regular interval. For each reference spectrum the position of the Raman bands at 390.9, 651.6, 797.2, 857.9, 1168.5, 1236.8, 1278.5, 1323.9, 1371.5 and 1561.5 cm^{-1} is determined by fitting a Lorentzian profile to the measured Raman band. The average difference between the measured and real Raman positions (avg. Δ-Raman shift) is reported for each reference spectrum. The Δ-Raman shift value is then combined with the extracted creation time of each reference spectrum file and a univariate spline is fitted through the Δ-Raman shift/creation time pairs (Figure 2). Each sample spectrum is then corrected by the Δ-Raman shift value that corresponds to its creation time.

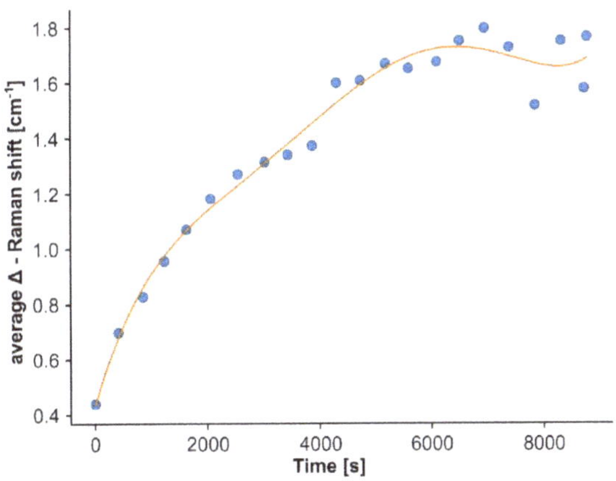

Figure 2. Temporal drift correction. The average deviation of 10 Raman band positions of 4-Acetamidophenol from the reported band positions [38] is plotted against the measurement time. It is apparent that the temporal drift is significant, and that the system stabilized after ca. 1.5 h.

The baseline is estimated by employing a method similar to the peak-clipping algorithm [39,40]. The Raman bands are clipped by iterative convolution of the spectrum with a Hanning window [41].

In this approach the baseline is represented by the convolution of the spectrum and a Hanning window and if an intensity value in the spectrum is higher than in the baseline, it is replaced by the baseline intensity value. Iterative application effectively removes the Raman bands (Figure 3). Iteration is terminated when a threshold value is met which is based on the decreasing difference between baselines of two consecutive iterations (Figure 3) and expressed by the squared sum of the difference. The baseline estimation depends on two parameters, namely the window size and threshold value. Both were determined empirically, and the window size is set to 0.05 times the number of datapoints in the spectrum, while the threshold value is fixed at 0.05.

The limit of detection (LOD) is given by a signal-to-noise ratio (SNR) of 3 [42]. Thus, spectra with a SNR ≤ 3 are excluded from the identification process, because they could lead to false positive results and unnecessarily extend computation time. SNR is defined as the average signal intensity above baseline divided by the standard deviation of the signal intensity [42]. This definition requires multiple accumulations, which is not advantageous for the method developed here because the total measuring time scales linearly with the number of accumulations per spectrum. Consequently, SNR must be estimated from a single accumulation and is defined here as the maximum intensity above baseline divided by the estimated noise intensity.

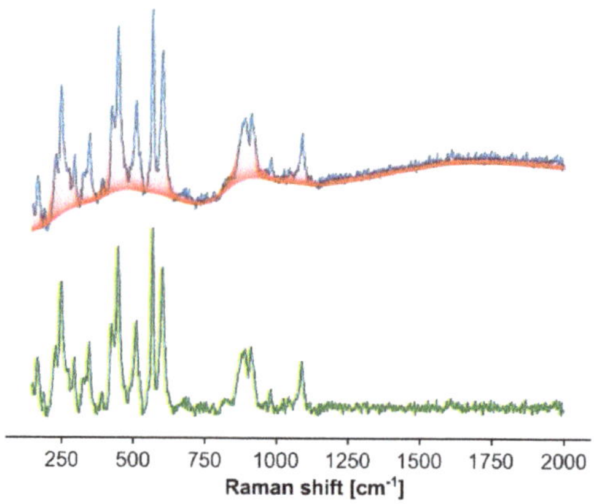

Figure 3. Baseline estimation. The iterative nature of the baseline estimation is visualized by the faint red lines which represent intermediate baseline estimates. The green spectrum is the baseline corrected spectrum.

Any part of the spectrum that contains no Raman signal can be used to estimate the noise intensity. Because the signal-free parts of a spectrum are a priori unknown, the spectrum is divided into segments of equal length (currently 15 pxls) and the spread of intensity values is computed in each segment. If a Raman band or part of it is present in a segment the intensity spread is comparatively high. Thus, the 25th percentile of all spread values is reported as the noise intensity estimate. Because the SNR defined here is not the true SNR, the LOD is conservatively set at SNR = 10 (Figure 4).

If the spectrum satisfies the LOD condition it is smoothed and scaled. Smoothing is achieved by convoluting the baseline corrected spectrum with a Hanning window of size 9 and the spectrum is scaled to the maximum value (Figure 4).

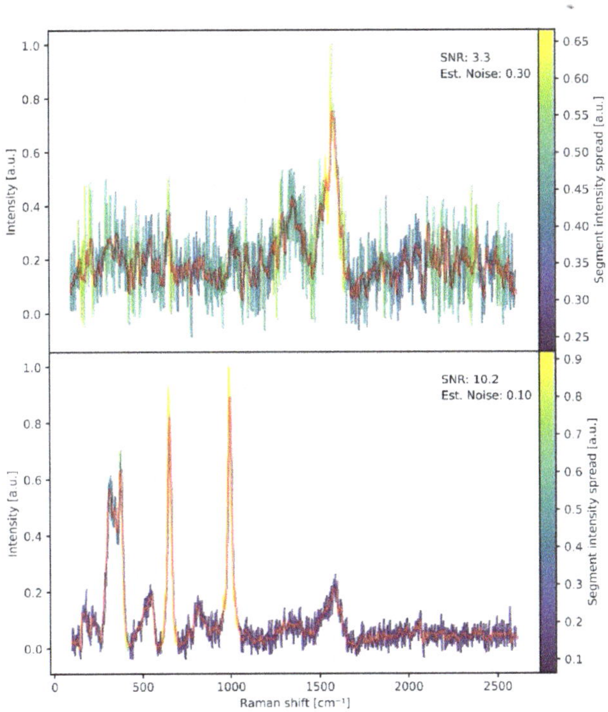

Figure 4. Noise estimation. Each vertical gray bar marks a segment of the spectrum and the color bar shows the intensity spread within each segment. Segments that contain a Raman band will have a large intensity spread, while segments without Raman bands will have a low intensity spread. The 25th percentile of all intensity spread values is reported as the estimated noise. The spectrum in the upper panel would be rejected and the spectrum in the lower panel would be accepted for further processing (LOD at SNR-threshold 10).

Pre-processing is finished by correcting for Raman bands that belong to the embedding medium. These are usually epoxy resins or UV-glues that show many Raman bands in the "finger-printing region" but also contain carbonyl groups and C=C bonds which give rise to Raman intensity between 1600–1800 cm^{-1}. Inorganic minerals usually do not show Raman bands in this spectral region. Thus, two bands between 1600 and 1800 cm^{-1} are selected in the drift corrected reference embedding medium spectra to derive a scaling factor (Figure 5). These peaks are then identified in the sample spectrum and the scaling factor specific to the sample spectrum is computed. The reference embedding medium spectrum is then multiplied by this scaling factor and subtracted from the sample spectrum (Figure 5).

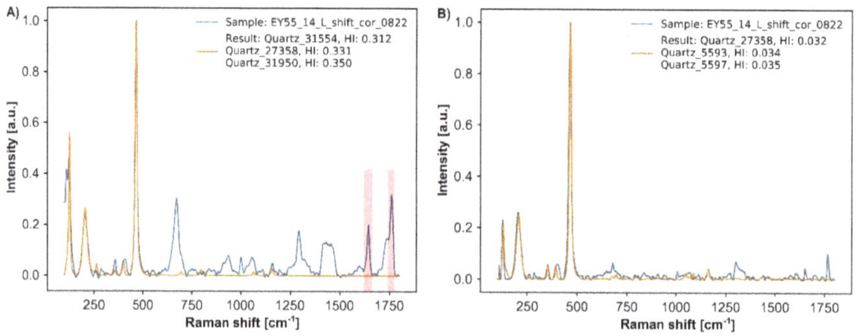

Figure 5. Embedding medium correction. The left panel (**A**) shows a composite spectrum of quartz and UV glue. Quartz is correctly identified, but the poor hit index of >0.3, due to the presence of UV-glue Raman bands, would lead to rejection of this result. A scaling factor for the UV-glue spectrum is computed based on the intensity ratio of the red shaded bands. A reference UV-glue spectrum is multiplied with the scaling factor and subtracted from the composite spectrum, leading to the corrected spectrum (right panel, **B**). After processing the hit index improves by an order of magnitude and the result would be accepted.

Phase identification is done by comparing each sample spectrum to each spectrum in the database. Currently, the database consists of the "LR" subgroup of the RRUFF database [43]. On 28.02.2018 this subgroup contained 9433 broad scans of lower resolution (~2 cm^{-1}) which are mainly unprocessed. This group was filtered for unprocessed spectra of minerals which have been verified by at least one other method except Raman spectroscopy. The remaining 8115 spectra were extended by additional 300 spectra of confirmed heavy minerals. The latter were chosen to reflect especially the influence of metamictization in zircon and to account for luminescence bands in several phases (e.g., zircon, apatite, monazite, etc.).

$$HI = 1 - \frac{(ref_m \cdot smpl_m)^2}{(ref_m \cdot ref_m) \times (smpl_m \cdot smpl_m)} \quad (1)$$

The "Hit Index" (HI, Equation (1)) is computed for each pair of sample and reference spectra, where ref_m refers to the mean-centered reference spectrum, $smpl_m$ refers to the mean-centered sample spectrum and the dot indicates vector dot products [44]. The HI is a correlation method with a success rate of ca. 80% (e.g., [45]) and ranges between 0 and 1, with 0 indicating a perfect match and 1 absolute disagreement. HI thresholds were empirically determined, and the identification results were differentiated into three groups: "Good hit" 0.0 to 0.15, "Medium hit" 0.15 to 0.3 and "No hit" for HI > 0.3. Spectra of the "Good hit" group are accepted, spectra of the "No hit" group are neglected and spectra of the "Medium hit" group should be reexamined, because these can contain correctly identified minerals, but often consist of mixed or poor quality spectra. After identification, the mineral names are sorted into a list of mineral groups approved by the International Mineralogical Association (IMA). The list (see "mineral dictionary" in the supplements) consists of mineral groups of the most common light minerals and mineral groups that contain heavy minerals, e.g., minerals with a density of 2.85 g/cm^3 or higher. The output of the evaluation routine is a .csv file that contains all results, three csv files which contain the results sorted by the HI group intervals and a .png file that shows the sample spectrum with the "best hit" reference spectrum (e.g., Figure 5B).

When the phase identification is complete, Raman spectra of relevant mineral groups are selected for spectral curve fitting. Curve fitting is essential to extract Raman band parameters such as position, height, width, and area. Of these parameters, position and width convey information on crystallographic structure and mineral composition (e.g., [46]). The latter can help to further differentiate sediment sources. Curve fitting is accomplished by using the automated iterative curve-fitting approach of [47] which allows batch processing of numerous spectra with curve-fitting protocols specific for the selected mineral group.

3. Application (Case Study)

The described approach was tested on modern sediments (Table 2) of the Fulda river, a medium-sized catchment (entire river length approx. 220 km) in central Germany. The primary goal of this case study was to investigate if the heavy minerals derived from two volcanic source areas within the catchment could be traced in the modern river sediment and discriminated against the dominating siliciclastic bedrock, and if any downstream modification of the volcanic assemblages could be determined.

The Fulda drainage area (Figure 6) contains a vast proportion of Lower Triassic siliciclastic sedimentary rocks of the Bunter sandstone formation and Middle Triassic carbonates of the Muschelkalk formation. In the lower reaches of the drainage area some Permian carbonates and evaporites of the Zechstein formation and few clastic sediments of the Rotliegend formation crop out. Major occurrences of Tertiary (Miocene, 19–10 Ma) volcanic rocks are found in the Vogelsberg volcanic complex (VVC) and in the northern Rhön area at the "Wasserkuppe" close to the Fulda spring. Some more volcanic spots together with Oligocene to Miocene sediments occur further downstream in the vicinity of the city of Kassel (Figure 6). The volcanic rocks in the Fulda drainage area are mainly olivine-bearing alkali basalts and tholeiitic basalts with minor trachytes (VVC and northern Rhön) and phonolites (only northern Rhön). Quaternary deposits are dispersed across the complete drainage area and are mainly composed of loess.

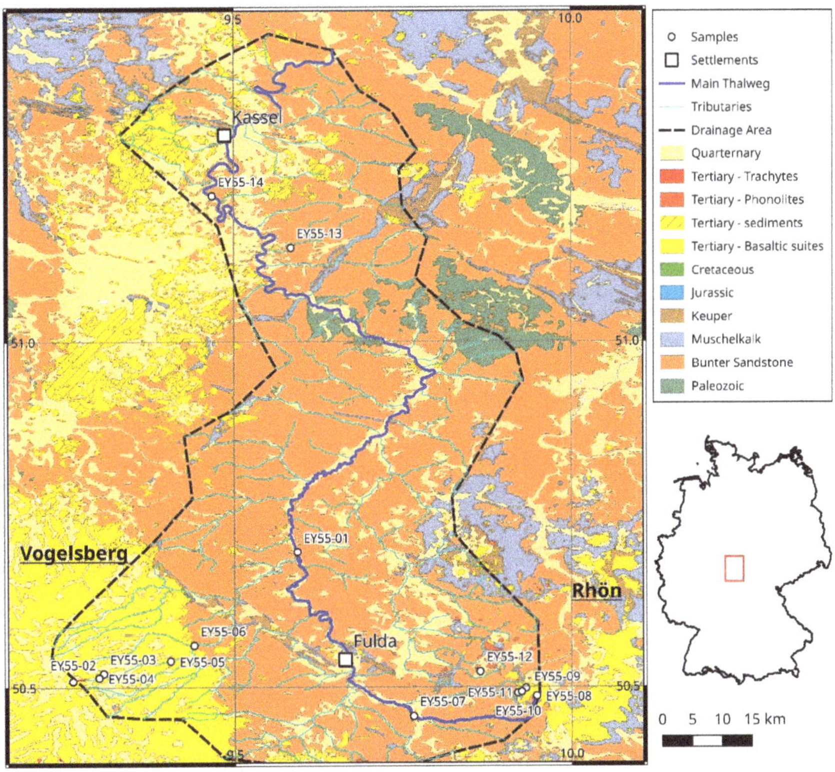

Figure 6. Geological map of the Fulda river drainage area and sample locations. Colors refer to the stratigraphic age.

A stable heavy-mineral assemblage is reported for the Bunter sandstone formation of central Germany which mostly consists of zircon and tourmaline in roughly equal proportions together with less TiO_2 minerals (rutile, anatase) and minor apatite; staurolite may occur in traces [48–51]. In general, the HM content is less than 1% and usually more than 50% of the heavy minerals are opaque phases. Similar composition and amount of heavy minerals are documented for the Permian clastic sediments [52]. The loess deposits contain higher proportions of heavy minerals (ca. 5%) and are mainly composed of amphibole, epidote and opaques in roughly equal proportions (ca. 20%) together with garnet, zircon, titanite, tourmaline, anatase (all ± 5%) and minor monazite, rutile and apatite (all <5%) and traces of spinel, sillimanite and kyanite [53]. The basalts of the VVC mainly deliver olivine and Ti–augite [54] as heavy minerals, while the heavy minerals of the trachyte consist predominantly of augitic clinopyroxene along with some titanite [55]. The volcanic rocks of the northern Rhön area provide a similar suite of heavy minerals; however, the phonolites contain significant amounts of aegirine–augite and aegirine [56].

3.1. Samples

In total 14 samples (Table 2) were collected across the Fulda river drainage area (Figure 6). The samples are divided into four groups, i.e., modern Fulda river sediments (EY55-01, -07, and -14; marked by red text in the figures), modern sediments of a Fulda tributary draining only the VVC (EY55-04, -05, and -06; marked by blue text in the figures), a modern sediment sample of a Fulda tributary draining only the Bunter sandstone formation (EY55-13; marked by violet text in the figures) and hard rock samples that depict potential source rocks (EY55-02, -03, -09, -10, -11, and -12; marked by green text in the figures). The modern Fulda river sediments were collected to reflect the downstream evolution of the heavy-mineral assemblage. EY55-07 is the most proximal sample and contains the assemblage without input from the VVC, EY55-01 is an intermediate sample containing the assemblage after the last Fulda tributary that drains the VVC and EY55-14 is the most distal sample. EY55-13 was sampled to reflect the heavy-mineral assemblage of the Bunter sandstone formation in the drainage area. Samples EY55-04 to -06 were collected to document the downstream evolution of VVC heavy-mineral assemblage.

3.2. Results

Table 3 lists all results related to the heavy-mineral assemblages determined by the method outlined in Section 2. The number of total spectra varies between 700 and 1200. Only transparent spectra with a SNR > 10 were evaluated. The difference between the number of spectra in the "Good hit" group and the number of spectra per heavy-mineral assemblage differs mainly due to the presence of light minerals, e.g., 668 "Good hit" spectra vs. 285 heavy-mineral spectra in fine sand sample EY55-14 (see "HM_result_files" in the supplements).

Table 3. A compilation of the results of spectral evaluation. trns: transparent, n.n.: no noise, opq: opaque, TS: total spectra, NS: noisy spectra, GH: good hit, MH: medium hit, NH: no hit, HR: hit rate, n: number of "good hit" heavy minerals. Amp: Amphibole, And: Andalusite, Ant: Anatase, Ap: Apatite, Brt: Baryte, Chm: Chamosite, Chr: Chromite, Cld: Chloritoid, Ep: Epidote, Grt: Garnet, Mag: Magnetite, Mnz: Monazite, Ol: Olivine, Opx: Orthopyroxene, Px: Pyroxene, Rt: Rutile, Sd: Siderite, Sil: Sillimanite, Spl: Spinel, St: Staurolite, Tur: Tourmaline, Ttn: Titanite, Zrn: Zircon.

Grain Size Fraction	Sample	trns	trns n.n.	opq	opq n.n.	TS	NS	GH	MH	NH	HR	Grt	Ol	Zrn	Ttn	Ep	Tur	Px	Amp	TiO2	Ap	Mnz	Other	n	
	EY55-01	849	722	260	218	1109	169	478	135	110	66	0.00	13.17	16.10	0.00	0.00	5.12	59.51	0.00	4.63	0.00	0.00	1.46	410	Ttn, Amp, Spl, Brt, Ap
	EY55-02	459	411	581	572	1040	57	319	34	58	78	0.00	0.00	0.00	0.00	0.00	0.00	91.42	0.00	0.00	7.84	0.00	0.75	268	Ol, TiO2,
	EY55-03	868	725	109	97	977	155	531	68	127	73	0.00	49.90	0.00	0.00	0.00	0.00	49.33	0.00	0.00	0.00	0.00	0.77	519	Ap
	EY55-04	845	601	350	314	1195	280	338	158	105	56	0.00	14.33	0.00	0.00	0.00	0.00	83.80	0.00	0.00	0.00	0.00	1.87	321	Zrn, Ttn, Ep, Ap, Mnz
	EY55-05	1013	776	162	147	1175	252	625	66	85	81	0.00	18.11	0.00	0.00	0.00	0.00	81.40	0.00	0.00	0.00	0.00	0.49	613	Ttn, TiO2, Spl
	EY55-06	829	675	371	329	1200	196	529	77	70	78	0.00	34.04	0.00	0.00	0.00	0.00	64.22	0.00	0.00	0.00	0.00	1.74	517	Zrn, Ep, Spl, Ap
fine sand	EY55-07	562	516	441	417	1003	70	415	56	45	80	0.00	11.14	31.65	1.27	0.00	4.30	49.11	0.00	1.52	0.00	0.00	1.01	395	Ep, Amp, Spl, Mnz
	EY55-09	1012	985	4	3	1016	28	941	16	28	96	0.00	94.36	0.00	0.00	0.00	0.00	5.53	0.00	0.00	0.00	0.00	0.11	940	Ap
	EY55-10	617	566	398	361	1015	88	232	102	232	41	0.00	0.00	0.00	12.72	0.00	0.00	76.32	0.00	1.75	8.33	0.00	0.88	228	Ol, Zrn, Ep, Amp
	EY55-11	1028	991	74	74	1102	37	888	93	10	90	0.00	23.87	0.00	0.00	0.00	0.00	76.13	0.00	0.00	0.00	0.00	0.00	888	Titano-Mag
	EY55-12	766	666	262	255	1028	107	203	245	218	30	0.00	0.00	0.00	4.49	0.00	0.00	88.46	0.00	3.85	3.21	0.00	0.00	156	Mag, Feldspathoides
	EY55-13	622	600	378	378	1000	22	552	37	11	92	0.00	26.19	49.72	0.00	0.00	10.44	12.52	0.00	0.00	0.00	0.00	1.14	527	Gr, Amp, Ttn, Ant
	EY55-14	728	706	272	258	1000	36	668	18	20	95	1.40	33.33	20.00	0.00	1.75	5.61	28.42	1.05	2.46	4.21	1.05	0.70	285	Ttn, Opx, Cld, Mica, Brt
	EY55-01	652	544	378	315	1030	171	350	112	82	64	4.28	13.90	20.32	1.07	6.42	2.14	41.71	0.00	5.88	0.00	3.74	0.53	187	Sil, St, Ap
	EY55-02	397	348	722	722	1119	49	290	33	25	83	0.00	0.00	0.00	0.00	0.00	0.00	91.44	0.00	0.00	7.78	0.00	0.78	257	Ol
	EY55-03	767	671	293	278	1060	111	561	33	77	84	0.00	42.31	0.00	0.00	0.00	0.00	56.78	0.00	0.00	0.00	0.00	0.92	546	Ap
	EY55-04	745	535	256	236	1001	230	281	140	114	53	0.00	2.25	0.00	0.00	1.87	0.00	91.39	1.12	0.00	0.00	0.00	3.37	267	Grt, Zrn, Ttn, TiO2, Ap
	EY55-05	745	507	290	274	1035	254	305	94	108	60	0.00	14.43	0.00	0.00	1.72	0.00	82.13	0.00	0.00	0.00	0.00	1.72	291	Zrn, Amp, Ap
	EY55-06	655	445	354	315	1000	249	250	81	114	56	1.27	19.92	0.00	0.00	3.81	0.00	72.03	1.27	0.00	0.00	0.00	1.69	236	Ttn, Tur, TiO2, Ap
coarse silt	EY55-07	551	467	449	391	1000	142	321	80	66	69	0.00	15.71	29.49	0.00	2.88	1.60	43.27	1.28	2.56	0.00	0.00	3.21	312	Grt, Ttn, Cld, Ap, Mnz
	EY55-09	954	933	47	46	1002	22	866	31	36	93	0.00	92.12	0.00	0.00	0.00	0.00	7.76	0.00	0.00	0.00	0.00	0.12	863	Chr
	EY55-10	908	603	92	90	1000	307	284	136	183	47	0.00	0.00	0.00	0.00	0.00	0.00	97.71	0.00	0.00	1.38	0.00	0.92	218	Ttn, Titano-Mag
	EY55-11	934	910	82	81	1016	25	729	137	44	80	0.00	20.86	0.00	0.00	0.00	0.00	79.01	0.00	0.00	0.00	0.00	0.14	724	Ap
	EY55-12	931	714	81	81	1012	216	232	254	228	32	0.00	0.00	0.00	0.00	0.00	0.00	97.03	0.00	0.00	0.00	0.00	2.97	101	Ttn, Ap, Mnz, Zeo
	EY55-13	667	617	333	333	1000	50	554	41	22	90	0.00	9.81	15.09	0	22.26	6.60	36.42	4.53	1.89	1.32	0.00	2.08	530	Grt, Ttn, Sil, St, Mnz
	EY55-14	778	716	222	218	1000	66	648	44	24	91	1.23	8.62	5.23	1.54	23.69	4.00	37.54	7.38	4.92	5.54	0.00	0.31	325	St, Cld, Chm, Sd, Mnz,

Table 3. *Cont.*

Grain Size Fraction	Sample	trns	trns n.n.	opq	opq n.n.	TS	NS	GH	MH	NH	HR	Grt	Ol	Zrn	Ttn	Ep	Tur	Px	Amp	TiO2	Ap	Mnz	Other	n	
	EY55-01	492	441	208	168	700	91	290	113	38	66	4.32	3.60	3.96	0.00	17.63	4.32	53.60	7.91	2.16	1.08	0.00	1.44	278	Ttn, And, St, Spl, Mnz
	EY55-02	391	386	265	265	656	5	323	42	20	84	0.00	0.00	0.00	0.00	0.00	0.00	96.85	0.00	0.00	3.15	0.00	0.00	317	
	EY55-03	545	503	161	161	706	42	387	53	63	77	0.00	18.95	0.00	0.00	0.00	0.00	77.89	0.00	0.00	3.16	0.00	0.00	380	
	EY55-04	486	378	214	214	700	108	214	97	67	57	0.00	2.46	1.48	0.00	5.42	0.00	86.21	1.48	0.00	0.00	0.00	2.96	203	Grt, Chm, TiO2, Ap, Mnz
	EY55-05	457	412	248	248	705	45	289	84	39	70	1.51	4.91	4.15	0.00	22.26	1.51	56.98	4.15	1.51	1.89	0.00	1.13	265	Ttn, Ky, St, Mnz
	EY55-06	448	418	164	163	612	31	260	112	46	62	0.00	2.37	1.58	0.00	5.14	0.00	85.77	1.19	0.00	3.16	0.00	0.79	253	Tur, TiO2
medium silt	EY55-07	435	397	271	269	706	40	264	81	50	67	0.00	1.54	0.00	0.00	2.70	0.00	94.21	0.00	0.00	0.00	0.00	1.54	259	Grt, Ttn, Amp, Ap
	EY55-09	1185	1131	15	15	1200	54	511	297	323	45	0.00	61.70	0.00	0.00	0.00	0.00	38.30	0.00	0.00	0.00	0.00	0.00	470	
	EY55-10	481	394	221	207	702	101	178	107	109	45	0.00	0.00	0.00	0.00	0.00	0.00	93.25	3.68	0.00	0.00	0.00	3.07	163	Ol, Ep, Rt, Ap, Mnz, Titano-Mag
	EY55-11	613	600	105	96	718	22	465	80	55	78	0.00	20.53	0.00	0.00	0.00	0.00	76.85	0.00	0.00	2.39	0.00	0.24	419	Ant, Titano-Mag
	EY55-12	638	559	84	72	722	91	229	177	133	42	0.00	0.00	0.00	0.00	0.00	0.00	98.96	0.00	0.00	0.00	0.00	1.04	192	Ol, Ap, Mag, Lorenzenite
	EY55-13	488	455	217	216	705	34	366	58	31	80	3.55	2.48	12.06	0.00	25.18	4.26	32.62	10.99	6.38	1.77	0.00	0.71	282	Ttn, And
	EY55-14	496	449	207	204	703	50	385	34	30	86	4.64	2.50	8.21	1.79	25.00	2.50	33.57	13.93	3.57	1.79	2.50	0.00	280	Sil, And, St, Chm, Sd

3.2.1. Heavy-Mineral Assemblages

The Fulda river sediment samples (EY55-01, 07, 14) and, interestingly, the sediments of the tributary draining only the Bunter sandstone formation (EY55-13) show a mixed provenance of the Bunter sandstone source dominated by zircon, tourmaline and rutile (ZTR) and the olivine and pyroxene rich basaltic source (Figure 7). Only the most distal sample (EY55-14) contains few garnets, epidotes, amphiboles, and apatite. Except for the latter, these phases are absent in all other samples of the fS grain size fraction. The Vogelsberg sediment samples (EY55-04, 05, 06) reflect the bimodal composition (olivine, pyroxene) of the basalts (EY55-03). The Vogelsberg trachyte (EY55-02) primarily delivers pyroxene and little apatite while the Rhön trachyte (EY55-10) additionally contains titanite. The Rhön phonolite (EY55-12) appears to be of similar composition as the Rhön trachyte, but contains many clinopyroxene grains of aegirin–augitic to aegirin composition which is not the case for the trachytes.

In all sediment samples the amount of olivine decreases with decreasing grain size while the content of epidote, amphibole and garnet increases (Figure 8). The decrease in olivine content towards finer grain size is also observed in the Vogelsberg and Rhön basalts (EY55-03, 11). Titanite even disappears in the silt fractions of the Rhön trachyte (EY55-10) and phonolite (EY55-12). Interestingly, zircon which is present in all Fulda river sediment samples in all grain size fractions, appears in the Vogelsberg sediments in the finest fractions only (Figure 7).

In the fS fraction the Ol/Px ratio of the Vogelsberg sediments (EY55-04, -06) shows a marked increase with increasing transport distance. The same is true for the modern Fulda sediments. The proximal (EY55-07) and intermediate (EY55-01) samples show similar Ol/Px ratios of less than one, while the distal sample (EY55-14) has as Ol/Px ratio of approximately one.

3.2.2. Semi-Quantitative Mineral Chemistry by Raman Spectroscopy: Olivine

The estimation of the content of the Mg-endmember (forsterite content) after [19] is based on the shift of two Raman bands in the olivine spectrum related to internal stretching vibrations of the SiO_4-group [57]. These bands are named DB1 and DB2 in [19] and occur at Raman shift values of ca. 820 cm^{-1} and ca. 850 cm^{-1}. A suite of olivines, identified as such by the outlined approach (Methods section), were also analyzed by EMPA. The forsterite (Fo) contents estimated by Raman and EMPA range between 55% and 85% and are in good agreement within an uncertainty of ±10% Fo content for most of the analyses (blue dots in Figure 9A). However, the Fo content of several olivines with Fo content around 60% to 65% according to EMPA is severely underestimated by Raman spectroscopy (green dots in Figure 9A). The main trend in DB1 and DB2 positions of all analyzed olivines reveals a range of Fo contents between ca. 55 to 100% (Figure 10), while the hard rock basalt samples (EY55-03, -09, -11) show a second trend (Figure 10, Figure 11 and Figure S3) that deviates from the Fo content related trendline. This second trend is no longer observed in the finest grain size fraction (mSi, Figure 12) and not in any of the sediment samples (Figure 10, Figures S2 and S4).

3.2.3. Semi-Quantitative Mineral Chemistry by Raman Spectroscopy: Pyroxene

Approaches after [20,58,59] allow estimation of the chemical composition of pyroxenes in terms of Ca, Mg, and Fe content by Raman spectroscopy. The molar ratios X^{Mg2+} ($Mg^{2+}/(Mg^{2+} + Ca^{2+} + Fe^{2+})$) and X^{Ca2+} ($Ca^{2+}/(Ca^{2+} + Mg^{2+} + Fe^{2+})$) can be determined using the procedure after [20] and this method was tested on pyroxenes selected from samples EY55-01, -02, -03, and -07. To this end the chemical compositions of pyroxenes, identified by Raman spectroscopy, were determined by EMPA. Comparing the compositional results based on the procedure after [20] to EMPA results indicates that X^{Mg2+} can be estimated within ±0.1 range by Raman spectroscopy and X^{Ca2+} is generally overestimated by Raman spectroscopy which in turn leads to an underestimation of X^{Fe2+} (Figure 13).

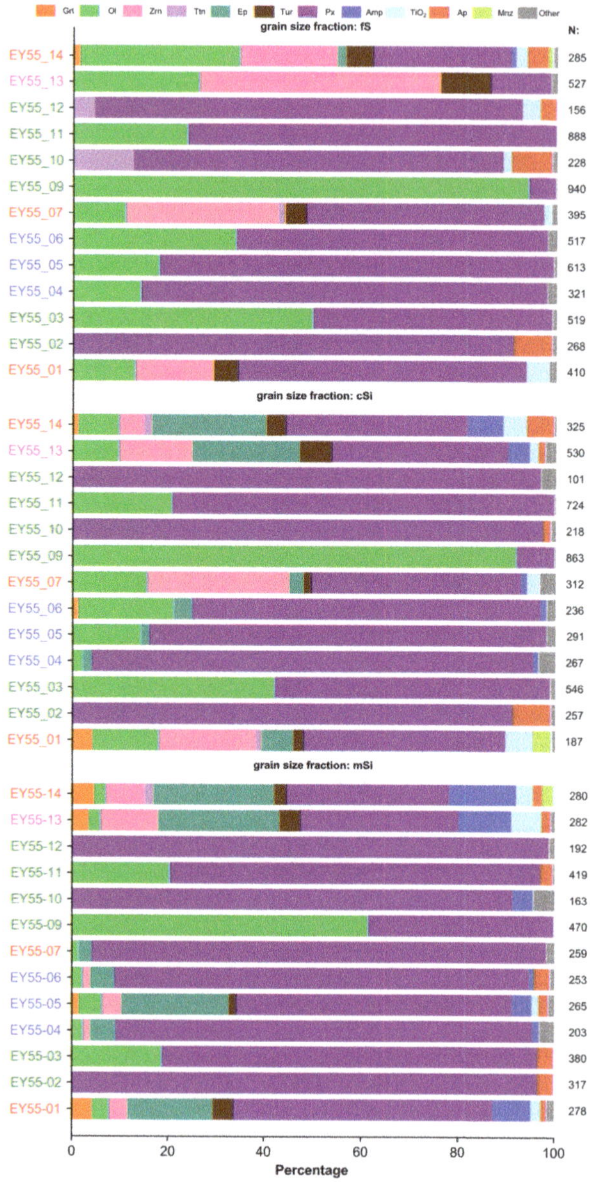

Figure 7. Heavy-mineral assemblage of the grain size fractions mSi, cSi, and fS. N is the number of "good hit" spectra of heavy minerals and sample names are color coded to reflect Fulda river sediment samples (red), sediment samples of the Vogelsberg volcanic complex (blue) and source rock samples (green).

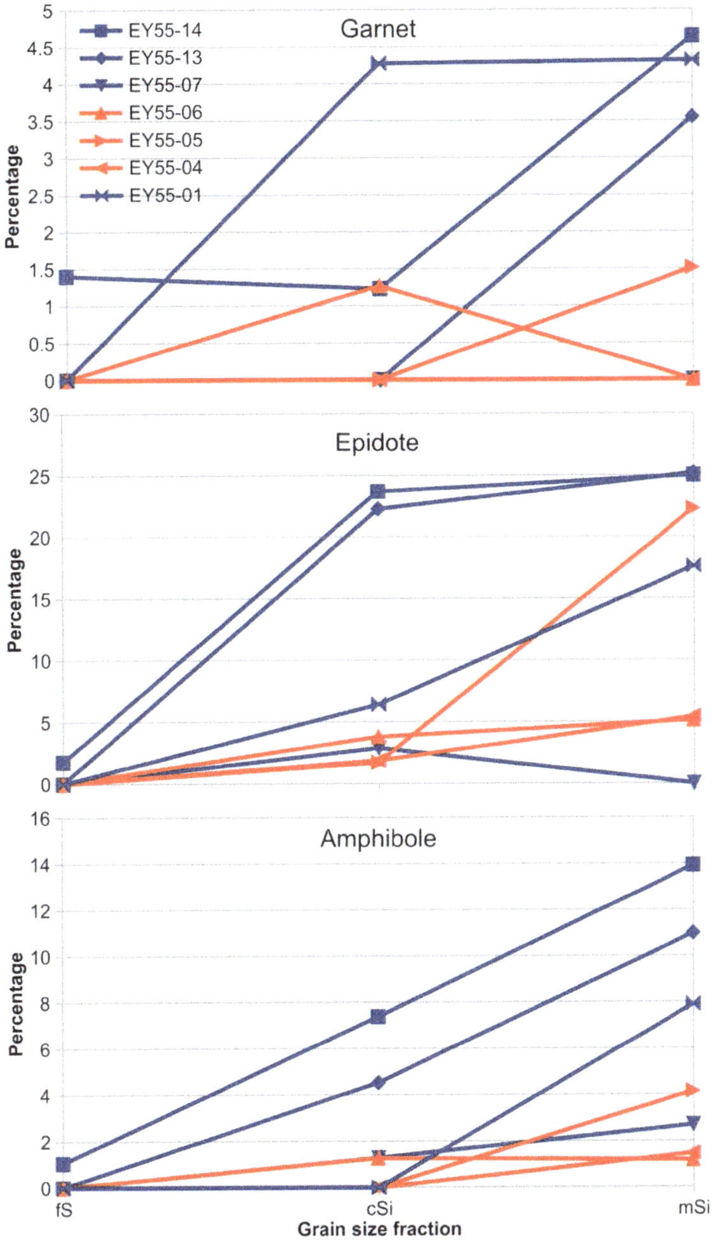

Figure 8. Grain size trends. The three panels show that the content of garnet, epidote, and amphibole, given in percentage of the heavy-mineral assemblage, increases in all sediment samples with decreasing grain size.

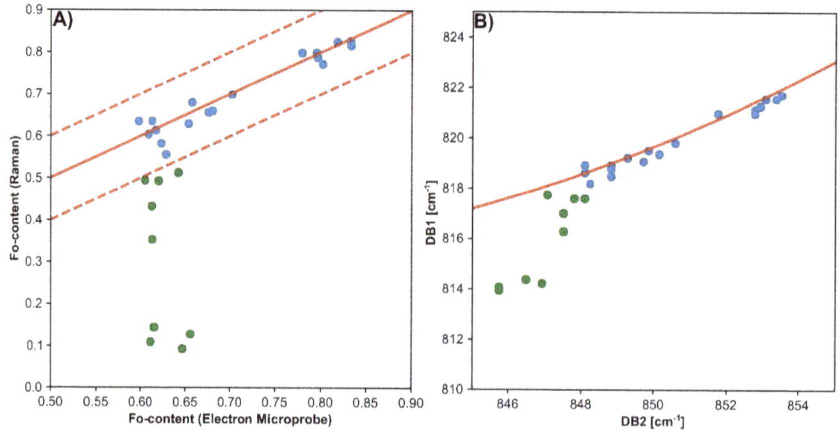

Figure 9. Fo content in olivines estimated by Raman spectroscopy. Panel (**A**) shows the forsterite content estimated from the Raman spectrum following the approach of [19] vs. electron microprobe data. The dashed lines mark the reported uncertainty of ± 0.1 for the Raman-based estimate. The green data points mark samples for which the Raman approach significantly underestimated the forsterite content. Panel (**B**) shows the trend in positional shift of the DB1 and DB2 Raman bands. The trend with increasing Fo content, as reported in [19], is indicated by the red line and the data points that produce Fo underestimates are shown in green.

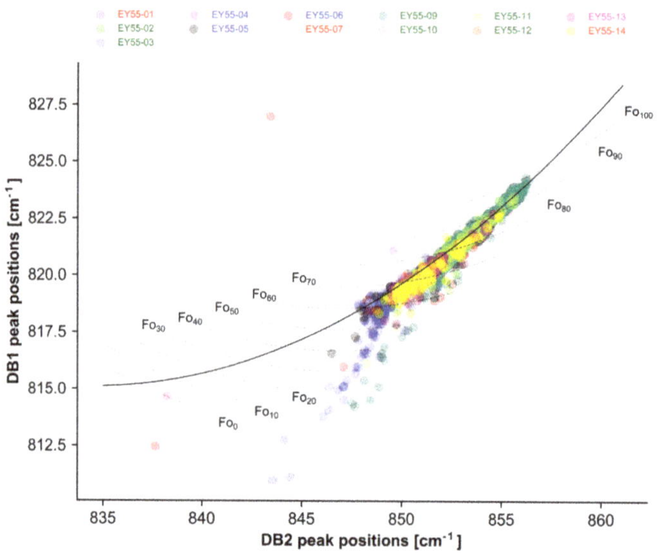

Figure 10. Fo content estimated by Raman spectroscopy for all samples of grain size fraction fS. The Fo content along the main trend (solid line) ranges between 55% to 100%. Only the hard rock samples (EY55-03, -09, -11, -12) contain olivines that show DB1 and DB2 positions deviating from the general Fo content trend line.

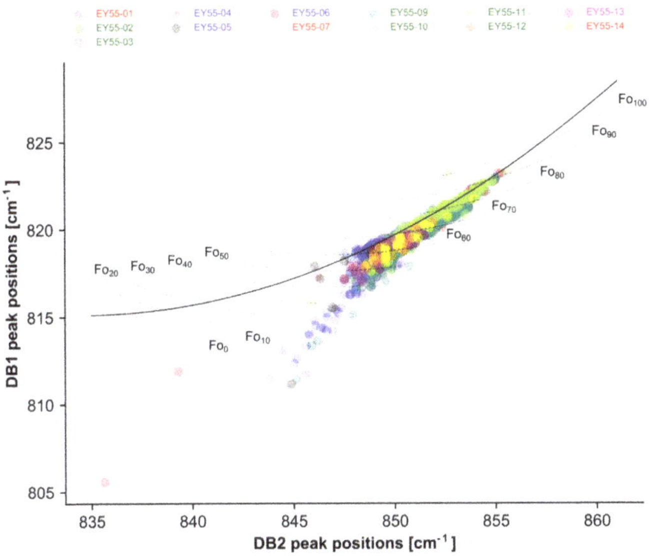

Figure 11. Fo content estimated by Raman spectroscopy for all samples of grain size fraction cSi. The Fo content along the main trend (solid line) ranges between 55% to 85%. Only the hard rock samples (EY55-03, -09, -11, -12) contain olivines that show DB1 and DB2 positions deviating from the general Fo content trend line.

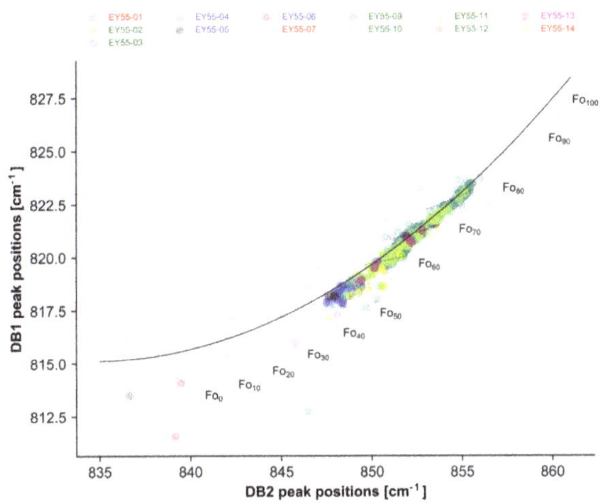

Figure 12. Fo content estimated by Raman spectroscopy for all samples of grain size fraction mSi. The Fo content along the main trend (solid line) ranges between 50% to 100%. No clear deviating trend as in the coarser grain size fractions (Figure 10; Figure 11) is observed.

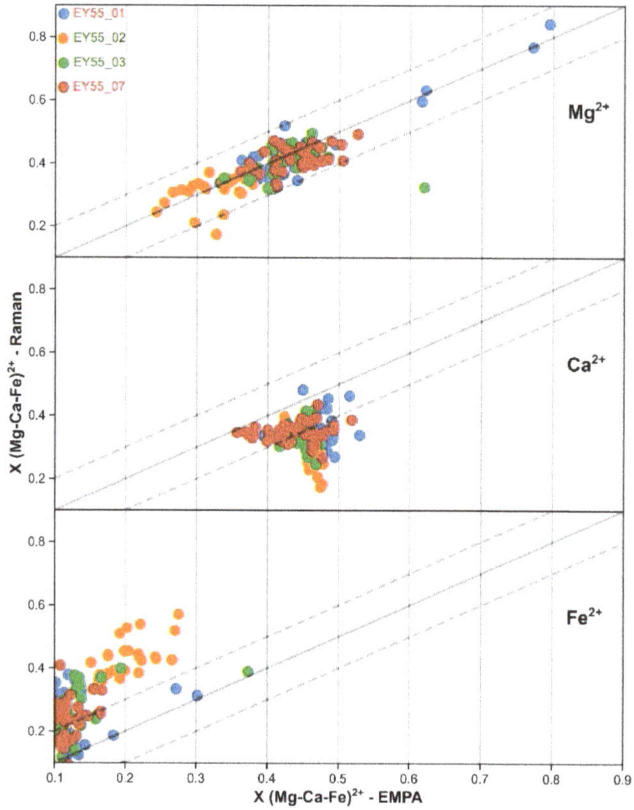

Figure 13. Estimation of X(Mg–Ca–Fe)$^{2+}$ by Raman spectroscopy and EMPA. Mg^{2+} and Ca^{2+} content was estimated following the approach of [20]. Mg^{2+} content is estimated reasonably well by this method, but Ca^{2+} content is generally underestimated. Due to the latter Fe^{2+} is overestimated.

Therefore, only X^{Mg2+} was estimated for all pyroxenes from all grain size fractions (Figure 14). The X^{Mg2+} values mostly range between 0.2 and 0.5. As expected, the Mg content of pyroxenes in sediments deposited within the Vogelsberg (EY55-04, -05, -06) is the same as in the sampled basalt (EY55-03) for all analyzed grain size fractions. However, this average value is slightly higher for the coarsest fraction (ca. 0.41 vs. 0.37, Figure 14). Sediment samples of the Fulda river directly down- and up-stream of the Vogelsberg (EY55-01, -07) also provide pyroxenes with X^{Mg2+} similar to the basalt sample. Pyroxenes of the sample EY55-11 (basalt, Rhön area) show X^{Mg2+} values equivalent to sample EY55-03 (basalt, Vogelsberg). The higher differentiated volcanic rocks (EY55-02, -10, -12) and distal Fulda river sediments (EY55-14) contain pyroxenes with relatively less Mg. Additionally, the distal sediment samples show narrower X^{Mg2+} distributions in the grain size fractions cSi and fS than all other samples. In the finest grain size fraction mSi X^{Mg2+} distributions are generally narrower and the contrast between samples is less pronounced.

Besides X^{Mg2+}, several Raman peak positions in the pyroxene spectrum can be used to differentiate pyroxene types. By plotting the positions of peak 2 against peak 3 in the spectral region R2 and R3 (Figure 15) after [20] quadrilateral and non-quadrilateral pyroxenes are readily distinguishable. From Figure 15 it follows that a group of pyroxenes with aegirine–augitic composition is present in the Rhön phonolites (EY55-12). Interestingly, this plot also reveals that a single aegirine–augite was detected in sample EY55-01 (Fulda sediment).

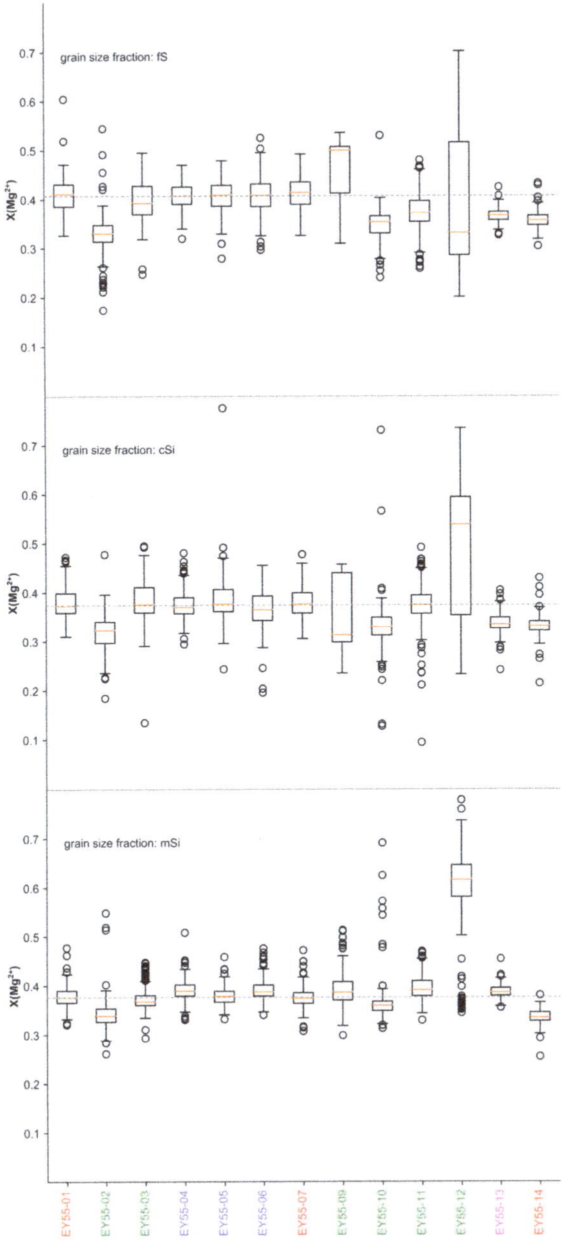

Figure 14. Boxplots of X(Mg^{2+}) for all samples and grain size fractions. The dashed line marks the average Mg^{2+}-content of basaltic clinopyroxenes of the Vogelsberg volcanic complex (samples EY55-03 to EY55-06) for the respective grain size fraction. The clinopyroxene phenocrysts (grain size fraction cSi and fS) of the higher differentiated rocks (EY55-02, -10, -12) show the lowest Mg^{2+}-contents, while the spread and average Mg^{2+}-content in the sediment samples decrease with increasing transport distance (proximal samples: EY55-01 and -07; distal sample: EY55-14). The distributions of Mg^{2+}-contents of the matrix clinopyroxenes (grain size fraction mSi) are generally very similar.

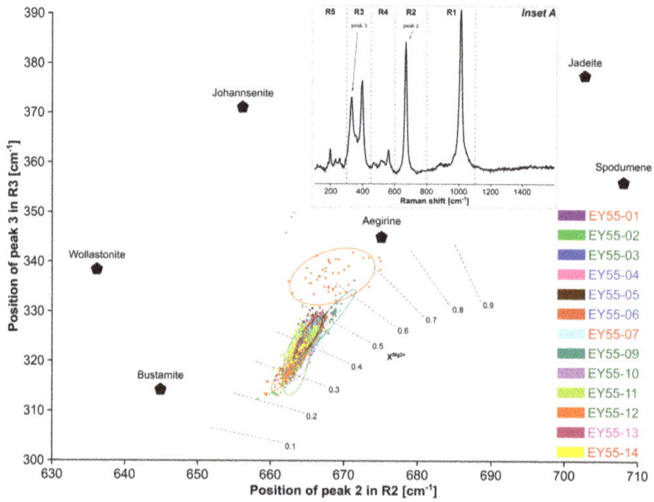

Figure 15. Variability of Raman peak positions for quadrilateral and non-quadrilateral pyroxenes after [20]. The positions of peak 2 and peak 3 in regions R2 and R3 (see inset A) off all clinopyroxenes (grain size fraction fS) are plotted together with average peak positions of non-quadrilateral pyroxenes. All data points plot along the Mg^{2+} trend of quadrilateral pyroxenes, except a group of pyroxenes of sample EY55-12 which plot close to the average aegirine values.

4. Discussion

We will first discuss the results with respect to various aspects in the frame of the case study (Chapter 4.1 to 4.3), followed by methodological aspects including potential disadvantages and future developments (Chapters 4.4 and 4.5).

4.1. HM Assemblage

The occurrence of significant amounts of epidote and amphibole in the silt-sized fractions of the sediment samples suggest a third source that is different in composition to the Lower Triassic Bunter sandstone formation and the Miocene volcanic rocks of the Vogelsberg and Rhön. Considering (i) the exclusive occurrence of epidotes and amphiboles in the silt-sized fractions, (ii) the occurrence of epidote in all sediment samples of the drainage area including the small creeks at the flanks of the Vogelsberg volcano but in none of the volcanic hardrocks, (iii) similar pattern of amphibole occurrence with very minor exceptions (e.g., not detected in mSi faction EY55-07, present in mSi fraction of one hardrock EY55-10; Figure 7), and (iv) the widespread occurrence of Quaternary loess deposits in the drainage area, especially at the flanks of the Vogelsberg and the Rhön (Figure 6), loess appears to be the most reasonable source for silt-sized epidote and amphibole. As outlined above, the loess deposits in the drainage area contain significant amounts of amphibole and epidote in similar proportions (~20%) along with fewer garnet and zircon contents (~5%; [53]). In the silt-sized fractions the maximum amphibole, epidote, garnet and zircon contents are 14%, 25%, 5% and 12%, respectively. Considering that some silt-sized zircon most likely derives from the Bunter sandstone units, these proportions are in generally good agreement with the results after [53] and substantiate the hypothesis of a loess source.

Most olivines in the detritus have a forsterite content ranging between 50% to 100% (see Section 4.2) which translates into a range of specific gravities from 3.35 to ca. 3.8 g/cm^2 [60] while the clinopyroxenes have an augitic to diopsidic composition (see Section 4.3) which translates into a specific gravity of ca. 3.34 g/cm^2 and ca. 3.28 g/cm^2 [61]. Thus, the olivines and clinopyroxenes should have roughly similar settling velocities and be deposited in the same grain size window. The Ol/Px ratio in the fS grain size fraction increases with increasing transport distance (Figure 7) which points to a reduction in pyroxene compared to olivine. Possible reasons for apparent reduction of clinopyroxene could be lowering of settling velocity due to differences in grain shape or differential disintegration rates. In both cases pyroxene grains are likely to be affected more. In case of grain shape dependent settling velocity (given similar density) the olivines will probably settle faster due to a more isometric shape, while the pyroxenes would settle slower due to a more prismatic shape (see, for instance, [62–64]). Abrasion experiments by [65] show that in a heterogeneous mixture of basaltic components, augitic clinopyroxene is significantly faster abraded than olivine and that the average sphericity, given as the ratio of short to long axis (with 1 being a perfect sphere), during abrasion is much lower for augitic pyroxene compared to olivine, e.g., 0.65 vs. 0.8 [65]. Thus, the apparent reduction in pyroxene content during transport is probably a combined effect of differential settling velocities, with the clinopyroxenes sinking slower than the olivines, and faster disintegration of the clinopyroxenes. This is corroborated by decreasing Ol/Px ratios with decreasing grain size, as observed for most of the sediment samples, especially the most distal one, EY55-14 (Figure 7).

Sample EY55-13 was intended to reflect the heavy-mineral composition of the Bunter sandstone formation since the sampled tributary only drains this formation. However, the content of olivine and pyroxene in the heavy-mineral assemblage (Table 3 and Figure 7) points to a significant basaltic source and, thus, this sample cannot be used to infer the general Bunter sandstone heavy-mineral endmember.

4.2. Olivine Composition

The underestimation of the Fo content of some olivine is related to a shift of the DB1 and DB2 positions, which is unrelated to chemical variability (Figure 9). Optical re-examination of the respective olivine grains revealed that these grains frequently show alteration rims of iddingsite (Figure S1), which is the product of oxidative weathering of olivine and comprises a mixture of goethite and smectite [66]. Since goethite is far less translucent than the accompanied olivine, it will absorb the laser light to a much greater extent which favors heating of the iddingsitizised grain.

To test whether the observed band shifts are related to heating during measurement, an altered olivine of sample EY55-03 was selected for a series of measurements, in which laser power was the only variable (Figure 16). The spectrum recorded at 1% laser power shows several spectral features at ~200, 300, 400, 700, 1300 cm^{-1} which likely belong to an iron hydroxide phase and the main olivine bands DB1 and DB2 are situated at 818 and 850 cm^{-1}, respectively. When increasing the laser power to 25% most spectral features disappear except DB1 and DB2, which broaden, equalize in intensity and shift to lower Raman shift values (Figure 16B). Re-measuring with 1% laser power reveals that DB1 and DB2 have shifted back to their original positions, regained their former relative intensities and their widths decreased again. Additionally, Raman bands that can be clearly related to hematite appear in the spectrum, which points to a thermally induced transformation of goethite to hematite during measurement. This transformation occurs at about 300 °C [67]. Considering this temperature as minimum, significant changes in spectral parameters of DB1 and DB2 can be explained by temperature-dependent anharmonic effects [68]. Thus, the second trend indicates the degree of heating, i.e., the larger the positional shift the higher the temperature during measurement. Because heating is to large proportion a function of absorption, the second trend theoretically reflects the amount of goethite present in the grain and therefore the degree of iddingsitization and hence the degree of olivine weathering. Interestingly, the trend related to iddingsitization in DB1 and DB2 is only observed in the crushed hard rock samples. This implies that the weathered rim is removed effectively already during erosion and early stages of fluvial transport. This conclusion is supported by the fact that sediment samples proximal to the basalt source (EY55-04, -05, -06) show nearly no iddingsitization trend in DB1 and DB2.

The primary Fo content trend probably reflects the crystallization related variability during cooling. Comparing only the basalt samples (Figure 10) shows that the Vogelsberg sample (EY55-03) contains more olivines with lower Fo content (~50–60%) than the Rhön basalts (EY55-09, -11; ~60-70%).

Olivines of the sediment samples (EY55-01, -07, -14) show similar Fo content ranges (Figure 10). This indicates that the Fo content does not allow differentiation between sources in this case.

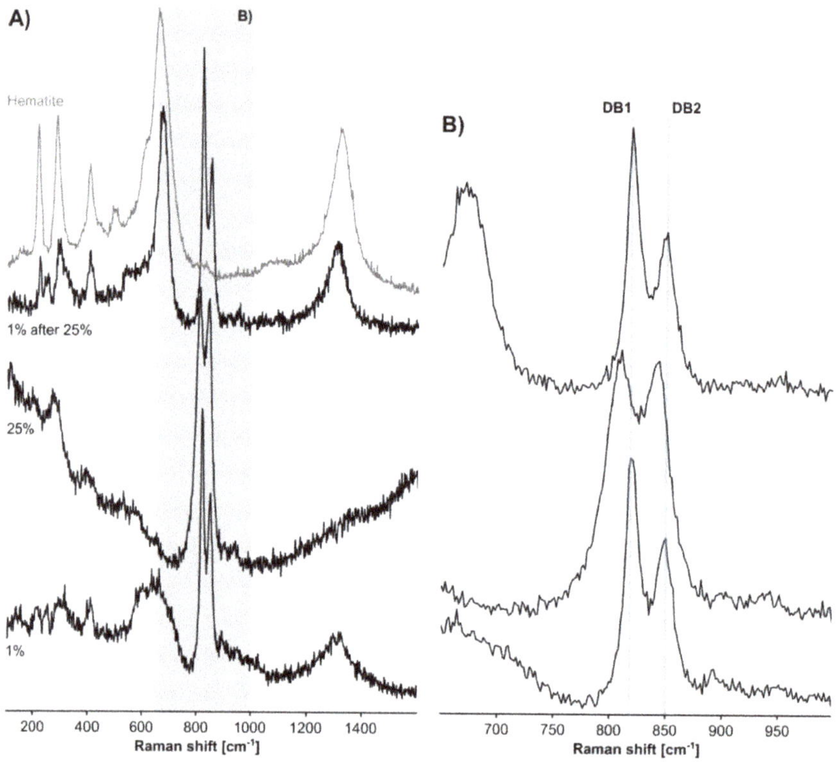

Figure 16. Altered olivine spectra. Panel (**A**) shows a progression of Raman spectra of olivine recorded at the same position. The bottom spectrum is a composite spectrum of olivine and probably goethite at 1% laser power. The middle spectrum is recorded at 25% laser power where only the olivine Raman bands remain. The upper spectrum is again recorded at 1% laser power and is a composite of olivine and hematite (gray hematite spectrum for comparison). The gray shaded area is zoomed in panel (**B**). Here the positions of the DB1 and DB2 bands in olivine are marked by gray lines to demonstrate that the positional shift in DB1 and DB2 is a temporary feature at 25% laser power.

4.3. Pyroxene Composition

Estimating X^{Mg2+} from the clinopyroxene Raman spectrum enables differentiation of quadrilateral pyroxenes from basic and more differentiated volcanic rocks. Here the basaltic clinopyroxenes (EY55-03, -09, -11) show median X^{Mg2+} values of 0.39 (EY55-03), 0.37 (EY55-11) and 0.5 (EY55-09). The elevated X^{Mg2+} value of sample EY55-09 is probably related to the presence of peridotite xenoliths that contain clinopyroxenes with higher clinoenstatite content of 49 to 63% [69] compared to the median X^{Mg2+} value of xenolith-free basalt in the same area (EY55-11). The clinopyroxenes derived from the trachytes and phonolites show median X^{Mg2+} values of 0.33 to 0.35 (EY55-02, -10, -12). With median X^{Mg2+} values of 0.41 it is obvious that most of the clinopyroxenes in the modern sediments proximal to the Vogelsberg and Rhön (EY55-01, -07) are sourced by the basalts. However, with a median X^{Mg2+} value of 0.36 the distal sediment sample (EY55-14) indicates significant contribution from lower X^{Mg2+} clinopyroxenes and, thus, a more differentiated source for the clinopyroxenes. Considering the amount of trachytes and phonolites that croup out in the Fulda drainage area (Figure 6), it is improbable that such small sources, if not extremely fertile, will significantly influence the clinopyroxene composition in the distal sediment samples. A more likely explanation than a different source, is a process that selectively alters the clinopyroxene suite in the sediments. The reduction in spread of X^{Mg2+} values from source rocks and proximal sediments to the distal sediments favors a process over a distinct source. Reduction of the X^{Mg2+} content points towards an increase in Fe which, in case of related decrease in Ca, favors a more augitic composition. This would suggest an enrichment of augitic clinopyroxenes in the distal sediment samples. Because many clinopyroxenes of the basalt source and proximal sediments have a diopsidic composition (Figure 17), differential weathering resistance of augitic and diopsidic clinopyroxene could explain the decrease in spread and average X^{Mg2+} values in the distal samples. In fact, [70] showed that the solubility of augite in water at a pH of 5–6 is about two orders of magnitude less than the solubility of diopside. Thus, it is likely that decrease of X^{Mg2+} in the distal sediment samples reflects the faster dissolution of diopside during fluvial transport.

Given the fact that there are several solid solution series in the pyroxene system, for instance diopside–hedenbergite or enstatite–ferrosilite, identification of intermediate compositions by comparing complete Raman spectral traces is not precise enough, e.g., the differentiation of diopsidic or augitic clinopyroxenes is often inconclusive (Figure 18). These subtle changes in composition are better reflected by shifts in Raman band positions which allow differentiation of groups of quadrilateral pyroxenes (Figure 15, EY55-09 vs. EY55-12). Still, based on the band positions it remains unclear if the clinopyroxene is a diopside or augite, but the information that they are distinct is gained (see EMPA data and Raman band positions in Figure 18). From Figure 15 it is evident that most band positions plots on the Mg^{2+}-compositional trend of quadrilateral pyroxenes, while a smaller group points towards a higher aegirine content. This smaller group is mainly made up of clinopyroxenes of sample EY55-12. The sample itself shows two distinct populations of diopsidic/hedenbergitic (low Mg^{2+}) clinopyroxenes and aegirin–augitic to aegirin-rich clinopyroxenes. Within the group of aegirin–augitc clinopyroxenes few spectra of other samples can be found, indicating that these samples (EY55-01, EY55-06) contain aegirin clinopyroxene. The fact that only one aegirin–augite spectrum was detected in the sediment samples, reflects that this mineral group is present in very small amounts in the detritus. This either points to a regionally confined or an infertile source. Only the phonolites of the Rhön (EY55-12) contain aegirin–augite to larger proportions and are considered the most likely source for the detected detrital aegirin–augite. Considering that the phonolites are relatively resistant to weathering in temperate climate conditions, the source is in this case both regionally confined and infertile. Due to the increased number of observations, inherent to the proposed method, chances are higher to observe the detrital signal of such small sources. This is a major advantage, considering that the number of observations can easily be increased to further reduce the counting-statistical uncertainty.

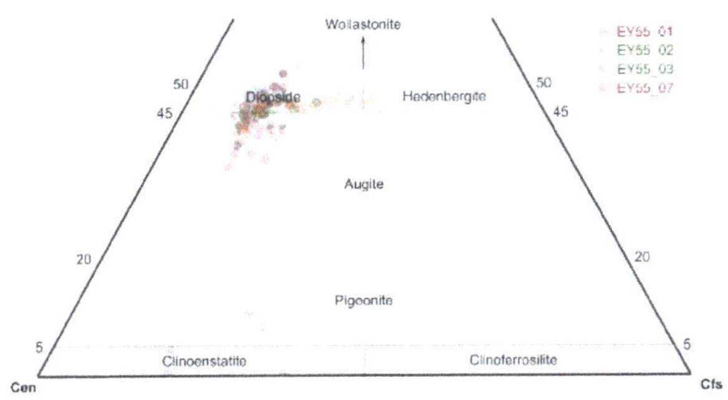

Figure 17. Composition of pyroxenes as derived by EMPA (grain size fraction fS).

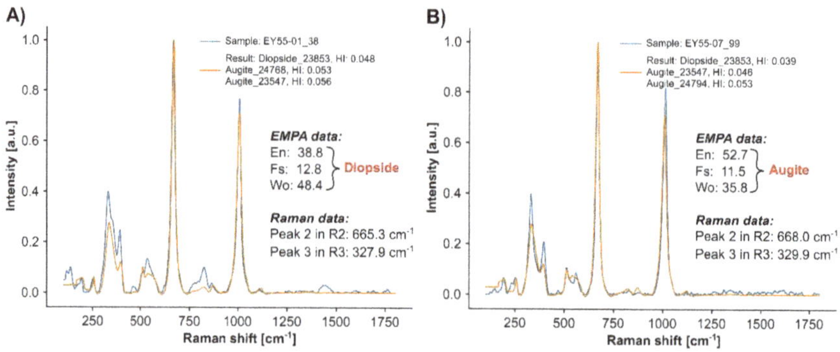

Figure 18. Results of Raman spectral comparison vs. chemical classification. Panel (**A**) No distinction between diopside and augite is possible based on the hit indices, while the EMPA data clearly identifies a diopsidic composition. Panel (**B**) No distinction between diopside and augite is possible based on the hit indices, while the EMPA data clearly identifies an augitic composition. Although the identification using the complete spectral trace is inconclusive, the positions of peaks 2 and 3 reflect the compositional differences between the augite and diopside.

4.4. Remarks on the Methodology and Outlook

Currently, the most severe bottleneck of the method, in terms of time, is the manual selection of measurement positions. Although user control avoids measurement positions on inclusions, grain boundaries, fractures, or polymineralic grains (lithic fragments), the time needed to select ca. 1000 grains takes about an hour. Considering that at least 1000 grains per sample should be selected and that 16–20 samples are mounted on one slide, it would take about 16 to 20 h to prepare the list of measurement positions. Thus, there is a strong need to automate this process as well. Obviously, image analysis is the method of choice; however the combination of transparent, light to dark colored to complete opaque grains in a transparent matrix is not readily solved by applying gray value threshold filters to separate the foreground from the background. We currently follow two different approaches to circumvent the problem. The first approach uses changing interference colors during rotation of perpendicular polarizers. In the other approach the pattern recognition capabilities of convoluted neural networks shall be used to identify the grains based on their shape and appearance.

The method can be further improved by reducing the number of low-quality spectra (i.e., HI values > 0.15). The majority of low-quality spectra fall into two categories, i.e., low SNR spectra and spectra affected by fluorescence. The low SNR spectra were most probably measured in an out-of-focus position. This is because the real sample surface is only approximated by interpolating through a set of support points and the local topography sometimes does not match this approximated surface. Considering that the axial resolution of confocal microscope is approximated by $1.4 \times \lambda/(NA)^2$ [71], where λ is the laser wavelength and NA the numerical aperture of the lens, the axial resolution of the typically used microscope objectives (100x, 0.8 NA; 50x, 0.5 NA) is, thus, ca. 1.2 μm and ca. 3 μm, respectively. Consequently, slight deviation of the local topography and approximated topography lead to a drastic reduction of Raman counts per second. Thus, it is crucial to stay in exact focus during the complete measurement series. This would decrease low SNR spectra and would speed up the measurements, because the maximum integration time could be lowered. This problem can be resolved with a dedicated focusing system that automatically readjusts the focus to the current measurement position. Such systems are commercially available. Spectra affected by fluorescence must be detected automatically, for instance by evaluating the ratio of integral intensities of the raw and baseline corrected spectrum, and marked for re-remeasuring with a different laser wavelength. This could be done while measuring. Recorded spectra will be directly evaluated/identified and marked for re-measurement with different settings (laser wavelength, accumulation time, etc.) depending on spectral quality (SNR, fluorescence).

The success rate of mineral identification using the correlation method is approximately 70% to 80% and depends significantly on the baseline estimation. If the misidentification is specific to certain phases needs to investigated, but is likely to be random. Nevertheless, the success rate must be improved and the approach after [45] will be implemented. This method is based on the discrimination capabilities of deep convolutional neural networks and appears to outmatch common comparison methods such as k-nearest neighbors (e.g., [72]), gradient boosting (e.g., [73]), random forest (e.g., [74]), support vector machines (e.g., [75,76]) or correlation (e.g., [44]) and does not require baseline estimation on which the other methods heavily rely [45].

The shortcomings outlined above can, however, be resolved and in its current state the method's efficiency already allows a drastic increase in the number of measured grains per sample and also the number of samples per project. This implies that future sedimentary provenance studies allow the processing of higher sample numbers, which in turn results in an increased spatial and, if stratigraphically well constrained, temporal resolution. This concept could be extended to heavy minerals extracted from shales, which constitute about two thirds of the sedimentary record and contain significant amounts of heavy minerals. To date they are often inferred from bulk chemical approaches [77]. Since the Raman-HMA approach enables systematic analysis of silt-sized minerals, sedimentary provenance analysis in shales opens a promising new field of research.

4.5. Potential Disadvantages

The potentially most severe drawback of the method is that it inherently relies on spectroscopic data while optical information is, at least, partly neglected, i.e., the Raman spectrum does not convey insight on, for instance, the shape (rounded vs. non-rounded), size, color, amount of inclusions or corrosion features [78]. The high-resolution images (Figure 1) allow deduction of most of the named properties. Nevertheless, this must be done by the user, which is why stringent pairing of optical and spectroscopic data is currently not realized. In the future, this is going to be remediated with the implementation of image analysis, as outlined in Section 4.4, which can then be used to infer, for instance, color and/or shape parameters for each automatically detected grain.

The number and type of (mineral) inclusions are not readily assessed by the Raman-HMA method, but convey important information on pressure and temperature of metamorphic source rocks (e.g., [79]) or sedimentary provenance (e.g., [14]). However, the inclusion inventory can be assessed by combining the imaging capabilities of modern Raman spectrometers (e.g., [80]) with the referenced mineral positions, created by applying the Raman-HMA method. Thus, inclusions in selected minerals (e.g., garnet) can be characterized automatically by creating hyperspectral cubes at the mineral locations. These huge datasets can then be quickly screened for indicative Raman bands of, for instance, coesite or diamond.

Another inconvenience of the method is that for technical reasons, wide grain size ranges cannot be embedded within the same heavy-mineral slide. Therefore, two or more grain size fractions of one sample need to be prepared. However, this drawback is counterbalanced by the high sample throughput of the method.

5. Conclusions

With the proposed method it is now possible to assess the heavy-mineral assemblage of silt- and sand-sized sediments rapidly and with strongly increased objectivity. This is a major advantage compared to microscopical identification, which is limited by grain size and affected by the operator's knowledge and experience of optical properties of heavy minerals. Additional discrimination potential of the Raman-HMA method arises from the capability to semi-quantitatively determine the compositions in solid solutions series, which has been shown by analyzing the forsterite content and X^{Mg2+} of detrital olivine and clinopyroxene, respectively. Since all measured minerals are spatially referenced, transfer to other devices (EMPA, LA-ICPMS) is easily achieved which enables multi-approach single-grain analysis. The feasibility of this approach has been demonstrated to work by means of EMPA and Raman spectroscopy, applied to a medium-sized catchment with three component endmember mixings of Triassic siliclastics, Cenozoic volcanics, and Pleistocene loess.

Supplementary Materials: The following are available online at http://www.mdpi.com/2075-163X/9/7/385/s1. The supplementary data consists of five figures, a zip archive, and Python file containing a dictionary object of the mineral groups. Figure S1: A weathered olivine grain of sample EY55-03. The orange to brown weathering rim is clearly visible and documents the degree of iddingsitization. Figure S2: Forsterite content estimated from Raman band positions DB1 and DB2. Only the band positions of olivines of the Fulda river sediments (grain size fraction fS) are shown to document that iddingsitization related trend is not recognizable in the sediment samples. Figure S3: Forsterite content estimated from Raman band positions DB1 and DB2. Only the band positions of olivines of the basalt hard rock samples (grain size fraction fS) are shown to document the iddingsitization related trend. Figure S4: Forsterite content estimated from Raman band positions DB1 and DB2. Only the band positions of olivines of the Vogelsberg sediments (grain size fraction fS) are shown to document that iddingsitization related trend is already removed after short transportation distances. Figure S5: Sample fixation. The lower left panel shows the "sample grid" and funnel used to fill the single compartments of the "sample grid". The "sample grid" is fixed on a flat and flexible silicon layer by double-sided adhesive foil. Each compartment is filled by a quartered-down amount of heavy minerals using the funnel. The heavy minerals are distributed across the compartment floor and unfixed grains are removed by turning the "sample grid" upside down and gently tapping on the respective compartment. After this procedure, the compartment is sealed, and the next compartment is filled. This is being repeated until all samples are fixed or compartments filled. After removing the "sample grid" a few drops of UV glue are placed on the samples and the abraded glass slide is pressed on the samples followed by curing with UV irradiation. The zoom-in (upper right) shows a glass slide with 16 samples fixed in cured UV glue. HM result files: The "HM_result_files.zip" is a compressed archive of the evaluated heavy-mineral Raman spectra sorted into three folders reflecting the grain size fractions fS, cSi, and mSi. The evaluation results of the "Good hit" spectra for each sample are stored in a txt file. These txt files contain the spectrum index, the identified mineral group, the identified mineral variety and the hit index of the best hit. Mineral dictionary: The python file "mineral_dict.py" is a simple txt file that contains a Python dictionary object that assigns various mineral varieties into their respective mineral groups.

Author Contributions: Conceptualization: N.K.L., H.v.E. and I.D.; Methodology: N.K.L., J.K., P.A.; Software: N.K.L.; Writing—Original Draft Preparation: N.K.L., H.v.E.; Writing—Review and Editing: N.K.L., J.K., P.A., I.D., H.v.E.; Supervision: N.K.L.; Project Administration: N.K.L., H.v.E.

Funding: This research received no external funding.

Acknowledgments: The authors would like to thank Mr Andreas Kronz (Department of Geochemistry, Georg-August University Göttingen) for his assistance during EMPA measurements and Mr Ontje Lünsdorf for his computer sciences advice. Three unknown reviewers are thanked for their criticism and insights which helped to improve the manuscript.

Conflicts of Interest: The authors declare no conflict of interest.

References

1. Weltje, G.J.; von Eynatten, H. Quantitative provenance analysis of sediments: Review and outlook. *Sediment. Geol.* **2004**, *171*, 1–11. [CrossRef]
2. Morton, A.C.; Hallsworth, C.R. Processes controlling the composition of heavy mineral assemblages in sandstones. *Sediment. Geol.* **1999**, *124*, 3–29. [CrossRef]
3. Garzanti, E.; Andò, S.; Vezzoli, G. Grain-size dependence of sediment composition and environmental bias in provenance studies. *Earth Planet. Sci. Lett.* **2009**, *277*, 422–432. [CrossRef]
4. Garzanti, E.; Andò, S.; Limonta, M.; Fielding, L.; Najman, Y. Diagenetic control on mineralogical suites in sand, silt, and mud (Cenozoic Nile Delta): Implications for provenance reconstructions. *Earth Science Rev.* **2018**, *185*, 122–139. [CrossRef]
5. von Eynatten, H.; Tolosana-Delgado, R.; Karius, V.; Bachmann, K.; Caracciolo, L. Sediment generation in humid Mediterranean setting: Grain-size and source-rock control on sediment geochemistry and mineralogy (Sila Massif, Calabria). *Sediment. Geol.* **2016**, *336*, 68–80. [CrossRef]
6. Mange, M.A.; Morton, A.C. Geochemistry of Heavy Minerals. In *Heavy Minerals in Use*, 1st ed.; Mange, M.A., Wright, D.T., Eds.; Elsevier: Amsterdam, The Netherlands, 2007; pp. 345–391.
7. von Eynatten, H.; Dunkl, I. Assessing the sediment factory: The role of single grain analysis. *EarthScience Rev.* **2012**, *115*, 97–120. [CrossRef]

8. Morton, A.C.; Hallsworth, C. Identifying provenance-specific features of detrital heavy mineral assemblages in sandstones. *Sediment. Geol.* **1994**, *90*, 241–256. [CrossRef]
9. Andò, S.; Garzanti, E. Raman spectroscopy in heavy-mineral studies. *Geol. Soc. London Spec. Publ.* **2013**, *386*, 395–412. [CrossRef]
10. Mange, M.A.; Maurer, H.F.W. *Schwerminerale in Farbe*; Ferdinand Enke Verlag Stuttgart: Stuttgart, Germany, 1991.
11. Fries, M.; Steele, A. Raman Spectroscopy and Confocal Raman Imaging in Mineralogy and Petrography. In *Confocal Raman Microscopy*; Dieing, T., Hollricher, O., Toporski, J., Eds.; Springer: Berlin/Heidelberg, Germany, 2010; p. 289.
12. Andò, S.; Vignola, P.; Garzanti, E. Raman counting: A new method to determine provenance of silt. *Rend. Lincei* **2011**, *22*, 327–347. [CrossRef]
13. Korsakov, A.V.; Hutsebaut, D.; Theunissen, K.; Vandenabeele, P.; Stepanov, A.S. Raman mapping of coesite inclusions in garnet from the Kokchetav Massif (Northern Kazakhstan). *Spectrochim. Acta Part A Mol. Biomol. Spectrosc.* **2007**, *68*, 1046–1052. [CrossRef]
14. Schönig, J.; Meinhold, G.; von Eynatten, H.; Lünsdorf, N.K. Tracing ultrahigh-pressure metamorphism at the catchment scale. *Sci. Rep.* **2018**, *8*, 2931. [CrossRef] [PubMed]
15. Schönig, J.; von Eynatten, H.; Meinhold, G.; Lünsdorf, N.K. Diamond and coesite inclusions in detrital garnet of the Saxonian Erzgebirge, Germany. *Geology* **2019**. [CrossRef]
16. Vermeesch, P. How many grains are needed for a provenance study? *Earth Planet. Sci. Lett.* **2004**, *224*, 441–451. [CrossRef]
17. Pullen, A.; Ibáñez-Mejía, M.; Gehrels, G.E.; Ibáñez-Mejía, J.C.; Pecha, M. What happens when n= 1000? Creating large-n geochronological datasets with LA-ICP-MS for geologic investigations. *J. Anal. At. Spectrom.* **2014**, *29*, 971–980. [CrossRef]
18. Smith, D.C. The RAMANITA© method for non-destructive and in situ semi-quantitative chemical analysis of mineral solid-solutions by multidimensional calibration of Raman wavenumber shifts. *Spectrochim. Acta Part A Mol. Biomol. Spectrosc.* **2005**, *61*, 2299–2314. [CrossRef] [PubMed]
19. Kuebler, K.E.; Jolliff, B.L.; Wang, A.; Haskin, L.A. Extracting olivine (Fo-Fa) compositions from Raman spectral peak positions. *Geochim. Cosmochim. Acta* **2006**, *70*, 6201–6222. [CrossRef]
20. Wang, A.; Jolliff, B.L.; Haskin, L.A.; Kuebler, K.E.; Viskupic, K.M. Characterization and comparison of structural and compositional features of planetary quadrilateral pyroxenes by Raman spectroscopy. *Am. Mineral.* **2001**, *86*, 790–806. [CrossRef]
21. Bersani, D.; Andò, S.; Vignola, P.; Moltifiori, G.; Marino, I.G.; Lottici, P.P.; Diella, V. Micro-Raman spectroscopy as a routine tool for garnet analysis. *Spectrochim. Acta Part A Mol. Biomol. Spectrosc.* **2009**, *73*, 484–491. [CrossRef]
22. Leissner, L.; Schlüter, J.; Horn, I.; Mihailova, B. Exploring the potential of Raman spectroscopy for crystallochemical analyses of complex hydrous silicates: I. Amphiboles. *Am. Mineral.* **2015**, *100*, 2682–2694. [CrossRef]
23. Lenz, C.; Nasdala, L.; Talla, D.; Hauzenberger, C.; Seitz, R.; Kolitsch, U. Laser-induced REE3+ photoluminescence of selected accessory minerals - An 'advantageous artefact' in Raman spectroscopy. *Chem. Geol.* **2015**, *415*, 1–16. [CrossRef]
24. Nasdala, L.; Wenzel, M.; Vavra, G.; Irmer, G.; Wenzel, T.; Kober, B. Metamictisation of natural zircon: Accumulaton versus thermal annealing of radioactivity-induced damage. *Contrib. Mineral. Petrol.* **2001**, *141*, 125–144. [CrossRef]
25. Hanchar, J.M.; Ruschel, K.; Nasdala, L.; Škoda, R.; Finger, F.; Kronz, A.; Többens, D.M.; Möller, A. A Raman spectroscopic study on the structural disorder of monazite–(Ce). *Mineral. Petrol.* **2012**, *105*, 41–55.
26. Heller, B.M.; Lünsdorf, N.K.; Dunkl, I.; Molnár, F.; von Eynatten, H. Estimation of radiation damage in titanites using Raman spectroscopy. *Am. Mineral.* **2019**, *104*, 857–868. [CrossRef]
27. Mezger, K.; Krogstad, E.J. Interpretation of discordant U-Pb zircon ages: An evaluation. *J. Metamorph. Geol.* **1997**, *15*, 127–140. [CrossRef]
28. Zack, T.; Moraes, R.; Kronz, A. Temperature dependence of Zr in rutile: Empirical calibration of a rutile thermometer. *Contrib. Mineral. Petrol.* **2004**, *148*, 471–488. [CrossRef]
29. Triebold, S.; Luvizotto, G.L.; Tolosana-Delgado, R.; Zack, T.; von Eynatten, H. Discrimination of TiO_2 polymorphs in sedimentary and metamorphic rocks. *Contrib. Mineral. Petrol.* **2011**, *161*, 581–596. [CrossRef]
30. Fleet, W.F. Petrological Notes on the Old Red Sandstone of the West Midlands. *Geol. Mag.* **1926**, *63*, 505–516. [CrossRef]

31. Moore, D.M.; Reynolds, R.C. *X-Ray Diffraction and the Identification and Analysis of Clay Minerals*, 2nd ed.; Oxford University Press: Oxford, UK, 1997.
32. Tributh, H.; Lagaly, G. Aufbereitung und Identifizierung von Boden- und Lagerstättentonen. 1. *GIT-Fachzeitschrift für das Lab.* **1986**, *30*, 524–529.
33. de Faria, D.L.A.; Lopes, F.N. Heated goethite and natural hematite: Can Raman spectroscopy be used to differentiate them? *Vib. Spectrosc.* **2007**, *45*, 117–121. [CrossRef]
34. Hanesch, M. Raman spectroscopy of iron oxides and (oxy)hydroxides at low laser power and possible applications in environmental magnetic studies. *Geophys. J. Int.* **2009**, *177*, 941–948. [CrossRef]
35. Besl, P.J.; McKay, N.D. A Method for Registration of 3-D shapes. *IEEE Trans. Pattern Anal. Mach. Intell.* **1992**, *14*, 239–256. [CrossRef]
36. Dörfer, T.; Bocklitz, T.; Tarcea, N.; Schmitt, M.; Popp, J. Checking and improving calibration of raman spectra using chemometric approaches. *Zeitschrift fur Phys. Chemie* **2011**, *225*, 753–764. [CrossRef]
37. Popp, J.; Tarcea, N. Raman data analysis. In *EMU Notes in Mineralogy, Volume 12—Applications of Raman Spectroscopy to Earth Sciences and Cultural Heritage*; Dubessy, J., Caumon, M.-C., Rull, F., Eds.; European Mineralogical Union & Mineralogical Society of Great Britain & Ireland: London, UK, 2012; p. 504.
38. McCreery, R. McCreery Group—National Institute for Nanotechnology, University of Alberta, 2015. Available online: https://www.chem.ualberta.ca/~{}mccreery/ramanmaterials.html (accessed on 24 May 2019).
39. Ryan, C.G.; Clayton, E.; Griffin, W.L.; Sie, S.H.; Cousens, D.R. SNIP, a statistics-sensitive background treatment for the quantitative analysis of PIXE spectra in geoscience applications. *Nucl. Inst. Methods Phys. Res. B* **1988**, *34*, 396–402. [CrossRef]
40. Morháč, M.; Matoušek, V. Peak clipping algorithms for background estimation in spectroscopic data. *Appl. Spectrosc.* **2008**, *62*, 91–106. [CrossRef] [PubMed]
41. Harris, F.J. On the Use of Windows for Harmonic Analysis with the Discrete Fourier Transform. *Proc. IEEE* **1978**, *66*, 51–83. [CrossRef]
42. McCreery, R.L. Raman Spectroscopy for Chemical Analysis. In *Volume 157 in Chemical Analysis—A Series of Monographs on Analytical Chemistry and Its Applications*; Winefordner, J.D., Ed.; Wiley-Interscience: Hoboken, NJ, USA, 2000.
43. Lafuente, B.; Downs, T.R.; Yang, H.; Stone, N. The power of databases: the RRUFF project. In *Highlights in Mineralogical Crystallography*; Armbruster, T., Danisi, R.M., Eds.; De Greuyter: Berlin, Germany, 2015; pp. 1–30.
44. Rodriguez, J.D.; Westenberger, B.J.; Buhse, L.F.; Kauffman, J.F. Quantitative evaluation of the sensitivity of library-based Raman spectral correlation methods. *Anal. Chem.* **2011**, *83*, 4061–4067. [CrossRef] [PubMed]
45. Liu, J.C.; Osadchy, M.; Ashton, L.; Foster, M.; Solomon, C.J.; Gibson, S.J. Deep convolutional neural networks for Raman spectrum recognition: A unified solution. *Analyst* **2017**, *142*, 4067–4074. [CrossRef]
46. Nasdala, L.; Smith, D.C.; Kaindl, R.; Ziemann, M.A. Raman spectroscopy: Analytical perspectives in mineralogical research. *EMU Notes Mineral.* **2004**, *6*, 281–343.
47. Lünsdorf, N.K.; Lünsdorf, J.O. Evaluating Raman spectra of carbonaceous matter by automated, iterative curve-fitting. *Int. J. Coal Geol.* **2016**, *160–161*, 51–62. [CrossRef]
48. Sindowski, K.-H. Schüttungsrichtungen und Mineral-Provinzen im westdeutschen Buntsandstein. *Geol. Jahrb.* **1957**, *73*, 277–294.
49. Okrajek, A. Sedimentpetrographische Untersuchung toniger und sandiger Lagen des Mittleren Buntsandsteins in Bohrungen und Tagesaufschlüssen Süd-Niedersachsens. *Beiträge zur Mineral. und Petrogr.* **1965**, *11*, 507–534.
50. Heim, D. Petrographische Beitrage zur Paläogeographie des Buntsandsteins. *Notizblatt des Hess. Landesamtes für Bodenforsch. zu Wiesbad.* **1966**, *94*, 235–258.
51. Brüning, U. Stratigraphie und Lithofazies des Unteren Buntsandsteins in Südniedersachsen und Nordhessen. *Geol. Jahrb.* **1986**, *90*, 125.
52. Pryor, W.A. Petrology of the Weissliegendes sandstones in the Harz and Werra-Fulda areas, Germany. *Geol. Rundsch.* **1971**, *60*, 524–552. [CrossRef]
53. Scheffer, F.; Meyer, B.; Kalk, E. Mineraluntersuchungen am Würm-Löß südniedersächsischer Lößfluren als Voraussetzung für die Mineralanalyse verschiedener Lößbodentypen. *Chem. Erde.* **1958**, *19*, 338–360.
54. Bogaard, P.J.F.; Wörner, G. Petrogenesis of Basanitic to Tholeiitic Volcanic Rocks from the Miocene Vogelsberg, Central Germany. *J. Petrol.* **2003**, *44*, 569–602. [CrossRef]

55. Reischmann, T.; Schraft, A. *Der Vogelsberg - Geotope im größten Vulkangebiet Mitteleuropas*; Hessisches Landesamt für Umwelt und Geologie: Wiesbaden, Germany, 2009.
56. Ficke, B. Petrologische Untersuchungen an tertiären basaltischen bis phonolitschen Vulkaniten der Rhön. *Mineral. Petrol.* **1961**, *7*, 337–436. [CrossRef]
57. Chopelas, A. Single crystal Raman spectra of forsterite, fayalite, and monticellite. *Am. Mineral.* **1991**, *76*, 1101–1109.
58. Huang, E.; Chen, C.H.; Huang, T.; Lin, E.H.; Xu, J.A. Raman spectroscopic characteristics of Mg-Fe-Ca pyroxenes. *Am. Mineral.* **2000**, *85*, 473–479. [CrossRef]
59. Tribaudino, M.; Montovani, L.; Bersani, D.; Lottici, P. Raman spectroscopy of (Ca,Mg)$MgSi_2O_6$ clinopyroxenes. *Am. Mineral.* **2012**, *97*, 1339–1347. [CrossRef]
60. Mursky, G.A.; Thompson, R.M. A specific gravity index for minerals. *Can. Mineral.* **1958**, *6*, 273–287.
61. Poldervaart, A. Correlation of physical properties and chemical composition in the plagioclase, olivine and orthopyroxene series. *Am. Mineral.* **1950**, *35*, 1067–1079.
62. Komar, P.D.; Reimers, C.E. Grain shape effects on settling rates. *J. Geol.* **1978**, *86*, 193–209. [CrossRef]
63. Le Roux, J.P. Settling velocity of ellipsoidal grains as related to shape entropy. *Sediment. Geol.* **1996**, *101*, 15–20. [CrossRef]
64. Garzanti, E.; Andò, S.; Vezzoli, G. Settling equivalence of detrital minerals and grain-size dependence of sediment composition. *Earth Planet. Sci. Lett.* **2008**, *273*, 138–151. [CrossRef]
65. Cornwall, C.; Bandfield, J.L.; Titus, T.N.; Schreiber, B.C.; Montgomery, D.R. Physical abrasion of mafic minerals and basalt grains: Application to martian aeolian deposits. *Icarus* **2015**, *256*, 13–21. [CrossRef]
66. Smith, K.L.; Milnes, A.R.; Eggleton, R.A. Weathering of Basalt: Formation of Iddingsite. *Clays Clay Miner.* **1987**, *35*, 418–428. [CrossRef]
67. Gialanella, S.; Girardi, F.; Ischia, G.; Lonardelli, I.; Mattarelli, M.; Montagna, M. On the goethite to hematite phase transformation. *J. Therm. Anal. Calorim.* **2010**, *102*, 867–873. [CrossRef]
68. Pine, A.S.; Tannenwald, P.E. Temperature dependence of Raman linewidth and shift in α-quartz. *Phys. Rev.* **1969**, *178*, 1424–1430. [CrossRef]
69. Franz, L.; Seifert, W.; Kramer, W. Thermal evolution of the mantle underneath the Mid-German Crystalline Rise: evidence from mantle xenoliths from the Rhone area (Central Germany). *Mineral. Petrol.* **1997**, *61*, 1–25. [CrossRef]
70. Brantley, S.L.; Chen, Y. Chemical weathering rates of pyroxenes and amphiboles. In *Chemical Weathering Rates of Silicate Minerals*; White, A.F., Brantley, S.L., Eds.; Mineralogical Society of America: Washington, DC, USA, 1995; p. 583.
71. Dubessy, J.; Caumon, M.-C.; Rull, F.; Sharma, S. Instrumentation in Raman spectroscopy: elementary theory and practice. In *EMU Notes in Mineralogy—Volume 12 Applications of Raman spectroscopy to Earth sciences and cultural heritage*; Dubessy, J., Caumon, M.-C., Rull, F., Eds.; European Mineralogical Union & Mineralogical Society of Great Britain & Ireland: London, UK, 2012; pp. 83–172.
72. Cover, T.M.; Hart, P.E. Nearest Neighbor Pattern Classification. *IEEE Trans. Inf. Theory* **1967**, *13*, 21–27. [CrossRef]
73. Mason, L.; Bartlett, P.; Baxter, J.; Frean, M. Boosting Algorithms as Gradient Descent. In *Advances in Neural Information Processing Systems 12*; MIT Press: Cambridge, MA, USA, 2000; p. 1098.
74. Breiman, L. Random Forests. *Mach. Learn.* **2001**, *45*, 5–32. [CrossRef]
75. Cortes, C.; Vapnik, V. Support-Vector Networks. *Mach. Learn.* **1995**, *297*, 273–297. [CrossRef]
76. Drucker, H.; Burges, C.J.C.; Kaufman, L.; Smola, A.; Vapnik, V.; Long, W.; Nj, B. Support Vector Regression Machines. In *Advances in Neural Information Processing Systems 9*; MIT Press: Cambridge, MA, USA, 1996; p. 1118.
77. Totten, M.W.; Hanan, M.A. Heavy minerals in shale. In *Heavy Minerals in Use*; Mange, M.A., Wright, D.T., Eds.; Elsevier: Amsterdam, The Netherlands, 2007; p. 1283.
78. Andò, S.; Garzanti, E.; Padoan, M.; Limonta, M. Corrosion of heavy minerals during weathering and diagenesis: A catalog for optical analysis. *Sediment. Geol.* **2012**, *280*, 165–178. [CrossRef]
79. Kohn, M.J. "Thermoba-Raman-try": Calibration of spectroscopic barometers and thermometers for mineral inclusions. *Earth Planet. Sci. Lett.* **2014**, *388*, 187–196. [CrossRef]
80. Zoubir, A. *Raman Imaging*; Springer: Berlin/Heidelberg, Germany, 2012; Volume 168.

minerals

Article

Composition of Amphiboles in the Tremolite–Ferro–Actinolite Series by Raman Spectroscopy

Danilo Bersani [1,*], Sergio Andò [2], Laura Scrocco [3], Paolo Gentile [2], Emma Salvioli-Mariani [3], Laura Fornasini [1] and Pier Paolo Lottici [1]

1 Department of Mathematical, Physical and Computer Sciences, University of Parma, Parco Area delle Scienze 7/A, 43124 Parma, Italy
2 Department of Earth and Environmental Sciences, University of Milano-Bicocca, Piazzale della Scienza 4, 20126 Milano, Italy
3 Department of Chemistry, Life Sciences and Environmental Sustainability, University of Parma, Parco Area delle Scienze 157/A, 43124 Parma, Italy
* Correspondence: danilo.bersani@unipr.it

Received: 12 June 2019; Accepted: 13 August 2019; Published: 16 August 2019

Abstract: Amphiboles are an important family of rock forming minerals, whose identification is crucial in provenance studies as well as in many other fields of geology, archaeology and environmental sciences. This study is aimed to find a quick way to characterize Ca-amphiboles in the tremolite ($Ca_2Mg_5Si_8O_{22}(OH)_2$)–ferro–actinolite ($Ca_2Fe_5Si_8O_{22}(OH)_2$) series. Raman spectroscopy is established as technique to perform non-destructive and quick analysis, with micrometric resolution, able to give the composition in terms of $Mg/(Mg + Fe^{2+})$ ratio. To exploit the method, a preliminary characterization is performed by Scanning Electron Microscopy coupled with Energy-dispersed X-ray Spectroscopy (SEM-EDS). Two independent methods to evaluate the composition from the Raman data (aiming to an accuracy of about 5%), using the low-wavenumbers part of the spectrum and the OH stretching bands, are developed. The application of the proposed method to micro-Raman mappings and the possible use of handheld Raman spectroscopy to have compositional information on Ca-amphiboles are discussed.

Keywords: Raman spectroscopy; amphiboles; tremolite; actinolite; heavy minerals

1. Introduction

Amphiboles are a widespread family of rock forming minerals with general formula ($A_{0-1}B_2C_5T_8O_{22}W_2$); they are the most common heavy minerals in Alpine–Himalayan orogenic belts formed by continental collision [1]. Amphiboles are formed under a wide range of pressures, temperatures and chemical environments, and their composition reflects the crystallization conditions providing important information on the metamorphic evolution of source areas [2]. Some chemical properties of detrital minerals are directly revealed by optical features that are quickly identified and recorded during grain counting [3]. Amphiboles, and in particular detrital amphiboles, thus represent accurate provenance tracers that can be profitably used in provenance studies. In the tremolite–ferro-actinolite series, $\square Ca_2Mg_5Si_8O_{22}(OH)_2$–$\square Ca_2Fe^{2+}{}_5Si_8O_{22}(OH)_2$, the members can be distinguished by $X = Mg/(Mg + Fe^{2+})$ ratio. The compositional range of tremolite extends from $\square Ca_2Mg_5Si_8O_{22}(OH)_2$ to $\square Ca_2Mg_{4.5}Fe^{2+}{}_{0.5}Si_8O_{22}(OH)_2$ ($0.9 \le X \le 1$) actinolite extends from $\square Ca_2Mg_{<4.5}Fe^{2+}{}_{>0.5}Si_8O_{22}(OH)_2$ to $\square Ca_2\ Mg_{2.5}Fe^{2+}{}_{2.5}Si_8O_{22}(OH)_2$ ($0.5 < X < 0.9$), and ferro-actinolite extends from $\square Ca_2Mg_{<2.5}Fe^{2+}{}_{>2.5}\ Si_8O_{22}(OH)_2$ to $\square Ca_2Fe^{2+}{}_5Si_8O_{22}(OH)_2$ $X < 0.5$ [4].

A great variety of techniques can be applied for the identification of single crystal in sediments: optical microscope studies [5], microprobe analysis [6], Raman spectroscopy [7], Fourier transform infrared spectroscopy and X-ray diffraction [8]. In provenance studies and heavy-mineral analysis of sediments, Raman discrimination of amphiboles is not yet widely applied [9]. Raman databases, mostly based on the OH stretching bands (3500–3800 cm^{-1} spectral range), are sparse in literature. In this paper we have checked the advantages of the Raman spectroscopic approach and discussed limitations applying this technique on real samples and with different spectrometers in the laboratory and in the field. This study is intended as a complementary tool to the classical optical studies performed by polarizing microscope and it aims to help to distinguish, in a quick and representative way, the different suites of Ca-amphiboles commonly found in modern sediments. An additional objective is the reduction of the bias introduced by the different skill of each operator, always present during the optical assessment of amphibole colour in heavy-mineral slides. When analyzing rocks in the amphibolite-facies, the distinction between common blue-green hornblende and rare actinolite is important. It can be assessed by a well established and time consuming chemical analysis by SEM-EDS, but it is not easy to be achieved only through the optical properties collected during observation with a mineralogical microscope and by different operators. Extinction angle varies with chemistry but its measurement is affected by the orientation of the mineral in grain mounts. Even the different grain size (15–500 micron), normally observed under the microscope, strongly influences the colors and pleochroism of amphiboles for the same chemistry. This can be a major issue, as an example, during the analysis of detrital amphiboles in marine scientific expeditions where the suite of amphiboles to be recognized over a big sample requires a huge effort to achieve a proper identification. The discrimination made only by traditional optical techniques between tremolite and cummingtonite or between actinolite and grunerite requires a careful identification of subtle differences in birefringence and extinction, needing a very expert eye, with a high risk of errors.

In this work, we took advantage from the collections of heavy minerals from sediments, available at the Department of Earth and Environmental Sciences (DISAT) in Milano-Bicocca (Milano, Italy). The huge number of slides (more than 5500) prepared in more than 15 years for heavy mineral analyses, from samples collected in different geological settings and countries all around the world [10], contain thousands of amphiboles embedded in Canada Balsam. For this study, we selected Alpine mineral samples in the tremolite–ferro–actinolite series, sometimes called nephrite series from the name of the green rock nephrite comprising massive cryptocrystalline felted calcic amphiboles of the homonymous series [11]. The characterization was performed by micro-Raman spectroscopy and Scanning Electron Microscopy coupled with Energy-dispersed X-ray Spectroscopy (SEM-EDS), to obtain the relationship between the Raman spectra and the $X = Mg/(Mg + Fe^{2+})$ ratio, with the aim to obtain a method for the estimation of the composition directly from the Raman spectrum. The attempt to obtain the composition of members of the nephrite series, and in particular the cation ratio X by means of micro-Raman spectroscopy is not new. We start from the fundamental work on the cation distribution on amphiboles of Wang et al. [12] and from the application of Raman micro-spectroscopy on the study of jades proposed by Chen et al. [11] to develop a quick method for the identification of nephrite using both low and high wavenumber regions of the spectrum. The use of the method to realize composition maps in zoned crystals of actinolite is shown.

In addition, we evaluated the possible use of mobile instrumentation. This can extend the application fields for the Raman analysis of the calcic amphiboles, for example, in the field archaeology and conservation science. Nephrite is the main component of the so-called "true-jade" (the other main type of jade has pyroxene nature, being composed by the clinopyroxene jadeite $Na(Al,Fe^{3+})Si_2O_6$). The identification of the composition of nephrite is then of great interest in archaeology and art history, giving information useful for provenance and authentication purposes. The use of a mobile Raman spectrometer can allow the study of archaeologic jade objects [13,14] in situ and in fast and completely non-invasive way. In addition, fibrous tremolite is considered an asbesto mineral, very dangerous for human health. The use of mobile Raman spectroscopy can allow a quick identification directly in the natural environment [15].

2. Materials and Methods

The analyzed samples are twenty minerals in the tremolite–ferro–actinolite series, coming from different localities in the Alps, supposed to cover all the compositional range from nearly pure tremolite to ferro-actinolite terms (Table 1). The samples were embedded in araldite resin and polished, in order to perform SEM-EDS analysis. On each sample, the same points were analyzed by SEM-EDS and micro-Raman spectroscopy according to the following strategy. We measured with SEM-EDS 8/10 points for each sample. Then we selected only homogeneous points in order to maintain the average compositional dispersion lower than 0.025 (taken as the uncertainty for SEM-EDS measurements). Then micro-Raman measurements (nearly 10) were performed as nearest as possible to the selected points. A new set of raw mineral specimens, in part overlapping the previous one, was used for the analysis with the portable Raman system.

Table 1. Provenance of the twenty amphiboles analyzed with SEM-EDS and micro-Raman spectroscopy.

Sample No.	Provenance
1	Villa Lake, Valle d'Ayas, Valle d'Aosta, Italy
2,3,4	Pineta di Primolo, Valmalenco, Lombardia, Italy
5,6,19	Miage Glacier, Val Veny, Valle d'Aosta, Italy
7,8	Valle Formazza, Piemonte, Italy
9,11,12	Alpe di Caneto, Valle Vigezzo, Piemonte, Italy
10	Val di Basso, Valle Vigezzo, Piemonte, Italy
13	Passo dei Laghi, Valle Devero, Piemonte, Italy
14	Tirolo, Alto-Adige, Italy
16, 21	Campolongo, Canton Ticino, Switzerland
17	Pizzo Rosso, Valle Aurina, Alto Adige, Italy
18	Alpe Rosso, Valle Vigezzo, Piemonte, Italy
20	Pizzo Bandiera, Valle Devero, Piemonte, Italy

Non-polarized micro-Raman spectra have been obtained on the mineral fragments embedded in the resin stab in nearly backscattering geometry with a Horiba LabRam apparatus (HORIBA Scientific, Kyoto, Japan), equipped with an Olympus microscope (Olympus, Tokyo, Japan) with 10×, 50×, ULWD 50× and 100× objectives and a motorized x–y stage. The 632.8 nm line of a He–Ne laser and the 473.1 nm line of a doubled Nd:YAG laser were used as excitation; the 632.8 nm line was mostly used to obtain high resolution spectra in the low wavenumber range (100–1200 cm^{-1}), whereas the 473.1 nm source was utilized for higher efficiency in the detection of the OH stretching signals of the OH groups in the range 3000–4000 cm^{-1}. Laser power (less than 1 mW on the sample) was controlled by means of density filters, to avoid heating effects. The minimum lateral resolution was about 1 μm (with the 100× objective), the depth resolution was set to few micrometers by means of a confocal hole. The spectral resolution was ~2 cm^{-1} (at 632.8 nm) and ~4 cm^{-1} (at 473.1 nm). The system was calibrated using the 520.6 cm^{-1} Raman band of silicon. In the high wavenumber range, the spectra were further calibrated using spectral lamps: in particular, for the OH stretching region, with the 473.1 nm laser, we used the main emission lines of Hg in fluorescent bulbs (546.07, 576.96 and 579.07 nm).

The spectra were collected using 100× and 50× ULWD (ultra long working distance) objectives. Several points for each samples were measured, with repeated acquisitions of 60 s each. A simple polynomial background subtraction was performed on each spectrum with LabSpec® software. The same program was used to perform spectral deconvolution, using Gauss-Lorentzian functions, to determine position, width and intensity (amplitude and area) of the Raman bands. With excitation at 632.8 nm, the position of the bands was obtained with an uncertainty better than 0.5 cm^{-1}.

Raman maps were obtained on a rectangular matrix of 9 × 11 points in an area of 40 × 75 square micrometers. The acquisition time for each point was 30 s. A gray tone image was obtained by associating the lightness to the position of the actinolite main peak in the range 660–680 cm^{-1}. A smoothing in the digital map was performed for a best comparison with the microscope photograph of the same area.

In addition, Raman spectra have been obtained on some raw samples, not embedded in resins, with a high resolution confocal Raman microscope (Renishaw inVia Reflex, Renishaw, Wotton-under-Edge, UK), equipped with a Leica DM2500 polarizing microscope (Leica Microsystems, Wetzlar, Germany), with a 50× LWD (long working distance) objective and motorized x–y stages. Spectra were obtained by the 532 nm line of a solid-state laser with a power of few mW at the sample, spectral resolution of ±0.5 cm^{-1} and acquisition time of 2 min.

Raman measurements were also made with a portable instrument: a handheld Enspectr RaPort® spectrometer, equipped with a 532 nm laser (max power 30 mW, used at half power) with a spectral resolution of ~8 cm^{-1}. The laser spot is approximatively 0.5 mm. The wide measurement range (100–4000 cm^{-1}) makes this tool a good choice for the in situ study of hydrous silicates [16,17]. The instrument has a weight of nearly 2 Kg and is operated by a laptop through USB 2.0 connection. Typical acquisition times are of the order of 60 s, but some dark samples required longer acquisition times.

SEM-EDS measurements have been performed on the mineral samples embedded in resin stab. Samples were analyzed for their major elements with a Tescan VEGA TS Univac 5136 XM scanning electron microscope coupled with an EDAX Genesis 4000 XMS Imaging 60 SEM-EDS at the University of Milano-Bicocca, Milano, Italy, under an electron beam accelerated at 20 kV, with 250 nm spot size and a current of 190 ± 1 pA, The quantification was obtained using natural standards (Astimex Scientific Ltd, Toronto, ON, Canada). Counting time was 100 sec plus deadtime (25%). ZAF matrix correction was applied. The Fe^{3+} content was calculated according to Hawthorne et al. [4] and end member according to IMA classification [4,18].

3. Results and Discussion

The chemical composition of the samples of the main set obtained by SEM-EDS is reported in Table 2. All the investigated samples belong to the nephrite series of calcic amphiboles, and range from pure tremolite (sample #21) to Fe-actinolite (sample #17). Some samples of tremolite contain small amount of iron (up to 3.26 wt % FeOt) and aluminium (up to 3.68 wt % Al_2O_3). The actinolite samples show a variable concentration of iron (5.27–13.16 wt % FeOt) with Fe^{3+} ranging from 0 to 0.325 apfu. Some actinolites have small content of chromium (up to 0.80 wt % Cr_2O_3).

Table 2. Representative compositions of the investigated amphiboles obtained by SEM-EDS. *Act* = actinolite, *Trm* = tremolite, *Fe-Act* = ferro-actinolite.

	1	2	3	4	5	6	7	8	9	10	11	12	13	14	16	17	18	19	20	21
SiO_2	54.5	55.3	53.9	55.3	54.7	56.3	55.6	57.3	59.1	58.5	59.2	56.1	59.0	56.6	57.1	51.4	54.5	54.8	54.7	59.8
TiO_2	–	0.1	0.2	0.1	0.1	–	–	–	–	–	–	–	–	–	0.1	–	–	–	–	–
Al_2O_3	2.4	1.6	2.4	0.6	3.1	1.7	2.8	3.7	–	0.8	–	2.7	–	2.3	0.4	5.5	2.2	0.9	4.8	–
FeOt	10.7	5.4	6.4	7.2	8.1	9.7	6.1	2.2	1.8	2.9	3.1	3.3	2.1	5.3	0.5	19.7	7.5	13.2	6.6	–
MnO	–	0.1	0.1	0.2	0.1	–	–	–	–	–	–	–	–	–	–	0.4	1.8	0.4	0.2	–
MgO	17.6	21.2	19.7	20.2	18.5	17.8	20.2	21.8	23.3	22.6	22.8	21.7	22.7	20.2	24.0	7.8	18.9	14.9	19.6	24.5
CaO	11.1	11.7	12.0	12.1	9.9	12.6	12.1	13.4	13.1	13.5	12.9	13.3	13.4	12.0	13.5	10.6	11.0	13.2	12.2	13.9
Na_2O	1.5	0.7	0.8	0.3	1.5	–	–	–	–	0.3	–	0.5	0.6	–	0.3	1.5	0.9	0.2	1.0	–
K_2O	–	0.1	0.2	0.2	0.1	–	–	–	–	–	–	–	–	–	0.1	0.5	0.2	0.2	–	–
Cr_2O_3	–	0.4	0.8	0.1	0.3	–	–	–	–	–	–	–	–	0.8	–	–	–	–	–	–
	97.8	96.6	96.5	96.3	96.4	98.1	96.8	98.4	97.3	98.6	98.0	97.6	97.8	97.2	96.0	97.4	97.0	97.8	99.1	98.2
Structural formula on the basis of 23 oxygens																				
Si	7.791	7.781	7.658	7.865	7.762	7.910	7.782	7.762	8.072	7.949	8.060	7.734	8.052	7.856	7.907	7.674	7.750	7.914	7.544	8.040
Ti	–	0.007	0.020	0.015	0.011	–	–	–	–	–	–	–	–	–	0.009	–	–	–	–	–
Al	0.414	0.264	0.400	0.102	0.518	0.275	0.465	0.587	–	0.131	–	0.436	–	0.383	0.062	0.962	0.369	0.160	0.785	–
Fe^{3+}	0.325	0.219	0.265	0.135	0.213	0.000	0.000	0.000	0.000	0.035	0.000	0.185	0.036	0.000	0.008	0.014	0.263	0.038	0.296	–
Fe^{2+}	0.957	0.422	0.492	0.720	0.759	1.149	0.718	0.250	0.212	0.298	0.351	0.190	0.204	0.623	0.054	2.479	0.631	1.553	0.462	–
Mn	–	0.012	0.018	0.023	0.015	–	–	–	–	–	–	–	–	–	–	0.047	0.219	0.048	0.023	–
Mg	3.752	4.439	4.180	4.283	3.915	3.720	4.210	4.396	4.749	4.569	4.635	4.459	4.619	4.181	4.964	1.732	4.013	3.204	4.018	4.913
Ca	1.551	1.762	1.825	1.843	1.511	1.903	1.811	1.948	1.896	1.967	1.890	1.970	1.962	1.785	2.000	1.701	1.676	2.039	1.799	2.007
Na	0.426	0.197	0.229	0.088	0.404	–	–	–	–	0.074	–	0.147	0.151	–	0.075	0.434	0.251	0.070	0.265	–
K	–	0.028	0.031	0.036	0.019	–	–	–	–	–	–	–	–	–	0.019	0.105	0.036	0.031	–	–
Cr	–	0.042	0.089	0.012	0.038	–	–	–	–	–	–	–	–	0.088	–	–	–	–	–	–
X	0.80	0.91	0.89	0.86	0.84	0.76	0.85	0.95	0.96	0.94	0.93	0.96	0.96	0.87	0.99	0.41	0.86	0.67	0.90	1.00
Mg_v	0.75	0.87	0.85	0.83	0.80	0.76	0.85	0.95	0.96	0.93	0.93	0.92	0.95	0.87	0.99	0.41	0.82	0.67	0.84	1.00
	Act	*Act*	*Act*	*Act*	*Act*	*Act*	*Act*	*Trm*	*Trm*	*Trm*	*Trm*	*Trm*	*Trm*	*Act*	*Trm*	*Fe-Act*	*Act*	*Act*	*Act*	*Trm*

The Raman spectrum of tremolite and actinolite between 100 and 1100 cm^{-1} (Figure 1) is characterized by the typical features of the spectra of amphibole minerals [19–21]. A possible schematic representation is the following: under 200 cm^{-1}, lattice vibrations; between 300 and 600 cm^{-1}, Mg–OH and Fe–OH vibrations, Si–O–Si bending motions and OH^- librations; between 650 and 750 cm^{-1}, the strong Si–O–Si symmetric stretching, over 750 cm^{-1}, the O–Si–O symmetric stretching and the O–Si–O and Si–O–Si asymmetric stretching motions.

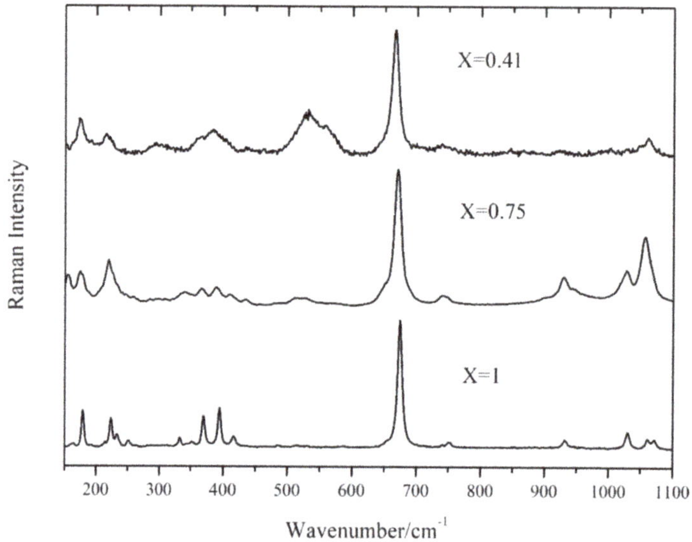

Figure 1. Raman spectra at low wavenumbers of different minerals of tremolite Fe-actinolite series with different X ratios.

As Mg is substituted by Fe^{2+}, the Raman bands broaden; the main band at nearly 670 cm^{-1} changes its width (FWHM) from 9 cm^{-1} for pure tremolite (X = 1) to 18 cm^{-1} for Fe-actinolite (X = 0.41). Some of the minor peaks are no more resolved and merge into broad features.

The main feature of the Raman spectra in the low-wavenumber region, at nearly 675 cm^{-1}, is the Si–O–Si symmetrical stretching (Figure 1) with A_g symmetry. This mode, when substituting Mg^{2+} with the heavier Fe^{2+}, downshifts from 675 cm^{-1} in pure tremolite to 667 cm^{-1} in Fe-rich actinolite (X = 0.41, at the border with Fe-actinolite). A linear trend of the wavenumber with composition is expected by the one-mode behavior usually shown by Si–O vibrations in chain silicates and in particular in amphiboles [22]. This trend is clearly evidenced in Figure 2. A linear interpolation allows to obtain a simple formula relating the position of the main A_g band with the X ratio as obtained by SEM-EDS data (see Table 2): $X = 0.066 \times (\nu - 659.3)$, where ν is expressed in cm^{-1} and 659.33 is the ν value for $X = 0$, as obtained from the interpolation. The least squares linear fitting procedure returned a value of the coefficient R^2 = 0.962 and the standard errors on the slope (0.003). The intercept is zero with a standard error of 0.04. The standard error on X obtained from the interpolation is 0.03.

Only two points (not used in the interpolation procedure) are markedly far from the straight line, related to two samples presenting specific analysis problems. The first one is related to sample #5, very small with a large compositional variability (e.g., Fe^{2+} amount ranges from 0.58 to 1.27 in different points) and showing a small amount of Mn, able to occupy M4 as well as M1, M2, M3 sites. The second point is related to sample #10, highly zoned with large variations in composition. For these two samples, the main source of error is the compositional gradient near the analysed points; a small error in the position can produce a big change in the X value or in the wavenumber of the main Raman peak giving useless results. The parameters obtained from the linear interpolation represent a first tool useful to determine the composition of minerals in the nephrite series from a quick Raman spectrum. This method requires a very good wavenumber calibration of the spectrometer, but it can be also used when noisy or weak spectra are obtained, being based only on the position of the most intense Raman peak.

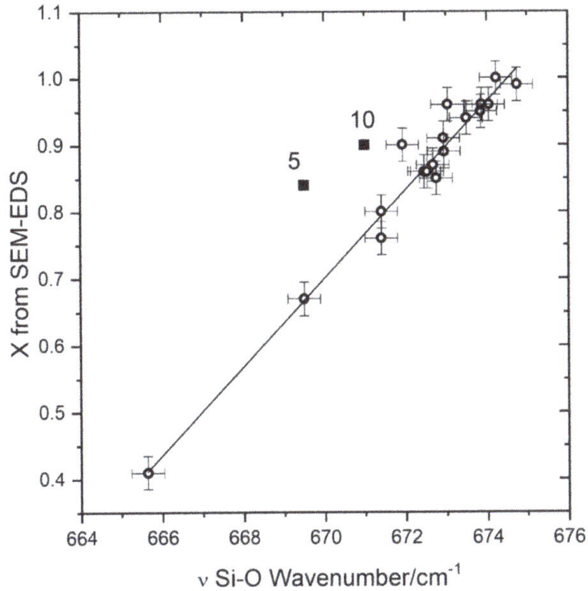

Figure 2. Relation between the wavenumber ν of the A_g main Raman peak at low wavenumber and the chemical composition $X = Mg/(Mg + Fe^{2+})$. The straight line represents the result of the linear interpolation: $X = 0.066 \times (\nu - 659.3)$. The numbered points are the outliers described in the text.

We verified the shift to lower frequencies of the main A_g peak with the increase of iron, by making a micrometric Raman map (10 × 5 points) on a zoned, needle-shaped, tremolite crystal coming from Alpe Rosso (Vigezzo Valley, Italy). As visible in the microscope image (Figure 3 top) the green color, due to the presence of iron, increases from left to right (nearly along the maximum elongation of the crystal). The grey color map (Figure 3 bottom) represents the position of the A_g main band: lower lightness means lower wavenumber (higher Fe amount). As expected, the intensity of green color in the picture is well paralleled by the dark parts of the Raman map.

In the high-wavenumbers region (3600–3700 cm^{-1}) a number of OH stretching bands (from 1 to 4) is present. The number and relative intensity of these bands depend on the X value. This is due to the fact that in amphiboles the OH groups are bonded to 3 sites indicated as C in the general chemical formula. Theses C-sites are usually occupied in nephrites by Mg^{2+} and Fe^{2+} ions. The frequency of the OH stretching vibrations is then influenced by the population of the 3 nearest C-sites [11,12,20,23]. The OH stretching mode shows a two-mode behaviour, meaning that for every different atomic configuration, a different Raman peak arises [20]. When only Mg is present (as in pure tremolite), only one peak (at 3675 cm^{-1}) is observed. As the amount of iron increases, the other peaks, related to different combinations of Fe^{2+} and Mg^{2+} in the sites close to OH, arise at lower wavenumbers (Figure 4). We built a simple statistical model, using the binomial distribution, based on the hypothesis that the 3 sites close to OH are randomly occupied by Fe^{2+} and Mg^{2+} ions, with a probability proportional to their relative amount. The area of the OH Raman bands at about 3675 cm^{-1} (OH surrounded by 3 Mg^{2+} ions) will be proportional to X^3, while the area of the band at 3660 cm^{-1} (OH surrounded by 2 Mg^{2+} and 1 Fe^{2+} ions) will be proportional to $3X^2(1 - X)$. The ratio between the two bands is then $A_{12} = X/[3(1 - X)]$. In this way we obtain another relation useful to estimate the X value from the Raman spectrum: $X = (A_{12})/(1/3 + A_{12})$.

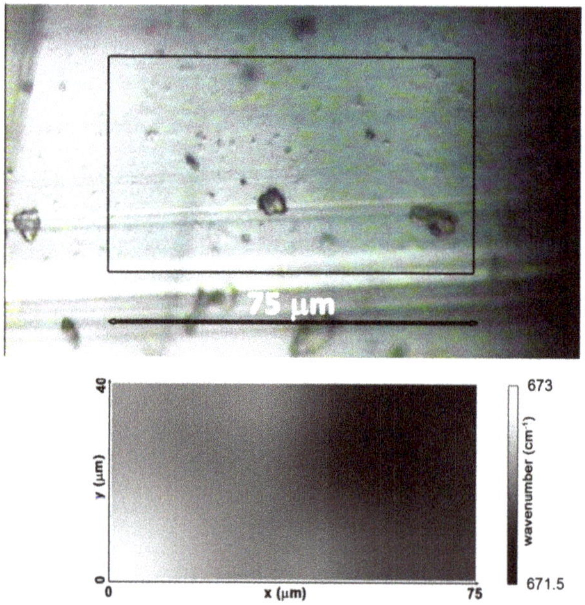

Figure 3. Microscope image (**top**) and Raman map (**bottom**) representing the shift of the main A_g band obtained on a zoned actinolite crystal.

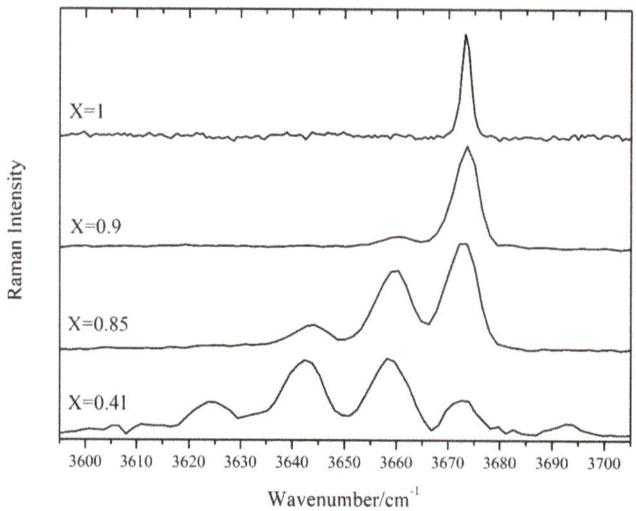

Figure 4. Raman spectra at high wavenumbers of minerals of tremolite Fe-actinolite series with different X ratios.

In Figure 5 the composition (*X* value) estimated from the areas of the OH bands and that measured by SEM-EDS is shown: a good agreement is observed, with a very slight tendency of the "OH bands" method to underestimate *X*. Only for a sample (#3), with composition far from ideal nephrite, with many extra ions, the obtained values differ more than 0.1. The standard error on X obtained with this method, calculated from the differences between Raman (OH bands) and SEM-EDS data is 0.04.

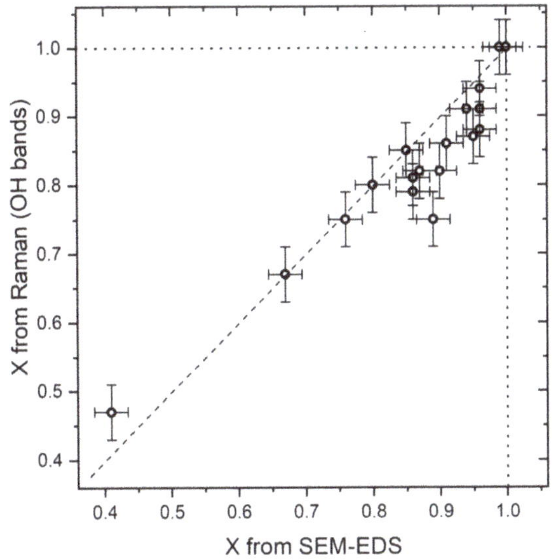

Figure 5. Comparison between the compositions (X ratios) of the tremolite Fe-actinolite samples estimated by the OH stretching bands and those obtained by SEM-EDS. The dashed lines (for X = 1.0 and for equal values) are drawn for visual aid.

The simple equation $X = (A_{12})/(1/3 + A_{12})$, gives results comparable with those obtained by the equation proposed by Chen et al. [11] but the latter is based on the intensity of all the OH features, while our calculation requires only the intensity of two peaks, usually the most intense ones in the Mg-rich samples and easy to measure. The choice to work with only two peaks was made in order to obtain compositional information even using noisy spectra, with just the most intense peaks emerging. This can be important when working on the field, with real samples, or maybe in a museum when the object can be analysed only with a contactless technique and for a short time. A disadvantage is that this method strongly relies on the hypothesis of a random distribution of the cations, that could be not true. However, the cross-check with the other methods (the low-wavenumber method presented before and the method proposed by Chen et al. [11]) indicate that the possible deviations from the randomness are not relevant. The evaluation of the composition (X value) from the ratio of the intensities of the OH stretching bands is not affected by the wavenumber calibration of the instrument and can be made independently of that obtained from the position of the main A_g peak.

The comparison between the two proposed methods is shown by Figure 6 where the X values obtained starting from the position of the main Raman band at nearly 670 cm^{-1} and from the area ratio of the OH Raman stretching bands are compared. The two methods give comparable results, but the X values obtained with the "OH stretching" method are in most cases slightly lower (the average difference is 0.05) than those obtained by the "Si–O stretching" one. This difference is probably due to a slight underestimation provided by the "OH stretching" method, as already suggested by Figure 5. Deviations of the real amphiboles from the theoretical behaviour expected by our statistical model as well as the non-flat spectral answer of the spectrometer can be between the causes of this difference. On the other hand, X values obtained by the "Si–O stretching" method have a better dispersion (see Figure 2), as expected from an empirical procedure. For that reason they can be considered more reliable, but only when a perfect calibration of the spectrometer is provided.

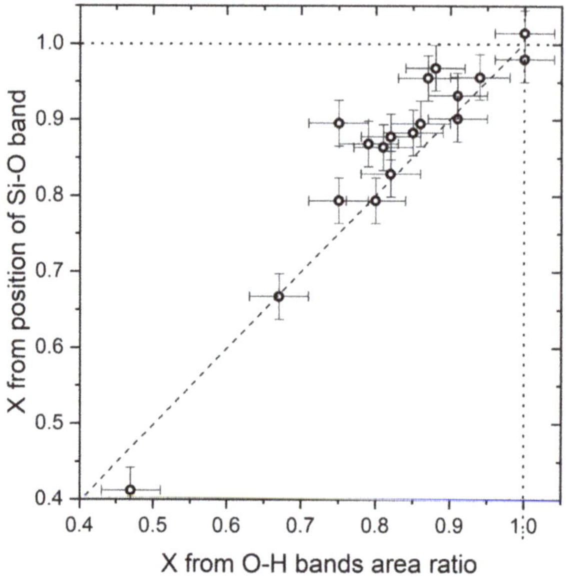

Figure 6. Comparison between the X values obtained starting from the position of the main Raman band at nearly 670 cm^{-1} and from the area ratio of the OH Raman stretching bands. The dashed lines (for $X = 1.0$ and for equal values) are drawn for visual aid.

To evaluate the possible effect of Fe^{3+} on the position of the Si–O stretching main band and on the area of the OH stretching bands, we replotted the graphs and recalculated the parameters using the total amount of iron (Fe_{total}) instead of Fe^{2+}. The obtained trends are very similar and the obtained parameters (including the estimated X values) show variations largely smaller than the calculated uncertainty. For that reason, and because the definition of tremolite and actinolite depends on the Mg/(Mg + Fe^{2+}) ratio, we decided to not include Fe^{3+} in the calculations.

The possibility to obtain compositional information on the X value in situ, for example directly on the outcrops or in a museum environment, using a compact portable instrument, was then evaluated. Due to the impossibility to analyze micrometric crystals or very small felted clusters, a partly different set of minerals was used. Of course, due to the large spot, possible thin zonations are averaged.

A comparison between some spectra obtained using the micro-Raman and the mobile Raman spectrometers is reported in Figure 7 (low wavenumber range) and Figure 8 (high wavenumber range).

It is clearly visible (Figure 7a–c) that in the low wavenumber range the spectra obtained with the handheld spectrometer are of good quality, often comparable with those obtained with the micro-Raman apparatus, despite the lower spectral resolution. The mobile instrument has a fixed geometry and cannot be calibrated with the required precision. For this reason the position of the Raman bands is not enough accurate to obtain the X composition from the wavenumber of the main A_g mode. However, when a reference line is present (e.g. the strong 546 nm line of a common fluorescent tube, appearing at 485 cm^{-1} in the spectrum), the difference in position of the main A_g mode between iron rich ("Serifos") and iron poor ("Valmalenco") terms is clearly detectable (Figure 7d).

In the high wavenumber range (Figure 8), the spectra obtained with the mobile spectrometer on the nephrite minerals are very weak and noisy, because the OH stretching bands are in a region near the limit of the spectral window measured by our handheld spectrometer, with a very bad signal to noise ratio. Even with very long acquisition times (several minutes) it is difficult to observe clean OH stretching peaks. Nevertheless, the distribution of the intensities between the various peaks well reflects that obtained with the micro-Raman spectrometer. Even if the quality of the spectra is not enough to give a quantitative estimation of *X*, it is still possible to distinguish between iron-rich and magnesium-rich terms of the series.

One of the macroscopic samples (Serifos) used for the test of the mobile Raman spectrometer show the spectral features of a very Fe-rich ferro-actinolite. For that reason we decided to measure its composition even with SEM-EDS, obtaining $X = 0.31$. The X value calculated from the position of the main Si–O stretching band measured with a micro-Raman spectrometer ($\nu = 664.2$ cm^{-1}) is $X = 0.32$, in very good agreement with SEM-EDS. To estimate the X value from the areas of the OH stretching bands in Fe-rich samples, it is possible to use the ratio (A_{43}) between the two strongest bands related to (Fe,Fe,Fe) and (Fe, Mg, Mg) configurations to calculate $1 - X = (A_{43})/(1/3 + A_{43})$. For the Serifos sample we obtained $X = 0.25$, not far from the value obtained by SEM-EDS. These results confirm the validity of both Raman-based methods in the Fe-rich region of the series.

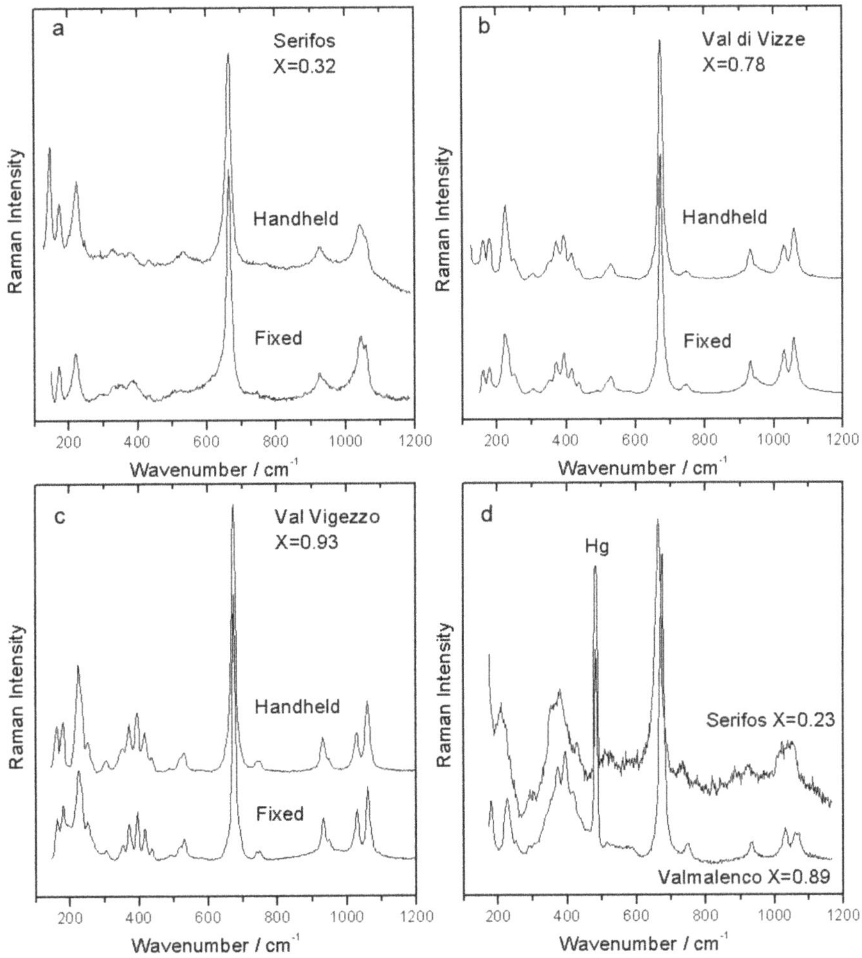

Figure 7. (**a–c**): comparison between spectra obtained on different samples in the low-wavenumbers region with fixed and handheld Raman spectrometers; (**d**): comparison between spectra obtained on Fe-rich and Fe-poor samples with the handheld spectrometer in the low-wavenumber region. Hg = Hg emission line of a fluorescent tube. The *X* values reported are obtained by micro-Raman (fixed Raman) data.

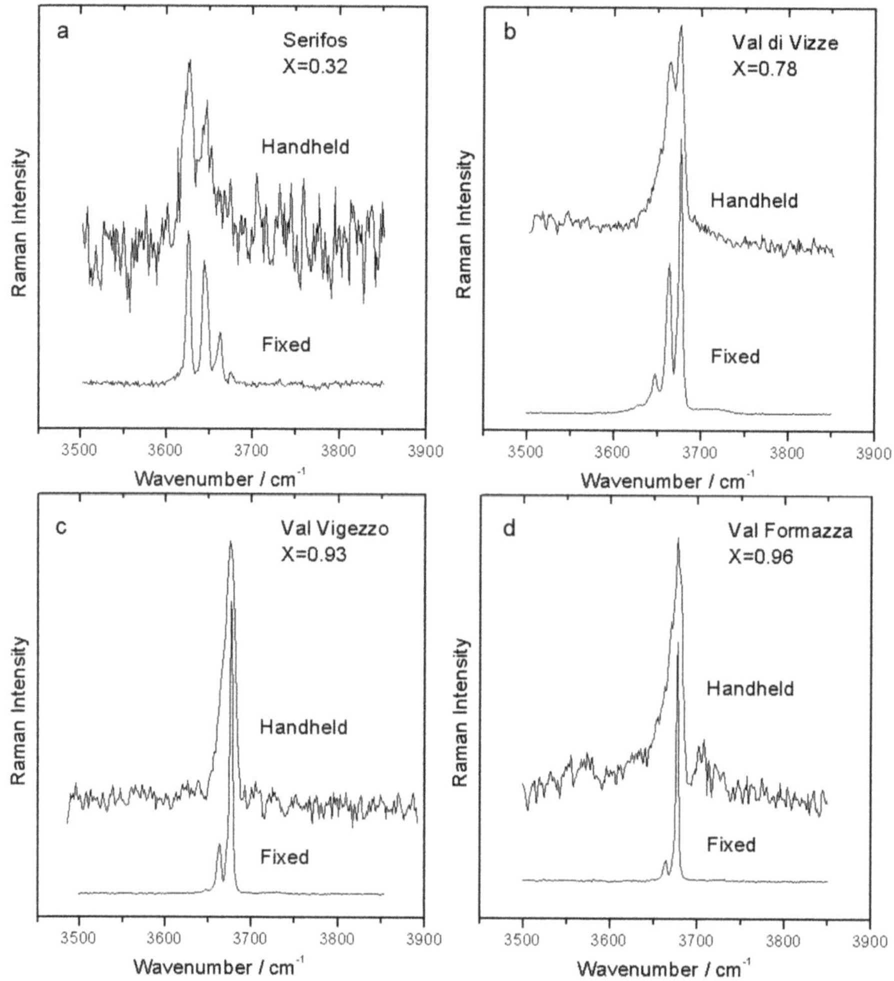

Figure 8. (**a**–**c**): comparison between spectra obtained on different samples in the OH-stretching region with fixed and handheld Raman spectrometers, (**d**): comparison between spectra obtained on Fe-rich and Fe-poor samples with the handheld spectrometer in the OH-stretching region. The *X* values reported in this figures are obtained by micro-Raman (fixed Raman) data.

4. Conclusions

We obtained two simple and independent methods to evaluate the composition of nephritic minerals from their Raman spectrum. The first one, based on the position of the main A_g band, shows a better sensitivity and linearity with the changes of the X value. On the other hand, the second method, based on the ratio of the areas of the two most intense OH stretching bands, is nearly insensitive to the spectrometer calibration. The two methods are consistent and can be chosen depending on the experimental setup and the quality of the signal in the different spectral regions.

The fast and easy to automatize calculation of the X value on Raman micro-maps allows to study compositional variations on microscopic scale directly on crystals, fibers or archaeological items, without requiring any preparation of the samples.

The identification of minerals of the tremolite Fe-actinolite series and the discrimination between Fe-rich and Mg-rich terms from their Raman spectra is immediate also for not skilled operators. This study shows that it is possible to obtain the most important information even using very compact (handheld) mobile Raman spectrometers; this result is very important to extend the use of Raman spectroscopy for the identification of Ca amphiboles not only to routine analysis, but also for *on-the-field* applications in sedimentology, petrography, archaeometry and environmental analysis.

Author Contributions: Conceptualization, D.B. and S.A.; Formal analysis, P.G. and E.S.-M.; Investigation, D.B., S.A., L.S. and P.G.; Methodology, D.B. and S.A.; Visualization, L.F.; Writing—original draft, D.B.; Writing—review & editing, S.A. and P.P.L.

Funding: This research received no external funding

Conflicts of Interest: The authors declare no conflict of interest.

References

1. Garzanti, E.; Doglioni, C.; Vezzoli, G.; Andò, S. Orogenic Belts and Orogenic Sediment Provenance. *J. Geol.* **2007**, *115*, 315–334. [CrossRef]
2. Andò, S.; Morton, A.; Garzanti, E. Metamorphic grade of source rocks revealed by chemical fingerprints of detrital amphibole and garnet. In *Sediment Provenance Studies in Hydrocarbon Exploration and Production;* Scott, R.A., Smyth, H.R., Morton, A.C., Richardson, N., Eds.; Geological Society London Special Publications: London, UK, 2014; Volume 386, pp. 351–371.
3. Garzanti, E.; Vezzoli, G.; Lombardo, B.; Andò, S.; Mauri, E.; Monguzzi, S.; Russo, M. Collision-orogen provenance (Western and Central Alps): Detrital signatures and unroofing Trends. *J. Geol.* **2004**, *112*, 145–164. [CrossRef]
4. Hawthorne, F.C.; Oberti, R.; Harlow, G.E.; Maresch, W.V.; Martin, R.F.; Schumacher, J.C.; Welch, M.D. Nomenclature of the amphibole supergroup. *Am. Miner.* **2012**, *97*, 2031–2048. [CrossRef]
5. Cascalho, J. Provenance of Heavy Minerals: A Case Study from the WNW Portuguese Continental Margin. *Minerals* **2019**, *9*, 355. [CrossRef]
6. Liang, W.; Garzanti, E.; Andò, S.; Gentile, P.; Resentini, A. Multimineral fingerprint of Transhimalayan and Himalayan sources of Indus-derived Thal Desert sand (Central Pakistan). *Minerals* **2019**, *9*, 457. [CrossRef]
7. Lünsdorf, N.K.; Kalies, J.; Ahlers, P.; Dunkl, I.; Von Eynatten, H. Semi-Automated Heavy-Mineral Analysis by Raman Spectroscopy. *Minerals* **2019**, *9*, 385. [CrossRef]
8. Hahn, A.; Vogel, H.; Andó, S.; Garzanti, E.; Kuhn, G.; Lantzsch, H.; Schüürman, J.; Vogt, C.; Zabel, M. Using Fourier transform infrared spectroscopy to determine mineral phases in sediments. *Sediment. Geol.* **2018**, *375*, 27–35. [CrossRef]
9. Andò, S.; Garzanti, E. Raman spectroscopy in heavy mineral studies. In *Sediment Provenance Studies in Hydrocarbon Exploration and Production*; Scott, R.A., Smyth, H.R., Morton, A.C., Richardson, N., Eds.; Geological Society London Special Publications: London, UK, 2014; Volume 386, pp. 395–412.
10. Garzanti, E.; Andò, S. Plate tectonics and heavy-mineral suites of modern sands. In *Heavy Minerals in Use*; Mange, M.A., Wright, D.T., Eds.; Elsevier: Amsterdam, The Netherlands, 2007; Volume 58, pp. 741–763.
11. Chen, T.-H.; Calligaro, T.; Pagès-Camagna, S.; Menu, M. Investigation of Chinese archaic jade by PIXE and μRaman spectrometry. *Appl. Phys. A* **2004**, *79*, 177–180. [CrossRef]
12. Wang, A.; Dhamelincourt, P.; Turrell, G. Raman microspectroscopic study of the cation distribution in amphiboles. *Appl. Spectrosc.* **1988**, *42*, 1441. [CrossRef]
13. Wang, R.; Zhang, W.-S. Application of Raman spectroscopy in the nondestructive analyses of ancient Chinese jades. *J. Raman Spectrosc.* **2011**, *42*, 1324–1329. [CrossRef]
14. Casadio, F.; Douglas, J.G.; Faber, K.T. Noninvasive methods for the investigation of ancient Chinese jades: An integrated analytical approach. *Anal. Bioanal. Chem.* **2007**, *387*, 791–801. [CrossRef] [PubMed]
15. Rinaudo, C.; Belluso, E.; Gastaldi, D. Assessment of the use of Raman spectroscopy for the determination of amphibole asbestos. *Miner. Magaz.* **2004**, *68*, 455–465. [CrossRef]
16. Jehlička, J.; Culka, A.; Bersani, D.; Vandenabeele, P. Comparison of seven portable Raman spectrometers: Beryl as a case study. *J. Raman Spectrosc.* **2017**, *48*, 1289–1299. [CrossRef]

17. Petriglieri, J.R.; Laporte-Magoni, C.; Gunkel-Grillon, P.; Tribaudino, M.; Bersani, D.; Sala, O.; Le Mestre, M.; Vigliaturo, R.; Bursi Gandolfi, N.; Salvioli-Mariani, E. Mineral fibres and environmental monitoring: A comparison of different analytical strategies in New Caledonia. *Geosci. Front. Ophiolites* **2019**, in press. [CrossRef]
18. Leake, B.E.; Wooley, A.R.; Arps, C.E.; Birch, W.D.; Gilbert, M.C.; Grice, J.D.; Hawthorne, F.C.; Kato, A.; Kisch, H.J.; Krivovichec, V.G.; et al. Nomenclature of amphibole. Report of the Subcommittee on Amphiboles of the International Mineralogical Association Commission on New Minerals and Mineral names. *Eur. J. Miner.* **1997**, *9*, 623–651. [CrossRef]
19. Apopei, A.I.; Buzgar, N. The Raman study of amphiboles. *An. Ştiinţifice Univ. Cuza Iaşi Geol.* **2010**, *1*, 57–83.
20. Leissner, L. Crystal Chemistry of Amphiboles Studied by Raman Spectroscopy. Master's Thesis, Mineralogisch-Petrographisches Institut, Universität Hamburg, Hamburg, Germany, 2014.
21. Leissner, L.; Schlüter, J.; Horn, I.; Mihailova, B. Exploring the potential of Raman spectroscopy for crystallochemical analyses of complex hydrous silicates: I. Amphiboles. *Am. Miner.* **2015**, *100*, 2682–2694. [CrossRef]
22. Andrut, M.; Gottschalk, M.; Melzer, S.; Najorka, J. Lattice vibrational modes in synthetic tremolite-Sr-tremolite and tremolite-richterite solid solutions. *Phys. Chem. Miner.* **2000**, *27*, 301–309. [CrossRef]
23. Sbroscia, M.; Della Ventura, G.; Iezzi, G.; Sodo, S. Quantifying the A-site occupancy in amphiboles: A Raman study in the OH-stretching region. *Eur. J. Miner.* **2018**, *30*, 429–436. [CrossRef]

minerals

MDPI

Article

Correlation of Hydrocarbon Reservoir Sandstones Using Heavy Mineral Provenance Signatures: Examples from the North Sea and Adjacent Areas

Andrew Morton [1,2,*] **and Paula McGill** [3]

1 HM Research Associates Ltd., Giddanmu, St Ishmaels, Pembrokeshire SA62 3TJ, UK
2 CASP, University of Cambridge, Madingley Rise, Cambridge CB3 0UD, UK
3 HM Research Norway A/S, Furuveien 7A, 4025 Stavanger, Norway; paula.hmresearch@gmail.com
* Correspondence: heavyminerals@hotmail.co.uk

Received: 18 October 2018; Accepted: 24 November 2018; Published: 3 December 2018

Abstract: Correlation of hydrocarbon reservoir sandstones is one of the most important economic applications for heavy mineral analysis. In this paper, we review the fundamental principles required for establishing correlation frameworks using heavy mineral data, and illustrate the applications of a wide variety of heavy mineral techniques using a number of case studies from hydrocarbon reservoirs in the North Sea and adjacent areas. The examples cover Triassic red-bed successions in the central North Sea and west of Shetland, which have been subdivided and correlated using provenance-sensitive ratio data and mineral morphologies; Middle Jurassic paralic sandstones in the northern North Sea, correlated using garnet geochemistry; Upper Jurassic deep water sandstones in the northern North Sea, discriminated using rutile geochemistry and detrital zircon age data; and the "real-time" application of the technique at well site in Devonian-Carboniferous fluvio-lacustrine sandstones of the Clair Field, west of Shetland.

Keywords: heavy minerals; correlation; North Sea; Jurassic; Triassic; Carboniferous; Devonian

1. Introduction

One of the most important practical economic applications of heavy mineral analysis is the correlation of sandstones that host reserves of oil and gas. Correlation is one of the key components required to build a petroleum reservoir model, because it places fundamental constraints on the understanding of the reservoir architecture. Correlation of reservoir sandstones is therefore crucial for optimising recovery of our ever-decreasing supply of hydrocarbons. Biostratigraphy, integrated with seismic and geophysical log interpretation, remains the principal tool for oilfield correlation [1]. However, there are many situations where biostratigraphic data lack sufficient resolution, for example due to unfavourable depositional conditions (such as non-marine or paralic settings or very rapid sedimentation in marine environments) or deep burial alteration causing degradation of the biostratigraphic signal.

In such cases, alternative methods are required [2,3]. In broad terms, three distinct approaches have been followed: sequence stratigraphic methods (such as core- and outcrop-based sedimentological criteria, geophysical log correlation, and seismic reflection data); direct or indirect dating (such as radiometric dating, magnetostratigraphy, and seawater $^{87}Sr/^{86}Sr$ analysis), and provenance-based analytical techniques. Heavy mineral correlation belongs to the third of these categories. It is one of a family of provenance-based tools that also includes chemostratigraphy [4,5], Sm-Nd isotope stratigraphy [6,7] and clay mineral stratigraphy [8,9].

2. Background

The successful application of any provenance-based correlation method depends on the presence of changes in sediment provenance or transport history during deposition of the succession in question. These changes are reflected in the mineralogy and geochemistry of the sediment, and can therefore be detected by one or more of the various analytical techniques. The main problem with provenance-based methods is that a number of other processes capable of affecting bulk mineralogy and geochemistry operate during the sedimentary cycle ([10] and references therein). These processes are:

1. weathering at source, prior to incorporation in the transport system
2. mechanical breakdown during transport
3. weathering during periods of alluvial storage on the floodplain
4. hydraulic processes during transport and final deposition
5. diagenesis during deep burial
6. weathering at outcrop

These processes can partially or totally mask the variations caused by changes in provenance, and reliable provenance-based correlations therefore crucially depend on filtering out the effects of these overprinting factors. The most important factors that bias the provenance signal are hydrodynamic processes during transport and deposition and diagenesis during deep burial [10,11]. Weathering is also an important control in some instances, but mechanical abrasion is not believed to be of great importance in most circumstances [10,12].

Heavy mineral analysis is especially well-placed to identify changes in provenance in clastic successions. This is (i) because of the long history of research into the controls on assemblage compositions, and (ii) because the knowledge gained has enabled recognition of parameters that either minimise the effects, or are entirely independent of, these factors. In addition, the method is especially well suited to application in the hydrocarbon industry because it deals with individual particles that constitute the sample, rather than the attributes of bulk samples. Consequently, data can be generated equally well from ditch cuttings (by far the most common sample type available) as from cores and sidewall cores, because the effects of components added to the drilling mud system (such as mud-weighting agents) can be eliminated.

3. History of Application

Almost immediately after it was recognised that heavy mineral assemblages provide important constraints on the provenance of sand and sandstone, e.g., [13–15], applications to the petroleum industry became commonplace. Examples of this early work includes studies by Reed [16] and Reed and Bailey [17] on the Tertiary of California; by Hedberg [18] on the Tertiary of the Maracaibo Basin, Venezuela; by Edson [19] on the Paleozoic of the mid-continental US; and by Bornhauser [20] and Cogen [21] on the Tertiary of the US Gulf Coast. The work by Reed [16] and Reed and Bailey [17] showed that high-resolution correlation is possible using a variety of heavy minerals, principally amphibole, pyroxene, epidote, zircon, and titanite. Hedberg [18] showed that heavy mineral data accurately identified an unconformity surface on the basis of the distribution of andalusite, sillimanite, kyanite, staurolite, garnet, apatite, and chloritoid, and that this surface could be traced for at least 35 km. Edson [19], by contrast, considered that correlation is possible even using the relatively impoverished heavy mineral suites found in Ordovician and Carboniferous sandstones of the mid-continental US. Bornhauser [20] and Cogen [21] demonstrated the existence of variations in amphibole, epidote, kyanite, and staurolite in the Tertiary of the US Gulf Coast, and used these to construct a series of heavy mineral zones. Key observations from the work of Cogen [21] were that (i) heavy mineral assemblages become less diverse with increasing burial depth, and (ii) the boundaries between the heavy mineral zones cross-cut stratigraphic boundaries.

Even at this early stage, warning signs concerning the instability of several detrital heavy minerals had started appearing, e.g., [22,23]. A key period in the application of heavy minerals to correlation issues was the early 1940's, with the virtually simultaneous publication of four papers that all demonstrated the extensive nature of heavy mineral dissolution in the subsurface [24–27]. This coincided with the observation that heavy mineral zonal boundaries cross-cut stratigraphic boundaries in the Tertiary of the US Gulf Coast [21], implying that heavy mineral stratigraphy may be suspect. Consequently, from this point onwards, applications of heavy minerals for stratigraphic purposes declined markedly, and very few papers on this topic appeared in the literature for several decades, a significant exception being the work of Feo-Codecido [28].

Interest in the application of heavy mineral analysis for hydrocarbon reservoir stratigraphy was rekindled during the 1980's. This was partly because of the increased understanding of heavy mineral behaviour during diagenesis in the subsurface e.g., [29–32], which led to the recognition of which minerals, and which mineral parameters, could be regarded as reliable provenance indicators in highly diagenetically-modified sandstones [33,34]. Concurrently, developments in microbeam analytical techniques had started to enable major element geochemical characterisation of stable or ultrastable heavy mineral components, such as garnet [35] and tourmaline [36]. At the same time, textural studies of individual stable heavy mineral components, notably apatite, tourmaline, and zircon, showed that variations in these parameters could be used for correlation purposes in biostratigraphically-barren hydrocarbon-bearing reservoir successions of the UK, such as the Devonian [37] and the Triassic [34,38]. More recently, continued developments in analytical techniques have increased the scope for the application of single-grain geochemistry for evaluation of changes in provenance, including trace element data from diagenetically-stable phases such as apatite and rutile e.g., [39,40], and U-Pb isotopic data from ultrastable phases such as zircon [41,42]. As a result, there is now a wide range of heavy mineral analytical methods available for correlation purposes. In this paper, we demonstrate the application of a variety of heavy mineral approaches to a number of different hydrocarbon reservoir successions in the North Sea and adjacent areas (Figure 1).

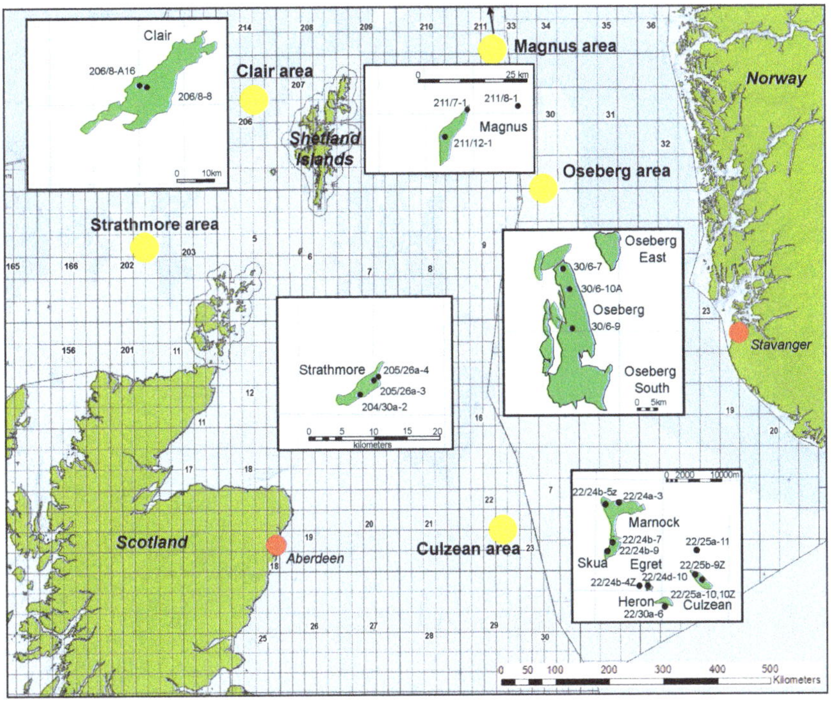

Figure 1. Location map of the North Sea and west of Shetland areas showing the oil and gas fields discussed in this paper. Numbers are the UK and Norwegian quadrants.

4. Correlation Using Conventional Heavy Mineral Assemblages: Strathmore Field, West of Shetland

Most Triassic sandstones in and around the British Isles were deposited in fluvial and aeolian environments under arid to semi-arid conditions [43]. Red-bed facies are therefore widespread and biostratigraphic controls are generally scarce. This is the situation in the Strathmore Field, which lies in the East Solan Basin, west of Shetland (Figure 1), and which hosts hydrocarbons in the Triassic [44]. Correlation of the Triassic succession in the Strathmore Field therefore has important economic implications. The Triassic, which is up to 1000 m thick, comprises two sand-rich formations, the Otter Bank Formation overlain by the Foula Formation (Figure 2). The Otter Bank Formation is underlain by the Otter Bank Shale Formation, dated as Griesbachian (basal Induan) on the basis of a distinctive palynological assemblage [44,45].

The Otter Bank Formation consists of braided sandy fluvial deposits overlain by interbedded fluvial and aeolian sabkha deposits. A very sparse palynological assemblage suggests an Early Triassic age [45], with palaeomagnetic data suggesting deposition in the Diererian-Smithian (mid Early Triassic). The Foula Formation comprises interbedded fluvial and aeolian sabkha deposits overlain by wholly fluvial deposits higher in the succession. Palynological constraints are poor but suggest that deposition began in the Ladinian and extended into the Carnian, and palaeomagnetic data are also consistent with a Ladinian age [45].

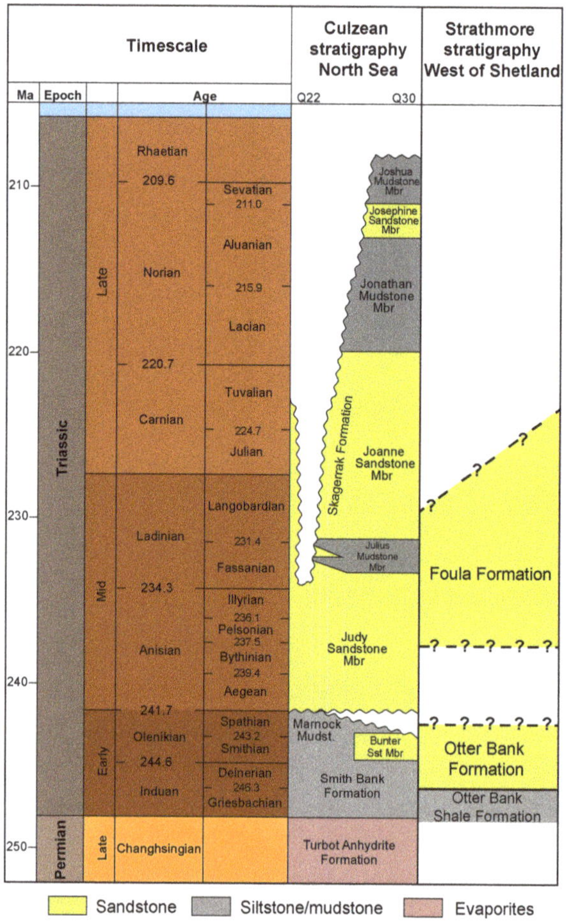

Figure 2. Triassic stratigraphy of the central North Sea and Strathmore Field areas, adapted from Goldsmith et al. [46,47], Swiecicki et al. [45], Herries et al. [44], and Mouritzen et al. [48].

Morton et al. [49] showed that the Triassic succession in three wells (204/30a-2, 205/26a-3, and 205/26a-4) from the Strathmore Field can be subdivided and correlated using conventional heavy mineral data (i.e., data collected by optical analysis using the petrographic microscope). The correlation depends on variations in three key provenance-sensitive ratio parameters, apatite:tourmaline (ATi), garnet:zircon (GZi), and rutile:zircon (RuZi), as defined by Morton and Hallsworth [33].

The heavy mineral assemblages in sandstones from 205/26a-3 and 205/26a-4 were characterised exclusively using core material. However, only three short cores were taken in 204/30a-2, and the heavy mineral stratigraphy for this well was mainly constructed using data from ditch cuttings.

In all three wells, the lower part of the succession has relatively low RuZi (ca. 20–30), but this rises to ca. 60 coincident with the boundary between the Otter Bank Formation and the Foula Formation as defined by a change on the gamma ray log (Figure 3). The increase in RuZi is mirrored by GZi, although the rise is less dramatic. Provenance-sensitive heavy mineral data therefore enable subdivision of the succession into heavy mineral unit O (below) and heavy mineral unit F (above).

These units can be further subdivided, principally using ATi and RuZi. ATi is consistently high in the Foula Formation but an upward-increasing trend in ATi defines a two-fold subdivision of the Otter Bank Formation into subunits O2 and O1 (Figure 3). Most of the Foula Formation has high RuZi, but two thin zones in 204/30a-2 have lower values (Figure 3). Variations in RuZi therefore enable subdivision of Unit F into five subunits, three with high RuZi (F1, F3, and F5) and two intervening intervals with low RuZi (F2 and F4). The two subunits with low RuZi are also characterised by slightly lower GZi. The lower of the two low-RuZi subunits (F2) can also be traced into the cored well 205/26a-4.

In summary, variations in RuZi, GZi, and ATi enabled the establishment of a high-resolution stratigraphic framework for the Triassic of the Strathmore Field, with correlations between the three analysed wells being clearly evident (Figure 3). The clear distinction between the units and subunits is also manifested by cross-plots of the ratio parameters (Figure 4). Foula Formation samples form well-defined clusters, with very high ATi and GZi and generally high RuZi. Otter Bank samples show greater variation, but display little overlap with the Foula Formation. They have uniformly low RuZi, and have generally lower ATi and GZi than the Foula Formation. The only area of overlap concerns subunit F2, subunit F4 and unit O, since all three intervals have similar RuZi, GZi, and ATi values (Figure 4). Further support for changes in provenance in the Triassic of the Strathmore Field are given by garnet geochemical and zircon U-Pb data [49], but the conventional heavy mineral data nevertheless provide a stand-alone correlation framework.

The Strathmore Field correlation study enabled a comparison between data acquired from core and ditch cuttings. Core data can be regarded as reliable since their position within the well is accurately known and they are generally uncontaminated or otherwise altered during the drilling process. Ditch cuttings samples, however, represent material composited across a range of depths (dependent on the sampling frequency), are subject to contamination by downhole caving and drilling mud additives, and are also subject to modification owing to the mechanical action of the drill bit.

There is evidence for contamination of ditch cuttings samples in the Triassic of Strathmore. Core samples contain virtually no epidote, calcic amphibole, or clinopyroxene, but in the cuttings, these minerals form up to 4.0%, 2.5%, and 0.5% of the assemblages [49]. The presence of minor amounts of these minerals is likely to be due either to caving of lithologies from higher in the well bore, or to their inherent presence as contaminants in the drilling mud system. The deviation in assemblage composition between core and cuttings illustrates the problems that can be associated with interpretation of heavy mineral data from ditch cuttings, and shows the advantages of using provenance-sensitive ratios to characterise sandstones that might be subject to contamination because of the drilling process. Provenance-sensitive ratios show little difference between core and ditch cuttings (Figure 4), the only deviation being that some cuttings samples in the Otter Bank Formation of 204/30a-2 have marginally lower ATi than Otter Bank core samples. This is due to the susceptibility of apatite to mechanical loss compared with the other heavy minerals because of its lower hardness, causing it to be preferentially lost through the grinding action of the drill bit [10].

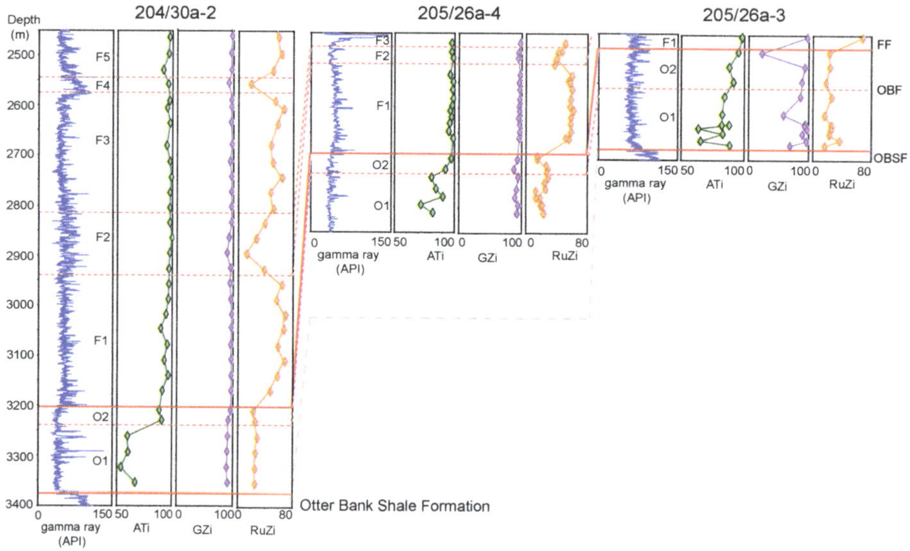

Figure 3. Correlation of the Triassic in Strathmore Fields wells 204/30a-2, 205/26a-3, and 205/26a-4 using heavy mineral provenance-sensitive ratio signatures. For well locations, see Figure 1. FF = Foula Formation, OBF = Otter Bank Formation, OSBF = Otter Bank Shale Formation

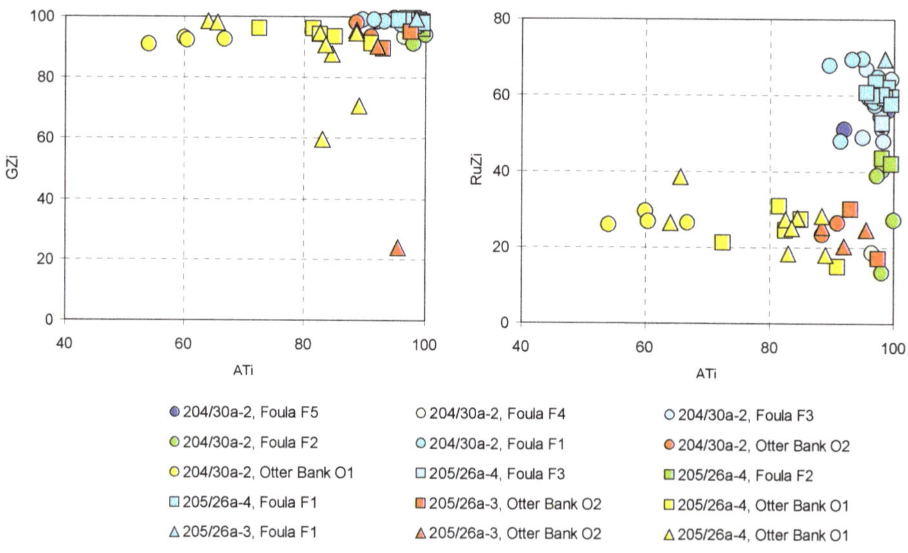

Figure 4. *Cont.*

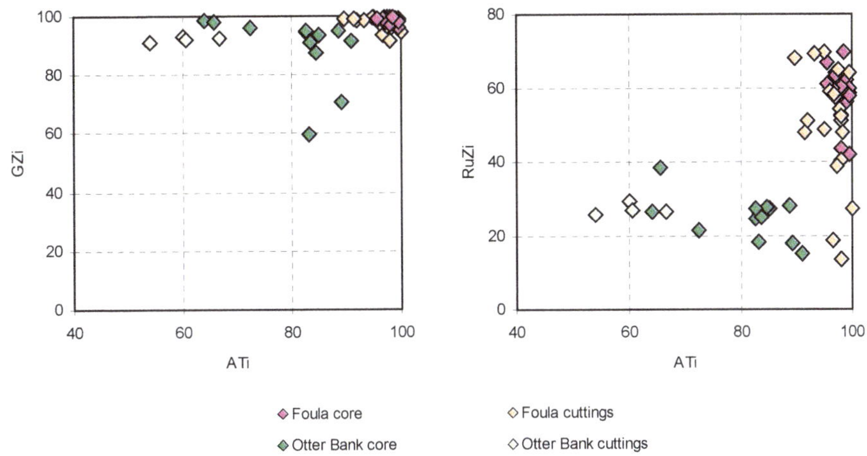

Figure 4. Crossplots of the provenance-sensitive heavy mineral ratios ATi, GZi, and RuZi showing (upper) discrimination of Foula and Otter Bank sandstones, and (lower) slight deviation in ATi between data from cores and ditch cuttings.

5. Correlation Using Heavy Mineral Morphology: Culzean Field Area, Central North Sea

The Triassic Skagerrak Formation, which hosts hydrocarbons in a number of fields in the central North Sea, has historically proved difficult to subdivide and correlate on biostratigraphic grounds. The Skagerrak Formation is interpreted to be a dryland terminal fluvial system, including arid terminal splay-like sheet flood deposits and more humid channel-confined and associated floodplain deposits [50]. The generally unfavourable red-bed lithologies, coupled with very deep burial, are responsible for the scarcity of palynological data over most of the region. Palynological data recovery is better in the southeastern part of the region (Quadrant 30), which enabled Goldsmith et al. [46,47] to subdivide the Skagerrak Formation into the Judy, Joanne, and Josephine Sandstone Members and intervening Julius, Jonathan, and Joshua Mudstone Members, and to assign stratigraphic ages to these members (Figure 2). However, recovery of age-diagnostic palynological floras has proved more elusive towards the northwest (Quadrant 22), and consequently heavy mineral stratigraphic studies have been conducted in this area in order to provide a correlation framework. These studies revealed that the assemblages are typically restricted in diversity [34,51], owing to extensive diagenetic modification of the detrital heavy mineral suite during deep burial. Mange-Rajetzky [34] noted that the monotonous nature of the assemblages precludes stratigraphic subdivision and correlation at heavy mineral species level, but observed the existence of distinct stratigraphic patterns in heavy mineral morphology (Figure 5), which she used to establish a semi-regional correlation framework. The initial work by Mange-Rajetzky [34] has been recently extended into the newly-discovered Culzean Field (Figure 1) and adjacent wells [48]. The Culzean discovery well (22/25a-9Z) is especially significant because it yielded relatively rich and age-diagnostic palynoflora assemblages, enabling the heavy mineral stratigraphy in the Quadrant 22 area to be tied into the biostratigraphically-based subdivision of the Skagerrak Formation into the various sandstone and mudstone members defined in wells from Quadrant 30 [46,47].

Mange-Rajetzky [34] used variations in morphology of apatite, tourmaline, and zircon in the template well 22/24b-5Z (Figure 5) to establish three main zones (SB, M, and A), with further subdivision of zone M into four subzones (q, d, c, and b). The study by Mouritzen et al. [48] concentrated on apatite morphology, since this phase is by far the most abundant component of the heavy mineral assemblages. Variations in the apatite roundness index (ARi) faithfully match the zones and subzones identified by Mange-Rajetzky [34] in well 22/24b-5Z (Figure 6). Another parameter that proved useful is zircon:apatite (ZAi), which is very low at the base of the succession but shows distinct increases higher in the stratigraphy. However, this parameter is heavily influenced by hydrodynamic conditions at the time of deposition, since it compares the relative abundance of apatite and zircon, which have contrasting densities (zircon being considerably denser than apatite). The overall upward increase in this parameter reflects the general upward-coarsening of the Triassic succession. Therefore, while variations in this parameter are stratigraphically useful on a local scale, they are unlikely to be applicable regionally. The reproducibility of the ARi and ZAi trends between the Marnock Field well 22/24b-5Z and the Culzean Field well 22/25a-9Z is illustrated in Figure 6. The consistency of these patterns enabled Mouritzen et al. [48] to define 6 heavy mineral zones, and by integrating the biostratigraphic information they were able to identify which of the key mudstone marker horizons found in Quadrant 30 [46,47] are present in the Quadrant 22 area.

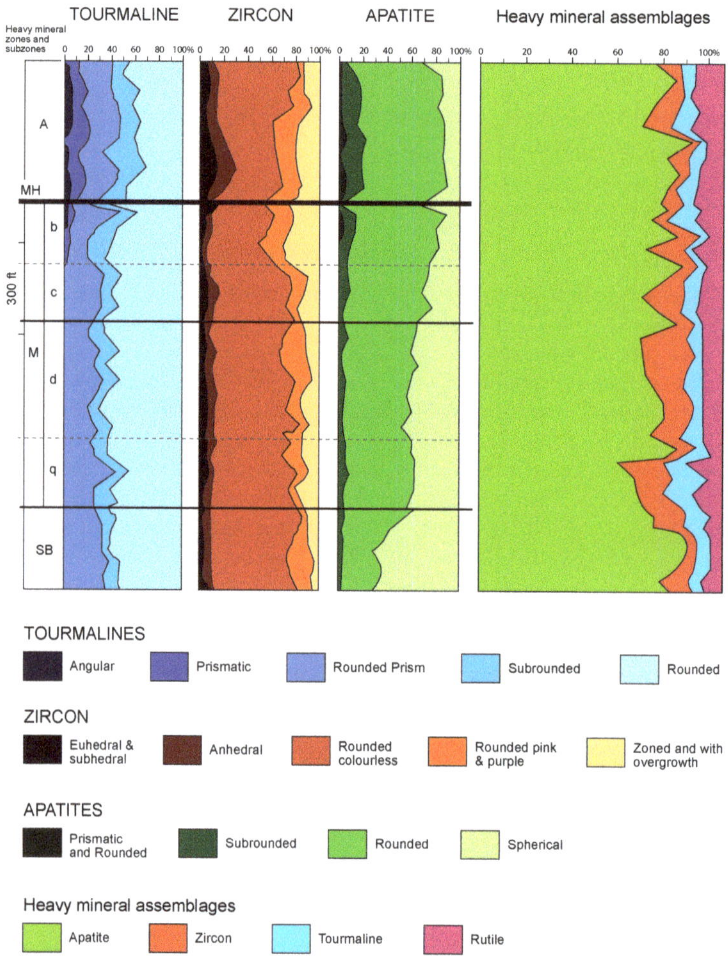

Figure 5. Stratigraphic subdivision of the Triassic Skagerrak Formation of well 22/24a-5Z (see Figure 1) using morphological variations of tourmaline, zircon, and apatite [34]. Note that the conventional heavy mineral data from the same interval show no systematic variation.

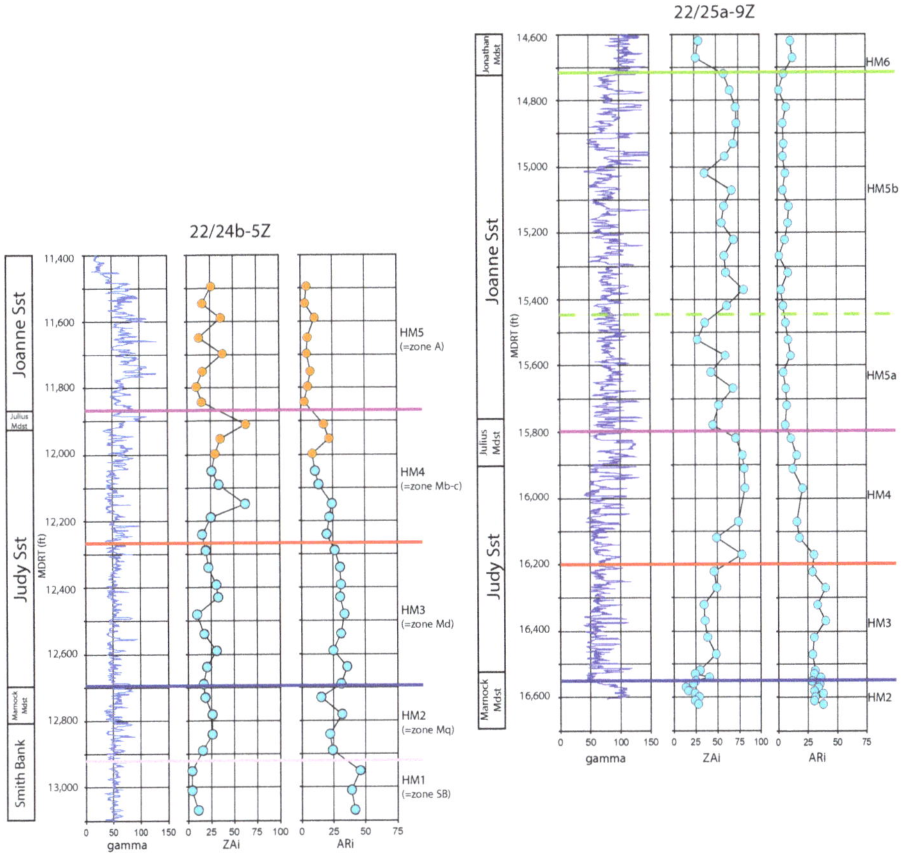

Figure 6. Heavy mineral zonation of the Triassic Skagerrak Formation in Marnock Field well 22/24a-5 and Culzean Field well 22/25a-9Z using apatite roundness index (ARi) and zircon:apatite index (ZAi), after Mouritzen et al. [48]. For 22/24b-5Z, note the correspondence between the Mouritzen et al. zonation [48] and the earlier subdivision by Mange-Rajetzky [34], as shown in Figure 5.

The integrated heavy mineral and biostratigraphic correlation of the Skagerrak Formation in Quadrant 22 demonstrates that the majority of the reservoir succession in the Culzean wells is equivalent to the Joanne Sandstone Member and the overlying Jonathan Mudstone Member (Figure 7). The presence of the underlying Julius Mudstone Member and Judy Sandstone Member is proved by penetrations in the discovery well 22/25a-9Z (which also extends down into the top of the Marnock Mudstone) and in 22/25a-11. The wells from Egret (22/24d-10) and Heron (22/30a-6) yielded the same heavy mineral stratigraphy (Jonathan Mudstone Member, Joanne Sandstone Member and Julius Mudstone Member). By contrast, the successions in wells further to the northwest are largely assigned to the Judy Sandstone Member together with the underlying Marnock Mudstone and Smith Bank Formation. Younger Skagerrak units (Julius Mudstone Member and the lower part of the Joanne Sandstone Member) were identified in only two cases (22/24b-5Z and 22/24b-7).

In previous correlations of the Skagerrak Formation [50,52], the Egret and Heron wells (22/24d-10 and 22/30a-6) are considered to be directly correlative with Marnock and Skua wells (22/24b-5Z and 22/24b-7), with the reservoir succession being correlated with the Judy Sandstone Member. The Mouritzen et al. [48] correlation offers a different interpretation, whereby the successions in Egret and Heron are interpreted as equivalent to the Joanne Sandstone Member, whereas the Marnock and Skua wells contain both Judy and Joanne sandstones. The new correlation therefore indicates significant modification to the understanding of Triassic stratigraphy in Quadrant 22, and in particular that the Joanne Sandstone Member and Jonathan Mudstone Member are more widely distributed than was previously recognised.

6. Correlation Using Garnet Geochemistry: Oseberg Field, Northern North Sea

The potential value of single-grain major element geochemistry of detrital garnet populations for discriminating sandstones supplied by different transport systems, and for identifying sediment source areas, was first demonstrated by Morton [35]. This paper was followed not long afterwards by the first application of garnet geochemistry in hydrocarbon reservoir subdivision and correlation [53]. The oilfield concerned is Oseberg (Figure 1), where hydrocarbons are hosted by the Middle Jurassic Brent Group, comprising (from base to top) the Oseberg, Rannoch, Etive, Ness, and Tarbert formations [54,55]. The Oseberg Formation, which is the main reservoir, represents a local fan delta that prograded from east to west. The Rannoch-Etive, Ness, and Tarbert formations respectively represent the delta front-shoreline, delta plain, and abandonment phase of the Brent delta that migrated northwards across the Oseberg Field. When planning production strategy, it is crucial that the understanding of sandstone distribution is maximised, so that sand body geometry and channel orientation, which control permeability anisotropy, can be evaluated.

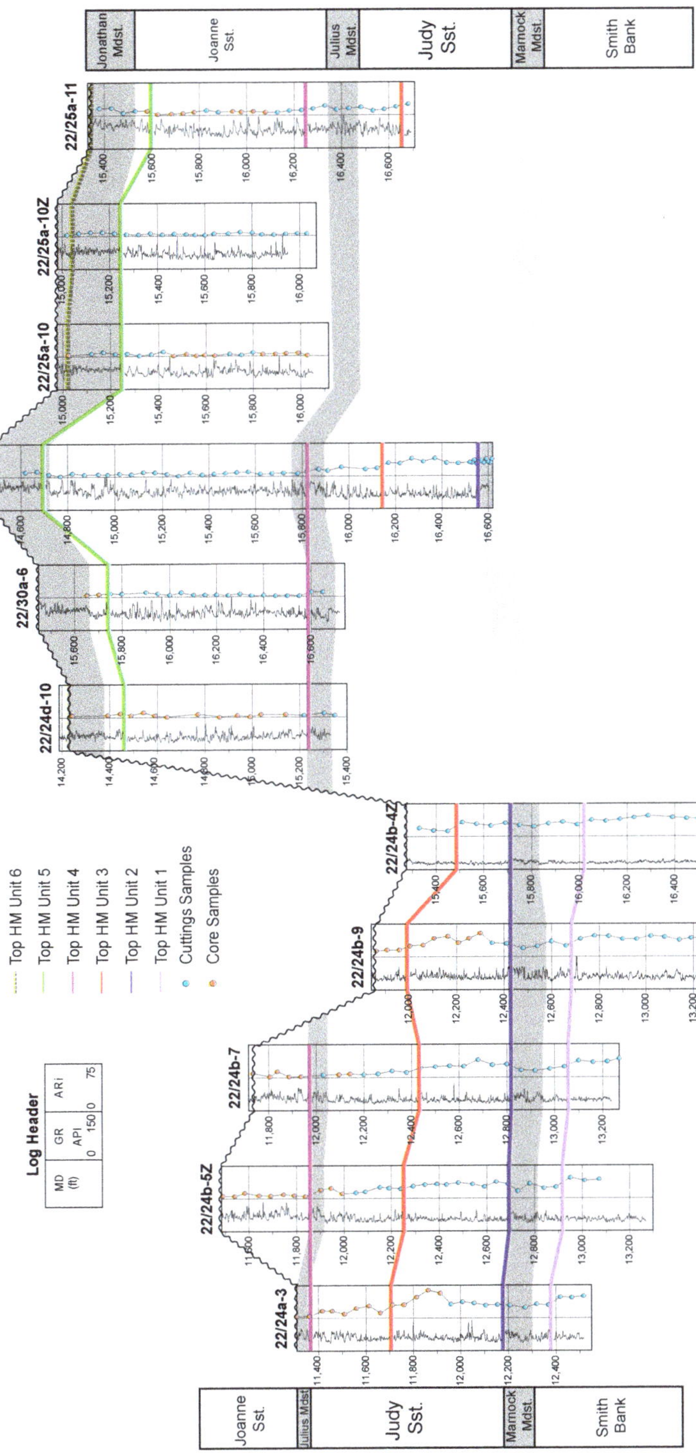

Figure 7. Semi-regional correlation of the Triassic in UK quadrant 22, central North Sea, using variations in apatite morphology [48]. For well locations, see Figure 1.

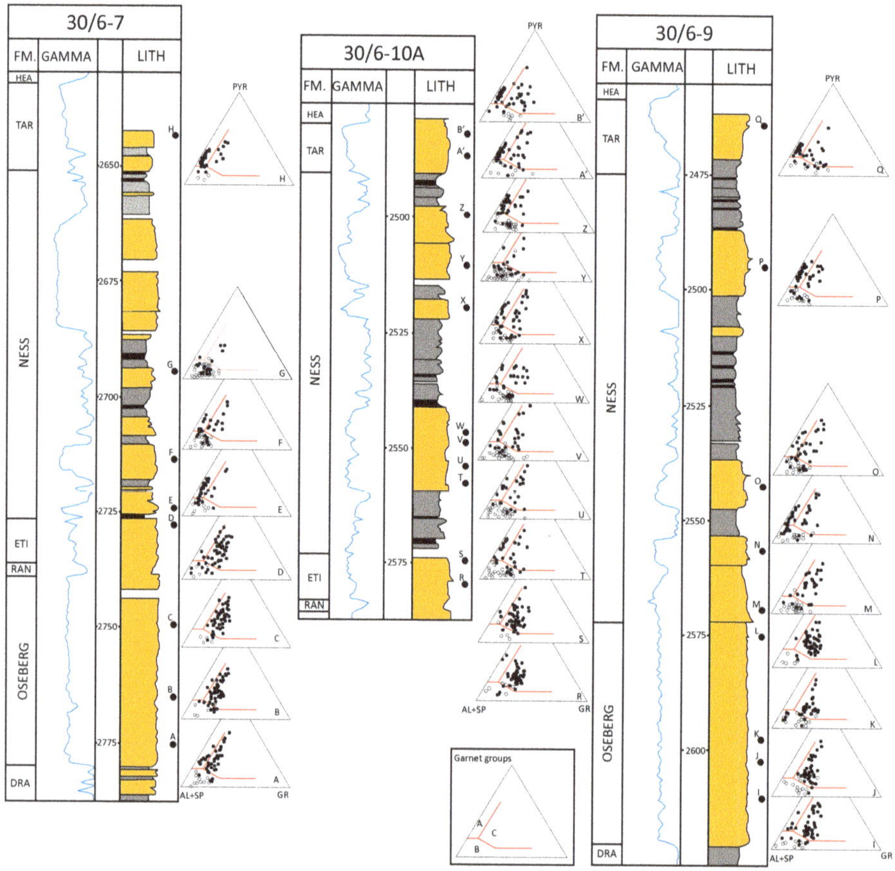

Figure 8. Stratigraphic breakdown and correlation of the Brent Group in Oseberg Fields well 30/6-7, 30/6-9, and 30/6-10A using garnet geochemical signatures. AL = almandine, SP = spessartine, GR = grossular, PYR = pyrope. Garnet fields A, B, and C are as defined by Morton et al. [56] and Mange and Morton [57]. Yellow = sandstone, grey = mudstone, black = coal.

Three wells were used in the garnet stratigraphy study, 30/6-7, 30/6-9, and 30/6-10A [53,58]. Garnet populations in the Brent Group of these Oseberg Field wells show considerable heterogeneity (Figure 8), with representation of varying numbers of Type A (low-Ca, high-Mg), Type B (low-Mg, variable-Ca), and Type C (high-Mg, high-Ca) types as defined by Mange and Morton [57]. Type C garnets dominate throughout the Oseberg, Rannoch, and Etive formations, with the base of the Ness being marked by a major reduction in this garnet type and by a concurrent increase in Types B and A. The Ness Formation has garnet assemblages dominated by Type B with generally subordinate Type A, and the overlying Tarbert Formation has assemblages consistently dominated by Type A with subordinate Types B and C (Figure 8).

This case study therefore demonstrates the existence of laterally-consistent differences in garnet provenance in the Oseberg-Etive, Ness, and Tarbert intervals across the Oseberg Field, showing that the garnet geochemical method is a useful guide to stratigraphy at formation level. Possibly the most important outcome, however, is the identification of a change in garnet geochemistry between samples M and L, within what appears to be a massive sand development in the lower part of the 30/6-9 well (Figure 8). The garnet assemblages in samples I, J, K, and L are dominated by Type C garnets, consistent with a stratigraphic location within the Oseberg Formation. The overlying samples from the sand body (M and N), however, have assemblages dominated by Types B and A, corresponding to a stratigraphic position within the Ness. This observation indicates that the Etive, Ness, and top part of the Oseberg formations in 30/6-9 were eroded by a channel sand at the base of the Ness. The identification of a Ness channel sand in direct contact with the Oseberg fan-delta sands has important implications for production geology, since the two have markedly different geometries and, by association, permeability anisotropies.

A major challenge identified by Johnsen et al. [55] is to map the more isolated fluvial channel sandstones of the Ness Formation, which account for about 25% of the oil in place but only 10% of the reserves. The garnet data offer some important insights into the relationships between the Ness sandstones in the three analysed wells. For example, it is evident that some of the sand bodies are complex, since they have internal variations in garnet provenance characteristics. This is illustrated by samples Y and Z, from 30/6-10A, which have contrasting garnet assemblages, one dominated by Type B and the other by much higher contents of Type A. The garnet data therefore indicate that the samples are from different channel sands and that the analysed sand body is composite, with a sand-on-sand contact. By contrast, other sand bodies have uniform mineralogy: for example, samples T, U, V, and W in 30/6-10A, which were taken at relatively close spacing within a sand-rich interval, show virtually no difference in garnet assemblage compositions. Although intra-well correlations cannot be unequivocally assigned, the garnet data can be used to suggest likely correlations and to rule out other possibilities. For example, the sand body sampled at point G in 30/6-7 could correlate with that at point Y in 30/6-10A, since both garnet assemblages almost exclusively comprise Type B garnets. By contrast, the basal Ness sand body in 30/6-7 (sample E) cannot correlate with the basal Ness in 30/6-10A (samples T, U, V, and W) since the two samples have markedly different proportions of Type A garnets.

7. Correlation Using Rutile Geochemistry and Zircon Geochronology: Magnus Field, Northern North Sea

Deep-water basin floor fan sandstones assigned to the Magnus Sandstone Member (Kimmeridgian-Volgian) and the underlying Ptarmigan Sandstone Member (Oxfordian-Kimmeridgian) form the reservoir for the Magnus oilfield in the northern North Sea (Figures 1 and 9). Since a high-resolution biostratigraphic correlation framework exists for the reservoir succession [59], alternative correlation schemes have not been considered a priority. However, there is no consensus of opinion on the source area for the Magnus sandstones. Fraser et al. [60] state that they were derived from an uplifted part of the Margareta Spur to the north of the Magnus Field, and De'Ath and Schuyleman [61] indicate the sandstones were introduced by a submarine fan system from the NW. The Ptarmigan Sandstone Member is believed to represent the initial products of the submarine fan system that subsequently led to deposition of the Magnus Sandstone Member [61]. Heavy mineral data have been acquired from three wells in and adjacent to the Magnus Field (211/7-1, 211/8-1 and 211/12-1, Figure 1), and in this example we consider the value of these data for reservoir subdivision and correlation.

The heavy mineral suites are low in diversity, typical for North Sea sandstones buried to in excess of 2900 m. Four minerals (apatite, garnet, rutile, and zircon) account for 93.0–99.5% (mean 96.7%) of the assemblages, with anatase, monazite, and tourmaline being present consistently but in minor amounts. Other phases (notably chrome spinel and chloritoid) are found in minor amounts and in only a small number of samples. Provenance-sensitive heavy mineral ratios show comparatively little variation both within and between the two sandstone members (Figure 10).

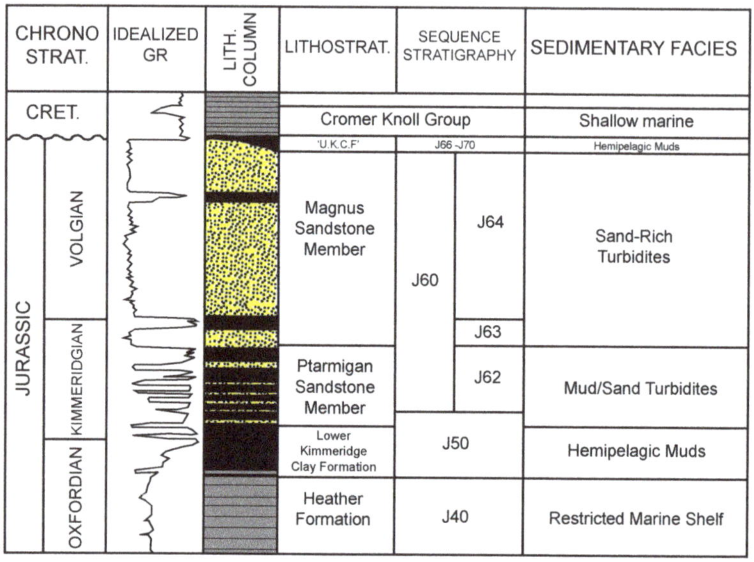

Figure 9. Stratigraphy of the reservoir interval in the Magnus Field, northern North Sea [59,60,62]. Yellow = sandstone, grey = mudstone, black = high-gamma (organic-rich) mudstone.

The only parameter to display significant variations is GZi, but the range shown by this parameter is clearly related to burial depth (Figure 10). Therefore, the observed variations in GZi are attributable to garnet dissolution during diagenesis, and the parameter cannot be regarded as a reliable indicator of provenance characteristics. The evidence for advanced garnet dissolution also precludes the use of garnet geochemistry as a provenance or correlation tool, since increasing garnet dissolution leads to biasing of garnet geochemical populations due to the relative instability of Ca-rich garnets compared with Ca-poor garnets ([10] and references therein). There are only two subtle hints of a difference in provenance within the conventional heavy mineral data set. First, the Ptarmigan Sandstone Member has very slightly higher chrome spinel:zircon index (CZi) values (0.5–4.3%) compared with the Magnus Sandstone Member (0.0–0.5%). Second, chloritoid is present in very small proportions (0.5%) in two of the four Ptarmigan Sandstone samples, but was not found in any of the 15 Magnus Sandstone samples. Using such scanty data to propose a difference in provenance and to subdivide the reservoir succession would be hazardous.

Despite the lack of clear evidence for any differences in provenance in the conventional data set, there are major differences in both rutile trace element geochemistry and detrital zircon geochronology. Rutile data (Figure 11) show that metapelites were the main rutile source for the Magnus Sandstone Member, with 84–92% of rutiles having metapelitic signatures. Furthermore, the Magnus Sandstone contains abundant granulite-facies rutiles, which form 39–51% of the populations. By contrast, metamafic rutiles are significantly more abundant in the Ptarmigan Sandstone, forming 45% of the assemblages, and granulite-facies grains are much less abundant (10–13%). Differentiation of the two sandstone members is reinforced by the zircon age data, which show that the Magnus and Ptarmigan units have markedly different spectra (Figure 12). The three analysed Magnus samples have bimodal populations, with a well-defined Early Paleozoic group (59–62% of zircons with <10% discordance) and a broader Neoarchaean group (27–35%), and only minor representation of the intervening Proterozoic (3–16%). The Ptarmigan population, by contrast, almost entirely comprises Proterozoic zircons (95%), with hardly any Early Paleozoic or Archaean zircons (4% and 1%, respectively). This example shows the potential for using rutile trace element and zircon geochronology data for reservoir subdivision and correlation in the absence of any clear evidence from basic heavy mineral data and where application of garnet geochemistry is not possible due to burial diagenesis.

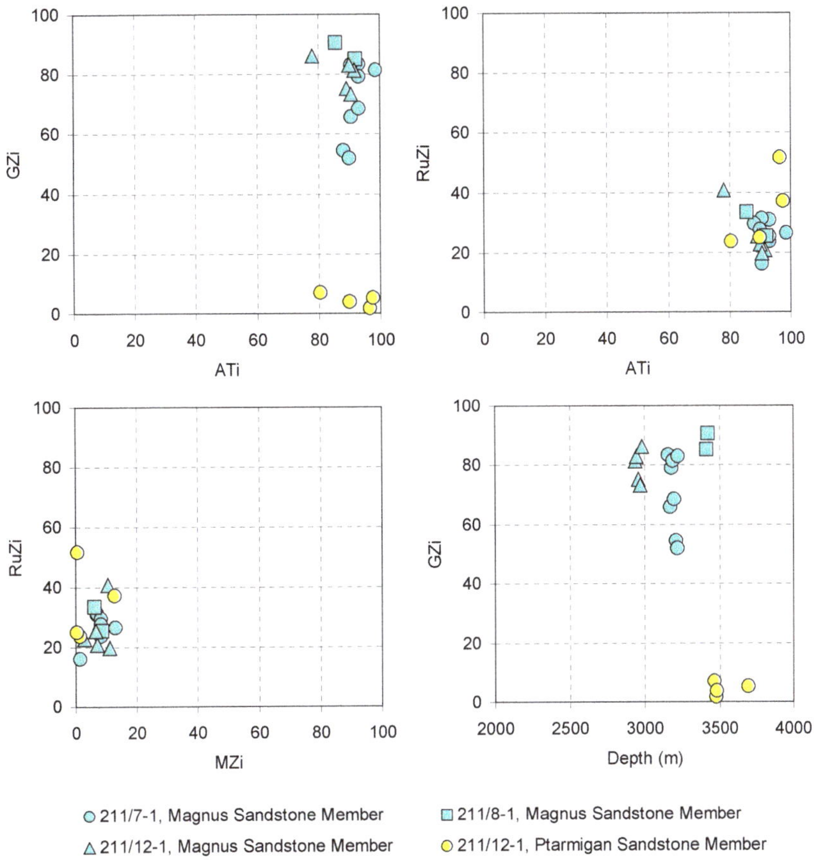

Figure 10. Provenance-sensitive heavy mineral ratios for the Magnus Sandstone and Ptarmigan Sandstone members in the Magnus Field (see Figure 1). Note that there is no clear discrimination between the two sandstones in terms of ATi, RuZi, and MZi values. GZi is much lower in the Ptarmigan sandstone but this unit is more deeply buried and the lower GZi is likely to be a function of greater diagenetic modification.

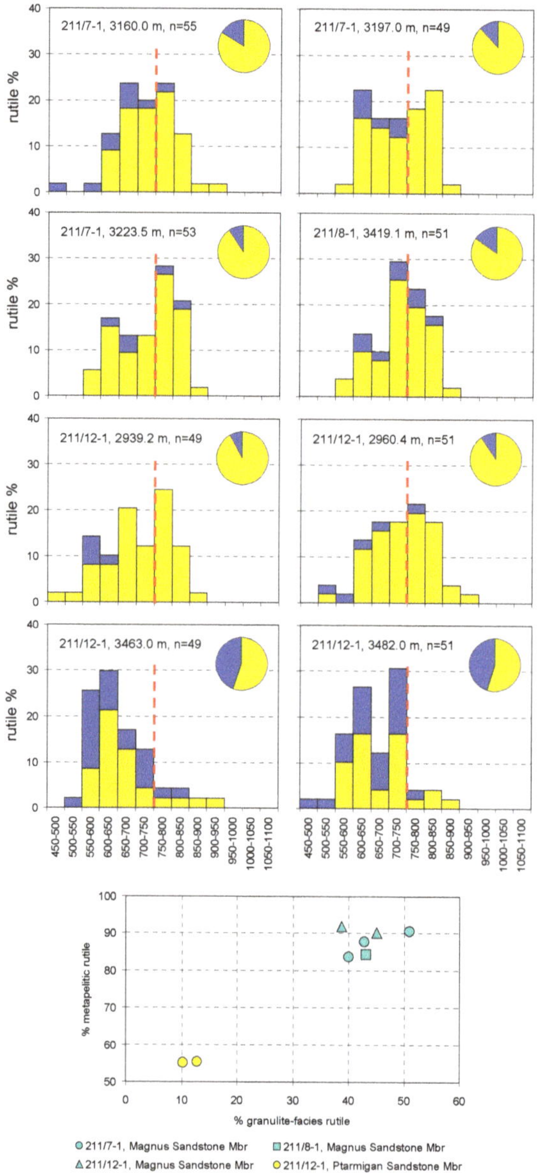

Figure 11. Discrimination of Magnus and Ptarmigan sandstones using rutile trace element geochemistry. Magnus rutile populations are dominated by metapelitic types and include a high proportion that formed in high-grade granulite facies conditions, whereas Ptarmigan sandstones have much higher proportions of metamafic types and lack evidence for high-grade metamorphism. Categorisation of metamafic and metapelitic rutiles uses Cr and Nb contents as defined by Meinhold et al. [63]. Crystallisation temperatures were determined using Zr contents following Watson et al. [64].

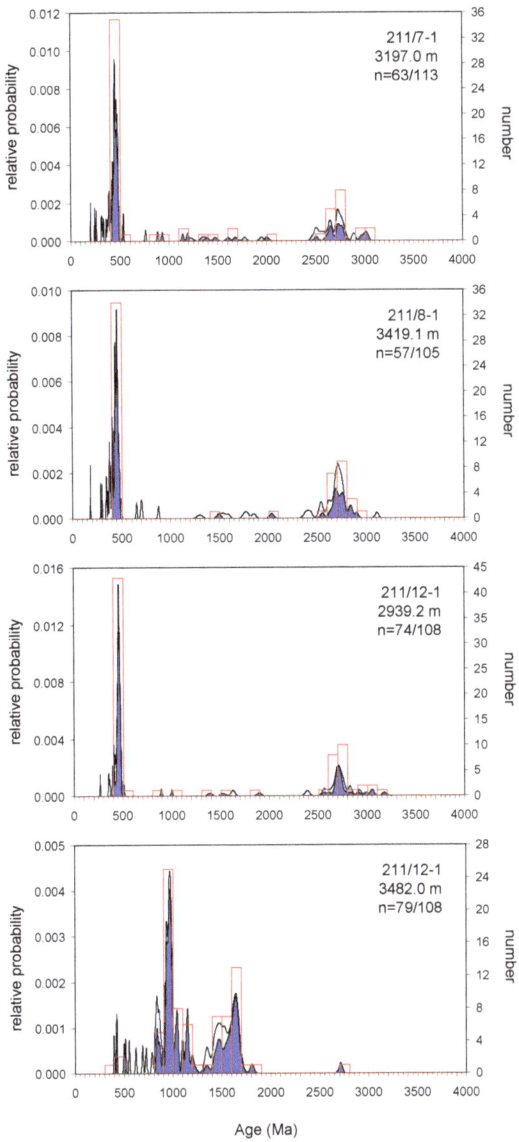

Figure 12. Probability-density plots of detrital zircon ages in three Magnus Sandstone Member samples (211/7-1, 3197.0 m; 211/8-1, 3418.1 m; 211/12-1, 2938.2 m) and one Ptarmigan Sandstone Member sample (211/12-1, 3482.0 m), confirming the rutile geochemical evidence for a difference in provenance. 'n' is number of zircons with <10% discordance in the total zircon population.

The rutile and zircon data provide some constraints on the provenance of these enigmatic sandstones. The source of the Magnus Sandstone Member can be reconstructed as comprising high-grade metasedimentary rocks at upper amphibolite and granulite facies, as indicated by the garnet-rich heavy mineral assemblages and rutile geochemical data. The zircon data indicate the source area was heavily affected by the Early Paleozoic Caledonian orogenic cycle. The Early Paleozoic zircons fall in the 410–490 Ma range, corresponding to the age of late tectonic granites, the alkali suite and late granites in northern Scotland [65], and to the Grampian and Scandian orogenic events [66]. The older, predominantly Archaean, ages are interpreted as reflecting the detrital zircon populations in the metasedimentary source rocks. Similar bimodal zircon populations, in association with heavy mineral assemblages indicating derivation from high-grade metasedimentary rocks, have been found elsewhere in the northern North Sea, such as the Late Triassic-Early Jurassic Statfjord Formation in the Brent Field [41]. The Early Carboniferous Upper Clair Group in the Clair Field (Figure 1), on the western side of the Shetland Platform, also has a bimodal zircon population with Early Paleozoic and Neoarchaean groups in association with garnet- and staurolite-rich heavy mineral suites evidently of metasedimentary provenance, although in this case the Early Paleozoic group is less common than the Neoarchaean [67]. The occurrence of metasedimentary-sourced sandstones with comparable zircon age spectra in sandstones located both to the east and west of the Shetland Platform suggests the source lay on the intervening platform area, although derivation from the north, as suggested by Fraser et al. [60], cannot be ruled out. The contrast in rutile data and zircon age spectra in the Ptarmigan sandstone compared with the Magnus sandstone indicates the two units have a different provenance. The Ptarmigan zircon data can be matched with a Shetland Isles provenance, in particular with the Dalradian metasediments that outcrop on the islands. The largest zircon peak in the Ptarmigan spectrum (920–990 Ma) is broadly equivalent to the timing of metamorphism of the Dalradian at 938 $\pm$ 8 Ma [68], and the second largest group (c. 1450–1700 Ma) is also common as a detrital component in the Shetland Dalradian [69]. The presence of chloritoid and chrome spinel is also suggestive of a source on the Shetland Isles, since both minerals are characteristic of this area. Chrome spinel is a major component of the Unst ophiolite [70], and chloritoid is abundant in modern stream sediments from this area [58].

8. Application of Heavy Mineral Stratigraphy at the Well Site: Clair Field, West of Shetland

One of the most exciting applications of heavy mineral stratigraphy for the hydrocarbons industry sector is its 'real-time' use for geosteering of high angle wells. This application is a logical extension of the technique given the established correlative value of heavy mineral analysis in post-well studies. High-angle (horizontal or near-horizontal) wells are commonly used to exploit hydrocarbon reserves, since a large number of high-angle wells can be drilled from a single location. Successful exploitation generally depends on ensuring the well bore remains in the productive parts of the reservoir (the so-called 'pay zone'). Geosteering involves the use of geological parameters to alter the trajectory of wells during drilling to ensure the well bore remains within the pay zone, and requires continual monitoring of the geology during drilling operations. This is normally achieved using a combination of cuttings description by the wellsite geologist, geophysical logging while drilling (LWD), and high-resolution biostratigraphy. However, in reservoirs with insufficient biostratigraphic control, heavy mineral analysis can play an important role. Heavy mineral stratigraphy has been carried out at the well site in a number of fields on the UK continental shelf, such as Ross and Hannay [71].

However, the most important application to date has been on the Clair Field, west of Shetland (Figure 1), where the technique was first pioneered in the 1990's [72]. The reservoir succession in the Clair Field consists of fractured Devonian to Carboniferous fluvio-lacustrine clastic sediments, and is largely barren of microfossils and palynomorphs owing to the adverse depositional environment. Initial heavy mineral work demonstrated the important potential role for heavy minerals to establish a stratigraphic framework for the Clair Field [37].

As discussed earlier in this paper, the basic prerequisite for all successful heavy mineral correlation schemes is that there were variations in provenance and transport history during deposition of the reservoir succession. For application at the well site, it is also crucial that the parameters used in the scheme can be determined from ditch cuttings, since production wells are rarely cored. Furthermore, the variations must be laterally extensive across the area under development, and finally, it must be possible to acquire the data in a timely fashion in order to be able to influence drilling decisions.

The first step in the process involved creation of a correlation scheme based on meaningful geological criteria. The fully cored well 206/8-8 (Figure 13) was used for this purpose. Variations in a number of key parameters enable subdivision of the succession into two major lithostratigraphic units, the Lower Clair Group and the Upper Clair Group [37,73]. The Lower Clair Group has been further subdivided into Units I–VI, with the Upper Clair Group comprising Units VII–X. Some of these units have been further subdivided on the basis of internal trends in one or more of the parameters.

The criteria used in the Clair Field correlation scheme are the apatite:tourmaline index (ATi), apatite roundness index (ARi), garnet:zircon index (GZi), rutile:zircon index (RuZi), and staurolite:zircon index (SZi), as defined by Morton and Hallsworth [33], Morton et al. [72], and Morton et al. [73]. In addition, the abundance of the unstable minerals epidote and titanite, and the associated unstables:tourmaline index (UTi), have proved useful locally in parts of the field. Variations in these parameters are due to a combination of provenance, transport history, and diagenetic factors. Provenance variations result from the interplay between an axial sediment transport system and sediment of local derivation from an adjacent basement high, together with longer-term variations in sediment composition supplied by the axial system and a fundamental shift in provenance at the Lower Clair Group–Upper Clair Group boundary [37,73,74]. The highest ARi values are found in Units III and V (Figure 13), which display the greatest aeolian influences in the succession. Unit III shows evidence for wind modification in the form of sand flats [75] and Unit V contains wind ripple laminated sandstones [37]. Variations in ARi therefore reflect the extent of aeolian transport. Hence, variations in ATi, ARi, GZi, RuZi, and SZi all have geological significance. Variations in abundance of the unstable minerals epidote and titanite (and UTi) are partly controlled by stratigraphy, with their presence in Unit VI and absence in Unit V (Figure 13) being related to the difference in reservoir quality between the two [73,74]. However, differences in burial depth means that variations in abundance of unstable minerals and UTi are not consistent across the field, and these parameters therefore have only local correlative value.

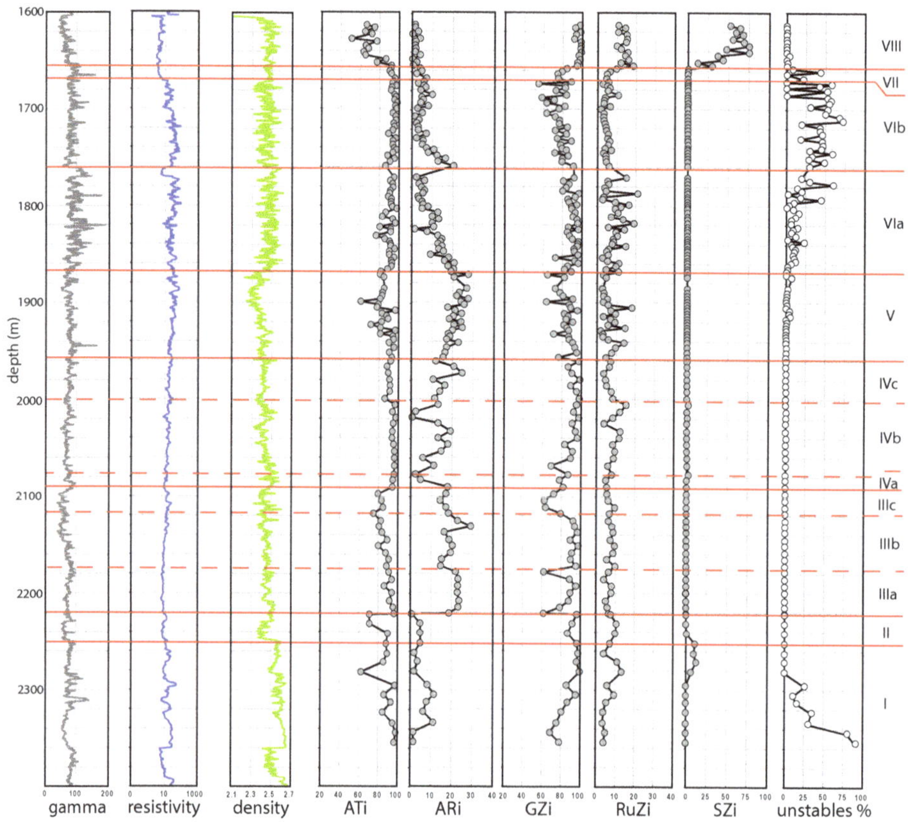

Figure 13. Heavy mineral stratigraphic framework for the Devonian-Carboniferous Clair Group reservoir succession in the Clair field, west of Shetland (Figure 1), illustrated using the fully cored well 206/8-8. Stratigraphic units (I, II, etc.) were originally defined by Allen and Mange-Rajetzky [37] and subsequently revised by Morton et al. [73] and Morton and Milne [74].

Having established the existence of a high-resolution heavy mineral stratigraphic breakdown of the 206/8-8 succession, the next step was to demonstrate that the same events can be seen in an uncored well where only ditch cuttings samples are available. This was successfully tested using well 206/8-9Y, which is located immediately adjacent to the cored well 206/8-8 [73]. All the events seen in the cored well proved to be readily identifiable in the uncored well [73], with a slight blurring of some of the boundaries owing to the inevitable contamination associated with the drilling mud system. With this successful outcome, further work established that the same events can be seen in other parts of the field, indicating that the stratigraphic framework can be applied across the entire area [73].

The final step in the feasibility study was to undertake a pilot study at the well site in order to determine whether it was possible to acquire data quickly enough for the information to be useful in decision making. During the pilot studies on wells 206/8-10Z and 206/8-11Z, heavy mineral data were acquired in less than 2 h from receipt of sample, and owing to the relatively slow penetration rates associated with Clair drilling, heavy mineral information was generally acquired ahead of the LWD [72,74]. In consequence, heavy mineral stratigraphic analysis has been undertaken on a real-time basis at the well site for virtually all appraisal and development wells since production drilling first started in 2005 [74].

Heavy mineral data have been used in the decision-making process in a variety of situations, including picking of casing and coring points, whether to maintain or alter well trajectory, and when to terminate drilling. Some of the formation tops have only subtle geophysical log expressions, and since heavy mineral data establish trends and predict formation changes before they are encountered, they are critical in aiding geosteering decisions. Heavy mineral analysis has also been used to monitor stratigraphy and to pick formation tops when logging tools have failed, allowing drilling to continue and thereby avoiding the need to change the bottom-hole assembly. In this paper, we review the study of well 206/8-A16, originally discussed in the paper by Morton and Milne [74] on the application of heavy mineral stratigraphy to Clair Field appraisal and development wells drilled since 2005.

Well 206/8-A16 was drilled into a fault terrace close to a major fault that separates two large-scale structural blocks (Core and Ridge areas) of the Clair Field [74]. Seismic resolution is poor in the area because of extensive faulting, leading to considerable uncertainty concerning the predicted pre-drill stratigraphy. Another pre-drill uncertainty was the possibility that the extensive faulting might be associated with widespread mineralisation, and this might affect log responses. For a number of reasons, therefore, wellsite heavy mineral analysis was considered to be critical in understanding the stratigraphy while drilling, as well as helping to pick the casing point in the uppermost part of Unit VI.

The heavy mineral stratigraphy of the well and the interpreted well track are shown in Figures 14 and 15, respectively. The stratigraphy encountered in the shallower parts of the succession was essentially as predicted. A basal Cretaceous sandstone with relatively high GZi, high RuZi, and low SZi overlies the Upper Clair Group Unit IX identified on the basis of very high GZi and SZi. The subsequent increase in unstable mineral abundances and UTi values marks the top of Upper Clair Group Unit VIII. Entry into Unit VII was diagnosed by decreasing GZi, SZi, and RuZi and an increase in ATi. There was an LWD tool failure during drilling in Unit VII, but because heavy mineral stratigraphic analysis was available at well site, there was no need to make a three-day trip to change the bottom-hole assembly, and the $9^5/_8$ inch casing was successfully set at the top of Lower Clair Group Unit VIb.

Almost immediately after drilling commenced on the $8\frac{1}{2}$ inch hole section, the predicted complexity in Lower Clair Group stratigraphy became apparent. The first sample analysed had Unit VIb characteristics as expected, but this was followed immediately by a sharp increase in GZi suggesting entry into Unit VIa. The anomalously short penetration of Unit VIb was interpreted as the result of faulting, a view supported by the influx of abundant fracture-fill mineral phases (notably pyrite). The abundance of fracture-fill minerals persisted for a considerable distance along hole, indicating that the well track was following an extensively mineralised fault corridor. The mineralisation led to anomalous geophysical log responses, notably much higher resistivity than normal (compare Figure 14 with the typical Clair Field log responses in Figure 13). This caused serious complications in interpretation of the LWD data, thereby increasing the reliance on heavy minerals for diagnosing stratigraphy in the well. Anomalous log responses persisted through Unit VIa, Unit V and the relatively thin Unit IV, but the entry into Unit III was associated with a return to more typical log responses and a reduction in mineralisation (Figure 14), suggesting the well exited the strongly faulted zone at this point.

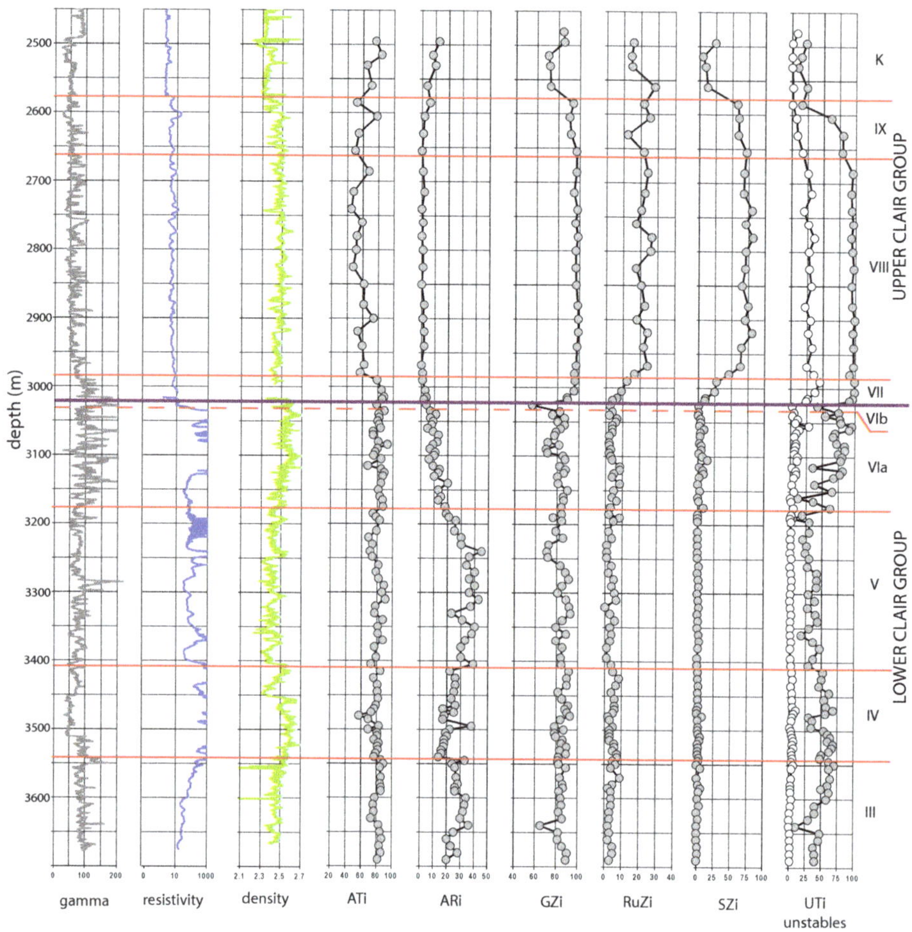

Figure 14. Variations in heavy mineral parameters acquired at the well site during drilling of the high-angle well 206/8-A16, Clair Field, west of Shetland, together with the interpreted stratigraphy. Note the highly anomalous resistivity log responses in units IV, V, and VIa, caused by extensive mineralisation. Well location is shown in Figure 1 and the geological interpretation along the well track is shown in Figure 15.

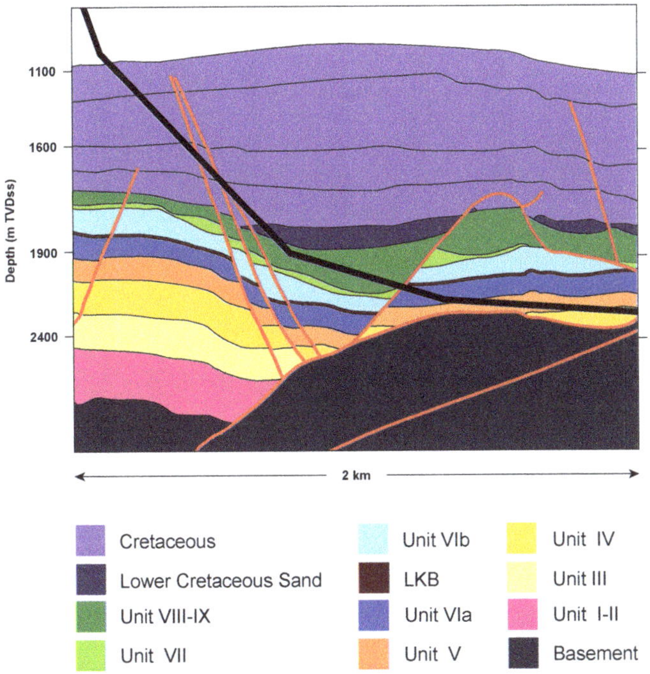

Figure 15. Interpreted schematic geological section along the path of well 206/8-A16 illustrating the geology encountered down to Unit IV, adapted from Morton and Milne [74]. Continued drilling led to penetration of Unit III before TD (see Figure 14).

9. Concluding Remarks

Correlation of hydrocarbon reservoir sandstones is one of the most important practical applications for heavy mineral analysis. Successful correlation schemes depend on recognition of changes in sediment provenance and transport history in the interval under investigation. Heavy mineral assemblages are uniquely placed to identify such changes since they are highly sensitive indicators of provenance, but care must be taken to ensure that heavy mineral variations are not related to differences in hydrodynamic conditions at the time of deposition, or to deep burial diagenesis. A wide variety of parameters are available to identify changes in provenance and to create correlation frameworks, including provenance-sensitive heavy mineral ratio data, mineral morphologies, major element mineral chemistry, trace element mineral chemistry, and single-grain dating, notably of zircon. This paper illustrates the application of all these techniques in the North Sea and adjacent areas, where the majority of such work has been undertaken. Nevertheless, the potential for applications elsewhere is obvious, as demonstrated by work on such disparate areas as the Middle East [76,77], Myanmar [78], Papua New Guinea [79], and Trinidad [80,81].

Although the technique is most widely applied in reservoirs with poor biostratigraphic control, it has important applications in all clastic reservoirs because it provides information on the actual reservoir, rather than the bounding mudstones: in this sense, it effectively correlates reservoir volumes [11]. Furthermore, although not the primary aim of heavy mineral correlation studies, continued applications within sedimentary basins generate a large data base that can be used to constrain sediment provenance, transport, and dispersal patterns and entry points [11].

Author Contributions: Writing-original draft preparation by A.M., revision by P.M., and editing by A.M. and P.M. Figures—A.M. and P.M.

Funding: This review article received no external funding.

Acknowledgments: We are grateful to GdF Suez E&P Ltd. for allowing release of the rutile and zircon data information discussed herein, and to the reviewers for their helpful comments on a previous version of this paper.

Conflicts of Interest: The authors declare no conflicts of interest.

References

1. Jones, R.W.; Simmons, M.D. (Eds.) *Biostratigraphy in Production and Development Geology*; Geological Society, Special Publication: London, UK, 1999; p. 152.
2. Dunay, R.E.; Hailwood, E.A. (Eds.) *Non-Biostratigraphical Methods of Dating and Correlation*; Geological Society, Special Publication: London, UK, 1995; p. 89.
3. Ratcliffe, K.T.; Zaitlin, B.A. (Eds.) *Application of Modern Stratigraphic Techniques: Theory and Case Histories*; SEPM Special Publication: Broken Arrow, OK, USA, 2010; p. 94.
4. Pearce, T.J.; Besly, B.M.; Wray, D.S.; Wright, D.K. Chemostratigraphy: A method to improve interwell correlation in barren sequences—A case study using onshore Duckmantian/Stephanian sequences (West Midlands, U.K.). *Sediment. Geol.* **1999**, *124*, 197–220. [CrossRef]
5. Ratcliffe, K.T.; Wilson, A.; Payenberg, T.; Rittersbacher, A.; Hildred, G.V.; Flint, S.S. Ground truthing chemostratigraphic correlations in fluvial systems. *Bull. Am. Assoc. Pet. Geol.* **2015**, *99*, 155–180.
6. Mearns, E.W.; Knarud, R.; Raestad, N.; Stanley, K.O.; Stockbridge, C.P. Samarium-neodymium isotope stratigraphy of the Lunde and Statfjord Formations of Snorre oil field, northern North Sea. *J. Geol. Soc.* **1989**, *146*, 217–228. [CrossRef]
7. Dalland, A.; Mearns, E.W.; McBride, J.J. The application of samarium-neodymium (Sm-Nd) provenance ages to correlation of biostratigraphically barren strata: A case study of the Statfjord Formation in the Gullfaks oilfield, Norwegian North Sea. In *Non-Biostratigraphical Methods of Dating and Correlation*; Dunay, R.E., Hailwood, E.A., Eds.; Geological Society, Special Publications: London, UK, 1995; Volume 89, pp. 201–222.
8. Jeans, C.V. Clay mineral stratigraphy in Palaeozoic and Mesozoic red bed facies, onshore and offshore UK. In *Non-Biostratigraphical Methods of Dating and Correlation*; Dunay, R.E., Hailwood, E.A., Eds.; Geological Society of London Special Publications: London, UK, 1995; Volume 89, pp. 31–55.
9. Pay, M.D.; Astin, T.R.; Parker, A. Clay mineral distribution in the Devonian-Carboniferous sandstones of the Clair Field, west of Shetland, and its significance for reservoir quality. *Clay Miner.* **2000**, *35*, 151–162. [CrossRef]
10. Morton, A.C. Value of heavy minerals in sediments and sedimentary rocks for provenance, transport history and stratigraphic correlation. In *Quantitative Mineralogy and Microanalysis of Sediments and Sedimentary Rocks*; Mineralogical Association of Canada Short Course Series; Sylvester, P., Ed.; Mineralogical Association of Canada: Québec, QC, Canada, 2012; Volume 42, pp. 133–165.
11. Hurst, A.; Morton, A. Provenance models: The role of sandstone mineral-chemical stratigraphy. In *Sediment Provenance Studies in Hydrocarbon Exploration and Production*; Scott, R.A., Smyth, H., Morton, A., Richardson, N., Eds.; Geological Society of London, Special Publications: London, UK, 2014; Volume 386, pp. 7–26.
12. Garzanti, E.; Andò, S.; Vezzoli, G.; Lustrino, M.; Boni, M.; Vermeesch, P. Petrology of the Namib Sand Sea: Long-distance transport and compositional variability in the wind-displaced Orange Delta. *Earth-Sci. Rev.* **2012**, *112*, 173–189. [CrossRef]
13. Artini, E. Intorno alla composizione mineralogica delle sabbie di alcuni fiumi del Veneto, con applicazione ai terreni di transporto. *Riv. Miner. Cristal. Ital.* **1898**, *19*, 33–94. (In Italian)
14. Thomas, H.H. A contribution to the petrography of the New Red Sandstone in the west of England. *Q. J. Geol. Soc. Lond.* **1909**, *65*, 229–244. [CrossRef]
15. Gilligan, A. The petrography of the Millstone Grit of Yorkshire. *Q. J. Geol. Soc. Lond.* **1920**, *75*, 251–293. [CrossRef]
16. Reed, R.D. Role of heavy minerals in the Coalinga Tertiary formations. *Econ. Geol.* **1924**, *19*, 730–749. [CrossRef]
17. Reed, R.D.; Bailey, J.P. Subsurface correlation by means of heavy minerals. *Bull. Am. Assoc. Pet. Geol.* **1927**, *11*, 359–368.

18. Hedberg, H.D. Some aspects of sedimentary petrography in relation to stratigraphy in the Bolivar Coast fields of the Maracaibo Basin, Venezuela. *J. Paleontol.* **1928**, *2*, 32–42.
19. Edson, F.C. Heavy minerals as a guide in stratigraphic studies. *Am. Mineral.* **1932**, *17*, 429–436.
20. Bornhauser, M. Heavy mineral associations in Quaternary and late Tertiary sediments of the Gulf Coast of Louisiana and Texas. *J. Sediment. Petrol.* **1940**, *10*, 125–135. [CrossRef]
21. Cogen, W.M. Heavy-mineral zones of Louisiana and Texas Gulf coast sediments. *Bull. Am. Assoc. Pet. Geol.* **1940**, *24*, 2069–2101.
22. Edelman, C.H.; Doeglas, D.J. Reliktstructuren detritischer pyroxene und amphibole. *Tschermaks Mineral. Petrogr. Mitteil.* **1932**, *42*, 482–490. (In German)
23. Edelman, C.H.; Doeglas, D.J. Über umwandlungserscheinungen an detritischem staurolith und anderen mineralien. *Tschermaks Mineral. Petrogr. Mitteil.* **1934**, *44*, 225–234. (In German) [CrossRef]
24. Bramlette, M.N. The stability of minerals in sandstone. *J. Sediment. Petrol.* **1941**, *11*, 32–36. [CrossRef]
25. Pettijohn, F.J. Persistence of heavy minerals and geologic age. *J. Geol.* **1941**, *49*, 610–625. [CrossRef]
26. Smithson, F. The alteration of detrital minerals in the Mesozoic rocks of Yorkshire. *Geol. Mag.* **1941**, *78*, 97–112. [CrossRef]
27. Boswell, P.G.H. The stability of minerals in sedimentary rocks (Presidential Address). *Q. J. Geol. Soc. Lond.* **1942**, *97*, lvi–lxxv.
28. Feo-Codecido, G. Heavy mineral techniques and their application to Venezuelan stratigraphy. *Bull. Am. Assoc. Pet. Geol.* **1956**, *40*, 984–1000.
29. Morton, A.C. Stability of detrital heavy minerals in Tertiary sandstones of the North Sea Basin. *Clay Miner.* **1984**, *19*, 287–308. [CrossRef]
30. Milliken, K.L. Loss of provenance information through subsurface diagenesis in Plio-Pleistocene sediments, northern Gulf of Mexico. *J. Sediment. Petrol.* **1988**, *58*, 992–1002.
31. Milliken, K.L.; Mack, L.E. Subsurface dissolution of heavy minerals, Frio Formation sandstones of the ancestral Rio Grande Province, South Texas. *Sediment. Geol.* **1990**, *68*, 187–199. [CrossRef]
32. Mange, M.A.; Maurer, H.F.W. *Heavy Minerals in Colour*; Chapman and Hall: London, UK, 1992; p. 147.
33. Morton, A.C.; Hallsworth, C.R. Identifying provenance-specific features of detrital heavy mineral assemblages in sandstones. *Sediment. Geol.* **1994**, *90*, 241–256. [CrossRef]
34. Mange-Rajetzky, M.A. Subdivision and correlation of monotonous sandstone sequences using high resolution heavy mineral analysis, a case study: The Triassic of the Central Graben. In *Non-Biostratigraphical Methods of Dating and Correlation*; Dunay, R.E., Hailwood, E.A., Eds.; Geological Society of London Special Publications: London, UK, 1995; Volume 89, pp. 23–30.
35. Morton, A.C. A new approach to provenance studies: Electron microprobe analysis of detrital garnets from Middle Jurassic sandstones of the northern North Sea. *Sedimentology* **1985**, *32*, 553–566. [CrossRef]
36. Henry, D.J.; Guidotti, C.V. Tourmaline as a petrogenetic indicator mineral: An example from the staurolite-grade metapelites of NW Maine. *Am. Mineral.* **1985**, *70*, 1–15.
37. Allen, P.A.; Mange-Rajetzky, M.A. Sedimentary evolution of the Devonian-Carboniferous Clair Field, offshore northwestern UK: Impact of changing provenance. *Mar. Pet. Geol.* **1992**, *9*, 29–52. [CrossRef]
38. Mange, M.; Turner, P.; Ince, D.; Pugh, J.; Wright, D. A new perspective on the zonation and correlation of barren strata: An integrated heavy mineral and palaeomagnetic study of the Sherwood Sandstone Group, East Irish Sea Basin and surrounding areas. *J. Pet. Geol.* **1999**, *22*, 325–348. [CrossRef]
39. Morton, A.C.; Yaxley, G. Detrital apatite geochemistry and its application in provenance studies. In *Sediment Provenance and Petrogenesis: Perspectives from Petrography and Geochemistry*; Special Paper; Arribas, J., Critelli, S., Johnsson, M.J., Eds.; Geological Society of America: Boulder, CO, USA, 2007; Volume 420, pp. 319–344.
40. Zack, T.; von Eynatten, H.; Kronz, A. Rutile geochemistry and its potential use in quantitative provenance studies. *Sediment. Geol.* **2004**, *171*, 37–58. [CrossRef]
41. Morton, A.C.; Claoué-Long, J.; Berge, C. Factors influencing heavy mineral suites in the Statfjord Formation, Brent Field, North Sea: Constraints provided by SHRIMP U-Pb dating of detrital zircons. *J. Geol. Soc. Lond.* **1996**, *153*, 911–929.
42. Blum, M.D.; Milliken, K.T.; Pecha, M.A.; Snedden, J.W.; Frederick, B.C.; Galloway, W.E. Detrital-zircon records of Cenomanian, Paleocene, and Oligocene Gulf of Mexico drainage integration and sediment routing: Implications for scales of basin-floor fans. *Geosphere* **2017**, *13*, 1–37. [CrossRef]

43. Fisher, M.J.; Mudge, D.C. Triassic. In *Petroleum Geology of the North Sea: Basic Concepts and Recent Advances*; Glennie, K.W., Ed.; Blackwell Science Ltd.: Oxford, UK, 1998; pp. 212–244.
44. Herries, R.; Poddubiuk, R.; Wilcockson, P. Solan, Strathmore and the back basin play, west of Shetland. In *Petroleum Geology of Northwest Europe: Proceedings of the 5th Conference*; Fleet, A.J., Boldy, S.A.R., Eds.; Geological Society: London, UK, 1999; pp. 693–712.
45. Swiecicki, T.; Wilcockson, P.; Canham, A.; Whelan, G.; Homann, H. Dating, correlation and stratigraphy of the Triassic sediments in the West Shetlands area. In *Permian and Triassic Rifting in Northwest Europe*; Boldy, S.A.R., Ed.; Geological Society, Special Publications: London, UK, 1995; Volume 91, pp. 57–85.
46. Goldsmith, P.J.; Hudson, G.; Van Veen, P. Triassic. In *The Millenium Atlas: Petroleum Geology of the Central and Northern North Sea*; Evans, D., Graham, C., Armour, A., Bathurst, P., Eds.; Geological Society: London, UK, 2003; pp. 105–127.
47. Goldsmith, P.J.; Rich, B.; Standring, J. Triassic correlation and stratigraphy in the southern Central Graben, UK North Sea. In *Permian and Triassic Rifting in Northwest Europe*; Boldy, S.A.R., Ed.; Geological Society, Special Publications: London, UK, 1995; Volume 91, pp. 123–143.
48. Mouritzen, C.; Farris, M.A.; Morton, A.; Matthews, S. Integrated Triassic stratigraphy of the greater Culzean area, UK central North Sea. *Pet. Geosci.* **2017**, *24*, 197–207. [CrossRef]
49. Morton, A.C.; Herries, R.; Fanning, C.M. Correlation of Triassic sandstones in the Strathmore Field, west of Shetland, using heavy mineral provenance signatures. In *Heavy Minerals in Use, Developments in Sedimentology*; Mange, M., Wright, D.T., Eds.; Elsevier: Amsterdam, The Netherlands, 2007; Volume 58, pp. 1037–1072.
50. McKie, T.; Jolley, S.J.; Kristensen, M.B. Stratigraphic and structural compartmentalization of dryland fluvial reservoirs: Triassic Heron Cluster, Central North Sea. In *Reservoir Compartmentalization*; Jolley, S.J., Fisher, Q.J., Ainsworth, R.B., Vrolijk, P.J., Delisle, S., Eds.; Geological Society, Special Publications: London, UK, 2010; Volume 347, pp. 165–198.
51. Jeans, C.V.; Reed, S.J.B.; Xing, M. Heavy mineral stratigraphy in the UK Trias: Western Approaches, onshore England and the Central North Sea. In *Petroleum Geology of Northern Europe: Proceedings of the 4th Conference*; Parker, J.R., Ed.; Geological Society: London, UK, 1993; pp. 609–624.
52. McKie, T.; Audretsch, P. Depositional and structural controls on Triassicreservoir performance in the Heron Cluster, ETAP, Central North Sea. In *Petroleum Geology: North-West Europe and Global Perspectives–Proceedings of the 6th Petroleum Geology Conference*; Doré, A.G., Vining, B.A., Eds.; Geological Society: London, UK, 2005; pp. 285–298.
53. Hurst, A.R.; Morton, A.C. An application of heavy-mineral analysis to lithostratigraphy and reservoir modelling in the Oseberg Field, northern North Sea. *Mar. Pet. Geol.* **1988**, *5*, 157–169. [CrossRef]
54. Graue, E.; Helland-Hansen, W.; Johnsen, J.; Lømo, L.; Nøttvedt, A.; Rønning, K.; Ryseth, A.; Steel, R.J. Advance and retreat of Brent delta system, Norwegian North Sea. In *Petroleum Geology of North West Europe*; Brooks, J., Glennie, K.W., Eds.; Graham and Trotman: London, UK, 1987; pp. 915–937.
55. Johnsen, J.R.; Rutledal, H.; Nilsen, D.E. *Jurassic Reservoirs Field Examples from the Oseberg and Troll Fields: Horda Platform Area*; Norwegian Petroleum Society Special Publications; Elsevier: Amsterdam, The Netherlands, 1995; Volume 4, pp. 199–234.
56. Morton, A.C.; Hallsworth, C.R.; Chalton, B. Garnet compositions in Scottish and Norwegian basement terrains: A framework for interpretation of North Sea sandstone provenance. *Mar. Pet. Geol.* **2004**, *21*, 393–410. [CrossRef]
57. Mange, M.A.; Morton, A.C. Geochemistry of heavy minerals. In *Heavy Minerals in Use, Developments in Sedimentology*; Mange, M., Wright, D.T., Eds.; Elsevier: Amsterdam, The Netherlands, 2007; Volume 58, pp. 345–391.
58. Morton, A.C.; Stiberg, J.P.; Hurst, A.; Qvale, H. Use of heavy minerals in lithostratigraphic correlation, with examples from the Brent sandstones of the northern North Sea. In *Correlation in Hydrocarbon Exploration*; Collinson, J., Ed.; Graham and Trotman: London, UK, 1989; pp. 217–230.
59. Morris, P.H.; Payne, S.N.J.; Richards, D.P.J. Micropalaeontological biostratigraphy of the Magnus Sandstone Member (Kimmeridgian-Early Volgian), Magnus Field, UK North Sea. In *Biostratigraphy in Production and Development Geology*; Jones, R.W., Simmons, M.D., Eds.; Geological Society, Special Publications: London, UK, 1999; Volume 152, pp. 55–73.

60. Fraser, S.; Robinson, A.M.; Johnson, H.D.; Underhill, J.R.; Kadolsky, D.G.A.; Connell, R.; Johannessen, P.; Ravnås, R. Upper Jurassic. In *The Millenium Atlas: Petroleum Geology of the Central and Northern North Sea*; Evans, D., Graham, C., Armour, A., Bathurst, P., Eds.; Geological Society: London, UK, 2003; pp. 157–189.
61. De'ath, N.G.; Schuyleman, S.F. The geology of the Magnus oilfield. In *Petroleum Geology of the Continental Shelf of Northwest Europe*; Illing, L.V., Hobson, G.D., Eds.; Heyden: London, UK, 1981; pp. 342–351.
62. Shepherd, M. The Magnus Field, Blocks 211/7a, 12a, UK North Sea. In *United Kingdom Oil and Gas Fields, 25 Years Commemorative Volume*; Abbotts, I.L., Ed.; Geological Society, Memoir: London, UK, 1991; Volume 14, pp. 153–157.
63. Meinhold, G.; Anders, B.; Kostopoulos, D.; Reischmann, T. Rutile chemistry and thermometry as provenance indicator: An example from Chios Island, Greece. *Sediment. Geol.* **2008**, *203*, 98–111. [CrossRef]
64. Watson, E.B.; Wark, D.A.; Thomas, J.B. Crystallization thermometers for zircon and rutile. *Contrib. Miner. Petrol.* **2006**, *151*, 413–433. [CrossRef]
65. Stephenson, D. Introduction. In *Caledonian Igneous Rocks of Great Britain*; Geological Conservation Review Series; Stephenson, D., Bevins, R.E., Millward, D., Stone, P., Parsons, I., Highton, A.J., Wadsworth, W.J., Eds.; Joint Nature Conservation Committee: Peterborough, UK, 2000; Volume 17, pp. 3–26.
66. Chew, D.M.; Strachan, R.A. The Laurentian Caledonides of Scotland and Ireland. In *New Perspectives on the Caledonides of Scandinavia and Related Areas*; Corfu, F., Gasser, D., Chew, D.M., Eds.; Geological Society, Special Publications: London, UK, 2014; Volume 390, pp. 45–91.
67. Schmidt, A.S.; Morton, A.C.; Nichols, G.J.; Fanning, C.M. Interplay of proximal and distal sources in Devonian-Carboniferous sandstones of the Clair Basin, West of Shetland, revealed by detrital zircon U-Pb ages. *J. Geol. Soc. Lond.* **2012**, *169*, 691–702. [CrossRef]
68. Cutts, K.A.; Hand, M.; Kelsey, D.E.; Wade, B.; Strachan, R.A.; Clark, C.; Netting, A. Evidence for 930 Ma metamorphism in the Shetland Islands, Scottish Caledonides: Implications for Neoproterozoic tectonics in the Laurentian-Baltica sector of Rodinia. *J. Geol. Soc. Lond.* **2009**, *166*, 1033–1047. [CrossRef]
69. Strachan, R.A.; Prave, A.R.; Kirkland, C.L.; Storey, C.D. U-Pb detrital zircon geochronology of the Dalradian Supergroup, Shetland Islands, Scotland: Implications for regional correlations and Neoproterozoic–Palaeozoic basin development. *J. Geol. Soc. Lond.* **2013**, *170*, 905–916. [CrossRef]
70. Prichard, H.M.; Lord, R.A. The Shetland Ophiolite: Evidence for a supra-subduction zone origin and implications for platinum-group element mineralization. In *Mineral Deposits within the European Community*; Boissonnas, L., Omenetto, P., Eds.; Springer: Berlin, Germany, 1988; pp. 289–302.
71. Morton, A.C. The role of heavy mineral analysis during drilling of high angle wells. In *Heavy Minerals in Use, Developments in Sedimentology*; Mange, M., Wright, D.T., Eds.; Elsevier: Amsterdam, The Netherlands, 2007; Volume 58, pp. 1123–1142.
72. Morton, A.C.; Spicer, P.J.; Ewen, D.F. Geosteering of high-angle wells using heavy mineral analysis: The Clair Field, West of Shetland. In *Horizontal Wells: Focus on the Reservoir*; Carr, T.R., Mason, E.P., Feazel, C.T., Eds.; American Association of Petroleum Geologists, Methods in Exploration: Tulsa, OK, USA, 2003; Volume 14, pp. 249–260.
73. Morton, A.C.; Hallsworth, C.R.; Kunka, J.; Laws, E.; Payne, S.; Walder, D. Heavy mineral stratigraphy of the Clair Group (Devonian) in the Clair Field, west of Shetland, UK. In *Application of Modern Stratigraphic Techniques: Theory and Case Histories*; Ratcliffe, K.T., Zaitlin, B.A., Eds.; SEPM Special Publication: Broken Arrow, OK, USA, 2010; Volume 94, pp. 183–199.
74. Morton, A.; Milne, A. Heavy mineral stratigraphic analysis on the Clair Field, UK West of Shetlands—A unique real-time solution for redbed correlation while drilling. *Pet. Geosci.* **2012**, *18*, 115–127. [CrossRef]
75. Nichols, G.J. Sedimentary evolution of the Lower Clair Group, Devonian, west of Shetland: Climate and sediment supply controls on fluvial, aeolian and lacustrine deposition. In *Petroleum Geology: North-West Europe and Global Perspectives—Proceedings of the 6th Petroleum Geology Conference*; Doré, A.G., Vining, B., Eds.; Geological Society: London, UK, 2005; pp. 957–967.
76. Knox, R.W.O.B.; Cocker, J.D.; Filatoff, J. Heavy mineral stratigraphy of the Unayzah Formation and Basal Khuff Clastics (Carboniferous to Permian) of Central Saudi Arabia. *GeoArabia* **2010**, *15*, 17–80.
77. Knox, R.W.O.B.; Franks, S.G.; Cocker, J.D. Stratigraphic evolution of heavy-mineral provenance signatures in the sandstones of the Wajid Group (Cambrian to Permian), southwestern Saudi Arabia. *GeoArabia* **2007**, *12*, 65–96.

78. Naing, T.T.; Bussien, D.A.; Winkler, W.H.; Nold, M.; von Quadt, A. Provenance study on Eocene–Miocene sandstones of the Rakhine Coastal Belt, Indo-Burman Ranges of Myanmar: Geodynamic implications. In *Sediment Provenance Studies in Hydrocarbon Exploration and Production*; Scott, R.A., Smyth, H.R., Morton, A.C., Richardson, N., Eds.; Geological Society, Special Publications: London, UK, 2014; Volume 386, pp. 195–216.
79. Morton, A.C.; Humphreys, B.; Manggal, G.; Fanning, C.M. Provenance and correlation of Upper Jurassic and Lower Cretaceous reservoir sandstones in Papua New Guinea using heavy mineral analysis. In *Petroleum Exploration and Development in Papua New Guinea: Proceedings of the Fourth PNG Petroleum Convention*; Buchanan, P.G., Grainge, A.M., Thornton, R.C.N., Eds.; PNG Chamber of Mines and Petroleum: Port Moresby, Papua New Guinea, 2000; pp. 187–203.
80. Pindell, J.; Kennan, L.; Wright, D.; Erikson, J. Clastic domains of sandstones in central/eastern Venezuela, Trinidad, and Barbados: Heavy mineral and tectonic constraints on provenance and palaeogeography. In *the Origin and Evolution of the Caribbean Plate*; James, K., Lorente, M.A., Pindell, J., Eds.; Geological Society, Special Publications: London, UK, 2009; Volume 328, pp. 743–797.
81. Vincent, H.; Wach, G. The detrital record of Cretaceous to Pliocene sandstones across the NE South American margin. *J. Geol. Soc. Lond.* **2017**, *174*, 434–452. [CrossRef]

minerals

MDPI

Article

Multimineral Fingerprinting of Transhimalayan and Himalayan Sources of Indus-Derived Thal Desert Sand (Central Pakistan)

Wendong Liang *, Eduardo Garzanti *, Sergio Andò, Paolo Gentile and Alberto Resentini

Laboratory for Provenance Studies, Department of Earth and Environmental Sciences, University of Milano-Bicocca, Piazza della Scienza 4, 20126 Milano, Italy

* Correspondence: w.liang@campus.unimib.it (W.L.); eduardo.garzanti@unimib.it (E.G.)

Received: 16 July 2019; Accepted: 24 July 2019; Published: 26 July 2019

Abstract: As a Quaternary repository of wind-reworked Indus River sand at the entry point in the Himalayan foreland basin, the Thal Desert in northern Pakistan stores mineralogical information useful to trace erosion patterns across the western Himalayan syntaxis and the adjacent orogenic segments that fed detritus into the Indus delta and huge deep-sea fan throughout the Neogene. Provenance analysis of Thal Desert sand was carried out by applying optical and semi-automated Raman spectroscopy on heavy-mineral suites of four eolian and 11 fluvial sand samples collected in selected tributaries draining one specific tectonic domain each in the upper Indus catchment. In each sample, the different types of amphibole, garnet, epidote and pyroxene grains—the four dominant heavy-mineral species in orogenic sediment worldwide—were characterized by SEM-EDS spectroscopy. The chemical composition of 4249 grains was thus determined. Heavy-mineral concentration, the relative proportion of heavy-mineral species, and their minerochemical fingerprints indicate that the Kohistan arc has played the principal role as a source, especially of pyroxene and epidote. Within the western Himalayan syntaxis undergoing rapid exhumation, the Southern Karakorum belt drained by the Hispar River and the Nanga Parbat massif were revealed as important sources of garnet, amphibole, and possibly epidote. Sediment supply from the Greater Himalaya, Lesser Himalaya, and Subhimalaya is dominant only for Punjab tributaries that join the Indus River downstream and do not contribute sand to the Thal Desert. The detailed compositional fingerprint of Thal Desert sand, if contrasted with that of lower course tributaries exclusively draining the Himalaya, provides a semi-actualistic key to be used, in conjunction with complementary provenance datasets and geological information, to reconstruct changes in paleodrainage and unravel the relationship between climatic and tectonic forces that controlled the erosional evolution of the western Himalayan-Karakorum orogen in space and time.

Keywords: amphibole; garnet; epidote; pyroxene; provenance tracers; varietal studies; mineral chemistry; semi-automated Raman counting; Ladakh-Kohistan arcs; Himalaya; Nanga Parbat; Karakorum; Indus river

1. Introduction

Heavy minerals provide detailed information on the geology of source areas, which is particularly useful in the study of modern sand unmodified by diagenesis [1]. Subtler distinctions, however, may be required in provenance analysis wherever several different potential sources of sediment consist of similar lithological assemblages shedding similar heavy-mineral assemblages. This is often the case in orogenic sediment containing transparent-heavy-mineral suites typically dominated by amphibole, garnet, epidote, and pyroxene in various proportions [2–4]. In this case, distinctive geochemical signatures of single groups of detrital minerals can be used as a genetic tool to trace their provenance ("varietal studies"; [5]).

After the pioneering study dedicated to tourmaline by Krynine [6], and since modern geochemical techniques were applied on garnet [7], single-mineral analyses have been frequently used and proven to be an efficient means to trace sediment provenance (e.g., [8–12]). More and more sophisticated geochemical and geochronological methods are being applied with the aim to fingerprint the source of an increasing number of target minerals (e.g., [13]). Single-mineral studies have the advantage that fractionation by physical processes during erosion, transport and deposition, and by chemical processes during weathering and diagenesis, can generally be held as minimal. On the other hand, the information obtained from single-mineral datasets needs to be deciphered by correcting for the generally strong differences in mineral fertility of different potential source rocks [14,15]. This thorny fertility problem is best tackled when several mineral groups are investigated ("multimineral fingerprinting"; [16,17]), because provenance signals carried by different minerals are expected to differ, reflecting their different abundance in different source-rock domains. Emphasizing this crucial point is one of the goals of this article, which focuses on Transhimalayan and Himalayan sources of detritus transported by the Indus River across northern Pakistan to pinpoint the provenance of heavy minerals contained in eolian sand of the Thal Desert. This small dune field is located in central-northern Pakistan, confined between the Indus River in the west and the course of its major Punjab tributaries in the east (Figure 1). The overall petrographic, mineralogical, and geochemical signatures of Thal dunes indicate that the contribution of Himalayan-derived Punjab tributaries is negligible [18]. The Thal Desert, therefore, can be safely considered as representing a relict Quaternary repository of wind-reworked alluvial-fan sediment originally deposited by the upper Indus at the entry point in the Himalayan foreland basin. The detailed compositional fingerprint of Thal Desert sand, if contrasted with that of Punjab tributaries exclusively draining the Himalayan belt, thus provides an additional actualistic key to trace changes in erosion patterns within the huge catchment that has fed detritus to the Indus delta and deep-sea fan throughout the late Neogene [19,20]. To this goal, our dataset complements a previous work on major and trace elements in amphibole [21] and integrates the geochemical study of detrital garnet in sand of the middle Indus course and its Punjab tributaries [22]. We chose to focus on the chemical composition of detrital amphibole, garnet, epidote, and pyroxene because these four minerals, all solid-solution series, represent the four dominant species in orogenic sediments worldwide [23]. Other studies investigating provenance of Indus sediments focused on Pb isotopes in detrital K-feldspar and bulk-sediment Nd and Sr isotope fingerprints [24–26], zircon U-Pb or mica $^{39}Ar/^{40}Ar$ geochronology and apatite fission-track or (U–Th)/He thermochronology [27–29], sand petrography, heavy minerals, ^{10}Be cosmogenic nuclides [18,30], and clay mineralogy [31]. Such multi-technique approaches have shed new light on the relative role played by the interacting climatic and tectonic forces that controlled the erosional evolution of the western Himalayan-Karakorum orogen.

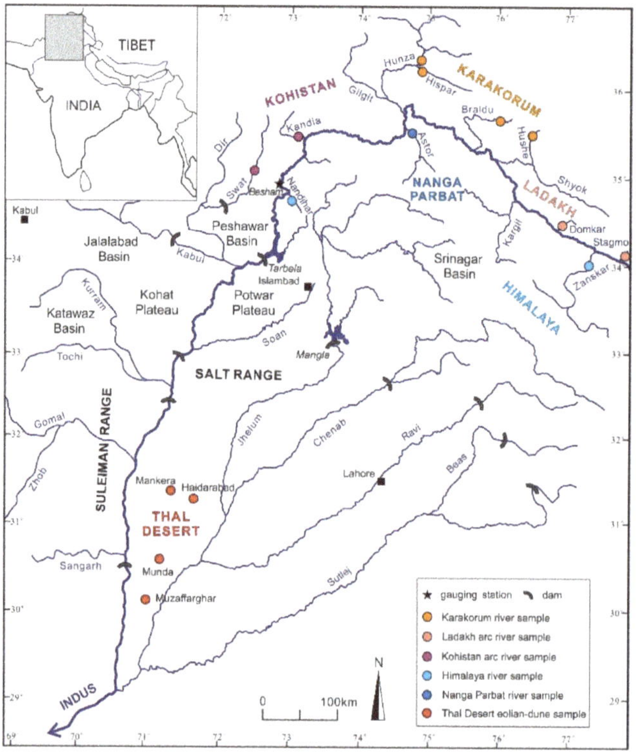

Figure 1. The Indus drainage system and sample locations in northern Pakistan.

2. The Indus River Catchment and the Thal Desert

The Indus River, sourced from the central southern Tibetan Plateau, flows in its upper course along the suture zone and Transhimalayan forearc basin, while receiving detritus from both the Ladakh arc in the north and the northern side of the Himalayan belt in the south [30]. Next, it cuts a deep gorge through the western Himalayan syntaxis, where very rapid erosion rates generate large amounts of detritus from the Karakorum belt, the Nanga Parbat crystalline massif, and the Kohistan arc [32–35]. Farther downstream, it flows across the Himalayan belt and Potwar Plateau [36] where it is joined by the Kabul River draining the Hindukush belt [37], crosses the Salt Range, and eventually reaches the foreland basin where it flows southward, confined between the front of the Sulaiman Range in the west and the Thal Desert in the east (Figure 1).

The upper Indus River is mainly fed by melting of ice and snow, and sediment flux consequently increases by two to three orders of magnitudes during the summer [38]. The annual suspended load of the Indus River, estimated as ~14×10^6 t upstream of the Shyok confluence (Figure 1), increases rapidly downstream owing to major contributions from the Shyok River (~23×10^6 t/a), Karakorum tributaries ($\leq 100 \times 10^6$ t, ~18×10^6 of which from the Hunza River), the Astor River draining Nanga Parbat (~2×10^6 t), and diverse tributaries draining the Kohistan arc, summing up to ~176×10^6 t at the Besham gauging station [39]. The annual sediment load reaching the Tarbela Dam, which was closed in 1974 in northern Pakistan, has been estimated at 200×10^6 t [40,41] or even at 287×10^6 t [42].

The hydrology of the Indus River has been so intensely regulated since the 1930s that most of its sediment has been trapped in artificial reservoirs and canals, and the annual flux in the lower course has been reduced to ~50×10^6 t [42]. The Indus Waters Treaty signed in 1960 gave rights to the entire flow of the Indus, Jhelum, and Chenab Rivers to Pakistan, and of the Ravi, Beas, and Sutlej Rivers to India.

Subsequently, all Punjab tributaries have been dammed and linked by canals to irrigate the arid lowlands and compensate for lost waters in eastern Pakistan. Water discharge dropped sharply, and flow in the Ravi and Sutlej rivers ceased altogether except during monsoon floods. The Mangla Dam, completed in 1967, reduced sediment load of the Jhelum River from 45 to < 0.5×10^6 t/a [43,44]. Among Indus tributaries draining the Sulaiman Range in western Pakistan, the Gomal River (basin area 36,000 km^2) is characterized by extreme concentration of suspended solids (42 g/L) and high sediment load (30×10^6 t/a), followed by the Kurram River (3×10^6 t/a; [42]). Other rivers are minor and mostly flow during flash floods.

2.1. Karakorum Belt

The composite Karakorum belt includes the Northern Karakorum sedimentary domain, the Central Karakorum batholith, and the Southern Karakorum metamorphic belt (Figure 2; [45,46]). In the Northern Karakorum zone, an Ordovician to Upper Cretaceous sedimentary succession lying non-conformably onto crystalline basement is exposed [47]. Black slates in the north, intruded by calc-alkaline gabbro-diorite, granodiorite, granite, and tonalite, contain andalusite, chloritoid, and epidote [48]. The Central Karakorum batholith comprises mid-Cretaceous granitoids intruded before the India-Asia collision and containing amphibole with residual clinopyroxene and accessory titanite, epidote, allanite, apatite, zircon, and opaque minerals [49]. Post-collisional leucogranites were intruded between 13 and 25 Ma (e.g., Baltoro batholith; [50]). The Southern Karakorum belt includes migmatitic domes undergoing rapid erosional exhumation and displays a northeastward increase in metamorphic grade from structurally lower phyllites to staurolite-, kyanite-, and eventually sillimanite-bearing metasedimentary rocks at the top. Impure dolomitic marbles containing diopside and corundum, and amphibolites with hornblende and garnet also occur [51–53].

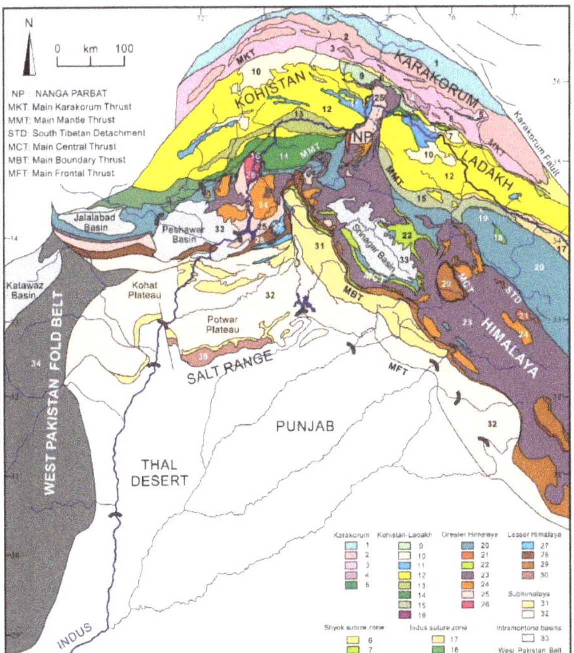

Figure 2. Geological map of the Indus catchment in northern Pakistan (modified from Pêcher et al. [34]). Geological units, from north to south: **Karakorum belt**: 1: northern sedimentary belt; 2: axial batholith and other granitoid rocks; 3: southern metamorphic belt; 4: felsic gneiss; 5: Masherbrum Greenstone Complex. **Shyok suture zone**: 6: mostly terrigeneous strata; 7: mélange zone (mainly volcanic rocks);

8: ultramafic rocks (Shyok and Dobani-Dassu lineament). **Kohistan and Ladakh arcs**: 9: Paleogene Chalt (Kohistan) and Khardung (Ladakh) volcanic rocks, Turmik volcaniclastic rocks; 10: undifferentiated volcano-sedimentary group; 11: metasedimentary rocks; 12: plutonic rocks; 13: gabbronorite (Chilas complex); 14: southern amphibolite; 15: Dras volcano-sedimentary group; 16: ultramafic rocks (Jijal complex). **Indus suture zone**: 17: Indus Group; 18: Spontang ophiolite; 19: imbricate thrust units with blueschist. **Tethys** and **Greater Himalaya**: 20: Paleozoic-Eocene sedimentary rocks; 21: Miocene leucogranite; 22: Permian Panjal Traps; 23: Greater Himalayan neometamorphic rocks; 24: Paleozoic intrusives; 25: mainly Paleoproterozoic orthogneiss; 26: Besham metaigneous rocks. **Lesser Himalaya**: 27: Paleozoic-Eocene strata; 28: upper nappe (mostly Mesoproterozoic metasedimentary rocks); 29: lower nappe (mostly Neoproterozoic and Paleozoic metasedimentary rocks); 30: Salt Range (Neoproterozoic to Eocene Indian margin strata). **Sub-Himalaya**: 31: Muree and Subathu formations (Cenozoic); 32: Siwalik Group (Neogene); 33: Peshawar and Srinagar Quaternary intramontane basins. **West Pakistan Belt**: 34: Sulaiman Range.

2.2. Ladakh and Kohistan Arcs

The Ladakh and Kohistan batholiths expose a complete section of mantle to upper crustal igneous rocks representing the dissected remnants of magmatic arcs fed by northern subduction of Neotethyan lithosphere during the Cretaceous to earliest Paleogene. The arcs are delimited by the Shyok ophiolitic suture in the north, generally ascribed to Upper Cretaceous (pre-Campanian) collision with the Karakorum block [54–58], and the Indus ophiolitic suture in the south, closed when India collided with Asia during the Paleocene [59,60].

The Kohistan arc is composed of six main units from bottom to top (south to north): (1) Jijal ultramafic-mafic complex yielding garnet, amphibole, clinopyroxene, and minor olivine, orthopyroxene, spinel and zoisite; (2) Kamila amphibolite; (3) Chilas ultramafic-mafic complex, containing orthopyroxene and clinopyroxene with minor olivine, magnetite, ilmenite, hornblende, and spinel; (4) Kohistan batholith, yielding mainly hornblende and locally clinopyroxene; (5) Jurassic-Cretaceous metavolcanic and metasedimentary rocks of the Jaglot and Chalt Groups, and the Aptian-Albian volcano-sedimentary Yasin Group [61–63].

The Ladakh batholith consists of a suite of Cretaceous to Paleogene mafic to felsic rocks (olivine norite to granite) yielding hornblende, augite, titanite, apatite, epidote, and zircon [64,65]. The batholith is non conformably overlain by Upper Cretaceous to Paleogene strata of the Indus Group [66,67]. Rocks exposed along the Indus suture also include Lower Cretaceous carbonates, ophiolitic mélange, and blueschists [68,69].

2.3. Himalayan Belt and Nanga Parbat Massif

The Himalayan Range formed as a consequence of continental collision between the Indian passive margin and the Asian active margin at ~60 Ma [70,71]. The orogenic belt consists of a series of southward propagating thrust sheets, which resulted in crustal thickening starting from the Eocene [72,73]. The Neoproterozoic to Eocene Tethys Himalayan succession consists of siliciclastic and carbonate rocks originally deposited onto the northern continental margin of India [74,75]. The Greater Himalaya, including slate intruded by Ordovician granitoids and sillimanite-bearing metasedimentary rocks at the top, represents the axial crystalline backbone of the range [33,76]. It is delimited in the north by the South Tibetan Detachment system lined with Miocene tourmaline-bearing leucogranite intrusions [77] and by the Main Central Thrust in the south [78]. Lesser Himalayan and Sub-Himalayan rocks exposed farther south include, respectively, Paleoproterozoic basement and Mesoproterozoic to Cenozoic cover strata displaying southward decreasing metamorphic grade [79,80] and orogen-derived Cenozoic molasse [81–83]. These rocks are drained by Punjab tributaries and shed detritus that contributes to Indus River load only downstream of the Thal Desert.

Only the Indus River cuts across the western Himalayan syntaxis, where the N/S-elongated crustal-scale Nanga Parbat antiform exposing Precambrian Indian gneissic basement overprinted by Himalayan metamorphism is bounded to the north by the Karakorum belt and flanked to the west and east by the Kohistan and Ladakh arcs. In the Nanga Parbat massif, sillimanite-bearing gneisses are structurally overlain by kyanite-bearing schists [84,85]. Leucogranite intrusions yielding tourmaline, apatite, zircon, monazite, and garnet [86] are as young as 1.4 Ma. Cooling ages of 5 Ma or even 1 Ma in the core of the dome [87] testify to ultra-rapid exhumation and very fast fluvial incision [88,89], with high denudation rates of 3–5 mm/year [90–92]. Upstream of the entry point in the foreland basin, the Indus River traverses the Potwar Plateau, where Himalayan-derived molassic Cenozoic rocks are widely exposed [93,94], and finally cuts across the Salt Range, including Paleozoic to Paleogene strata detached over uppermost Neoproterozoic/Cambrian salt and uplifted during the latest Miocene [95].

2.4. Thal Desert

The Thal Desert, a triangular region located in central northern Pakistan between ~30° and 32°30′ N and between ~71° and 72°E (Figure 2), is characterized by arid to semi-arid subtropical climate. This desert occupies the Sind-Sagar or Thal Doab (*doab* = land between two rivers, from *do* = two and *ab* = water in Urdu and Farsi), the region extending between the course of the Indus River in the west and the Punjab in the east, the fertile region crossed by the Himalayan rivers Jhelum, Chenab, Ravi, Beas, and Sutlej (*punjab* = five waters, from *panj* = five and *ab* = water).

The Thal Desert is delimited by the Salt Range foothills in the north, whereas the Indus floodplain is bounded by the Sulaiman Range in the west (Figure 1). The desert area is covered by low sand dunes (1–2 m in height) or rolling sand plains alternating with narrow valleys of cultivable land, and is underlain by Quaternary fluvial and eolian deposits more than 350 m thick in the south and even thicker in the central part of the desert [96]. The underlying alluvium mostly consists of laterally continuous fine to coarse sand, with minor gravel and isolated mud lenses. The coarsest deposits occur in the north close to the Salt Range, but otherwise the distribution of grain size is irregular, reflecting deposition by the constantly shifting paleo-Indus River.

3. Methods

3.1. Sampling

The sample set considered in this study includes four eolian-dune sand samples collected in February 2001 from the Thal Desert, along with 11 sand samples collected during 2001 and 2011 from active river bars in 11 tributaries draining each a different geological domain in the upper Indus River catchment ("first-order sampling scale" of [97]). These samples were accurately selected from a much larger sample set, described elsewhere [18,30,98], as the best suited to represent end-member sources of detritus derived from the Karakorum belt (upper Hushe, upper Braldu, upper Hunza, and Hispar samples), the Ladakh (Stagmo and Domkar samples) and Kohistan arcs (Kandia and Swat samples), the Nanga Parbat massif (Astor sample), and the Himalayan belt (Zanskar and Nandihar samples). The Thal dune samples are upper very fine to lower fine and well to moderately sorted sands (3.05–2.67 Φ, 0.43–0.84 σ_Φ); fluvial samples are upper very fine to lower medium and moderately-well to moderately sorted sands (3.20–1.51 Φ, 0.63–0.97 σ_Φ) (Supplementary Table S1).

3.2. Heavy Mineral Analyses

For each of the 15 selected samples, heavy minerals were separated with sodium polytungstate (density ~ 2.90 g/cm^3) from a split aliquot of the 63–250 μm or 32–500 μm fraction obtained by sieving, recovered by partial freezing with liquid nitrogen, and mounted on a glass slide. A polished thin section was also prepared, and mineralogical composition was determined by both counting under the microscope of ≥200 transparent heavy minerals in the glass slide and by semi-automated analysis of the polished thin section with a Raman spectrometer [99].

Heavy-mineral concentration, calculated as the volume percentage of total (HMC) and transparent (tHMC) heavy minerals in the bulk sample, ranges from poor (tHMC < 1), moderately poor

(1 ≤ tHMC < 2) and moderately rich (2 ≤ tHMC < 5), to rich (5 ≤ tHMC < 10), very rich (10 ≤ tHMC < 20) and extremely rich (20 ≤ tHMC < 50) [100,101].

3.3. Sources of Bias

As a consequence of choosing a 2Φ to 4Φ-wide size window for analysis in order to reduce technical problems during separation, mounting on the glass slide and identification under the microscope caused by detrital grains with great size differences, heavy minerals occurring in the finest tail of the size distribution (3% ± 2% of each bulk sample) and in the coarse tail (32% ± 22% of each bulk sample) were discarded. The sediment fraction considered for analysis ranged between 58% ± 21% for fluvial samples to 85% ± 9% for the better sorted Thal eolian-dune samples. The analytical bias thus introduced can be considered as minor, because the fine tail was almost entirely included whereas the coarse tail is strongly depleted in heavy minerals as the concentration of denser grains drops rapidly in the coarser classes of sediments deposited by tractive currents [102,103].

Another potential source of bias is represented by hydraulic-sorting processes, which may concentrate different minerals in distinct depositional sub-environments based on their size, density, and shape [104–106]. An efficient way to test for heavy-mineral enrichment or depletion in sediment samples is provided by chemical analyses, which readily reveal anomalous concentrations of chemical elements such as rare earth elements (REE) or zirconium preferentially hosted in ultradense minerals [101]. Among the four Thal dune samples, heavy-mineral enrichment is apparent for the Muzaffarghar sample S1470 containing much more Zr than the Munda sample S1474 (524 vs. 106 ppm), whereas the other two samples S1462 and S1463 have Zr concentration very close to the Upper Continental Crust standard (UCC; 195–213 vs. 190–193 ppm in the UCC; [107,108]) (Supplementary Table S1). Among river sands, only the Hispar sample shows high concentration of Zr, Th, and REEs relative to all other samples (Zr 395 vs. 110–186 ppm; Th 52 vs. 3–14 ppm; La 117 vs. 12–55 ppm and Y 41 vs. 12–27 ppm), suggesting hydraulic enrichment in heavy minerals. The heavy-mineral spectrum of samples systematically showing anomalous concentrations in these elements is expected to be enriched in denser heavy minerals such as garnet relative to low-density heavy minerals such as amphibole.

3.4. Microchemical Analyses

The same polished thin sections, in a photographic image of which all grains were properly identified and numbered, were carbon-coated and analyzed by Energy Dispersive X-ray Spectroscopy (EDS) under the scanning electron microscope (SEM) to obtain quantitative chemical information on the four most common detrital minerals in orogenic sediments (i.e., amphibole, garnet, epidote, and pyroxene). Microchemical analyses were carried out at the Department of Earth and Environmental Sciences, University of Milano-Bicocca (Milano, Italy), using a TESCAN TS5136XM with an electronic microprobe EDAX GENESIS 4000 XMS Imaging 60 SEM, voltage 20 KeV, detection time 20 s, spot size 250 nm and absorption current 190 ± 1 pA measured in Faraday cup, medium heating, take off angle 45°, working distance 23 mm. In each thin section, we counted ~100 grains for each mineral group (or all of those present in case we did not find enough). In the four Thal Desert samples, 400 amphibole, 395 epidote, 317 pyroxene, and 280 garnet grains were analyzed, thus allowing identification even of small detrital populations [109]. Overall, we analyzed 1504 amphibole, 1129 epidote, 861 pyroxene, and 755 garnet grains in the 15 selected samples. Information on sample locations, the result of heavy-mineral analyses, and the complete geochemical dataset including the percentages of each mineral variety in each sample are provided as Supplementary Material (Tables S1–S10). Statistical techniques used to illustrate our dataset include multidimensional scaling, which produces a map of points in which samples with similar mineralogical signature cluster closely together and dissimilar samples plot far apart [110] and the biplot [111], which allows us not only to discriminate among multivariate observations (data points) but also to visualize the mutual relationships among an even large number of variables (rays). The length of each ray is proportional to the variance of the corresponding parameter in the dataset, whereas if the angle between two rays is close to 0°, 90° or 180°, then the corresponding elements are directly correlated, uncorrelated or inversely correlated, respectively.

3.5. Amphibole Chemistry

The general chemical formula of the amphibole supergroup is $AB_2C_5T_8O_{22}W_2$, where A, B, and C are cations and W anions (A = □, Na, K, Ca, Pb, Li; B = Na, Ca, Mn^{2+}, Fe^{2+}, Mg, Li; C = Mg, Fe^{2+}, Mn^{2+}, Al, Fe^{3+}, Mn^{3+}, Ti^{4+}, V^{3+}, Li; T = Si, Al, Ti^{4+}, Be; W = (OH), F, Cl, O^{2-}; [112,113]). Following the recommendation of the International Mineralogical Association, amphibole minerals are divided into two groups based on the dominant anions at site W, i.e., $(OH, F, Cl)^-$ versus oxo-amphiboles. The $(OH, F, Cl)^-$ group is further subdivided into eight subgroups based on B cations. An Excel spreadsheet developed by Locock [114] was used to calculate the chemical formula and classify detrital amphiboles. The $Fe^{3+}/\Sigma Fe$ and $Mn^{3+}/\Sigma Mn$ ratios were calculated based on charge balance by normalizing the formula to one or more sets of cation sums because the valence state of Fe and Mn was not measured. All amphibole grains were considered to be monoclinic because only a few (<2%) orthorhombic amphibole grains were detected with Raman spectroscopy. For amphibole with W = 2 (OH, F, Cl), sufficient OH content was calculated to reach two (OH, F, Cl) per formula unit because H_2O+ was not measured and OH could not be estimated [114].

3.6. Garnet Chemistry

The general formula of garnet contains eight cations and 12 anions: $X_3Y_2Z_3\Phi_{12}$, where X = Na, Mg, Ca, Mn^{2+}, Fe^{2+}, Y; Y = Mg, Al, Si, Sc, Ti, V, Cr, Mn^{3+}, Fe^{2+}, Fe^{3+}, Zr, Sn; Z = Al, Si, Fe^3. The Excel spreadsheet developed by Locock [115], which considers 15 different garnet varieties and 14 end-members was used to calculate the molar proportion of garnet end-members from chemical data. The iron was entered as FeO_{tot} in the spreadsheet and the amounts of Fe^{2+} and Fe^{3+} were calculated by stoichiometric constraints because the proportion of FeO vs. Fe_2O_3 was not determined. Mn^{3+} was calculated only for compositions that cannot charge balance with Fe^{3+} alone.

Garnet, common in orogenic sediments derived from metasedimentary rocks, is a particularly valuable provenance tracer because it displays a wide range of major-element compositions and resists diagenetic dissolution better than epidote, amphibole, and pyroxene [116–118].

Different types of detrital garnets can be empirically distinguished according to their provenance by the use of the Fe + Mn–Mg–Ca ternary plot [5,119]. Type A garnet (high Mg, low Ca) is mainly shed by granulite-facies metasedimentary rocks, charnockites, and intermediate-felsic igneous rocks. Type B garnet (low Mg, variable Ca) is derived from either intermediate-felsic igneous rocks (sub-type Bi; $X_{Mg} < 20\%$, $X_{Ca} < 10\%$) or amphibolite-facies metasedimentary rocks (sub-type Bii). Type C garnet (high Mg, high Ca) is preferentially contained in high-grade metabasite (sub-type Ci) or ultramafic rocks such as pyroxenite and peridotite (sub-type Cii; $X_{Mg} > 40\%$, $X_{Ca} > 10\%$), and type D garnet (low Mg, very high Ca) in metasomatic rocks (skarn), very low-grade metabasite, and high-grade calc-silicate rocks.

The different origins of detrital garnet are also highlighted by the use of the Mn–Mg–Ca diagram, based on the observation that Mg^{2+} progressively substitutes for Mn^{2+} and Fe^{2+} in pyralspite garnet with increasing metamorphic temperature, whereas Ca^{2+} increases at increasing pressures [120,121].

3.7. Epidote Chemistry

For the classification of epidote-group minerals we used the Windows™ program WinEpclas developed by Yavuz and Yildirim [122] and based on the nomenclature recommended by the Comission on New Minerals and Mineral Names of the International Mineralogical Association. The structural formula of monoclinic epidote-group minerals can be expressed as A1A2M1M2M3 $[T_2O_7][TO_4](O4)(O10)$, where A1 = Ca, Mn^{2+}; A2 = Ca, Sr, Pb, Ce^{3+}, $(REE)^{3+}$; M1 = Mg, Fe^{2+}, Mn^{2+}, Al, Fe^{3+}, V^{3+}, Mn^{3+}, Cr^{3+}; M2 = Al, Fe^{3+}; M3 = Mg, Fe^{2+}, Mn^{2+}, Al, Fe^{3+}, V^{3+}, Mn^{3+}, Cr^{3+}; T = Si; O4 = O^{2-}, F^-; and O10 = OH^-, O^{2-} [123]. The normalization scheme based on the $\Sigma(A + M + T) = 8.0$ determines the mineral species on the basis of the dominant cations at sites A1, A2, M1, M2, and M3, and of anions at sites O4 and O10.

Zoisite, the orthorhombic polymorph of clinozoisite, cannot be distinguished chemically from its monoclinic polymorph clinozoisite, and is thus not considered as a distinct species by WinEpclas software.

3.8. Pyroxene Chemistry

The general formula of orthorhombic or monoclinic pyroxene can be expressed as ABZ_2O_6, where A = Ca, Fe^{2+}, Li, Mg, Mn^{2+}, Na, Zn; B = Al, Cr^{3+}, Fe^{2+}, Fe^{3+}, Mg, Mn^{2+}, Sc, Ti, V^{3+}; Z = Al, Si. Composition of detrital pyroxene was calculated on the basis of six oxygen atoms in the chemical formula, using the software developed by Sturm [124]. The nomenclature follows the rules set in Morimoto et al. [125]. The prefixes "aluminian" or "sodian" are added for clinopyroxene with $Al^{3+} > 0.1$ atoms per formula unit (a.p.f.u.) or $Na^{+} > 0.1$ a.p.f.u., respectively. The prefix "subsilicic" is added if Si^{4+} is < 1.75 a.p.f.u. Most pyroxene grains belong to the Quad chemical group (i.e., plot in the classical pyroxene quadrilateral, part of the Ca–Mg–Fe classification triangle; [125]). The J parameter is twice Na a.p.f.u.; the Q parameter is Ca + Mg + Fe^{2+} a.p.f.u.

4. Heavy Mineral Sources

Transparent heavy-mineral suites in all analyzed samples mostly consist (84% on average) of amphibole (47% ± 17%), epidote (17% ± 9%), pyroxene (12% ± 9%), and garnet grains (9% ± 8%).

Heavy-mineral concentration results to be much higher in river sand derived from the Kohistan arc (19%–44%) than from the Ladakh arc (5%–20%), and higher in river sand derived from the Nanga Parbat massif (6%–17%) than from both Karakorum belt and Greater Himalaya (3%–9%). Heavy mineral concentration is remarkably high in dune sand of the Thal Desert (12%–26%; Table 1).

Table 1. Heavy-mineral assemblages in river sands of the Upper Indus catchment (end-member sources) and eolian dunes of the Thal Desert (sediment sink) performed by semi-automated Raman spectroscopy.

Sample	River/Dune	Domain	HMC	tHMC	Zrn	Tur	Rt	Ttn	Ap	Amp	Cpx	Opx	Ol	Zo
S1749	Hushe	Karakorum	4.8	2.5	2	0.2	1	9	6	60	5	0	0.5	0
S1748	Braldu	Karakorum	6.5	4.5	0.9	1	3	8	6	42	11	0	0.5	0.1
S1437	Hunza	Karakorum	2.9	1.5	0.5	2	9	10	5	45	8	0.2	0	0.2
S1438	Hispar	Karakorum	8.7	6.7	2	1	1	10	3	26	6	0	0.2	0
S4426	Stagmo	Ladakh arc	12.6	12.0	0.4	0.4	0.4	6	0.9	82	6	0	0	0
S4430	Domkar	Ladakh arc	9.7	8.1	2	0.3	1	7	1	73	6	0.3	0	0
S1439	Kandia	Kohistan arc	44.2	33.4	0	0	0.5	7	0.5	51	4	0	0.2	2
S1440	Swat	Kohistan arc	31.4	27.5	0.2	0	2	1	0.6	49	37	4	0	0
S4419	Zanskar	G. Himalaya	4.8	4.6	1	5	2	8	8	21	11	0	0.2	0
S1426	Nandihar	G. Himalaya	4.8	4.0	0.6	8	1	5	3	32	13	2	0	0.6
S1432	Astor	Nanga Parbat	17.9	16.9	0.2	1	2	1	0.3	64	6	0	0.2	0.6
S1462	Mankera	Thal Desert	21.2	15.3	0	1	1	4	0.3	40	10	1	0.5	0.5
S1463	Haidarabad	Thal Desert	24.2	18.6	0	0.4	2	6	1	36	11	3	0.4	0.9
S1470	Muzaffarghar	Thal Desert	26.4	17.7	0.3	0.8	2	4	1	35	13	2	1	0.7
S1474	Munda	Thal Desert	12.3	10.0	0.3	1	1	4	1	40	13	2	0.3	0.3

Sample	River/Dune	Domain	Czo	&Ep	Grt	Cld	St	And	Ky	Sil	&HM	Tot	%Amp	%Grt
S1749	Hushe	Karakorum	1	11	4	0	0	0	0	0.2	0.6	100.0	1.5	0.1
S1748	Braldu	Karakorum	5	11	7	1	0.4	0.8	0	0.4	1	100.0	1.9	0.3
S1437	Hunza	Karakorum	6	13	1	0	0.2	0.3	0	0	0.5	100.0	0.7	0.0
S1438	Hispar	Karakorum	9	15	26	0.2	0.5	0	0	0	0.2	100.0	1.7	1.7
S4426	Stagmo	Ladakh arc	0	3	0	0	0	0	0	0	0	100.0	9.9	0.0
S4430	Domkar	Ladakh arc	0.5	7	1	0	0	0	0	0	0.8	100.0	5.9	0.1
S1439	Kandia	Kohistan arc	22	10	2	0	0	0	0	0	0	100.0	17.1	0.7
S1440	Swat	Kohistan arc	2	4	0.6	0	0	0	0	0	0	100.0	13.5	0.2
S4419	Zanskar	G. Himalaya	0.3	11	21	0	0.3	0	0.5	9	1	100.0	1.0	1.0
S1426	Nandihar	G. Himalaya	1	2	22	0	2	0	8	0.6	0.4	100.0	1.3	0.9
S1432	Astor	Nanga Parbat	5	9	9	0	0.2	0.2	0.8	0.2	0.2	100.0	10.8	1.6
S1462	Mankera	Thal Desert	15	9	14	0.3	0.8	0	1	0.2	0.7	100.0	6.1	2.1
S1463	Haidarabad	Thal Desert	11	13	14	0	0.6	0.2	0.4	1	0	100.0	6.6	2.5
S1470	Muzaffarghar	Thal Desert	14	11	11	0	0.5	0	0.8	0.5	0.5	100.0	6.3	2.0
S1474	Munda	Thal Desert	11	12	10	0	0.6	0	1	0.3	0.1	100.0	4.0	1.0

On average, over 700 transparent heavy minerals were counted per sample (ranging from 275 for S4426 to 1300 for S1748; Supplementary Table S2). HMC = heavy mineral concentration; tHMC = transparent heavy mineral concentration; Zrn = zircon; Tur = tourmaline; Rt = rutile; Ttn = titanite; Ap = apatite; Amp = amphibole; Cpx = clinopyroxene; Opx = orthopyroxene; Ol = olivine; Zo = zoisite; Czo = clinozoisite; &Ep = allanite and other epidote-group minerals; Grt = garnet; Cld = chloritoid; St = staurolite; And = andalusite; Ky = kyanite; Sil = sillimanite; &HM = other transparent heavy minerals (monazite, anatase, brookite, prehnite, axinite, gahnite, barite, vesuvianite). Percentages of amphibole, garnet, epidote, and pyroxene on bulk samples are given in the four last columns to the right.

4.1. Karakorum

The studied river sands derived from the Karakorum contain moderately poor to rich transparent-heavy-mineral suites dominated by amphibole, with subordinate epidote, clinopyroxene, titanite and rare clinozoisite, apatite, and garnet. The Hispar River, draining mid-crustal rocks rapidly exhumed in metamorphic domes of the Southern Karakorum Belt, carries the richest transparent-heavy-mineral suite containing subequal amounts of garnet and amphibole (Table 1).

Amphibole grains in sand of the upper Hushe River, which largely drains granitoid rocks, are mainly pargasite (52%), hastingsite (26%), and hornblende (Figure 3A). Garnet grains mainly plot in field Bi of the Fe + Mn–Mg–Ca plot and in the low P/T field of the Ca–Mg–Mn plot (Figure 4A), reflecting provenance from intermediate-felsic igneous rocks. Epidote-group minerals include abundant REE-rich allanite (11% on average) and detrital pyroxene is mainly diopside with subordinate augite with a low wollastonite (Wo) value and negligible orthopyroxene (Figures 5A and 6A).

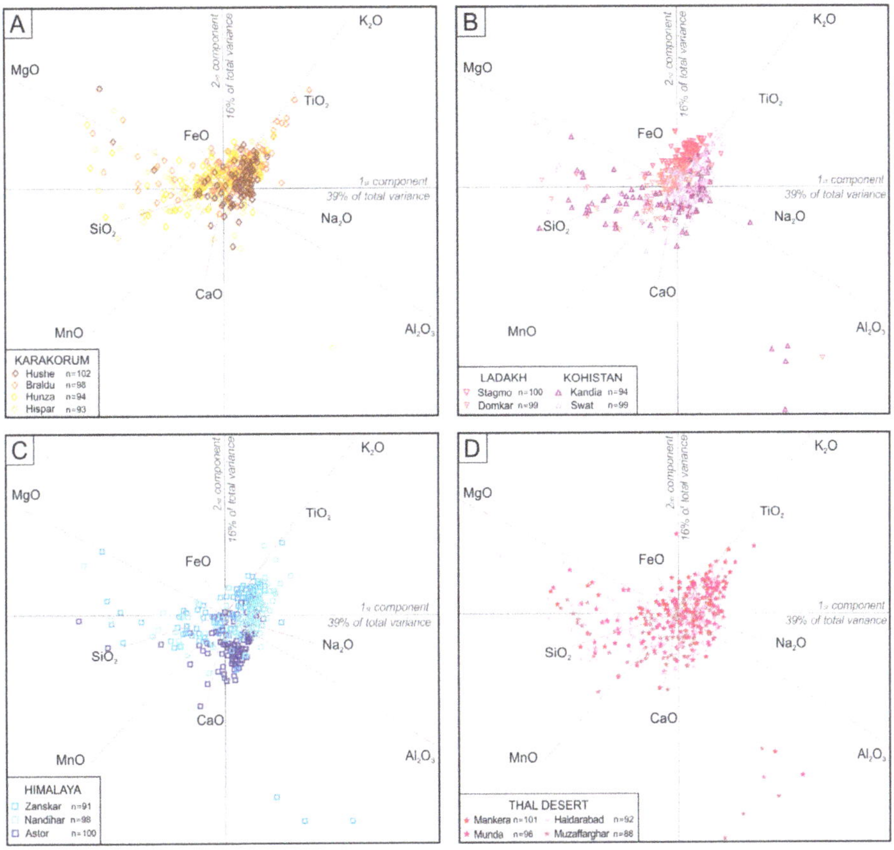

Figure 3. Chemical composition of detrital amphibole in river and eolian sands of northern Pakistan. All data were plotted in one single biplot [111], next separated into four panels to allow comparison between end-member sources (**A** = Karakprum, **B** = Ladakh and Kohistan arcs, **C** = Himalaya) and the sediment sink (**D** = Thal Desert).

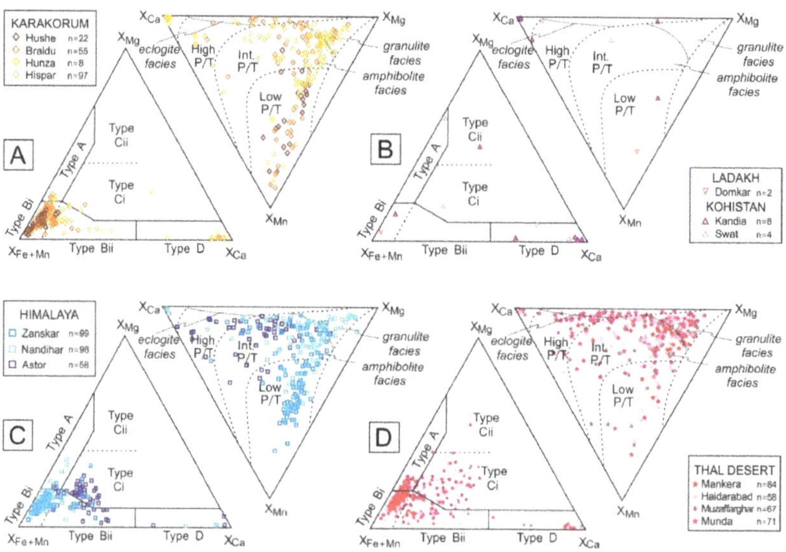

Figure 4. Chemical composition of detrital garnet in river sands of northern Pakistan (**A** = Karakorum, **B** = Ladakh and Kohistan arcs, **C** = Himalaya) and in the (**D**) Thal Desert. (Fe + Mn–Mg–Ca plot after Mange and Morton [5]; Ca–Mg–Mn plot after Win et al. [121]). XFe, XMg, XCa, XMn = molecular proportions of Fe^{2+}, Mg, Ca, and Mn.

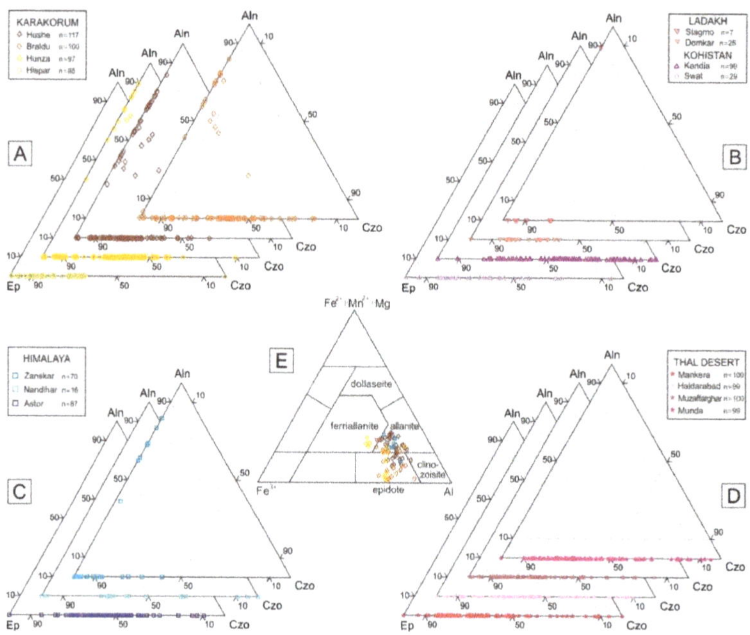

Figure 5. Chemical composition of detrital epidote-group minerals in river sands of northern Pakistan and in the Thal Desert. (**A**–**D**) The proportion of clinozoisite, allanite, and epidote were calculated as Al (atoms per formula unit (a.p.f.u.)) − 2, if Al (a.p.f.u.) > 2 (otherwise the proportion was taken as zero), as rare earth elements (REE) (a.p.f.u.), and as 1-allanite-clinozoisite, respectively [126]. Aln: allanite; Ep: epidote; Czo: clinozoisite. (**E**) Classification of REE-bearing epidote grains after Kartashov [127].

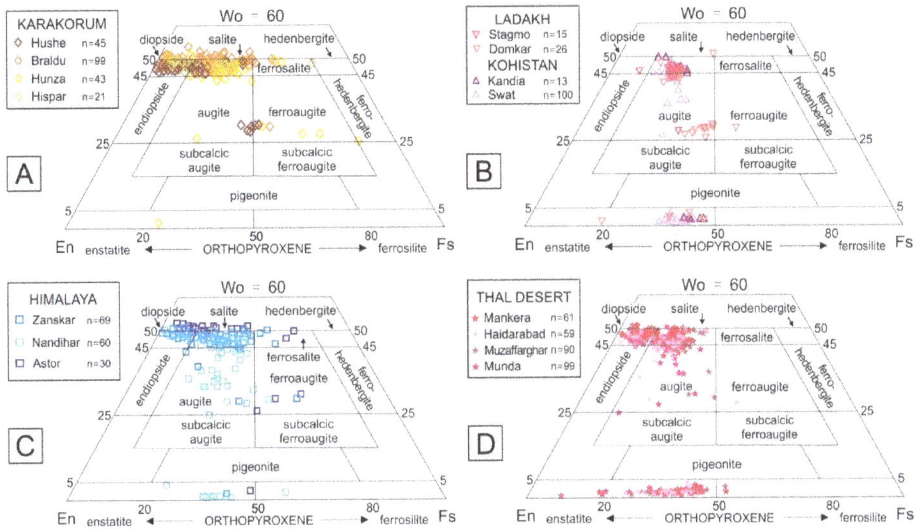

Figure 6. Chemical composition of detrital pyroxene in river sands of northern Pakistan (**A** = Karakorum, **B** = Ladakh and Kohistan arcs, **C** = Himalaya) and in the (**D**) Thal Desert. Pyroxene quadrilateral from Poldervaart and Hess [128] and Morimoto et al. [125]. Wo: Wollastonite ($Ca_2Si_2O_6$); En: enstatite ($Mg_2Si_2O_6$); Fs: ferrosilite ($Fe_2Si_2O_6$).

Sand of the upper Braldu River draining the axial part of the Karakorum belt contains mainly hornblende prevailing over pargasite and actinolite, and mainly Bi with subordinate Bii garnet plotting in the low and intermediate P/T fields (Figure 4A). Epidote dominates over allanite and clinozoisite. Pyroxene is mostly diopside, largely derived from upper-amphibolite facies metasedimentary rocks (99%; Figure 6A). Similar mineralogical signatures characterize upper Hunza and Hispar sands, mainly derived from the Northern Karakorum sedimentary domain and from the Southern Karakorum belt, respectively. Hispar sand however lacks allanite (Figure 5A), whereas upper Hunza sand contains only a few garnet grains, dominantly of type D (Figure 4A), and some ferroaugite grains (Figure 6A).

4.2. Ladakh and Kohistan Arcs

Stream sands derived from the Ladakh arc contain rich to very rich, amphibole-dominated transparent heavy-mineral suites including minor titanite, clinopyroxene, epidote, and only rare garnet (Figure 4B). Amphibole grains are mostly hornblende (~81% on average; Figure 3B), epidote is dominant (allanite is rare in Stagmo sand and absent in Domkar sand; Figure 5B), and detrital pyroxene is mainly diopside and augite (Figure 6B). Domkar sand includes a few orthopyroxene grains.

The very rich to extremely rich transparent heavy-mineral suites shed from the Kohistan arc are more varied (Table 1). Kandia sand yields mainly pargasite, hornblende, and actinolite among the amphibole group, abundant epidote-group minerals (mostly clinozoisite; Figure 5B), mostly high-Ca garnet of type D (Figure 4B), and only a few diopside and orthopyroxene grains (Figure 6B). In Swat sand, common detrital amphibole is mainly hornblende (42%), with minor pargasite and hastingsite, and rare actinolite and tschermakite. Clinopyroxene (diopside, minor augite) is abundant and orthopyroxene minor (Figure 6B). Epidote-group minerals are represented by epidote and clinozoisite (Figure 5B). The rare garnet grains are high in Ca and Mg (Figure 4B).

4.3. Greater Himalaya

Rivers draining amphibolite-facies metamorphic rocks of the Greater Himalaya carry moderately rich transparent heavy-mineral suites including amphibole, garnet, clinopyroxene and epidote, with minor titanite, tourmaline, apatite, sillimanite, and kyanite (Table 1).

The Zanskar River, sourced from the topmost part of the Greater Himalaya and cutting across the Tethys Himalaya, carries pargasite and hornblende with minor hastingsite, mainly Bi garnet with a few Bii and Ci grains (Figure 4C), and dominant epidote with minor allanite (Figure 5C). Diopside accounts for the vast majority of pyroxene grains (Figure 6C).

Similar amphibole varieties characterize Nandihar river sand, which contains mainly Bi garnet with minor A, Bii, and D grains (Figure 4C), epidote-clinozoisite but no allanite (Figure 5C), and a higher proportion of augite (30%) and orthopyroxene (25%) (Figure 6C).

4.4. Nanga Parbat

The very rich transparent-heavy-mineral suite of Astor River sand chiefly derived from the Nanga Parbat massif is dominated by amphibole with subordinate epidote-group minerals, garnet, and clinopyroxene (Table 1). Detrital amphibole is mainly hornblende with common tschermakite (14%) and minor pargasite (Figure 3C). Garnet grains are mainly Ci (57%) and minor Bii types (Figure 4C). Epidote and clinozoisite occur whereas allanite is lacking (Figure 5C), and detrital pyroxene is dominantly diopside with rare augite and orthopyroxene (Figure 6C).

5. Heavy-Mineral Provenance Tracers in Thal Desert Sand

Transparent-heavy-mineral suites of Thal Desert sand reflect the mineralogy of their diverse magmatic and metamorphic sources. As each source-rock domain contributes detrital species to the sediment load of the upper Indus River in different proportions, depending not only on exposure area and erosion rate but also on the different mineral concentrations (fertilities), every detrital mineral is expected to carry a distinct provenance signal (Figure 7). A source-rock domain may contribute one mineral (e.g., amphibole) in large proportion but another mineral (e.g., garnet) in negligible proportion; as a consequence, that domain will be over-represented in the detrital-amphibole spectrum but hardly seen in the detrital-garnet spectrum.

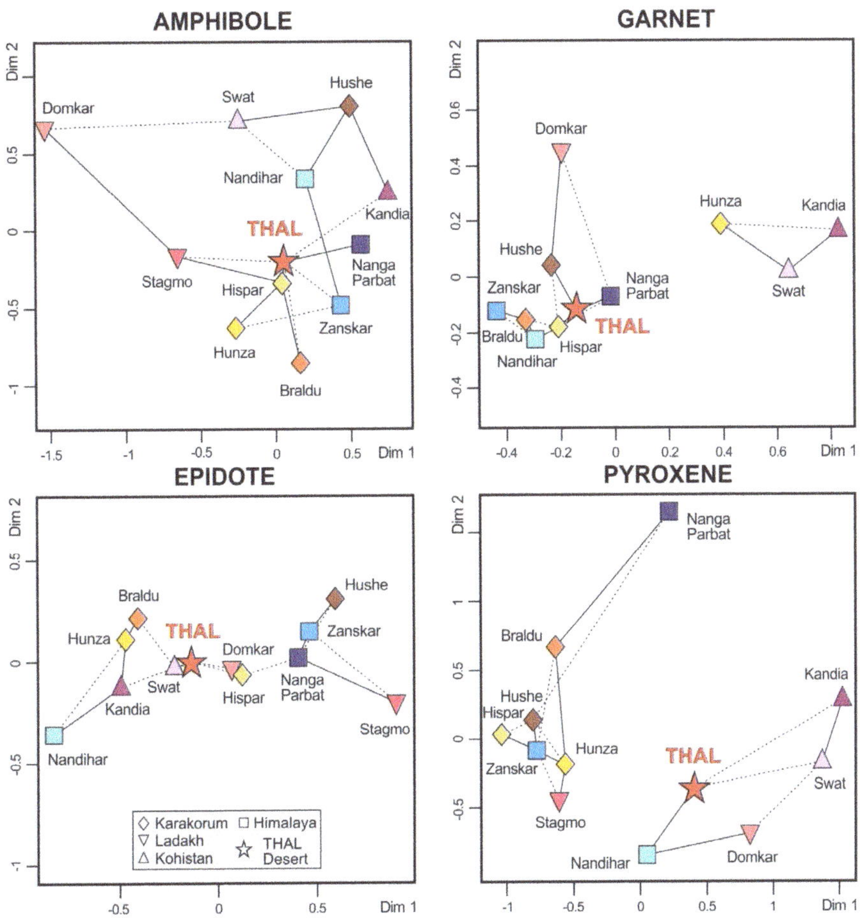

Figure 7. Multidimensional scaling maps based on the chemical signatures of the four studied mineral groups in river sands of northern Pakistan and in the Thal Desert (Supplementary Tables S7–S10). The four Thal dune samples are considered as subsamples of the same unitary population. Solid and dashed lines link closest and second-closest neighbors, respectively. The higher "stress" values (poorer fit) obtained for the amphibole (11.2%) and pyroxene maps (8.9%) than for the garnet (2.4%) and epidote maps (0.8%) largely reflect the higher number of varieties identified for amphibole (32) and pyroxene (39) than for epidote and garnet (six each). The four maps—plotted using the provenance package of Vermeesch et al. [129]—differ because different minerals are contained in markedly different proportions (fertilities) in different source-rock domains. Similarities among mineralogical spectra indicate the Kohistan arc as the main supplier of epidote and pyroxene, whereas amphibole and garnet were largely derived also from the Karakorum (Southern Karakorum gneiss domes drained by the Hispar River) and Himalaya (Nanga Parbat massif). The shape and color of sample symbols are same with those of Figures 3–6.

5.1. The Thal Desert as a Quaternary Sediment Sink

The very rich transparent-heavy-mineral suites of Thal Desert sand (tHMC 15.4 ± 3.9) mainly consist of amphibole, with common epidote, clinozoisite, clinopyroxene, and garnet. Detrital amphibole includes mainly hornblende, subordinate pargasite, actinolite (11% on average), hastingsite, and minor tschermakite (up to 5%; Figure 3D). Detrital garnet mainly consists of Bi grains with minor Ci (23%), Bii (18%), A (13%), and a few D grains (Figure 4D). Epidote-group minerals are mainly clinozoisite (54% on average) and epidote (Figure 5D). Detrital pyroxene is mainly diopside with common orthopyroxene (32% on average) and minor augite (Figure 6D).

The four studied dune samples are compositionally homogeneous (Table 1) and can thus be considered as subsamples of the same unitary population. Minor differences, however, are observed for instance between the adjacent Mankera and Haidarabad samples (Figure 1). Sample S1462 yielded more hornblende and more Bii than Ci garnets (20% vs. 18%), whereas sample S1463 yielded more hastingsite (27%), less Bii than Ci garnets (12% vs. 26%), and a lower diopside/augite ratio (Figures 8 and 9).

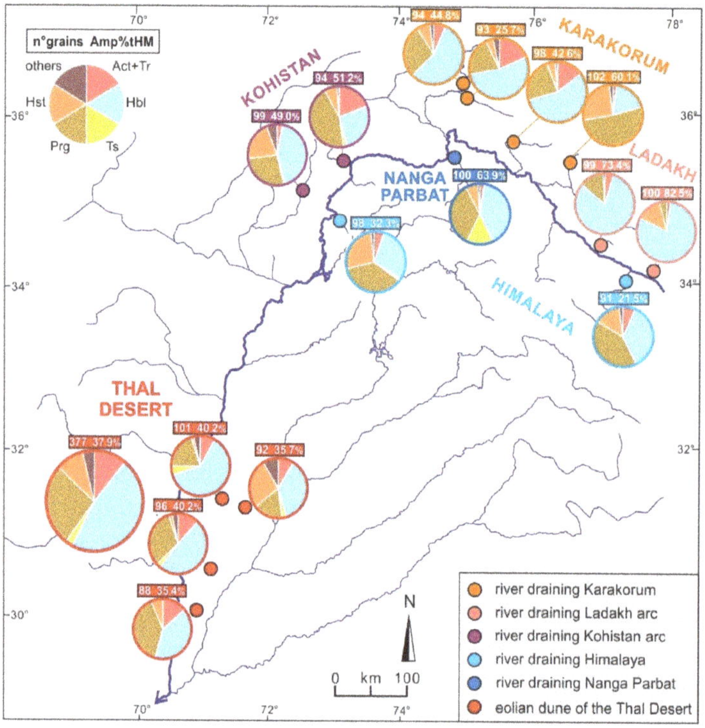

Figure 8. Calculated proportions of different varieties of detrital amphibole found in river and eolian sands of northern Pakistan. Act: actinolite; Hbl: hornblende; Hst: hastingsite; Prg: pargasite; Tr: tremolite; Ts: tschermakite.

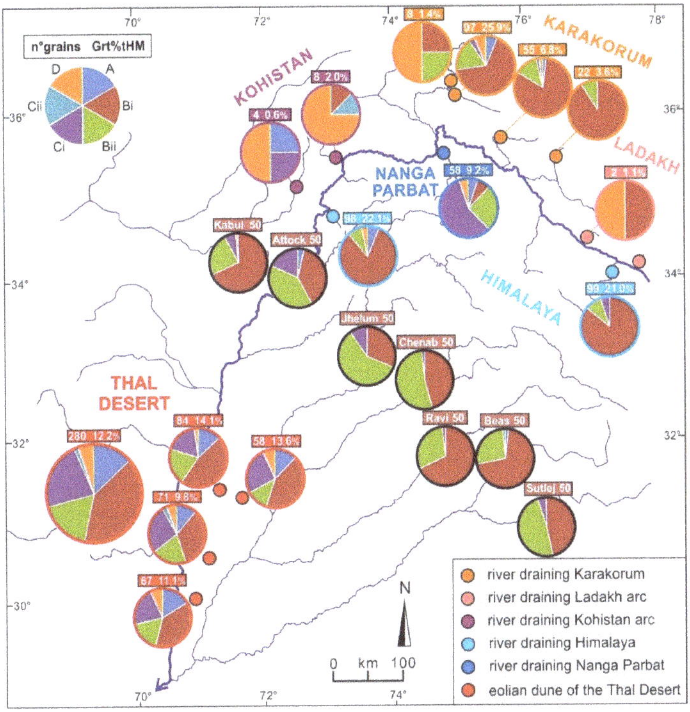

Figure 9. Calculated proportions of different varieties of detrital garnet (as defined in Mange and Morton, [5]) found in river and eolian sands of northern Pakistan. Data for Kabul, Indus, Jhelum, Chenab, Ravi, Beas, and Sutlej sands, circled in black, are from Alizai et al. [22]. Garnet types are explained in the text (Section 3.6).

5.2. *Heavy Mineral Concentration and Provenance Estimates*

A fundamental parameter in provenance analysis is represented by heavy-mineral concentration (HMC; [100,101]), which depends originally on the mineralogy and on the average density of parent rocks. The denser a rock is, the greater amount of dense minerals it contains and therefore can shed. Heavy-mineral concentration in sediments, however, can be modified even by an order of magnitude or more by hydraulic sorting during erosion, transport and sedimentation [106], or by chemical processes including weathering in soils and intrastratal dissolution during burial diagenesis [118,130]. Only in the absence of such environmental and diagenetic bias can terrigenous detritus be considered as produced purely by physical comminution and the mineralogy of daughter sand held to faithfully reflect the mineralogy of parent rocks. Under this strict assumption, the concentration (fertility) of each mineral can be determined for any specific source by the mineralogical analysis of daughter sand [15].

Our mineralogical dataset, integrated by data from Garzanti et al. [18] and Munack et al. [30], indicates that erosion in the diverse tectonic domains of the upper Indus catchment generate different amounts of heavy minerals. This depends principally on arc versus continental protoliths and crustal level exposed to erosion in each domain, because continental crust is more felsic and therefore less dense than arc crust, and because the Earth's crust is markedly stratified by density [131].

The very high transparent-heavy-mineral concentration in all of our four Thal dune-sand samples, which are not systematically enriched in densest minerals by selective-entrainment effects (Supplementary Table S1), points by itself to major heavy-mineral supply especially from dense mafic rocks exposed in deep tectonostratigraphic levels of the Kohistan arc and minor heavy-mineral supply from either the Karakorum or the Greater Himalaya. Similarity analysis [132] indicates that heavy-mineral suites resembling more closely those of Thal Desert dunes are those of Braldu and Hispar sands derived from the Central-Southern Karakorum, whereas the least similar are those of stream sand derived from the Ladakh arc. Forward-mixing calculations based on heavy-mineral data shown in Table 1 (mathematical method explained in Weltje [133] and Garzanti et al., [134]) confirm the Kohistan arc and the Central-Southern Karakorum as major sources of sediment for the Thal dunes.

5.3. The Amphibole Signal

The composite amphibole population of Thal Desert sand includes hornblende as well as other species identified in sands carried by diverse mountain tributaries of the Indus River, pointing to mixing from several sources (Figure 8). The relatively high amount of actinolite suggests contribution from Karakorum and/or Kohistan, and the presence of tschermakite indicates significant supply from the Nanga Parbat.

The massive appearance of blue-green hornblende in Upper Miocene foreland-basin strata of northern Pakistan was used as an indicator of rapid exhumation of the Kohistan arc [135]. Geochemical data from Lee et al. [21] confirm the Kohistan arc as a major source of amphibole, whereas the Nanga Parbat massif together with the Himalayan belt and the Ladakh arc in the uppermost catchment were held to be minor contributors. Major supply from the Kohistan arc was principally ascribed to high fertility, whereas the Southern Karakorum Belt was identified as the dominant source of bulk sediment also based on Nd isotope fingerprints [24].

5.4. The Garnet Signal

As the studied rivers draining the Ladakh and Kohistan arcs carry little garnet, most of which are Ca-rich type-D grains, the arcs cannot be considered as significant sources for garnet (Figure 9). However, Thal Desert dunes contain common high-Mg Ci and A garnet grains, which may have been derived not only from Nanga Parbat and the Greater Himalaya, respectively, but also from granulite-facies metagabbros and metasedimentary rocks exposed in the southern part of the Kohistan arc drained by the Indus River. These high-grade rocks of the lower arc crust may in fact contain up to 20%–30% garnet [136,137]. High-Mn garnets plotting in the low P/T field are sporadic in Thal dunes, but common in all river sands derived from the Karakorum belt and the Greater Himalaya (Figure 4), which argues against dominant garnet contribution from these sources.

Geochemical data from Alizai et al. [22] indicate that garnet in Kabul sand has intermediate signatures between those of the Karakorum belt and the Kohistan arc, and that Punjab Rivers draining the Himalayan belt carry mainly Bi and Bii garnets with minor Ci grains (Figure 9). Type D garnet occurs in all geological domains drained by the Indus River upstream of the Thal Desert but not in Punjab tributaries (Figure 9), which do not contribute significant amounts of sediment to the Thal Desert. Mg-rich garnet derived from the Kohistan arc and subordinately from the Karakorum belt, characteristic of Indus sand, are still the mark of Thar Desert dunes in southern Pakistan, chiefly representing the wind-reworked alluvial fan of the Indus River [22].

5.5. The Epidote Signal

Although epidote has been generally used as a provenance tracer based on isotopic fingerprints [138,139], major element geochemistry also provides critical information. Most important, allanite grains were not detected in all four Thal sand samples, which precludes significant contribution from the Northern and Central Karakorum drained by the upper Hunza, upper Braldu, and upper Hushe rivers, and by the upper part of the Greater Himalaya drained by the Zanskar River (Figure 10). The abundance of clinozoisite favors instead major contribution from the Kohistan arc and possibly from the Southern Karakorum, drained by the Hispar which is the only Karakorum river that does not carry allanite. Subordinate supply from the Nanga Parbat massif, mainly shedding epidote, and from the Greater Himalaya in Pakistan, dominantly shedding clinozoisite, cannot be ruled out. Although zoisite cannot be identified from geochemical data by WinEpclas software [122], Raman spectroscopy revealed its abundance in both Thal Desert and Kandia River sands (Table 1), confirming the Kohistan arc as a major source of detrital epidote for Thal dunes.

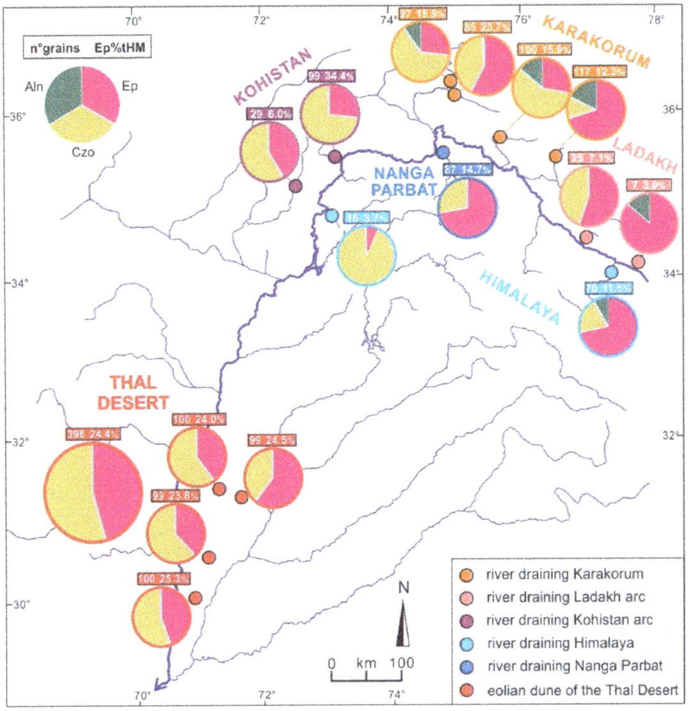

Figure 10. Calculated proportions of different varieties of detrital epidote-group minerals found in river and eolian sands of northern Pakistan. Note that Thal dunes, as well as sand from Kohistan and Nanga Parbat, lack allanite. Among Karakorum rivers, only the Hispar does not carry allanite, which singles out the Southern Karakorum as the only domain potentially representing a major source of epidote within the belt. Aln: allanite; Czo: clinozoisite; Ep: epidote.

5.6. The Pyroxene Signal

The abundance of orthopyroxene in Thal dunes points to dominant contribution from the Kohistan arc, with minor to negligible additional contributions from other sources (Figure 11). In fact, the Ladakh arc (Wo < 30), the Karakorum belt (upper Hushe sand; Wo ~30), and the Greater Himalaya (Nandihar sand; Wo < 40) shed mostly Ca-poor augite (Figure 6), whereas augite grains in Thal Desert dunes are mostly Ca-rich (Wo > 40). Detrital ferroaugite is negligible in Thal dunes, whereas it occurs in sand derived from Nanga Parbat, Greater Himalaya, and Karakorum belt (upper Hunza sand) (Figure 6).

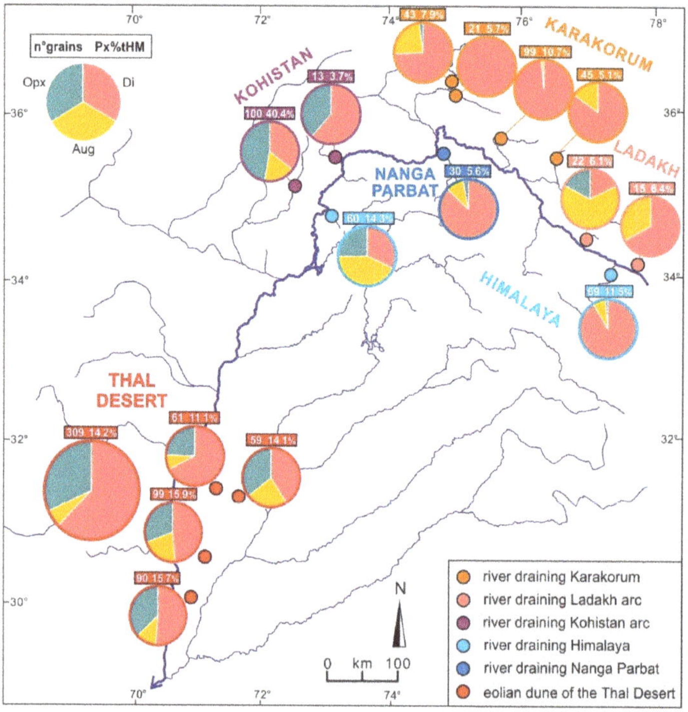

Figure 11. Calculated proportions of different varieties of detrital pyroxene found in river and eolian sands of northern Pakistan. Aug: augite (25 < Wo < 45); Di: diopside (Wo > 45); Opx: orthopyroxene (Wo < 5); Wo: wollastonite.

6. Conclusions

Varietal studies of heavy minerals have long been proven to provide crucial information on sediment provenance. The present study focuses on the chemical composition of detrital amphibole, garnet, epidote, and pyroxene because these solid-solution series are the four dominant minerals in orogenic sediments worldwide. The rich minerochemical dataset produced is intended as a basis useful to discriminate among the diverse sources of detritus within the upper part of the Indus River catchment in northern Pakistan, upstream of its entry point in the Punjab foreland basin. Therefore, the thorough quantitative description of mineralogical signatures of Thal Desert dune sand, representing a relict sink of sediment entirely derived from the upper Indus River in the Quaternary, offers a complementary way to trace erosion patterns across the western Himalayan syntaxis and adjacent orogenic segments.

High-resolution analysis of Thal Desert dune sand indicates that the Kohistan arc has played the principal role as a source of heavy minerals, especially as pyroxene and epidote are concerned. The similarity among mineralogical spectra suggests that the Southern Karakorum gneiss domes undergoing fast exhumation and the Nanga Parbat massif were major suppliers of amphibole and garnet, reflecting high erosion rates in the western Himalaya syntaxis. Among other Himalayan domains, a minor amount of heavy minerals was supplied by the Greater Himalaya, whereas detritus from the Lesser Himalaya and Subhimalaya becomes significant in Indus sand only in southern Pakistan, downstream of the confluence with Punjab tributaries. The contrast between mineralogical fingerprints of Thal Desert sand, entirely derived from geological domains exposed around the western Himalayan syntaxis, and those of detritus carried by Punjab tributaries, which drain the Himalayan belt exclusively, can be exploited to assess how the relative contributions from these different parts of the Himalayan-Karakorum orogen to the Indus delta to huge deep-sea fan have changed through time. Such a clear differentiation between Transhimalayan and Himalayan sources of detritus provides a semi-actualistic key that can be used, together with complementary compositional datasets and geological information, to make a step forward in the understanding of the erosional evolution of the Himalayan orogen and of landscape changes in the Punjab foreland basin as controlled by the complex interplay between climatic and tectonic forces in the recent and less recent past.

Supplementary Materials: The following are available online at http://www.mdpi.com/2075-163X/9/8/457/s1, Table S1: Sampling sites (including textural and geochemical information), Table S2: Heavy-mineral data obtained by semi-automated Raman spectroscopy, Table S3: Chemical composition of detrital amphiboles, Table S4: Chemical composition of detrital garnets, Table S5: Chemical composition of detrital epidote-group minerals, Table S6: Chemical composition of detrital pyroxenes, Table S7: Percentages of amphibole varieties in each sample, Table S8: Percentages of garnet varieties in each sample, Table S9: Percentages of epidote varieties in each sample, Table S10: Percentages of pyroxene varieties in each sample.

Author Contributions: Project design, E.G.; Data collection: W.L., P.G., S.A.; Semi-automated Raman counting and statistical analysis: A.R.; Writing and editing, W.L., E.G.

Funding: This study was supported financially by PRIN-MIUR (Progetto di Rilevante Interesse Nazionale, Ministero dell'Istruzione, dell'Università e della Ricerca) to E.Garzanti (2015EC9PJ5); MIUR – Dipartimenti di Eccellenza 2018–2022, Department of Earth and Environmental Sciences, University of Milano-Bicocca and State Scholarship Fund organized by China Scholarship Council (Grant No. 201606450012).

Acknowledgments: Heartfelt thanks to Giacomo Ghielmi and Filippo Lazzati, who collected Thal desert and Indus tributary sands in 2001, and to Mike Searle, Peter Clift, Jan Blöthe, and Henry Munack who provided additional samples from the upper Indus catchment. Careful constructive reviews by Carita Augustsson and other two reviewers are very gratefully acknowledged.

Conflicts of Interest: The authors declare no conflict of interest.

References

1. Mange, M.A.; Wright, D.T. *Heavy Minerals in Use. Developments in Sedimentology Series*; Elsevier: Amsterdam, The Netherlands, 2007; Volume 58, 1283p.
2. Garzanti, E.; Vezzoli, G.; Lombardo, B.; Ando, S.; Mauri, E.; Monguzzi, S.; Russo, M. Collision-orogen provenance (western Alps): Detrital signatures and unroofing trends. *J. Geol.* **2004**, *112*, 145–164. [CrossRef]
3. Garzanti, E.; Resentini, A.; Vezzoli, G.; Ando, S.; Malusa, M.G.; Padoan, M.; Paparella, P. Detrital fingerprints of fossil continental-subduction zones (Axial Belt Provenance, European Alps). *J. Geol.* **2010**, *118*, 341–362. [CrossRef]
4. Garzanti, E.; Al-Juboury, A.I.; Zoleikhaei, Y.; Vermeesch, P.; Jotheri, J.; Akkoca, D.B.; Obaid, A.K.; Allen, M.B.; Ando, S.; Limonta, M. The Euphrates-Tigris-Karun river system: Provenance, recycling and dispersal of quartz-poor foreland-basin sediments in arid climate. *Earth Sci. Rev.* **2016**, *162*, 107–128. [CrossRef]
5. Mange, M.A.; Morton, A.C. Geochemistry of heavy minerals. In *Heavy Minerals in Use. Developments in Sedimentology Series*; Mange, M.A., Wright, D.T., Eds.; Elsevier: Amsterdam, The Netherlands, 2007; Volume 58, pp. 345–391. [CrossRef]
6. Krynine, P.D. The tourmaline group in sediments. *J. Geol.* **1946**, *54*, 65–87. [CrossRef]
7. Morton, A.C. A new approach to provenance studies: Electron microprobe analysis of detrital garnets from Middle Jurassic sandstones of the northern North Sea. *Sedimentology* **1985**, *32*, 553–566. [CrossRef]
8. Morton, A.C. Geochemical studies of heavy minerals and their application to provenance research. In *Developments in Sedimentary Provenance Studies*; Special Publication 57; Morton, A.C., Todd, S.P., Haughton, P.D.W., Eds.; Geological Society of London Publications: London, UK, 1991; pp. 31–45. [CrossRef]
9. von Eynatten, H.; Gaupp, R. Provenance of Cretaceous synorogenic sandstones in the Eastern Alps: Constraints from framework petrography, heavy mineral analysis and mineral chemistry. *Sediment. Geol.* **1999**, *124*, 81–111. [CrossRef]
10. Meinhold, G. Rutile and its applications in earth sciences. *Earth Sci. Rev.* **2010**, *102*, 1–28. [CrossRef]
11. Andò, S.; Morton, A.; Garzanti, E. Metamorphic grade of source rocks revealed by chemical fingerprints of detrital amphibole and garnet. In *Sediment Provenance Studies in Hydrocarbon Exploration and Production*; Special Publication 386; Scott, R.A., Smyth, H.R., Morton, A.C., Richardson, N., Eds.; Geological Society of London Publications: London, UK, 2014; pp. 351–371. [CrossRef]

12. Malusà, M.G.; Wang, J.; Garzanti, E.; Liu, Z.C.; Villa, I.M.; Wittmann, H. Trace-element and Nd-isotope systematics in detrital apatite of the Po river catchment: Implications for provenance discrimination and the lag-time approach to detrital thermochronology. *Lithos* **2017**, *290*, 48–59. [CrossRef]
13. von Eynatten, H.; Dunkl, I. Assessing the sediment factory: The role of single grain analysis. *Earth Sci. Rev.* **2012**, *115*, 97–120. [CrossRef]
14. Moecher, D.P.; Samson, S.D. Differential zircon fertility of source terranes and natural bias in the detrital zircon record: Implications for sedimentary provenance analysis. *Earth Planet. Sci. Lett.* **2006**, *247*, 252–266. [CrossRef]
15. Malusà, M.G.; Resentini, A.; Garzanti, E. Hydraulic sorting and mineral fertility bias in detrital geochronology. *Gondwana Res.* **2016**, *31*, 1–19. [CrossRef]
16. Garzanti, E.; Limonta, M.; Vezzoli, G.; An, W.; Wang, J.; Hu, X. Petrology and multimineral fingerprinting of modern sand generated from a dissected magmatic arc (Lhasa River, Tibet). In *Tectonics, Sedimentary Basins, and Provenance: A Celebration of William, R. Dickinson's Career*; Special Paper 540; Ingersoll, R.V., Lawton, T.F., Graham, S.A., Eds.; Geological Society of America: Boulder, CO, USA, 2018; pp. 197–221. [CrossRef]
17. Guo, R.; Hu, X.M.; Garzanti, E.; Lai, W.; Yan, B. Magmatic and metamorphic events in Himalaya and Tibet traced by geochronological and geochemical fingerprinting of detrital zircon, monazite, rutile, and titanite. *Earth Sci. Rev.* **2019**, in press.
18. Garzanti, E.; Vezzoli, G.; Ando, S.; Paparella, P.; Clift, P.D. Petrology of Indus River sands: A key to interpret erosion history of the Western Himalayan Syntaxis. *Earth Planet. Sci. Lett.* **2005**, *229*, 287–302. [CrossRef]
19. Clift, P.D.; Shimizu, N.; Layne, G.; Blusztajn, J.S.; Gaedicke, C.; Schluter, H.-U.; Clark, M.K.; Amjad, S. Development of the Indus Fan and its significance for the erosional history of the Western Himalaya and Karakoram. *Geol. Soc. Am. Bull.* **2001**, *113*, 1039–1051. [CrossRef]
20. Clift, P.D.; Giosan, L.; Carter, A.; Garzanti, E.; Galy, V.; Tabrez, A.R.; Pringle, M.; Campbell, I.H.; France-Lanord, C.; Blusztajn, J. Monsoon control over erosion patterns in the western Himalaya: Possible feed-back into the tectonic evolution. In *Monsoon Evolution and Tectonic-Climate Linkage in Asia*; Special Publications 342; Clift, P.D., Tada, R., Zheng, H., Eds.; Geological Society of London Publications: London, UK, 2010; pp. 185–218. [CrossRef]
21. Lee, J.I.; Clift, P.D.; Layne, G.; Blum, J.; Khan, A.A. Sediment flux in the modern Indus River inferred from the trace element composition of detrital amphibole grains. *Sediment. Geol.* **2003**, *160*, 243–257. [CrossRef]
22. Alizai, A.; Clift, P.D.; Still, J. Indus Basin sediment provenance constrained using garnet geochemistry. *J. Asian Earth Sci.* **2016**, *126*, 29–57. [CrossRef]
23. Garzanti, E.; Andò, S. Plate tectonics and heavy-mineral suites of modern sands. In *Heavy Minerals in Use. Developments in Sedimentology Series*; Mange, M.A., Wright, D.T., Eds.; Elsevier: Amsterdam, The Netherlands, 2007; Volume 58, pp. 741–763. [CrossRef]
24. Clift, P.D.; Lee, J.I.; Hildebrand, P.; Shimizu, N.; Layne, G.D.; Blusztajn, J.; Blum, J.D.; Garzanti, E.; Khan, A.A. Nd and Pb isotope variability in the Indus River System: Implications for sediment provenance and crustal heterogeneity in the Western Himalaya. *Earth Planet. Sci. Lett.* **2002**, *200*, 91–106. [CrossRef]
25. Alizai, A.; Clift, P.D.; Giosan, L.; Van Laningham, S.; Hinton, R.; Tabrez, A.R.; Danish, M.; The Edinburgh Ion Microprobe Facility (EIMF). Pb isotopic variability in the modern-Pleistocene Indus River system measured by ion microprobe in detrital K-feldspar grains. *Geochim. Cosmochim. Acta* **2001**, *75*, 4771–4795. [CrossRef]
26. Jonell, T.N.; Li, Y.; Blusztajn, J.; Giosan, L.; Clift, P.D. Signal or noise? Isolating grain size effects on Nd and Sr isotope variability in Indus delta sediment provenance. *Chem. Geol.* **2018**, *485*, 56–73. [CrossRef]
27. Clift, P.D.; Campbell, I.H.; Pringle, M.S.; Carter, A.; Zhang, X.; Hodges, K.V.; Khan, A.A.; Allen, C.M. Thermochronology of the modern Indus River bedload: New insight into the controls on the marine stratigraphic record. *Tectonics* **2004**, *23*, TC5013. [CrossRef]
28. Campbell, I.H.; Reiners, P.W.; Allen, C.M.; Nicolescu, S.; Upadhyay, R. He-Pb double dating of detrital zircons from the Ganges and Indus Rivers: Implication for quantifying sediment recycling and provenance studies. *Earth Planet. Sci. Lett.* **2005**, *237*, 402–432. [CrossRef]
29. Alizai, A.; Carter, A.; Clift, P.D.; Van Laningham, S.; Williams, J.C.; Kumar, R. Sediment provenance, reworking and transport processes in the Indus River by U–Pb dating of detrital zircon grains. *Glob. Planet. Chang.* **2011**, *76*, 33–55. [CrossRef]

30. Munack, H.; Korup, O.; Resentini, A.; Limonta, M.; Garzanti, E.; Blöthe, J.H.; Scherler, D.; Wittmann, H.; Kubik, P.W. Postglacial denudation of western Tibetan Plateau margin outpaced by long-term exhumation. *Geol. Soc. Am. Bull.* **2014**, *126*, 1580–1594. [CrossRef]
31. Alizai, A.; Hillier, S.; Clift, P.D.; Giosan, L.; Hurst, A.; Van Laningham, S.; Macklin, M. Clay mineral variations in Holocene terrestrial sediments from the Indus Basin. *Quat. Res.* **2012**, *77*, 368–381. [CrossRef]
32. Treloar, P.J.; Petterson, M.G.; Jan, M.Q.; Sullivan, M.A. A re-evaluation of the stratigraphy and evolution of the Kohistan arc sequence, Pakistan Himalaya: Implications for magmatic and tectonic arc-building processes. *J. Geol. Soc.* **1996**, *153*, 681–693. [CrossRef]
33. DiPietro, J.A.; Pogue, K.R. Tectonostratigraphic subdivisions of the Himalaya: A view from the west. *Tectonics* **2004**, *23*, TC5001. [CrossRef]
34. Pêcher, A.; Seeber, L.; Guillot, S.; Jouanne, F.; Kausar, A.; Latif, M.; Majid, A.; Mahéo, G.; Mugnier, J.L.; Rolland, Y.; et al. Stress field evolution in the northwest Himalayan syntaxis, northern Pakistan. *Tectonics* **2008**, *27*, TC6005. [CrossRef]
35. Burg, J.P. The Asia–Kohistan–India collision: Review and discussion. In *Arc-Continent Collision*; Brown, D., Ryan, P.D., Eds.; Springer: Berlin/Heidelberg, Germany, 2011; pp. 279–309. [CrossRef]
36. Khan, I.A.; Bridge, J.S.; Kappelman, J.; Wilson, R. Evolution of Miocene fluvial environments, eastern Potwar plateau, northern Pakistan. *Sedimentology* **1997**, *44*, 221–251. [CrossRef]
37. Hildebrand, P.R.; Noble, S.R.; Searle, M.P.; Waters, D.J.; Parrish, R.R. Old origin for an active mountain range: Geology and geochronology of the eastern Hindu Kush, Pakistan. *Geol. Soc. Am. Bull.* **2001**, *113*, 625–639. [CrossRef]
38. Ferguson, R.I. Sediment load of the Hunza River. In *The International Karakoram Project*; Miller, K.J., Ed.; Cambridge University Press: Cambridge, UK, 1984; Volume 2, pp. 581–598.
39. Ali, K.F.; de Boer, D.H. Factors controlling specific sediment yield in the upper Indus River basin, northern Pakistan. *Hydrol. Process.* **2008**, *22*, 3102–3114. [CrossRef]
40. Einsele, G.; Hinderer, M. Terrestrial sediment yield and the lifetimes of reservoirs, lakes, and larger basins. *Geol. Rundsch.* **1997**, *86*, 288–310. [CrossRef]
41. Tate, E.L.; Farquharson, F.K. Simulating reservoir management under the threat of sedimentation: The case of Tarbela Dam on the River Indus. *Water Resour. Manag.* **2000**, *14*, 191–208. [CrossRef]
42. Rehman, S.S.; Sabir, M.A.; Khan, J. Discharge characteristics and suspended load from rivers of Northern Indus Basin, Pakistan. *Geol. Bull. Univ. Peshawar* **1997**, *30*, 325–326.
43. Milliman, J.D.; Quraishee, G.S.; Beg, M.A.A. Sediment discharge from the Indus River to the ocean: Past, present, and future. In *Marine Geology and Oceanography of the Arabian Sea and Coastal Pakistan*; Haq, B.U., Milliman, J.D., Eds.; Van Nostrand Reynolds: New York, NY, USA, 1984; pp. 65–70.
44. Meadows, A.; Meadows, P.S. *The Indus River: Biodiversity, Resources, Humankind*; Oxford University Press: Oxford, UK, 1999.
45. Searle, M.P.; Khan, M.A.; Fraser, J.E.; Gough, S.J.; Jan, M.Q. The tectonic evolution of the Kohistan-Karakoram collision belt along the Karakoram Highway transect, north Pakistan. *Tectonics* **1999**, *18*, 929–949. [CrossRef]
46. Hildebrand, P.R.; Searle, M.P.; Khan, Z.; Van Heijst, H.J. Geological evolution of the Hindu Kush, NW Frontier Pakistan: Active margin to continent-continent collision zone. In *Tectonics of the Nanga Parbat Syntaxis and the Western Himalaya*; Special Publication 170; Khan, M.A., Treloar, P.J., Searle, M.P., Jan, M.Q., Eds.; Geological Society of London Publications: London, UK, 2000; pp. 277–293. [CrossRef]
47. Gaetani, M.; Garzanti, E.; Jadoul, F.; Nicora, A.; Tintori, A.; Pasini, M.; Khan, K.S.A. The north Karakorum side of the Central Asia geopuzzle. *Geol. Soc. Am. Bull.* **1990**, *102*, 54–62. [CrossRef]
48. Zanchi, A.; Gaetani, M. The geology of the Karakoram range, Pakistan: The new 1: 100 000 geological map of Central-Western Karakoram. *Ital. J. Geosci.* **2011**, *130*, 161–262. [CrossRef]
49. Crawford, M.B.; Searle, M.P. Field relationships and geochemistry of pre-collisional (India-Asia) granitoid magmatism in the central Karakoram, northern Pakistan. *Tectonophysics* **1992**, *206*, 171–192. [CrossRef]
50. Searle, M.P.; Parrish, R.R.; Thow, A.V.; Noble, S.R.; Phillips, R.J.; Waters, D.J. Anatomy, age and evolution of a collisional mountain belt: The Baltoro granite batholith and Karakoram Metamorphic Complex, Pakistani Karakoram. *J. Geol. Soc.* **2010**, *167*, 183–202. [CrossRef]
51. Searle, M.P.; Tirrul, R. Structural and thermal evolution of the Karakoram crust. *J. Geol. Soc.* **1991**, *148*, 65–82. [CrossRef]

52. Rolland, Y.; Mahéo, G.; Guillot, S.; Pêcher, A. Tectono-metamorphic evolution of the Karakorum metamorphic complex (Dassu-Askole area, NE Pakistan): Exhumation of mid-crustal HT–MP gneisses in a convergent context. *J. Metamorph. Geol.* **2001**, *19*, 717–737. [CrossRef]
53. Palin, R.M.; Searle, M.P.; Waters, D.J.; Horstwood, M.S.A.; Parrish, R.R. Combined thermobarometry and geochronology of peraluminous metapelites from the Karakoram metamorphic complex, North Pakistan; New insight into the tectonothermal evolution of the Baltoro and Hunza Valley regions. *J. Metamorph. Geol.* **2012**, *30*, 793–820. [CrossRef]
54. Treloar, P.J.; Rex, D.C.; Guise, P.G.; Coward, M.P.; Searle, M.P.; Windley, B.F.; Petterson, M.G.; Jan, M.Q.; Luff, I.W. K-Ar and Ar-Ar geochronology of the Himalayan collision in NW Pakistan: Constraints on the timing of suturing, deformation, metamorphism and uplift. *Tectonics* **1989**, *8*, 881–909. [CrossRef]
55. Gaetani, M.; Jadoul, F.; Erba, E.; Garzanti, E. Jurassic and Cretaceous orogenic events in the North Karakoram: Age constraints from sedimentary rocks. In *Himalayan Tectonics*; Special Publications 74; Treloar, P.J., Searle, M.P., Eds.; Geological Society of London Publications: London, UK, 1993; pp. 39–52. [CrossRef]
56. Robertson, A.H.; Collins, A.S. Shyok Suture Zone, N Pakistan: Late Mesozoic–Tertiary evolution of a critical suture separating the oceanic Ladakh Arc from the Asian continental margin. *J. Asian Earth Sci.* **2002**, *20*, 309–351. [CrossRef]
57. Rehman, H.U.; Seno, T.; Yamamoto, H.; Khan, T. Timing of collision of the Kohistan-Ladakh Arc with India and Asia: Debate. *Isl. Arc* **2011**, *20*, 308–328. [CrossRef]
58. Borneman, N.L.; Hodges, K.V.; Soest, M.C.; Bohon, W.; Wartho, J.A.; Cronk, S.S.; Ahmad, T. Age and structure of the Shyok suture in the Ladakh region of northwestern India: Implications for slip on the Karakoram fault system. *Tectonics* **2015**, *34*, 2011–2033. [CrossRef]
59. Garzanti, E.; Baud, A.; Mascle, G. Sedimentary record of the northward flight of India and its collision with Eurasia (Ladakh Himalaya, India). *Geodin. Acta* **1987**, *1*, 297–312. [CrossRef]
60. Najman, Y.; Jenks, D.; Godin, L.; Boudagher-Fadel, M.; Millar, I.; Garzanti, E.; Horstwood, M.; Bracciali, L. The Tethyan Himalayan detrital record shows that India-Asia terminal collision occurred by 54 Ma in the Western Himalaya. *Earth Planet. Sci. Lett.* **2017**, *459*, 301–310. [CrossRef]
61. Jan, M.Q.; Howie, R.A. The mineralogy and geochemistry of the metamorphosed basic and ultrabasic rocks of the Jijal complex, Kohistan, NW Pakistan. *J. Petrol.* **1981**, *22*, 85–126. [CrossRef]
62. Jagoutz, O.; Müntener, O.; Ulmer, P.; Pettke, T.; Burg, J.P.; Dawood, H.; Hussain, S. Petrology and mineral chemistry of lower crustal intrusions: The Chilas Complex, Kohistan (NW Pakistan). *J. Petrol.* **2007**, *48*, 1895–1953. [CrossRef]
63. Dhuime, B.; Bosch, D.; Garrido, C.J.; Bodinier, J.L.; Bruguier, O.; Hussain, S.S.; Dawood, H. Geochemical architecture of the lower-to middle-crustal section of a paleo-island arc (Kohistan Complex, Jijal–Kamila area, northern Pakistan): Implications for the evolution of an oceanic subduction zone. *J. Petrol.* **2009**, *50*, 531–569. [CrossRef]
64. Honegger, K.; Dietrich, V.; Frank, W.; Gansser, A.; Thöni, M.; Trommsdorff, V. Magmatism and metamorphism in the Ladakh Himalayas (the Indus-Tsangpo suture zone). *Earth Planet. Sci. Lett.* **1982**, *60*, 253–292. [CrossRef]
65. Weinberg, R.F.; Dunlap, W.J. Growth and deformation of the Ladakh Batholith, Northwest Himalayas: Implications for timing of continental collision and origin of calc-alkaline batholiths. *J. Geol.* **2000**, *108*, 303–320. [CrossRef] [PubMed]
66. Garzanti, E.; Van Haver, T. The Indus clastics: Forearc basin sedimentation in the Ladakh Himalaya (India). *Sediment. Geol.* **1988**, *59*, 237–249. [CrossRef]
67. Henderson, A.L.; Najman, Y.; Parrish, R.; BouDagher-Fadel, M.; Barford, D.; Garzanti, E.; Andò, S. Geology of the Cenozoic Indus Basin sedimentary rocks: Paleoenvironmental interpretation of sedimentation from the western Himalaya during the early phases of India-Eurasia collision. *Tectonics* **2010**, *29*, TC6015. [CrossRef]
68. Anczkiewicz, R.; Burg, J.-P.; Hussain, S.S.; Dawood, H.; Ghazanfar, M.; Chaudhry, M. Stratigraphy and structure of the Indus Suture in the lower Swat, Pakistan, NW Himalaya. *J. Asian Earth Sci.* **1998**, *16*, 225–238. [CrossRef]
69. Mahéo, G.; Fayoux, X.; Guillot, S.; Garzanti, E.; Capiez, P.; Mascle, G. Relicts of an intra-oceanic arc in the Sapi-Shergol mélange zone (Ladakh, NW Himalaya, India): Implications for the closure of the Neo-Tethys Ocean. *J. Asian Earth Sci.* **2006**, *26*, 695–707. [CrossRef]
70. DeCelles, P.; Kapp, P.; Gehrels, G.; Ding, L. Paleocene-Eocene foreland basin evolution in the Himalaya of southern Tibet and Nepal: Implications for the age of initial India-Asia collision. *Tectonics* **2014**, *33*, 824–849. [CrossRef]

71. Hu, X.; Garzanti, E.; Wang, J.; Huang, W.; An, W.; Webb, A. The timing of India-Asia collision onset—Facts, theories, controversies. *Earth Sci. Rev.* **2016**, *160*, 264–299. [CrossRef]
72. Ratschbacher, L.; Frisch, W.; Liu, G.; Chen, C. Distributed deformation in southern and western Tibet during and after the India-Asia collision. *J. Geophys. Res. Solid Earth* **1994**, *99*, 19917–19945. [CrossRef]
73. Searle, M.; Corfield, R.I.; Stephenson, B.E.N.; McCarron, J.O.E. Structure of the North Indian continental margin in the Ladakh–Zanskar Himalayas: Implications for the timing of obduction of the Spontang ophiolite, India-Asia collision and deformation events in the Himalaya. *Geol. Mag.* **1997**, *134*, 297–316. [CrossRef]
74. Gaetani, M.; Garzanti, E. Multicyclic history of the Northern India continental margin (Northwestern Himalaya)(1). *AAPG Bull.* **1991**, *75*, 1427–1446.
75. Sciunnach, D.; Garzanti, E. Subsidence history of the Tethys Himalaya. *Earth Sci. Rev.* **2012**, *111*, 179–198. [CrossRef]
76. Pognante, U.T.; Lombardo, B. Metamorphic evolution of the High Himalayan crystallines in SE Zanskar, India. *J. Metamorph. Geol.* **1989**, *7*, 9–17. [CrossRef]
77. Herren, E. Zanskar shear zone: Northeast-southwest extension within the Higher Himalayas (Ladakh, India). *Geology* **1987**, *15*, 409–413. [CrossRef]
78. Steck, A. Geology of the NW Indian Himalaya. *Eclogae Geol. Helv.* **2003**, *96*, 147–196. [CrossRef]
79. Greco, A.; Spencer, D.A. A section through the Indian plate, Kaghan valley, NW Himalaya, Pakistan. In *Himalayan Tectonics*; Special Publications 74; Treloar, P.J., Searle, M.P., Eds.; Geological Society of London Publications: London, UK, 1993; pp. 221–236. [CrossRef]
80. Vannay, J.C.; Grasemann, B.; Rahn, M.; Frank, W.; Carter, A.; Baudraz, V.; Cosca, M. Miocene to Holocene exhumation of metamorphic crustal wedges in the NW Himalaya: Evidence for tectonic extrusion coupled to fluvial erosion. *Tectonics* **2004**, *23*, TC1014. [CrossRef]
81. Bossart, P.; Ottiger, R. Rocks of the Murree Formation in northern Pakistan: Indicators of a descending foreland basin of late Paleocene to middle Eocene age. *Eclogae Geol. Helv.* **1989**, *82*, 133–165.
82. Najman, Y.; Garzanti, E. Reconstructing early Himalayan tectonic evolution and paleogeography from Tertiary foreland basin sedimentary rocks, northern India. *Geol. Soc. Am. Bull.* **2000**, *112*, 435–449. [CrossRef]
83. White, N.; Pringle, M.; Garzanti, E.; Bickle, M.; Najman, Y.; Chapman, H.; Friend, P. Constraints on the exhumation and erosion of the High Himalayan slab, NW India, from foreland basin deposits. *Earth Planet. Sci. Lett.* **2002**, *195*, 29–44. [CrossRef]
84. Zeitler, P.K.; Chamberlain, C.P.; Smith, H.A. Synchronous anatexis, metamorphism, and rapid denudation at Nanga Parbat (Pakistan Himalaya). *Geology* **1993**, *21*, 347–350. [CrossRef]
85. Chamberlain, C.P.; Zeitler, P.K.; Barnett, D.E.; Winslow, D.; Poulson, S.R.; Leahy, T.; Hammer, J.E. Active hydrothermal systems during the recent uplift of Nanga Parbat, Pakistan Himalaya. *J. Geophys. Res. Solid Earth* **1995**, *100*, 439–453. [CrossRef]
86. Zeitler, P.K.; Chamberlain, C.P. Petrogenetic and tectonic significance of young leucogranites from the northwestern Himalaya, Pakistan. *Tectonics* **1991**, *10*, 729–741. [CrossRef]
87. Schneider, D.A.; Zeitler, P.K.; Kidd, W.S.F.; Edwards, M.A. Geochronologic constraints on the tectonic evolution and exhumation of Nanga Parbat, western Himalaya syntaxis, revisited. *J. Geol.* **2001**, *109*, 563–583. [CrossRef]
88. Burbank, D.W.; Leland, J.; Fielding, E.; Anderson, R.S.; Brozovic, N.; Reid, M.R.; Duncan, C. Bedrock incision, rock uplift and threshold hillslopes in the northwestern Himalayas. *Nature* **1996**, *379*, 505–510. [CrossRef]
89. Shroder, J.F.; Bishop, M.P. Unroofing of Nanga Parbat Himalaya. In *Tectonics of the Nanga Parbat Syntaxis and the Western Himalaya*; Special Publication 170; Khan, M.A., Treloar, P.J., Searle, M.P., Jan, M.Q., Eds.; Geological Society of London Publications: London, UK, 2000; pp. 163–179. [CrossRef]
90. Whittington, A.G. Exhumation overrated at Nanga Parbat, northern Pakistan. *Tectonophysics* **1996**, *260*, 215–226. [CrossRef]
91. Moore, M.A.; England, P.C. On the inference of denudation rates from cooling ages of minerals. *Earth Planet. Sci. Lett.* **2001**, *185*, 265–284. [CrossRef]
92. Zeitler, P.K.; Meltzer, A.S.; Koons, P.O.; Craw, D.; Hallet, B.; Chamberlain, C.P.; Kidd, W.S.; Park, S.K.; Seeber, L.; Bishop, M.; et al. Erosion, Himalayan geodynamics, and the geomorphology of metamorphism. *GSA Today* **2001**, *11*, 4–9. [CrossRef]

93. Johnson, N.M.; Stix, J.; Tauxe, L.; Cerveny, P.F.; Tahirkheli, R.A. Paleomagnetic chronology, fluvial processes, and tectonic implications of the Siwalik deposits near Chinji village, Pakistan. *J. Geol.* **1985**, *93*, 27–40. [CrossRef]
94. Najman, Y.; Garzanti, E.; Pringle, M.; Bickle, M.; Stix, J.; Khan, I. Early-Middle Miocene paleodrainage and tectonics in the Pakistan Himalaya. *Geol. Soc. Am. Bull.* **2003**, *115*, 1265–1277. [CrossRef]
95. Burbank, D.W.; Beck, R.A.; Mulder, T. The Himalayan foreland basin. In *The Tectonic Evolution of Asia (World and Regional Geology)*; Yin, A., Harrison, T.M., Eds.; Cambridge University Press: Cambridge, UK, 1996; pp. 149–188.
96. Nickson, R.T.; McArthur, J.M.; Shrestha, B.; Kyaw-Myint, T.O.; Lowry, D. Arsenic and other drinking water quality issues, Muzaffargarh District, Pakistan. *Appl. Geochem.* **2005**, *20*, 55–68. [CrossRef]
97. Ingersoll, R.V. Actualistic sandstone petrofacies: Discriminating modern and ancient source rocks. *Geology* **1990**, *18*, 733–736. [CrossRef]
98. Garzanti, E. The Himalayan foreland basin from collision onset to the present: A sedimentary-petrology perspective. In *Himalayan Tectonics: A Modern Synthesis*; Special Publication 483; Treloar, P., Searle, M.P., Eds.; Geological Society of London Publications: London, UK, 2019. [CrossRef]
99. Andò, S.; Garzanti, E. Raman spectroscopy in heavy-mineral studies. In *Sediment Provenance Studies in Hydrocarbon Exploration and Production*; Special Publication 386; Scott, R.A., Smyth, H.R., Morton, A.C., Richardson, N., Eds.; Geological Society of London Publications: London, UK, 2013; pp. 395–412. [CrossRef]
100. Garzanti, E.; Andò, S. Heavy-mineral concentration in modern sands: Implications for provenance interpretation. In *Heavy Minerals in Use*; Developments in Sedimentology; Mange, M.A., Wright, D.T., Eds.; Elsevier: Amsterdam, The Netherlands, 2007; Volume 58, pp. 517–545. [CrossRef]
101. Garzanti, E.; Andò, S. Heavy Minerals for Junior Woodchucks. *Minerals* **2019**, *9*, 148. [CrossRef]
102. Rubey, W.W. The size-distribution of heavy minerals within a water-laid sandstone. *J. Sediment. Petrol.* **1933**, *3*, 3–29. [CrossRef]
103. Garzanti, E.; Andò, S.; Vezzoli, G. Settling-equivalence of detrital minerals and grain-size dependence of sediment composition. *Earth Planet. Sci. Lett.* **2008**, *273*, 138–151. [CrossRef]
104. Briggs, L.I.; McCulloch, D.S.; Moser, F. The hydraulic shape of sand particles. *J. Sediment. Petrol.* **1962**, *32*, 645–656. [CrossRef]
105. Komar, P.D. The entrainment, transport and sorting of heavy minerals by waves and currents. In *Heavy Minerals in Use*; Developments in Sedimentology Series; Mange, M.A., Wright, D.T., Eds.; Elsevier: Amsterdam, The Netherlands, 2007; Volume 58, pp. 3–48. [CrossRef]
106. Garzanti, E.; Andò, S.; Vezzoli, G. Grain-size dependence of sediment composition and environmental bias in provenance studies. *Earth Planet. Sci. Lett.* **2009**, *277*, 422–432. [CrossRef]
107. Taylor, S.R.; McLennan, S.M. The geochemical evolution of the continental crust. *Rev. Geophys.* **1995**, *33*, 241–265. [CrossRef]
108. Rudnick, R.L.; Gao, S. Composition of the continental crust. In *Treatise on Geochemistry, The Crust*; Rudnick, R.L., Holland, H.D., Turekian, K.K., Eds.; Elsevier Pergamon: Oxford, UK, 2003; Volume 3, pp. 1–64. [CrossRef]
109. Vermeesch, P. How many grains are needed for a provenance study? *Earth Planet. Sci. Lett.* **2004**, *224*, 441–451. [CrossRef]
110. Vermeesch, P.; Garzanti, E. Making geological sense of 'Big Data' in sedimentary provenance analysis. *Chem. Geol.* **2015**, *409*, 20–27. [CrossRef]
111. Gabriel, K.R. The biplot graphic display of matrices with application to principal component analysis. *Biometrika* **1971**, *58*, 453–467. [CrossRef]
112. Hawthorne, F.C.; Oberti, R.; Harlow, G.E.; Maresch, W.V.; Martin, R.F.; Schumacher, J.C.; Welch, M.D. Nomenclature of the amphibole supergroup. *Am. Mineral.* **2012**, *97*, 2031–2048. [CrossRef]
113. Oberti, R.; Cannillo, E.; Toscani, G. How to name amphiboles after the IMA2012 report: Rules of thumb and a new PC program for monoclinic amphiboles. *Period. Mineral.* **2012**, *81*, 257–267. [CrossRef]
114. Locock, A.J. An Excel spreadsheet to classify chemical analyses of amphiboles following the IMA 2012 recommendations. *Comput. Geosci.* **2014**, *62*, 1–11. [CrossRef]
115. Locock, A.J. An Excel spreadsheet to recast analyses of garnet into end-member components, and a synopsis of the crystal chemistry of natural silicate garnets. *Comput. Geosci.* **2008**, *34*, 1769–1780. [CrossRef]

116. Morton, A.C.; Hallsworth, C. Stability of detrital heavy minerals during burial diagenesis. In *Heavy Minerals in Use*; Developments in Sedimentology Series; Mange, M.A., Wright, D.T., Eds.; Elsevier: Amsterdam, The Netherlands, 2007; Volume 58, pp. 215–245. [CrossRef]
117. Andò, S.; Garzanti, E.; Padoan, M.; Limonta, M. Corrosion of heavy minerals during weathering and diagenesis: A catalog for optical analysis. *Sediment. Geol.* **2012**, *280*, 165–178. [CrossRef]
118. Garzanti, E.; Ando, S.; Limonta, M.; Fielding, L.; Najman, Y. Diagenetic control on mineralogical suites in sand, silt, and mud (Cenozoic Nile Delta): Implications for provenance reconstructions. *Earth Sci. Rev.* **2018**, *185*, 122–139. [CrossRef]
119. Morton, A.; Hallsworth, C.; Chalton, B. Garnet compositions in Scottish and Norwegian basement terrains: A framework for interpretation of North Sea sandstone provenance. *Mar. Petrol. Geol.* **2004**, *21*, 393–410. [CrossRef]
120. Teraoka, Y.; Suzuki, M.; Kawakami, K. Provenance of Cretaceous and Paleogene sediments in the median zone of Southwest Japan. *Bull. Geol. Surv. Jpn.* **1998**, *49*, 395–411. [CrossRef]
121. Win, K.S.; Takeuchi, M.; Tokiwa, T. Changes in detrital garnet assemblages related to transpressive uplifting associated with strike–slip faulting: An example from the Cretaceous System in Kii Peninsula, Southwest Japan. *Sediment. Geol.* **2007**, *201*, 412–431. [CrossRef]
122. Yavuz, F.; Yildirim, D.K. A Windows program for calculation and classification of epidote-supergroup minerals. *Period. Mineral.* **2018**, *87*, 269–285. [CrossRef]
123. Armbruster, T.; Bonazzi, P.; Akasaka, M.; Bermanec, V.; Chopin, C.; Gieré, R.; Heuss-Assbichler, S.; Liebscher, A.; Menchetti, S.; Pan, Y.; et al. Recommended nomenclature of epidote-group minerals. *Eur. J. Mineral.* **2006**, *18*, 551–567. [CrossRef]
124. Sturm, R. PX-NOM—An interactive spreadsheet program for the computation of pyroxene analyses derived from the electron microprobe. *Comput. Geosci.* **2002**, *28*, 473–483. [CrossRef]
125. Morimoto, N.; Fabries, J.; Ferguson, A.K.; Ginzburg, I.V.; Ross, M.; Seifert, F.A.; Zussman, J.; Aoki, K.; Gottardi, G. Nomenclature of pyroxenes. *Miner. Petrol.* **1988**, *39*, 55–76. [CrossRef]
126. Graser, G.; Markl, G. Ca-rich ilvaite–epidote–hydrogarnet endoskarns: A record of late-magmatic fluid influx into the Persodic Ilímaussaq Complex, South Greenland. *J. Petrol.* **2007**, *49*, 239–265. [CrossRef]
127. Kartashov, P.M. Classification diagram for REE-bearing members of the epidote group based on crystallochemical data. In Proceedings of the Workshop on Acessory Minerals, Warsaw, Poland, 25–26 September 2014; pp. 19–21.
128. Poldervaart, A.; Hess, H.H. Pyroxenes in the crystallization of basaltic magma. *J. Geol.* **1951**, *59*, 472–489. [CrossRef]
129. Vermeesch, P.; Resentini, A.; Garzanti, E. An R package for statistical provenance analysis. *Sediment. Geol.* **2016**, *336*, 14–25. [CrossRef]
130. Garzanti, E.; Padoan, M.; Andò, S.; Resentini, A.; Vezzoli, G.; Lustrino, M. Weathering and relative durability of detrital minerals in equatorial climate: Sand petrology and geochemistry in the East African Rift. *J. Geol.* **2013**, *121*, 547–580. [CrossRef]
131. Garzanti, E.; Andò, S.; Vezzoli, G. The Continental Crust as a Source of Sand (Southern Alps cross-section, Northern Italy). *J. Geol.* **2006**, *114*, 533–554. [CrossRef]
132. Garzanti, E.; Vezzoli, G.; Andò, S. Paleogeographic and paleodrainage changes during Pleistocene glaciations (Po Plain, Northern Italy). *Earth Sci. Rev.* **2011**, *105*, 25–48. [CrossRef]
133. Weltje, G.J. End-member modelling of compositional data: Numerical statistical algorithms for solving the explicit mixing problem. *Math. Geol.* **1997**, *29*, 503–549. [CrossRef]
134. Garzanti, E.; Resentini, A.; Vezzoli, G.; Andò, S.; Malusà, M.; Padoan, M. Forward compositional modelling of Alpine orogenic sediments. *Sediment. Geol.* **2012**, *280*, 149–164. [CrossRef]
135. Cerveny, P.F.; Johnson, N.M.; Tahirkheli, R.A.K.; Bonis, N.R. Tectonic and geomorphic implications of Siwalik group heavy minerals, Potwar plateau, Pakistan. In *Tectonics of the Western Himalayas*; Special Papers 232; Malinconico, L.L., Lillie, R.J., Eds.; Geological Society of America: Boulder, CO, USA, 1989; pp. 129–136.
136. Yamamoto, H. Contrasting metamorphic P-T-time paths of the Kohistan granulites and tectonics of the western Himalayas. *J. Geol. Soc.* **1993**, *150*, 843–856. [CrossRef]
137. Jagoutz, O.; Schmidt, M.W. The formation and bulk composition of modern juvenile continental crust: The Kohistan arc. *Chem. Geol.* **2012**, *298*, 79–96. [CrossRef]

138. Keane, S.D.; Morrison, J. Distinguishing magmatic from subsolidus epidote: Laser probe oxygen isotope compositions. *Contrib. Mineral. Petrol.* **1997**, *126*, 265–274. [CrossRef]
139. Spiegel, C.; Siebel, W.; Frisch, W.; Berner, Z. Nd and Sr isotopic ratios and trace element geochemistry of epidote from the Swiss Molasse Basin as provenance indicators: Implications for the reconstruction of the exhumation history of the Central Alps. *Chem. Geol.* **2002**, *189*, 231–250. [CrossRef]

minerals

MDPI

Article

Provenance of Bengal Shelf Sediments: 1. Mineralogy and Geochemistry of Silt

Laura Borromeo [1,*], Sergio Andò [1], Christian France-Lanord [2], Giovanni Coletti [1], Annette Hahn [3] and Eduardo Garzanti [1,*]

1 Laboratory for Provenance Studies, Department of Earth and Environmental Sciences, University of Milano-Bicocca, 20126 Milano, Italy; sergio.ando@unimib.it (S.A.); giovanni.coletti@unimib.it (G.C.)

2 Centre de Recherches Pétrographiques et Géochimiques, Université de Lorraine-CNRS, BP 20, 54501 Vandoeuvre-lès-Nancy, France; cfl@crpg.cnrs-nancy.fr

3 MARUM Center for Marine Environmental Sciences, University of Bremen, 28359 Bremen, Germany; anhahn@uni-bremen.de

* Correspondence: laura.borromeo@unimib.it (L.B.); eduardo.garzanti@unimib.it (E.G.)

Received: 30 May 2019; Accepted: 1 October 2019; Published: 18 October 2019

Abstract: This article illustrates a multi-technique frontier approach for the provenance study of silt-size sediments. The mineralogical composition of low-density and heavy-mineral fractions of four samples of fine to very coarse silt deposited on the Bengal shelf was analyzed separately for six different grain-size classes by combining grain counting under an optical microscope, Raman spectroscopy, and X-ray diffraction. The geochemical composition was determined on both bulk-sediment samples and on their <5-μm classes. Such a "multiple-window" approach allowed capturing the full mineralogical information contained in each sample, as well as the size-dependent intra-sample variability of all compositional parameters. The comparison between grain-size distributions obtained by different methods highlighted a notable fallacy of laser granulometry, which markedly overestimated the size of the finest mode represented by fine silt and clay. As a test case, we chose to investigate sediments of the Bengal shelf, where detritus is fed from the Meghna estuary, formed by the joint Ganga and Brahmaputra Rivers and representing the largest single entry point of sediment in the world's oceans. The studied samples show the typical fingerprint of orogenic detritus produced by focused erosion of collision orogens. Bengal shelf silt is characterized by a feldspatho-quartzose (F-Q) composition with a Q/F ratio decreasing from 3.0 to 1.7 with increasing grain size, plagioclase prevailing over K-feldspar, and rich transparent-heavy-mineral assemblages including mainly amphibole with epidote, and minor garnet and pyroxene. Such a detrital signature compares very closely with Brahmaputra suspended load, but mineralogical and geochemical parameters, including the anomalous decrease of the Q/F ratio with increasing grain size, consistently indicate more significant Ganga contribution for cohesive fine silt. The accurate quantitative characterization of different size fractions of Bengal shelf sediments represents an essential step to allow comparison of compositional signatures characterizing different segments of this huge source-to-sink system, from fluvial and deltaic sediments of the Himalayan foreland basin and Bengal shelf to the Bengal Fan.

Keywords: provenance analysis; sieving of fine silt; fallacy of laser granulometry; benthic foraminifera; heavy minerals; Raman spectroscopy; Ganga–Brahmaputra river system; Bay of Bengal; Himalayan orogen

"Vast bamboo jungles towered over the river's majestic banks, covering those muddy lowlands known to the world as the Sundarbans of the Ganges."

Emilio Salgari, The Two Tigers, Chapter 1.

1. Introduction

Silt represents a very major part of fluvial detrital fluxes and 50% or more of sedimentary covers on Earth [1–4]. Because of the diverse difficulties involved in sample treatment and analysis, however, mud and mudrocks are infrequently considered in provenance studies (e.g., [5,6]). Silt is a main component of offshore shelf deposits, which represent the *trait-d'union* between fluvial sediments largely carried in suspension and deep-sea-fan turbidites. Learning to extract provenance information routinely from silt-sized sediments, therefore, represents a major step forward in the understanding of the tectonic and erosional evolution of mountain belts.

The purpose of this article is to develop a standard procedure for provenance studies based on high-resolution analysis of silt-sized detrital minerals, choosing the Bengal shelf as a test case (Figure 1).

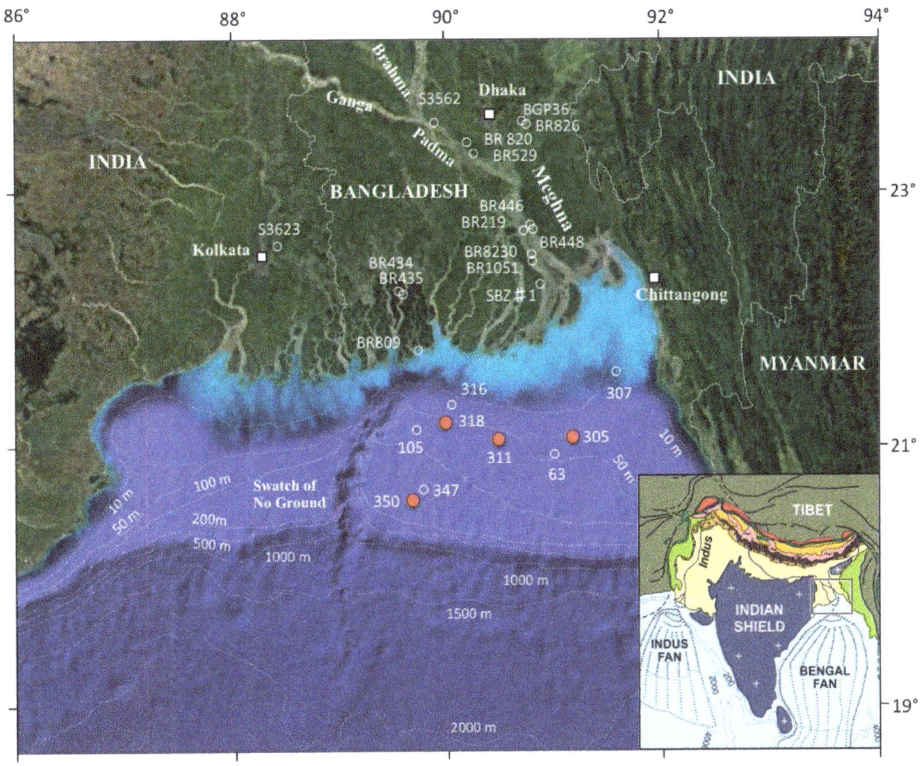

Figure 1. Location of vibrocore samples 305, 311, 318, and 350 (red dots), collected on the Bengal shelf during Sonne cruise 188-2 [7]. White circles indicate samples studied in the companion paper [8].

Himalayan-derived detritus is supplied to the Bengal shelf via the Meghna estuary of the Ganga and Brahmaputra rivers, which represents the largest single entry point of sediment in the world's oceans [9,10]. Sediment is partly deposited on the shelf, and partly bypasses it to generate turbiditic currents feeding the largest submarine fan on our planet [11,12]. The thorough investigation of the mineralogical variability associated with hydrodynamic sorting and other physical and chemical processes in shallow-marine environments represents the fundamental pre-requisite to link the orogenic provenance signatures of Ganga–Brahmaputra sediments with the deep-sea sedimentary record [13–16]. This article and the companion paper dedicated to Bengal shelf sand [8] are, thus, intended as complementary to previous studies carried out on both bedload and suspended-load sediments of the Ganga and Brahmaputra rivers in the proximal part of the sediment-routing system [17–27].

In order to carry out a quantitative mineralogical study of silt samples with similar accuracy and precision as currently reached for sand, we coupled traditional optical analyses under a petrographic microscope with micro-Raman spectroscopy and X-ray diffraction. For each of four silt samples cored from the Bengal shelf, we separately investigated six different size classes (from <5 μm to >63 μm) obtained by wet sieving. Such a "multiple-window" approach allows capturing the full mineralogical information contained in the sample, as well as the size-dependent intra-sample variability of all compositional parameters [28,29]. The grain-size distribution and geochemical composition of the samples was also determined by diverse techniques.

2. The Bengal Sediment System

The Ganga and Brahmaputra Rivers form the greatest system of sediment transport on Earth [11]. The Ganga River drains the northern part of the Precambrian Indian shield with its sedimentary and basaltic (Deccan Traps) covers, and all tectonic units of the Himalayan belt. These include the Paleozoic to Eocene sedimentary succession of the Tethys Himalaya [30,31], mainly amphibolite-facies metamorphic rocks of the Greater Himalaya [32,33], lower-grade metamorphic rocks and sedimentary strata of the Lesser Himalaya [34,35], and foreland-basin siliciclastic rocks of the Sub-Himalaya [36,37]. In addition to the Himalayan belt, the Brahmaputra River drains granitoid batholiths and sedimentary covers of the Lhasa block [38], the Transhimalayan forearc-basin [39] and Yarlung Tsangpo ophiolitic suture [40], and high-grade metamorphic rocks of the eastern Himalayan syntaxis [41]. The syntaxis provides a large part of the Brahmaputra sediment load, whereas the Indo-Burman Ranges and the Shillong Plateau supply subordinate additional detritus in the lower course [20,21]. The Ganga and Brahmaputra Rivers join ~200 km upstream of the coastline to form the Padma River (the name of the Ganga in Bangladesh) and debouch into the Bengal Sea in the tide-dominated estuary called Lower Meghna, because it also represents the lowermost tract of the Meghna River draining the Shillong Plateau and the Tripura (Indo-Burman) fold-belt. To the west of the estuary lies the Sundarban mangrove forest, formed during progressive eastward migration of the Ganga outlet during the late Holocene [9,42].

At least one billion tons of siliciclastic detritus are produced annually by erosion of the active Himalayan range, and then funneled through the Meghna estuary into the Indian Ocean. Most commonly cited figures range from 380 to 480 × 10^6 t/a for the Ganga River and from 650 to 680 × 10^6 t/a for the Brahmaputra, which implies ~40% of the sediment flux from the Ganga and ~60% from the Brahmaputra [25,27,43].

Rapid erosion and transport continuing for several tens of million years ended up in the colossal accumulation of 12.5 × 10^6 km^3 of orogen-derived sediments in the Bengal Fan [44]. In the long term, only ~10% of Himalayan detritus remained stored in the Ganga foreland basin [45] without reaching the sea. Despite its smaller catchment (~0.6 × 10^6 km^2 vs. ~1 × 10^6 km^2), the Brahmaputra carries a larger sediment load than the Ganga [46], and it is the big-river basin with the highest denudation rate on Earth [47].

Ganga–Brahmaputra sediments reaching Bangladesh appear to be subequally partitioned among the delta plain, the prograding subaqueous delta, and the deep-sea fan [48]. The subaerial and submarine delta combined cover 138,000 km^2, with total annual water and sediment fluxes from the Meghna estuary being around 970 km^3 and 1.7 × 10^6 t, respectively [49,50]. Most of the water and nearly all of the sediment are delivered during the summer monsoon season. Suspended load consists of fine to medium silt entrained close to the surface of the channel, passing to coarse and very coarse sandy silt at depth and to silty sand close to the channel bottom. Bedload is typically fine-grained well-sorted sand, and clay does not represent more than ~10%, even in surface load [19,25].

Offshore of the estuary mouth, a large subaqueous delta is prograding seaward. Wide sandy topsets extend down to water depths of ca. 15–30 m below sea level (b.s.l.) ~100 km from the coast, passing oceanward to delta-front foresets and beyond, at water depths around 80 m b.s.l., to muddy prodelta bottomsets [51–53].

2.1. Mineralogy of River Silt

Silt carried by the Ganga and Brahmaputra Rivers contains quartz and mica in similar proportions, associated with feldspar and heavy minerals. Quartz, feldspar, and heavy minerals steadily increase from shallow to deep suspended load relative to phyllosilicates. Composition, thus, changes from mica > quartz > feldspar in shallow suspended load to quartz > feldspar > mica close to the riverbed. The Ganga carries more quartz and less Ca-bearing plagioclase than the Brahmaputra. Ganga suspended load contains significant calcite and dolomite, and a moderately rich epidote–amphibole–garnet suite including clinopyroxene, tourmaline, titanite, rutile, zircon, kyanite, apatite, sillimanite, staurolite, and chloritoid; garnet increases progressively with depth relative to less dense amphibole and epidote. Brahmaputra suspended load contains virtually no calcite, very little dolomite, and a rich amphibole–epidote suite including garnet, clinopyroxene, minor titanite, tourmaline, apatite, rutile, sillimanite, zircon, kyanite, chloritoid, staurolite, hypersthene, monazite, and Cr-spinel [24].

2.2. Geochemistry of River Silt

Differences in the chemical composition of suspended load carried by the Ganga and Brahmaputra Rivers are difficult to determine accurately because of strong grain-size control. For a given grain size, Brahmaputra sediments are richer in Na, Sr, Cr, P, and possibly Co and Ni, reflecting higher abundance of plagioclase, amphibole, apatite, and Cr-spinel. Ganga sediments are instead richer in Ca, reflecting more abundant carbonate grains. Fine silt entrained close to the surface is strongly enriched in mica, clay minerals, oxy-hydroxides, and organic matter, and it is consequently characterized by higher concentration in chemical elements hosted in phyllosilicates, in heavy metals adsorbed on clays, and in constitutive water (Al, Fe, Mg, K, Ti, P, Mn, Rb, Cs, Be, Ba, V, Cr, Mo, W, Co, Ni, Cu, Zn, Ga, Ge, Pb, As, Sb, and Bi). Instead, because of progressive enrichment in faster-settling coarser, denser, and more spherical grains toward the bed, elements mostly hosted in tectosilicates (Si, Na, Ca, Sr) or ultradense minerals (Zr, Hf) are concentrated in deep suspended load. Other elements hosted in ultradense minerals but also associated with fine particles (Y, REE, Th, U, Ti, Nb, Ta, Cr, W, Mn, Fe, Cd, Ge, Sn, and P) reach maximum abundance in deepest suspended load, particularly above heavy-mineral-rich bedload, but are also enriched in surface load [19,24,25,27].

2.3. Clay Minerals

Because of high rates of physical erosion in the Himalayan orogen, the clay–mineral assemblage carried by the Ganga and Brahmaputra rivers is dominated by illite, with subordinate chlorite, smectite/montmorillonite, and kaolinite. Notable differences occur between Ganga and Brahmaputra clays. Because storage time and weathering intensity is notably higher in the Ganga than in the Brahmaputra floodplain [26,54], and because southern Ganga tributaries such as the Chambal River drain the Indian shield capped by Deccan Traps basaltic lavas [55], smectite/montmorillonite is common in Ganga clay (20–40%), whereas Brahmaputra clay is dominated by illite (70–80%) with subequal amounts of chlorite and kaolinite (~10% each), and negligible smectite (≤5%; [56–60]). The clay–mineral assemblage carried by the Meghna River is similar overall to that carried by the Brahmaputra, with more chlorite (~20%). Clay minerals in the Meghna estuary are dominated by illite (59–69%), associated with chlorite (16–20%), smectite/montmorillonite (5–16%), and kaolinite (7–10%; [60]). More variable data were reported from the Bengal shelf, with kaolinite reaching as high as 34–39% offshore of the Chittagong coast [61].

3. Analytical Methods

In order to quantitatively assess the mineralogical signature of offshore shelfal silt, evaluate the compositional differences with fluvial suspended load, and investigate hydraulic-sorting processes in shallow-marine environments, we analyzed in full detail four samples SO188-305VC 63-72, SO188-311VC 591-600, SO188-318VC 600, and SO188-350VC 600 (named for brevity 305, 311, 318, and 350 throughout the article) collected by a vibrocorer on the Bengal shelf during Sonne cruise 188-2 in July 2006 (Figure 1; [7]). Whereas sample 305 is a very coarse silt collected from upper foresets of the subaqueous delta, samples 311, 318, and 350 are fine silts collected from the lower foresets and bottomsets of the distal subaqueous delta. Further information on sampling sites is provided in Table S1 (Supplementary Materials).

The grain-size distribution of each sample was determined by both wet sieving and laser granulometry carried out with a Malvern Mastersizer 2000E particle size analyzer at the University of Milano-Bicocca. Mineralogical analyses were performed by combining traditional optical methods and Raman counting [62] separately for five size classes (5–10 μm, 10–15 μm, 15–32 μm, 32–63 μm, and >63 μm), which were obtained by wet sieving using nylon sieves (5-μm, 10-μm, and 15-μm mesh) and steel sieves (32-μm and 63-μm mesh). The 63-μm (4φ) sieve divides sand from mud, the 32-μm (5φ) sieve separates very coarse frictional silt from cohesive mud, and the 5-μm (≤8 φ) sieve filters out highly cohesive very fine silt and clay particles. Each size class was separated by centrifuging in sodium polytungstate into low-density (<2.90 g/cm^3, "light" LM) and high-density (>2.90 g/cm^3, "heavy" HM) mineral fractions, recovered by partial freezing with liquid nitrogen, and mounted on glass slides. Between 100 and 150 grains were counted on each slide. The mineralogical and chemical compositions of the <5-μm class were obtained by X-ray diffraction (XRD) and X-ray fluorescence (XRF), respectively.

The results of grain-size and mineralogical analyses are summarized in Tables 1 and 2. The complete grain-size, mineralogical, and geochemical datasets are provided in Tables S2–S7 (Supplementary Materials).

3.1. *Sieving of Cohesive Mud*

The finest grain-size classes (<5 μm, 5–10 μm, and 10–15 μm) of the four samples 305, 311, 318, and 350 were separated by wet sieving using small disposable hand-made nylon sieves (5-μm, 10-μm, and 15-μm meshes, by Sefar AG). The tissue nets were carefully glued on the base of a plexiglass cylinder 8 cm in diameter (Figure 2A). In order to avoid mesh clogging and facilitate water passage through the sieve, a small amount of each sample was sieved at each time. The finer particles passing through the sieve were recovered in beakers after waiting for the time sufficient to complete settling by all particles (several days and up to one or two weeks for finest classes; Figure 2B). In the meantime, the size class retained in the sieve was recuperated (Figure 2C). After all particles settled, the clean water in excess was siphoned with a small plastic tube, and the sediment fraction recovered and dried at 40 °C in the oven (Figure 2D). Such a laborious procedure, repeated for all sieve sizes, allowed us to obtain neatly separated grain-size classes for mineralogical analysis of cohesive mud.

3.2. *Optical Microscopy and Raman Spectroscopy*

Raman spectroscopy is a user-friendly technique well suited to identify detrital minerals down to the size of a few microns. The reliability of mineralogical analyses is greatly increased by coupling observations under the transmitted-light microscope with Raman spectroscopy, especially for finest size classes where the uncertainty of optical identification increases. In the mineralogical study of the >63-μm (sand), 32–63-μm (very coarse silt), 15–32-μm (coarse silt), and 10–15-μm (medium silt) classes of each sample, we routinely coupled optical microscopy and Raman spectroscopy, whereas both LM and HM fractions of the 5–10-μm class (fine silt) were studied with Raman spectroscopy only. On the HM fraction of each class, at least 200 transparent-heavy-mineral grains were counted on grain mounts under the microscope by the area method [63], and all grains of uncertain identification were checked systematically with Raman spectroscopy [64,65]. The definition of heavy minerals and the followed methodological protocol were according to Andò [66] and Garzanti and Andò [67]. For each sample and size class, heavy-mineral and transparent-heavy-mineral concentrations (HMC and tHMC indices) were calculated according to Garzanti and Andò [68].

Analysis of the low-density fraction LM of each size class required a preliminary inspection under the microscope to identify and determine by grain counting the percentage of altered turbid grains, organic matter, and phyllosilicates, which generate a weak Raman signal. Each silicate and carbonate grain encountered within the counted area was numbered on a photograph to make it retraceable for subsequent Raman spectroscopy analysis. More than 2000 spectra of quartz and feldspar grains were obtained and carefully identified.

For this study, we used four different micro-Raman spectrometers, equipped with red Helio-Neon red 633-nm, Argon-ion blue 488-nm, and solid-state green 532-nm lasers, all focused on the sample through a 50× long-working-distance objective (additional technical information is provided in Supplementary Materials). Software LABSPEC 5 was used to perform baseline subtraction, to eliminate the natural fluorescence of grains, and to assess peak positions using the pseudo-Voigt deconvolution of the Gaussian–Lorentzian function with an accuracy of 0.2 cm^{-1}. Detrital minerals in both LM and HM fractions were identified by comparison of their calibrated Raman peaks with reference spectra collected on in-house standards, reported in the literature for feldspars [69–72] and carbonates [73,74], or included in online databases [75–77]. For the HM fraction, we mainly referred to Huang et al. [78], Bersani et al. [79], and Andò and Garzanti [64].

Figure 2. Separating grain-size classes of cohesive mud by wet sieving. (**A**) The <15-μm fraction of sample 350 in the 10-μm nylon sieve. (**B**) Settling of the passing <10-μm fraction of sample 350 in beakers. (**C**) The obtained 10–15-μm class of sample 350. (**D**) The <10-μm fraction recovered after sieving at 10 μm.

3.3. Bulk Geochemistry

Bulk-sediment chemical analyses were carried out at *Service d'Analyse des Roches et des Minéraux* (SARM–CRPG, Nancy, France). Samples were firstly carefully washed using de-ionized water in order to remove sea salt, and then dried and powdered in an agate mortar. Element concentrations were measured by inductively coupled plasma optical emission spectrometry (ICP-OES) and mass spectrometry (ICP-MS) on bulk aliquots of ~100 mg of sediment after lithium metaborate fusion [80,81]. The relative uncertainty for major elemental concentration was <5%, except for Si, Al, and Fe, for which it was ~2%. For full information on analytical procedures and geostandards used, see Carignan et al. [81].

3.4. XRD and XRF Analyses

The mineralogical and chemical compositions of the <5-μm class, which includes both very fine silt and clay and was separated by settling velocity (Atterberg separation after Stokes' Law; [82]) were obtained at MARUM Center for Marine Environmental Sciences in Bremen. Chemical composition was measured using a PANalytical Epsilon3-XL XRF spectrometer equipped with a rhodium tube, several filters, and an SSD5 detector. Calibration was based on certified standard materials (GBW07309, GBW07316, MAG-1 [83]).

Detrital minerals were identified by standard powder X-ray diffraction (XRD) patterns. Analyses were carried out at Central Laboratory for Crystallography and Applied Material Sciences (ZEKAM, Department of Geosciences, University of Bremen). We used a Philips X'Pert Pro multipurpose diffractometer equipped with a Cu-tube (kα 1.541, 45 kV, 40 mA), a fixed divergence slit of $\frac{1}{4}$, a secondary Ni filter, and the X'Celerator detector system. Measurements were performed as a continuous scan from 3° to 85° 2θ, with a calculated step size of 0.016° 2θ (calculated time per step = 50 s). Minerals were identified by the Philips software *X'Pert HighScore*™ [84] and, for sheet silicates, by the freely available Apple Macintosh X-ray diffraction interpretation software *MacDiff 4.25* [85,86]). Quantification of the mineral assemblage was done via the *QUAX* full pattern method [87].

4. Data

4.1. Grain Size

Sieve analysis indicates that sample 305, cored in front of the Meghna estuary at a water depth of 30 m b.s.l. (Figure 1), includes 25 wt.% sand and 58 wt.% very coarse silt (32–63 μm; Figure 3A). The other samples cored in deeper waters to the west (311, 87 m b.s.l; 318, 59 m b.s.l.) and southwest (350; 129 m b.s.l.), instead, contain only <1 wt.% sand and ≥95 wt.% cohesive mud (<32 μm), mostly represented by very fine silt and clay (< 5 μm fraction 67–79 wt.%).

Laser granulometry emphasizes the bimodal character of all size distributions (Figure 3B). The moderately to poorly sorted (~1 σ_{φ}) coarse mode represents 92% of sample 305, where it apparently occurs in the very-fine-sand range (3.7 φ), whereas, in the other three samples, it apparently occurs in medium-coarse silt (~5 φ) and is either prominent (350) or subordinate (311 and 318). The fine mode, apparently occurring at ~7 φ and minor for sample 305, accounts for 72–73% of samples 311 and 318.

Grain-size data obtained on the four studied samples by wet sieving, which provides weight percentages, compare rather poorly with those obtained by laser granulometry, which provides volume percentages. Most important, squared-mesh sieves separate particles based on the length of their intermediate axis and, thus, let pass elongated and platy particles that may have larger volumes than spherical particles with the same intermediate diameter. On the contrary, laser granulometry detects particles in a fluid using a diffraction algorithm that preferentially considers the long axis in the estimate of particle size [88]. As a consequence, the size of non-spherical natural particles may be overestimated by 0.25–0.5 φ, an effect markedly enhanced for clay flakes (~2 μm particles being read as ~8 μm by the laser [89]). Sieve analysis demonstrates that the fine mode in all of our samples is represented by very fine silt and clay (Figure 3A) rather than by medium silt (Figure 3B), thus testifying how laser granulometry markedly underestimates the clay content of sediment samples [90,91].

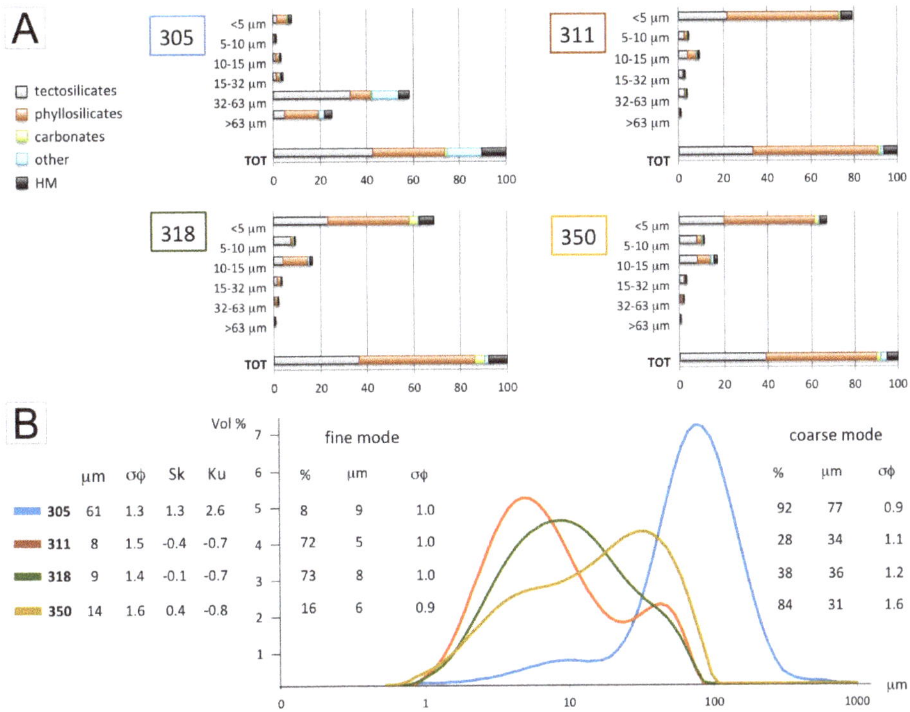

Figure 3. Grain-size and mineralogical data on Bengal shelf silt. (**A**) Grain size determined by wet sieving. Bulk-sediment mineralogy was analyzed by Raman spectroscopy coupled with optical microscopy for size classes >5 μm and by X-ray diffraction (XRD) for the <5-μm class. The "other" category includes rock fragments, altered grains, and organic matter. (**B**) Grain size determined by laser granulometry (μm = mean size in microns; σ_φ = sorting; Sk = skewness; Ku = kurtosis). The estimated percentages, mean grain size, and sorting of silt and sand modes are indicated.

4.2. Fossil Faunas

Fossils occurring in the >63-μm class of all mud-dominated samples, and most commonly in deeper-water offshore samples 311 and 350, include mainly benthic foraminifera (*Ammonia*, *Elphidium*, *Triloculina*, and various species of the Vaginulinidae, Uvigerinidae, Bolivinoidea, Nonionoidea, and Planorbulinoidea families and superfamilies) and a few planktonic foraminifera (mainly Globigerinidae) (Figure 4). Foraminifera are rare in the very-coarse-silt sample 305, collected from the topsets of the subaqueous delta in front of the Meghna estuary where terrigenous input is very high. Samples 311 and 318 are dominated by shallow-shelf dwellers like *Ammonia*, *Elphidium*, and *Triloculina*, whereas the most distal sample 350, collected in deeper water, is dominated by uvigerinids and bolivinids. Bolivinids, which as infaunal detritivore foraminifera commonly occur in deep, muddy, and poorly oxygenated environments rich in organic matter [92], are common in all fine silt samples collected in the distal foresets and bottomsets of the subaqueous delta.

Fossil faunas identified in the four studied silt samples were compared to those reported in the literature and investigated in the companion paper. A foraminiferal assemblage including *Elphidium*, *Asterorotalia*, *Ammonia*, some *Triloculina*, *Spiroloculina*, *Bolivina*, *Hoeglundina*, and no planktonic specimens was reported from core SO93-80 (water depth 78 m b.s.l. [93]). Mostly benthic foraminifera and fragments of mollusks and echinoids occur, associated with and commonly at the core of ooids, in very-fine-grained outer-shelf sand (sample SO347 in Figure 1; present water depth 126 m b.s.l.;

Figure 2I in Garzanti et al. [8]). Foraminifera are well preserved and include miliolids (e.g., *Triloculina*, *Quinqueloculina*, *Peneroplis*), rotaliids (e.g., *Cibicides*), textularids (e.g., *Textularia*), bolivinids, and rare globigerinids. The chambers of foraminiferal tests are commonly filled by glaucony, testifying to very slow accumulation rate [94]. This faunal assemblage indicates original deposition on a shallow-water inner shelf, followed by reworking, winnowing, and sediment starvation during the Holocene transgression [8]. The general lack of Indian Ocean sites in which ooids are forming at present [95,96] confirms the relict character of these palimpsest outer-shelf deposits [97].

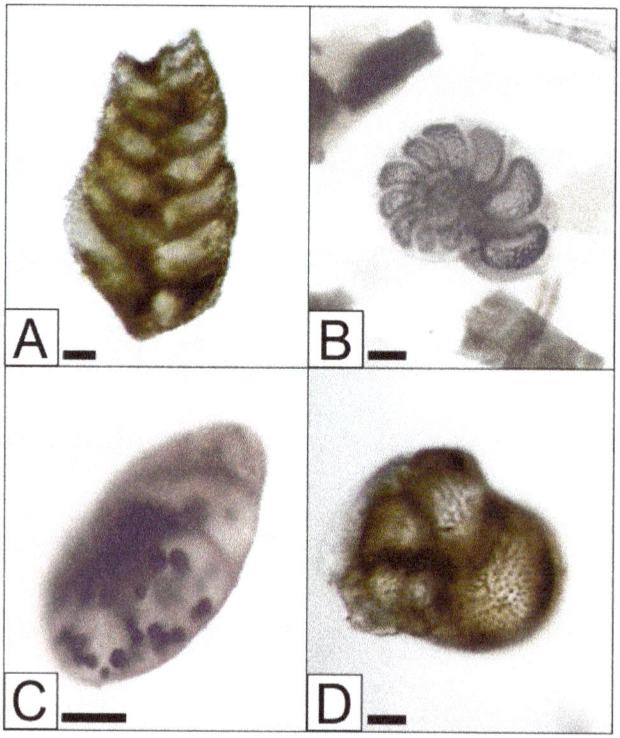

Figure 4. Foraminifera in Bengal shelf sediments (>63 μm class for all samples). (**A**) Bolivinoidea superfamily, Glaessner 1937 (sample 318); (**B**) Planorbulinoidea superfamily, Schwager 1877 (sample 350); (**C**) Vaginulinidae family, Reuss 1860 (sample 318); (**D**) Globigerinidae family, Parker and Jones 1862 (sample 350). Scale bar for all optical images = 30 μm.

4.3. Bulk-Sample Mineralogy

The four studied samples consist of about half phyllosilicates and clay minerals and one-fourth quartz, with the rest represented by feldspars, transparent and opaque heavy minerals, rock fragments, and minor carbonates (Figure 3A). The quartz/feldspar ratio tends to be slightly higher in the more distal and deeper-water feldspatho-quartzose (F-Q) samples 311 (Q/F 2.0) and 350 (Q/F 3.0) than in the shallower-water feldspar-rich feldspatho-quartzose samples 318 (Q/F 1.9) and 305 (Q/F 1.7; classification after Garzanti [98]) (Figure 5A). Among feldspars, albite prevails over K-feldspar, and Ca-bearing plagioclase is invariably subordinate. K-feldspar and Ca-bearing plagioclase tend to decrease in finer samples, whereas albite increases slightly. Among phyllosilicates, muscovite prevails over chlorite, and biotite is subordinate but tends to increase in coarser samples and coarser classes. Rock fragments increase in coarser classes for obvious geometrical reasons. Carbonates occur sporadically in all samples and all size classes, and include calcite, aragonite, and dolomite of both extrabasinal terrigenous and intrabasinal biochemical origin. Organic matter, soil particles, and vegetal fragments also occur, mainly in either finest or coarsest size classes.

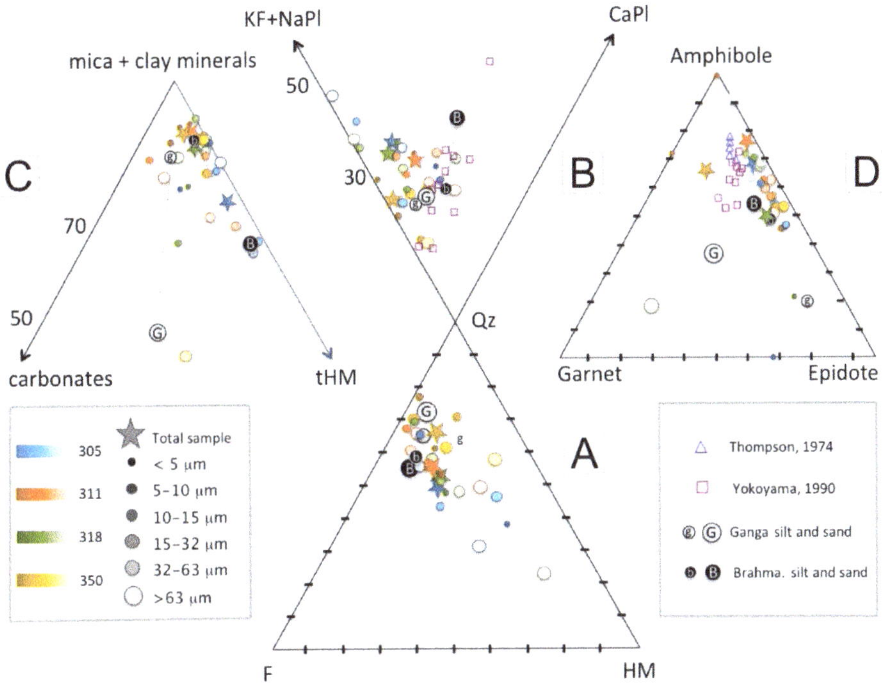

Figure 5. Triangular diagrams illustrating the mineralogical composition of the four studied samples, and of the six size-windows analyzed for each sample by combining grain counting under the microscope and Raman spectroscopy for classes >5 μm and XRD data for the <5-μm class. (**A**) Qz = quartz; F = total feldspars; HM = total heavy minerals; (**B**) KF + NaPl = alkali feldspars (including K-feldspar, albite, and ternary feldspars); CaPl = calcium-bearing plagioclase; (**C**) tHM = transparent heavy minerals; (**D**) relative abundance of the three most common groups of transparent heavy minerals in orogenic sediments. The average composition of Ganga (g = silt; G = sand) and Brahmaputra (b = silt; B = sand) sediments (from Garzanti et al. [24]), as well as mineralogical data on Bengal Fan sediments from Thompson [13] and Yokoyama et al. [15], are given for comparison.

Table 1. Multiple-window bulk-sediment mineralogical data for the four studied samples obtained by combining grain counting under the microscope with Raman spectroscopy for classes >5-µm and by X-ray diffraction (XRD) for the <5-µm class (in *italics*). Bulk-sample composition is in **bold**. GSZ = Grain size (in µm); Qz = quartz; KF= K-feldspar + ternary feldspar; Ab = albite; Ca-Pl = Ca-plagioclase; Rf = rock fragments; tHM = transparent heavy minerals; Op = opaque heavy minerals; Ms = muscovite; Bt = biotite; Chl = chlorite; Cc = calcite + aragonite; Do = dolomite; Alt = altered grains; OM = Organic Matter; & = others undetermined.

Sample	GSZ (µm)	Class (wt.%)	Qz	KF	Ab	Ca-Pl	Rf	tHM	Op	Ms	Bt	Chl	Clay	Cc	Do	Alt	OM	&	
305	<5	*7.9*	*14*	*1*	*5*		*0*	*9*	*7*	*20*	*0*	*16*	*18*	*7*	*0*	*0*	*0*	*1*	100.0
	5–10	1.1	40	4	6	7	0	5	0	15	11	6	0	0	0	6	2	0	100.0
	10–15	3.3	20	6	8	1	5	10	1	25	4	19	0	0	1	1	0	0	100.0
	15–32	4.3	23	2	5	1	12	19	0	19	0	18	0	0	1	0	0	0	100.0
	32–63	58.4	36	9	7	4	20	8	0	7	5	3	0	1	0	0	0	0	100.0
	>63	25.0	10	4	5	0	7	12	1	31	10	17	0	0	0	3	0	0	100.0
Total sample			**27**	**7**	**6**	**2.9**	**14**	**10**	**0.8**	**15**	**6**	**9**	**1**	**1**	**0.1**	**0.8**	**0.0**	**0.1**	100.0
311	<5	*79.1*	*18*	*2*	*8*		*0*	*6*	*1*	*6*	*0*	*10*	*49*	*1*	*0*	*0*	*0*	*0*	100.0
	5–10	4.4	41	5	9	4	0	1	0	19	4	5	0	4	0	4	3	0	100.0
	10–15	9.4	29	5	7	3	3	7	0	35	2	6	0	0	1	3	0	0	100.0
	15–32	2.6	50	6	10	11	0	6	0	11	1	3	0	0	1	2	0	0	100.0
	32–63	3.8	51	10	3	7	0	5	1	9	3	3	0	2	0	2	5	0	100.0
	>63	0.7	10	1	1	2	1	7	0	24	16	21	0	9	0	2	6	0	100.0
Total sample			**22**	**3**	**6**	**3.0**	**0**	**5**	**0.7**	**10**	**1**	**9**	**39**	**1**	**0.2**	**0.5**	**0.4**	**0.3**	100.0
318	<5	*68.8*	*21*	*2*	*10*		*0*	*9*	*0*	*19*	*0*	*11*	*22*	*1*	*5*	*0*	*0*	*1*	100.0
	5–10	9.2	58	3	14	3	0	3	0	9	0	1	0	0	3	3	3	0	100.0
	10–15	16.4	16	6	3	0	1	6	0	35	7	23	0	0	0	3	1	0	100.0
	15–32	3.4	31	3	9	3	0	7	0	21	19	8	0	0	0	0	0	0	100.0
	32–63	1.9	18	5	0	5	0	10	0	33	24	5	0	1	0	1	0	0	100.0
	>63	0.4	3	1	1	0	1	8	1	21	48	4	0	7	0	2	1	0	100.0
Total sample			**21**	**4**	**5**	**1.5**	**0**	**8**	**0.0**	**26**	**6**	**15**	**8**	**0**	**2.0**	**1.7**	**0.5**	**0.3**	100.4
350	<5	*67.1*	*21*	*2*	*6*		*0*	*4*	*1*	*13*	*0*	*10*	*40*	*2*	*1*	*0*	*0*	*1*	100.0
	5–10	11.0	56	3	13	0	0	1	0	12	2	7	0	2	0	1	4	0	100.0
	10–15	16.9	40	1	6	1	7	5	3	28	0	7	0	0	0	2	0	0	100.0
	15–32	2.9	56	5	10	6	1	3	0	8	6	1	0	0	1	2	0	0	100.0
	32–63	1.9	29	4	5	2	7	6	2	21	11	8	0	0	0	5	0	0	100.0
	>63	0.2	29	4	1	1	12	11	5	5	2	7	0	10	0	1	13	0	100.0
Total sample			**29**	**2**	**7**	**0.4**	**1**	**4**	**1.0**	**15**	**1**	**9**	**27**	**1**	**0.6**	**0.5**	**0.5**	**0.6**	100.0

Table 2. Multiple-window heavy-mineral data for the four studied samples obtained by combining grain counting under the microscope with Raman spectroscopy for classes >5 μm and by XRD for the <5 μm class. Bulk-sample composition is in **bold**; XRD data, in *italics*, are only indicative and were not considered in the calculation of bulk-sample composition; GSZ = Grain size (in μm); tHMC = transparent-heavy-mineral concentration; Zrn = zircon; Tur = tourmaline; Rt = rutile; TiOx = anatase + brookite; Ttn = titanite; Ap = apatite; Ep = epidote; Grt = garnet; Cld = chloritoid; St = staurolite; Ky = kyanite; Sil = sillimanite; Amp = amphibole; Px = pyroxene; &tHM = orthopyroxene, baryte, Cr-spinel, monazite, olivine, or xenotime. The weight percentage of each class was obtained by wet sieving; n.d. = not determined; tr. = in trace.

Sample	GSZ (μm)	Class (%)	tHMC	Zrn	Tur	Rt	TiOx	Ttn	Ap	Ep	Grt	Cld	St	Ky + Sil	Amp	Px	&tHM	
	<5	7.9	*9*	*2*	*0*	*0*		*0*	*3*	*43*	*20*	*0*	*0*	*0*	*30*		*2*	100
	5–10	1.1	5	1	1	2	3	3	2	28	4	1	0	0.4	50	4	0	100
	10–15	3.3	10	1	5	2	1	3	1	38	4	2	0	4	37	2	2	100
305	15–32	4.3	19	1	5	3	0	5	1	31	4	2	1	1	41	4	0	100
	32–63	58.4	8	0	2	1	1	5	2	25	5	2	0	2	52	4	0	100
	>63	25.0	12	0	0	0	0	0	7	7	0	0	0	0	73	13	0	100
Total sample			**10**	**0.1**	**2**	**0.7**	**0.3**	**3**	**4**	**20**	**3**	**1**	**0**	**1**	**58**	**7**	**0.1**	*100*
	<5	79.1	*6*	*4*	*0*	*0*		*0*	*6*	*0*	*0*	*0*	*0*	*0*	*90*		*0*	100
	5–10	4.4	1	0	1	2	1	2	1	30	4	1	0	0	56	2	0	100
	10–15	9.4	7	1	2	2	1	5	1	32	4	1	0	2	47	2	1	100
311	15–32	2.6	6	0	2	0	0	4	0	28	5	4	1	2	49	7	0	100
	32–63	3.8	5	1	1	1	0	3	3	28	2	3	1	1	51	7	0	100
	>63	0.7	7	4	6	4	0	0	0	33	6	0	6	4	39	0	0	100
Total sample			**5**	**0.9**	**2**	**1**	**0.4**	**4**	**1**	**31**	**4**	**2**	**0.4**	**1**	**48**	**3**	**1**	*100*
	<5	68.8	*9*	*0*	*0*	*13*		*0*	*0*	*41*	*9*	*0*	*0*	*0*	*37*		*0*	100
	5–10	9.2	3	1	2	1	0	3	1	21	3	1	0	0	62	4	0	100
	10–15	16.4	6	1	2	2	1	5	3	20	5	2	0	1	61	0	1	100
318	15–32	3.4	7	1	1	0	0	0	1	27	3	1	1	1	62	4	0	100
	32–63	1.9	10	1	1	0	0	3	2	24	2	3	0	1	57	7	1	100
	>63	0.4	8	7	7	0	0	10	3	13	43	0	0	0	13	3	0	100
Total sample			**8**	**0.7**	**2**	**1**	**0.3**	**4**	**2**	**21**	**4**	**1**	**0.1**	**1**	**60**	**2**	**1**	*100*
	<5	67.1	*4*	*0*	*0*	*14*		*0*	*0*	*0*	*23*	*0*	*0*	*0*	*55*		*8*	100
	5–10	11.0	1	4	6	3	1	3	1	34	5	2	0	0	33	6	1	100
	10–15	16.9	5	2	4	2	3	3	2	29	4	4	0	2	45	1	1	100
350	15–32	2.9	3	1	3	3	0	3	2	34	4	7	0	0	45	1	0	100
	32–63	1.9	6	0	2	1	0	5	1	37	3	4	1	2	46	2	0	100
	>63	0.2	11	n.d.	n.d.	n.d.	n.d.	n.d.	n.d.	n.d.	n.d.	n.d.	tr.	n.d.	n.d.	n.d.	n.d.	
Total sample			**4**	**2**	**4**	**2**	**2**	**3**	**2**	**31**	**4**	**4**	**0**	**2**	**44**	**1**	**1**	*100*

XRD analyses indicate that phyllosilicates make up half to two-thirds of the <5-μm class (Figure 3A). Clay minerals are mainly illite (44–67%), with subordinate chlorite (15–21%), smectite/montmorillonite (8–18% including illite–smectite mixed layers), and kaolinite (6–13%). On average, quartz represents 19% and feldspars 11% of the <5-μm class, with plagioclase dominant over K-feldspar. Pyrite, glauconite, opal or glass, Fe-oxyhydroxides, high-Mg calcite, ankerite, siderite, magnesite, zeolites, talc, baryte, gypsum, anhydrite, halite, and jarosite were also locally detected.

4.4. Heavy Minerals

Rich assemblages are mostly represented by amphibole (hornblende, actinolite), with subordinate epidote and minor garnet, the classic "triad" of transparent heavy minerals diagnostic of orogenic Himalayan provenance [99]. Titanite, clinopyroxene, tourmaline, apatite, chloritoid, rutile, zircon, kyanite, and sillimanite occur in all samples (Table 2), whereas orthopyroxene, olivine, Cr-spinel, monazite, xenotime, anatase, and baryte were only occasionally recorded.

Transparent-heavy-mineral concentration tends to increase from deeper-water samples 311 and 350 (tHMC 4.7 ± 1.0) to samples 318 and 305 (tHMC 8.5 ± 1.8), suggesting that heavy minerals are preferentially retained close to shore. Deeper-water samples 311 and 350 have a distinctly lower amphibole/epidote ratio (1.5 ± 0.1) than samples 318 and 305 (2.9 ± 0.1), indicating that amphibole is not preferentially winnowed offshore, being mainly associated with the coarse mode, whereas epidote is associated with the fine mode. These trends are also compatible with greater Ganga contribution to finer-grained distal muds.

Ferromagnesian inosilicates display variable stages of corrosion. Hornblende is more affected by weathering than actinolite, and ~70% of clinopyroxene grains are corroded and mostly etched to skeletal [100]. Epidote, garnet, and titanite are also commonly corroded.

4.5. Intra-Sample Mineralogical Variability

The adopted multiple-window approach allows us to investigate in detail how different minerals are distributed among the different grain-size classes of each sample (intra-sample variability) as a function of their density and shape [101,102]. Although inter-sample variability is relatively minor (i.e., the four studied samples show rather homogeneous bulk mineralogy), intra-sample variability is notable. Platy micas tend to concentrate in the coarse tail of the size distribution because of their low settling velocity. Clay minerals are the main component of the <5-μm class because of their fine crystal size.

Heavy minerals display a minimum in the 5–10-μm class—where phyllosilicates are also low and tectosilicates relatively enriched—and tend to increase in coarser classes (Figure 6). Ultradense and fast-settling Ti-oxides tend to concentrate in finer classes. The relative abundance of lower-density and slower-settling amphibole increases instead in the sandy coarse tail of the size distribution in sample 305, whereas the relative abundance of epidote is rather constant in all size classes (Figure 7). Other minerals do not show clear intrasample grain-size trends owing to mixing in different proportions of moderately to poorly sorted fine and coarse modes.

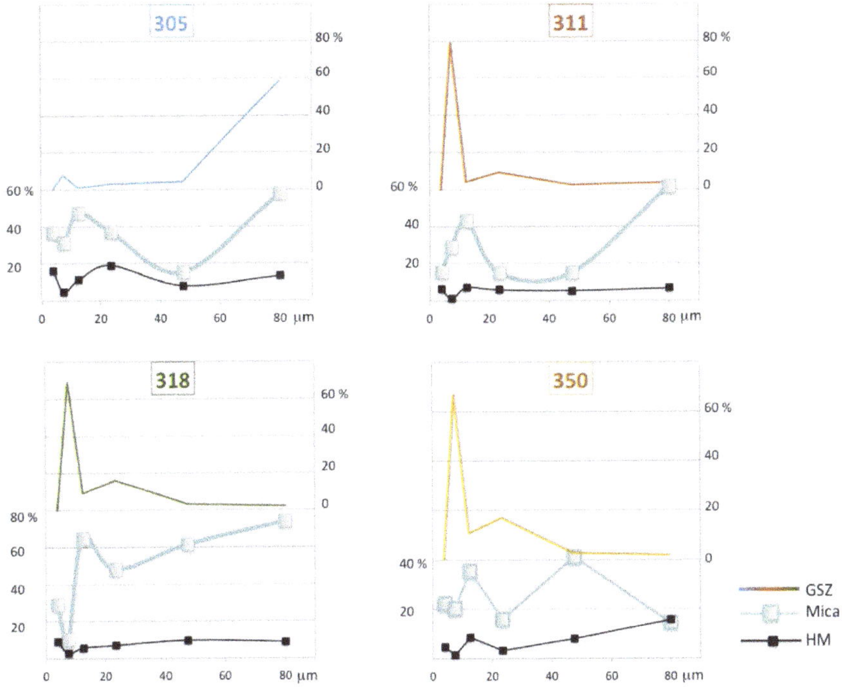

Figure 6. Intra-sample compositional variability. In most samples, mica (muscovite + biotite + chlorite) is concentrated in the coarse tail of the grain-size distribution (GSZ, determined by wet sieving). Heavy minerals (HM) show a minimum in the tectosilicate-rich 5–10-μm class and tend to increase in coarser classes. The lack of systematic grain-size control is ascribed to bimodal grain-size distribution as a result of mixing of fine and coarse modes in different proportions.

4.6. Geochemistry

The bulk-sample geochemical composition of Bengal shelf silt mirrors that of Himalayan-derived sediment supplied by the Meghna estuary (Figure 8). The Al/Si ratio, notoriously a proxy for grain size (e.g., [25]), ranges from 0.23 in the very-coarse-silt sample 305 to 0.38 in the fine-silt sample 311. Such values are characteristic of coarse silt and fine silt, respectively, whereas typical values for Bengal shelf sand are 0.11–0.17 [8]. The Ca content is below 1.6%, which implies low concentration of carbonates (<2%).

XRF data indicate that the composition of the <5-μm class is quite homogeneous in the four studied samples, although Ca decreases westward from higher values in samples 305 and 311 to the lowest value in the most distal sample 350. The Al/Si ratio ranges between 0.38 and 0.50, which are typical values for Himalayan-derived clay-rich mud [27].

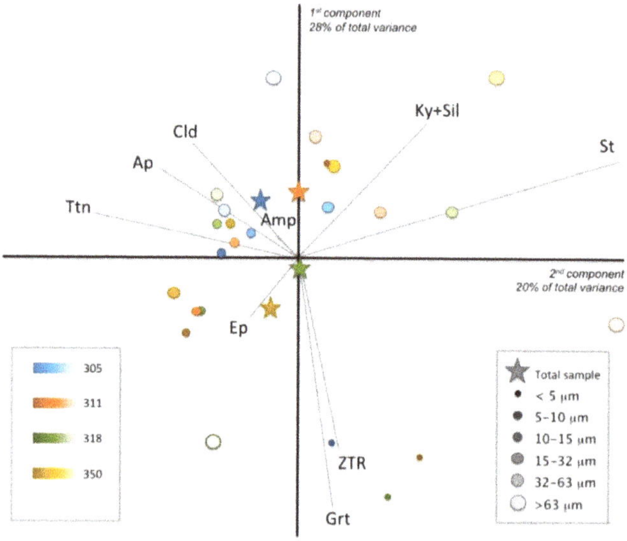

Figure 7. Inter-sample and intra-sample variability in transparent-heavy-mineral suites. The biplot [103] drawn using CoDaPack software by Comas-Cufí and Thió-Henestrosa [104]) displays multivariate observations (points) and variables (rays). The length of each ray is proportional to the variance of each element in the dataset. If the angle between two rays is close to 0°, 90°, and 180°, then the corresponding compositional parameters are directly correlated, uncorrelated, and inversely correlated, respectively. Ap = apatite; Cld = chloritoid; Ep = epidote; Grt = garnet; Ky+Sil = kyanite + sillimanite; St = staurolite; Ttn = titanite; ZTR = zircon + tourmaline + rutile.

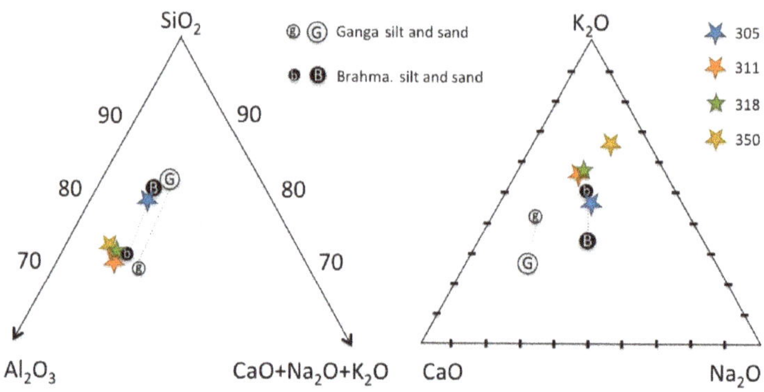

Figure 8. Triangular diagrams illustrating the bulk-sediment geochemistry of the four studied samples. The average composition of Ganga (g = silt; G = sand) and Brahmaputra (b = silt; B = sand) sediments is given for comparison. Relatively to the very-coarse-silt sample 305, fine-silt samples 311, 318, and 350 are notably enriched in Al and K, preferentially hosted in phyllosilicates.

5. Methodological Issues

In this section, we critically discuss the potential limitations of the proposed multi-technique approach with regard to the accuracy and precision of the obtained mineralogical dataset.

5.1. Quality Check of Sieving Results

Although wet sieving at 5-μm, 10-μm, and 15-μm mesh resulted to be challenging and time-consuming, the results were rewarding (Figure 2). The quality of the grain-size separation was systematically checked for both "light" (LM) and "heavy" (HM) mineral fractions under an optical microscope provided with a scale bar. Subsequently, every slide of the LM fraction (representing between 81% and 99% of each grain-size class) was mapped using a micro-camera also provided with a scale bar. Whereas no significant misfit was detected for samples 305 and 311, we did observe a few oversized grains—mainly phyllosilicates—in each class of samples 318 and 350. In order to minimize the error, we chose not to consider oversized grains during mineralogical analysis. As far as grain-size analysis is concerned, a more serious problem is associated with the inevitable loss of material in each passage, total loss being limited to 4% overall for very coarse silt (sample 305) but reaching between 9% and 15% for fine silt (samples 311, 318, and 350; Table S2, Supplementary Materials). After sieving, each net was scrutinized under the optical microscope to check the uniformity and potential deformation of mesh size (Figure 9).

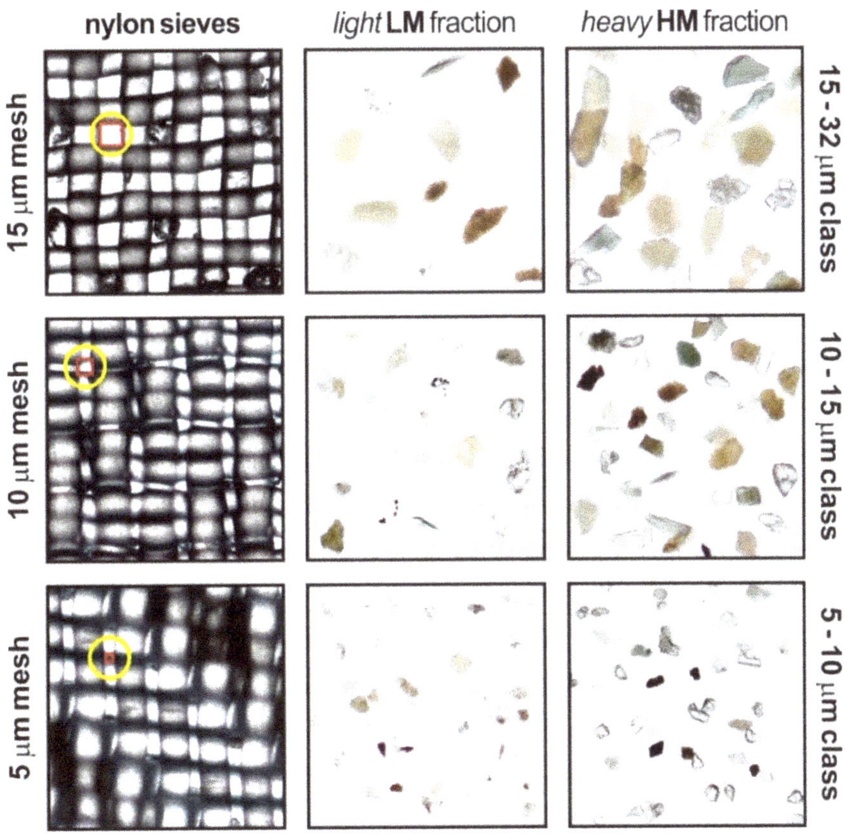

Figure 9. Images of the used nylon sieves (5-μm, 10-μm, and 15-μm mesh size) with the corresponding size classes obtained for sample 318, divided into low-density (<2.90 g/cm^3, "light" LM) and high-density (>2.90 g/cm^3 "heavy" HM) mineral fractions. Although the sieve meshes were slightly deformed and several grains got caught into them, after this laborious procedure, we managed to obtain neatly separated grain-size classes for mineralogical analysis of cohesive mud. Red square is included for scale.

5.2. Mineralogical Analysis of Cohesive Mud

Although Raman spectroscopy is proven to be an effective method to identify and count mineral grains down to the size of 1–2 μm only [105–107], this technique could not be applied to the <5-μm class of our samples. The finest sediment fraction mostly consists of clay minerals and very-fine-silt grains that display low Raman scattering, mainly because of the thin lamellar habit of phyllosilicates and, consequently, insufficient volume analyzed by the spectrometer. Moreover, most phyllosilicates in Bengal shelf mud are notably altered, which affects their mineral structure and chemical composition. This problem is attenuated for the 5–10-μm class, where only a few Raman spectra (≤2%) could not be determined owing to high background noise and a few other Raman spectra (≤2%) could be identified as phyllosilicates but not reliably assigned to a specific mineralogical species.

The mineralogy of the <5-μm class was, therefore, investigated by XRD, the most widely used technique for the analysis of clay minerals [108]. The results of the three methodologies used jointly (optical counting, Raman counting, and XRD), presented in Tables 1 and 2, are not perfectly comparable. The main difference between optical and Raman counting versus XRD is that the former techniques provide a punctual result—reflecting areal and, hence, volume percentages faithfully in case of point counting and only approximately in case of grain counting, which overestimates the volume percentage of smaller grains [109]—whereas XRD results are best-fit approximations of both composition and relative abundances. XRD data are expressed in weight percentages, rather than volume and frequency percentages as obtained with point and grain counting, and are influenced by the crystallographic parameters chosen for the diverse mineral phases. Moreover, the XRD analysis of the <5-μm class in bulk without separation of low-density and high-density particles poses a severe challenge in the case of Bengal shelf sediments, which contain a wide spectrum of detrital minerals derived from the great variety of rocks exposed in the vast Ganga and Brahmaputra catchments. Only the main mineral groups could, thus, be distinguished, with limited or no insight into rarer mineral phases, varieties within solid-solution series, or polymorphs. On the other hand, XRD analysis allowed us to obtain particularly valuable information on clay minerals, which could not be properly investigated by optical microscopy or Raman spectroscopy. Thus, even altered grains, which represent ≤6% of the >5-μm classes and could not be determined by optical or Raman methods, were identified.

6. Provenance

In this section, we draw inferences concerning the relative contributions from the Ganga and Brahmaputra Rivers to silt and clay deposited on the Bengal shelf. Provenance of mud in the subaerial tidal delta plain and subaqueous delta was investigated by Flood et al. [110] based on XRD mineralogy and XRF geochemistry, whereas Lupker et al. [27] and Goodbred et al. [42] used Sr and Nd isotopes, Sr concentrations, and major elements as provenance tracers. Grain-by-grain mineralogical analysis of fine silt poses diverse technical problems (e.g., difficult distinction of color and pleochroism under the optical microscope, weak Raman signal), which prevented us from achieving the same accuracy and precision obtained for sand. Moreover, sediment composition is controlled by several different factors including—in addition to provenance—grain size, hydraulic sorting, and chemical weathering.

6.1. Provenance of Silt

Detailed information obtained during previous work allowed us to discriminate between mineralogical signatures of suspended load carried by the Ganga and Brahmaputra Rivers (Table S8, Supplementary Material). The comparison between detrital modes of fluvial and shelfal sediments clearly indicates that silt deposited on the Bengal shelf has mineralogical signatures very close to Brahmaputra suspended load (Figure 10). The quartz/feldspar ratio and the proportions among different feldspar minerals are close to Brahmaputra silt (Figure 10A), and transparent-heavy-mineral suites are indistinguishable from those of both Brahmaputra and Meghna estuary silt (Figure 10B).

The Brahmaputra River is in fact characterized by a higher sediment flux and by a notably richer transparent-heavy-mineral suite than the Ganga [23,24].

Inspection of the biplot in Figure 10C, based on the entire spectrum of mineralogical parameters, allows us to distinguish provenance control (first principal component) from the grain-size effect (second principal component) on the mineralogy of fluvial suspended load. The grain-size trend is less clear for shelfal muds, which show full Brahmaputra affinity for the very-coarse-silt sample 305 and more intermediate character and greater Ganga affinity for fine-silt samples 311, 318, and 350 collected at deeper shelfal depths. The decrease in Q/F ratio with increasing grain size, which is atypical (e.g., [112,113]), is one clue suggesting greater contribution of finer mud fractions originally deposited by the Ganga River. Dominance of the Brahmaputra contribution in coarser silt and sand fractions and greater influence of the Ganga for cohesive mud may reflect the fact that the Brahmaputra River has greater flow velocity and competence than the Ganga River, and is, thus, able to entrain coarser particles at all channel depths [19,24,25]. Also, the very-coarse-silt sample 305 is deposited right offshore of the Meghna estuary, whereas fine-silt samples 311, 318, and 350 were collected in the western part of the shelf, where sediment contribution from the Ganga River and the subaerial delta may be more significant [42,110]. The marked difference between the coarser-grained substratum of proximal topsets and the fine-silt substratum of distal foresets and bottomsets is also reflected by foraminiferal assemblages, which are rare and oligotypic in sample 305 and progressively enriched in mud-dwelling species in samples 311, 318, and 350.

Further clues are provided by geochemical data (Figure 11). Simple best-fit calculations indicate that the chemical composition of very-coarse-silt sample 305 can be reproduced moderately well as a mixture of shallow and deep suspended load entrained in the Meghna estuary in proportions between 1:3 and 1:8, and best as a mixture composed one-third of Ganga shallow suspended load and two-thirds of Brahmaputra sediment with subequal proportions of shallow suspended load, deep suspended load, and bedload. The chemical composition of fine-silt samples 311, 318 and 350 can be reproduced moderately well as a mixture of shallow and deep suspended load entrained by the Padma River or Meghna estuary in subequal proportions, better as a mixture of ~40% Brahmaputra suspended load and ~60% Ganga clay, and best as a mixture of 62% very coarse Bengal shelf silt and 38% Ganga clay. The imperfect fit obtained in most simulations is caused for very coarse silt by a dearth in elements largely hosted in densest minerals (e.g., Y, REE, Th, Zr, Hf, and Cr), which can be ascribed to selective hydraulic segregation of fast-settling heavy minerals in the coastal setting and consequent depletion in offshore shelf sediments, and for fine silt by an excess in heavy metals, largely associated with the clay fraction. Additionally, shelf sediments appear systematically depleted relatively to fluvial sediments in Ca, Ba, P, and Sb (Figure 11). Provenance issues are discussed in greater detail in the companion paper [8], which focuses on sand but takes into account the entire grain-size spectrum of Bengal shelf sediments.

6.2. Provenance of Clay Minerals

As far as clay minerals are concerned, simple forward mixing calculations (mathematical approach described in Garzanti et al. [114] and Resentini et al. [115]), based on data summarized in Table 1 in Khan et al. [60], indicate that the Ganga River contributes ~40% and the Brahmaputra River ~60% of clay downstream of their confluence. Clay minerals fed into the Bengal shelf are supplied between 25% and 40% by the Ganga River and the remaining 60–75% by the Brahmaputra and Meghna Rivers. The clay–mineral assemblage identified by XRD analysis of the <5-μm class of the four studied mud-rich samples compares well with that determined by Khan et al. [60] on seven Meghna estuary samples, confirming that fine sediment deposited on the subaqueous Bengal delta is dominantly derived from the Meghna estuary and ultimately supplied ≤40% by the Ganga and ≥60% by the Brahmaputra and Meghna Rivers. Relatively abundant chlorite in our samples, in percentages similar to those reported from the Meghna River [60], suggests that clay–mineral contribution from the Tripura belt and eastern Bangladesh via the Meghna River may not be negligible.

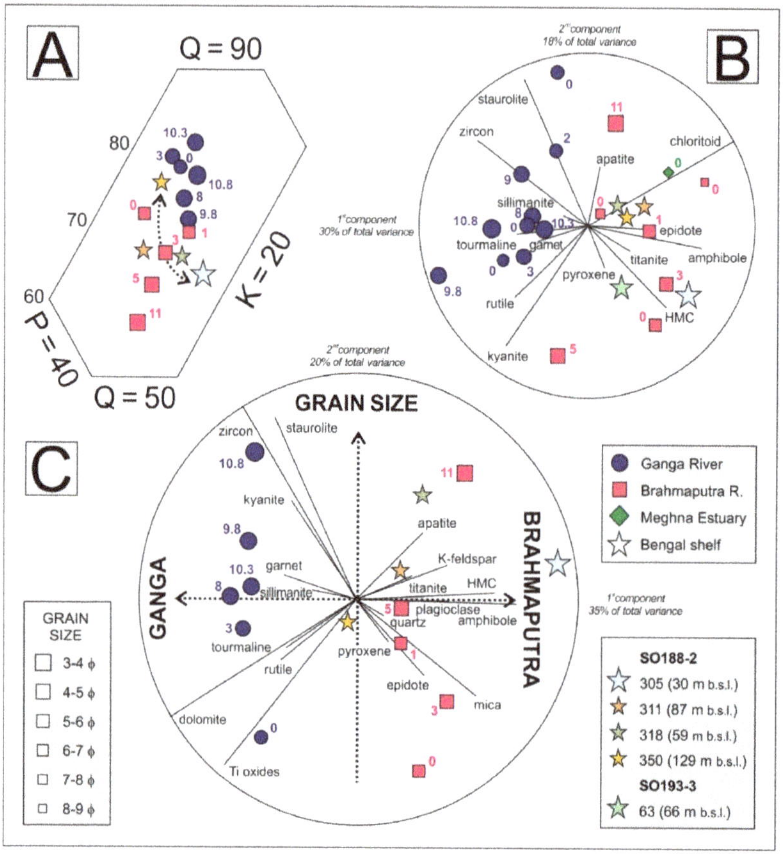

Figure 10. Detrital modes of Bengal shelf silt compared with Ganga and Brahmaputra suspended load (size of symbols proportional to sample grain size; depth of sampling within the river channel in meters indicated close to each symbol; data after Garzanti et al. [24]. Bengal shelf sediments display Brahmaputra signature, although the most distal and finest grained sample is more akin to the Ganga. (**A**) Ganga silt is distinguished by higher quartz/feldspar (Q/F) and lower plagioclase/feldspar (P/F) ratios than Brahmaputra silt (arrow indicates proximal/distal trend; QPK plot after Dickinson [111]). (**B**) The biplot [103] based on heavy-mineral data highlights the close affinity between heavy-mineral suites in Bengal Shelf silt and suspended load in the Brahmaputra River and Meghna estuary. (**C**) The biplot based on the entire mineralogical dataset (both LM and HM fractions) best highlights the different signatures of Ganga and Brahmaputra suspended load (first principal component, PC1) and the grain-size effect (PC2). Because of the presence of allochemical carbonate in shelf sediments, calcite grains were not included in the biplot.

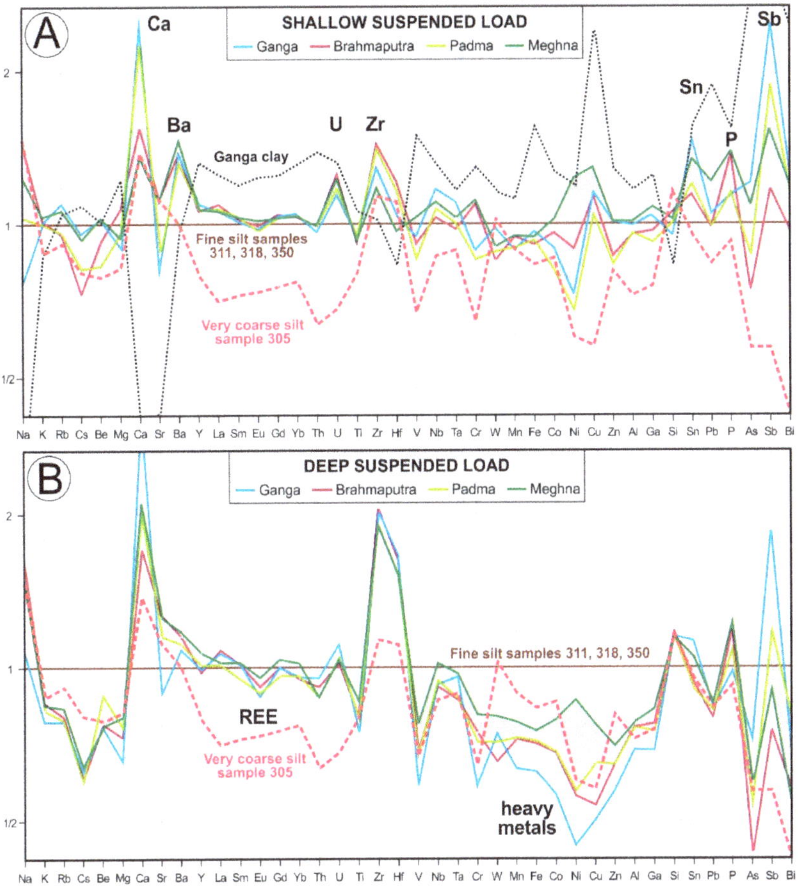

Figure 11. Geochemistry of Ganga, Brahmaputra, Padma, and Meghna suspended load entrained at shallow (≤2 m) and greater (>5 m) channel depth compared with Bengal shelf silt (elements arranged following the periodic table group by group; data normalized to the mean composition of fine silt samples 311, 318, and 350; Table S4, Supplementary Materials). (**A**) The composition of fine Bengal shelf silt compares broadly with shallow suspended load in the Padma River and Meghna estuary (gray field), in turn chiefly derived from the Ganga and Brahmaputra, but with notably less Ca, Ba, U, Zr, Sn, P, and Sb. Note that the composition of very coarse Bengal shelf silt (sample 305) is a quasi-perfect mirror image of that of Ganga clay (<2 μm; own data), which allows us to see fine Bengal shelf silt as a mixture of these two end members in subequal proportions. (**B**) The composition of very coarse Bengal shelf silt compares rather well with deep suspended load in the Padma River and Meghna estuary (gray field), even though with lower concentration of elements preferentially hosted in ultradense minerals (REE, Zr, Hf).

7. Conclusions

We developed a detailed procedure for the high-resolution, multiple-window analysis of the mineralogical and geochemical composition of mud-rich sediments. In order to achieve a similar accuracy and precision as that currently obtained for sand, several different techniques—including laser granulometry, optical microscopy, Raman spectroscopy, X-ray diffraction, ICP-OES/ICP-MS, and X-ray fluorescence—were applied either on the bulk sample or on the low-density LM and high-density HM fractions of six grain-size classes (<5 μm to >63 μm) of four Bengal shelf mud-rich samples separated by wet sieving. Grain-by-grain identification of transparent-heavy-mineral species down to 5 μm could be achieved by coupling optical observations under the microscope with Raman spectroscopy, whereas standard XRD and XRF methods were used to analyze the <5-μm class. Both inter-sample and intra-sample compositional variability could, thus, be investigated. Sieve analysis showed that the fine mode in all of our samples is represented by very fine silt and clay and, thus, confirmed that laser granulometry markedly underestimates the clay content of the analyzed sediment samples.

We chose the Bengal shelf as a test case, not only because this is part of the largest sediment system on Earth but also because a complete mineralogical and geochemical dataset was made available in the last two decades on different size classes of suspended load and bedload sediments carried by the Ganga and Brahmaputra rivers and funneled through the Meghna estuary. All of the studied silt samples and most of their size classes are feldspatho-quartzose. The Q/F ratio increases with decreasing grain size from 1.7 in the sand-bearing proximal sample 305 to 3.0 in the most distal sample 350; plagioclase prevails over K-feldspar. Rich transparent-heavy-mineral assemblages include mainly amphibole and subordinate epidote, and minor garnet and pyroxene. Mineralogical and geochemical fingerprints are typical of Himalayan-derived detritus and compare most closely to suspended load carried by the Brahmaputra River with more significant Ganga contribution for cohesive fine silt in distal foresets and bottomsets of the subaqueous delta. Sediment provenance and the potential modifications of the compositional signal owing to physical processes during transit from the Himalayan foreland basin to the Bengal remnant-ocean basin are discussed further in the companion paper [8]. Because the shelf represents the link between the huge Himalayan rivers and the largest turbiditic fan on Earth, a detailed quantitative characterization of Bengal shelf sediments represents an essential step to allow comparison of compositional signatures characterizing different segments of the entire source-to-sink system, from the fluvial and deltaic environment to the deep-sea fan.

Supplementary Materials: The following are available online at http://www.mdpi.com/2075-163X/9/10/640/s1; supplementary materials include information on technical characteristics of Raman spectrometers and on the location of the studied samples (Table S1), together with the laser granulometry (Table S2), bulk-mineralogy (Table S3), heavy-mineral (Table S4), XRD (Table S5), XRF (Table S6), and bulk-geochemistry (Table S7) datasets. Table S8 provides mineralogical data on Ganga and Brahmaputra suspended load (after Garzanti et al., 2011).

Author Contributions: Project design, E.G. and S.A.; writing and editing, E.G. and L.B.; grain-size and optical analyses, L.B. and S.A.; Raman spectroscopy analyses, L.B. and S.A.; geochemical analyses, C.F.-L.; XRD and XRF analyses of the <5-μm class, A.H.; fossil determination, G.C.; provenance interpretation, E.G.

Acknowledgments: The article benefited from careful reviews and constructive advice by two anonymous reviewers. We thank Alberto Paleari, Valentina Palanza, and Alessandro Lauria (Department of Material Sciences, University of Milano-Bicocca), Claudia Conti and Marco Realini (ICVCB-CNR), Danilo Bersani, Irene Aliatis, and Erica Lambruschi (Physics Department, University of Parma), and Massimo Placidi (Horiba Scientific) for fundamental help with the Raman spectroscopy analyses. Nicoletta Fusi assisted during laser granulometry analyses. We also thank Hermann Kudrass for kindly providing samples from Sonne cruise SO188-2. Funding was provided by Project MIUR-PRIN 2015EC9PJ5 "The subduction and exhumation of the continental lithosphere: their effects on the structure and evolution of the orogens" and MIUR—Dipartimenti di Eccellenza 2018–2022, Department of Earth and Environmental Sciences, University of Milano-Bicocca.

Conflicts of Interest: The authors declare no conflict of interest.

References

1. Blatt, H.; Jones, R.L. Proportions of exposed igneous, metamorphic, and sedimentary rocks. *Geol. Soc. Am. Bull.* **1975**, *86*, 1085–1088. [CrossRef]
2. Pettijohn, F.J. *Sedimentary Rocks*; Harper & Row: New York, NY, USA, 1975; 776p.
3. Garrels, R.M. Sediments cycling and diagenesis. *Geol. Surv. Bull.* **1986**, *1578*, 1–11.
4. Potter, P.E.; Maynard, J.B.; Depetris, P.J. *Mud and Mudstones: Introduction and Overview*; Springer Science & Business Media: Berlin, Germany, 2005; 297p.
5. Andò, S.; Aharonovich, S.; Hahn, A.; George, S.C.; Clift, P.D.; Garzanti, E. Integrating heavy-mineral, geochemical, and biomarker analyses of Plio-Pleistocene sandy and silty turbidites: A novel approach for provenance studies (Indus Fan, IODP Expedition 355). *Geol. Mag.* **2019**. [CrossRef]

6. Caracciolo, L.; Andò, S.; Vermeesch, P.; Garzanti, E.; McCabe, R.; Barbarano, M.; Paleari, C.; Rittner, M.; Pearce, T. A multidisciplinary approach for the quantitative provenance analysis of siltstone. Mesozoic Mandawa Basin, southeastern Tanzania. *Geol. Soc. Lond. Spec. Publ.* **2019**, *484*. [CrossRef]
7. Kudrass, H.R.; Spiess, V.; Bruns, A.; Ding, F.; Fekete, N.; Fenner, J.; France-Lanord, C.; Palamenghi, L.; Reinhardt, L.; Rühlemann, C.; et al. *Cruise Report SO 188-2 Chittagong (Bangladesh)—Penang (Malaysia) 06.07.2006—31.07.2006*; BGR: Hannover, Germany, 2007; pp. 1–100.
8. Garzanti, E.; Andò, S.; France-Lanord, C.; Limonta, M.; Borromeo, L.; Vezzoli, G. Provenance of Bengal Shelf Sediments. 2. Petrology and geochemistry of sand. *Minerals* **2019**, in press.
9. Allison, M.A.; Khan, S.R.; Goodbred, S.L.; Kuehl, S.A. Stratigraphic evolution of the late Holocene Ganges–Brahmaputra lower delta plain. *Sediment. Geol.* **2003**, *155*, 317–342. [CrossRef]
10. Kuehl, S.A.; Allison, M.A.; Goodbred, S.L.; Kudrass, H. The Ganges-Brahmaputra Delta. In *River Deltas—Concepts, Models, and Examples*; Giosan, L., Bhattacharya, J.P., Eds.; Society for Sedimentary Geology (SEPM) Special Publication NO. 83; Society for Sedimentary Geology (SEPM): Tulsa, OK, USA, 2005; pp. 413–434.
11. Goodbred, S.L.; Kuehl, S.A. Enormous Ganges-Brahmaputra sediment discharge during strengthened early Holocene monsoon. *Geology* **2000**, *28*, 1083–1086. [CrossRef]
12. Curray, J.R.; Emmel, F.J.; Moore, D.G. The Bengal Fan: Morphology, geometry, stratigraphy, history and processes. *Mar. Pet. Geol.* **2003**, *19*, 1191–1223. [CrossRef]
13. Thompson, R.W. Mineralogy of sands from the Bengal and Nicobar fans, Sites 218 and 211, Eastern Indian Ocean. *Initial Rep. DSDP* **1974**, *22*, 711–713.
14. Ingersoll, R.V.; Suczek, C.A. Petrology and provenance of Neogene sand from Nicobar and Bengal Fans, DSDP Sites 211 and 218. *J. Sedim. Petrol.* **1979**, *49*, 1217–1228.
15. Yokoyama, K. Mineralogy of silts from the Bengal Fan. *Proc. ODP Sci. Results* **1990**, *116*, 59–73.
16. France-Lanord, C.; Derry, L.; Michard, A. Evolution of the Himalaya since Miocene time: Isotopic and sedimentological evidence from the Bengal Fan. *Geol. Soc. Lond. Spec. Publ.* **1993**, *74*, 603–621. [CrossRef]
17. Galy, A.; France-Lanord, C.; Derry, L.A. The strontium isotopic budget of Himalayan rivers in Nepal and Bangladesh. *Geochim. Cosmochim. Acta* **1999**, *63*, 1905–1925. [CrossRef]
18. Galy, A.; France-Lanord, C. Higher Erosion rates in the Himalaya: Geochemical constraints on riverine fluxes. *Geology* **2001**, *29*, 23–26. [CrossRef]
19. Galy, V.; France-Lanord, C.; Beyssac, O.; Faure, P.; Kudrass, H.; Palhol, F. Efficient organic carbon burial in the Bengal fan sustained by the Himalayan erosional system. *Nature* **2007**, *450*, 407. [CrossRef]
20. Singh, S.K.; France-Lanord, C. Tracing the distribution of erosion in the Brahmaputra watershed from isotopic compositions of stream sediments. *Earth Planet. Sci. Lett.* **2002**, *202*, 645–662. [CrossRef]
21. Garzanti, E.; Vezzoli, G.; Andò, S.; France-Lanord, C.; Singh, S.K.; Foster, G. Sand petrology and focused erosion in collision orogens: The Brahmaputra case. *Earth Planet. Sci. Lett.* **2004**, *220*, 157–174. [CrossRef]
22. Garzanti, E.; Vezzoli, G.; Andò, S.; Lavé, J.; Attal, M.; France-Lanord, C.; DeCelles, P. Quantifying sand provenance and erosion (Marsyandi River, Nepal Himalaya). *Earth Planet. Sci. Lett.* **2007**, *258*, 500–515. [CrossRef]
23. Garzanti, E.; Andò, S.; France-Lanord, C.; Vezzoli, G.; Censi, P.; Galy, V.; Najman, Y. Mineralogical and chemical variability of fluvial sediments: 1. Bedload sand (Ganga–Brahmaputra, Bangladesh). *Earth Planet. Sci. Lett.* **2010**, *299*, 368–381. [CrossRef]
24. Garzanti, E.; Andò, S.; France-Lanord, C.; Censi, P.; Vignola, P.; Galy, V.; Lupker, M. Mineralogical and chemical variability of fluvial sediments. 2. Suspended-load silt (Ganga-Brahmaputra, Bangladesh). *Earth Planet. Sci. Lett.* **2011**, *302*, 107–120. [CrossRef]
25. Lupker, M.; France-Lanord, C.; Lavé, J.; Bouchez, J.; Galy, V.; Métivier, F.; Gaillardet, J.; Lartiges, B.; Mugnier, J.L. A Rouse-based method to integrate the chemical composition of river sediments: Application to the Ganga basin. *J. Geophys. Res. Earth Surf.* **2011**, *116*, 1–24. [CrossRef]
26. Lupker, M.; France-Lanord, C.; Galy, V.; Lavé, J.; Gaillardet, J.; Gajurel, A.P.; Guilmette, C.; Rahman, M.; Singh, S.K.; Sinha, R. Predominant floodplain over mountain weathering of Himalayan sediments (Ganga basin). *Geochim. Cosmochim. Acta* **2012**, *84*, 410–432. [CrossRef]
27. Lupker, M.; France-Lanord, C.; Galy, V.; Lavé, J.; Kudrass, H. Increasing chemical weathering in the Himalayan system since the Last Glacial Maximum. *Earth Planet. Sci. Lett.* **2013**, *365*, 243–252. [CrossRef]

28. Weltje, G.J. A quantitative approach to capturing the compositional variability of modern sands. *Sediment. Geol.* **2004**, *171*, 59–77. [CrossRef]
29. Garzanti, E.; Andò, S.; Vezzoli, G. Grain-size dependence of sediment composition and environmental bias in provenance studies. *Earth Planet. Sci. Lett.* **2009**, *277*, 422–432. [CrossRef]
30. Garzanti, E. Stratigraphy and sedimentary history of the Nepal Tethys Himalayan passive margin. *J. Asian Earth Sci.* **1999**, *17*, 805–827. [CrossRef]
31. Sciunnach, D.; Garzanti, E. Subsidence history of the Tethys Himalaya. *Earth Sci. Rev.* **2012**, *25*, 179–198. [CrossRef]
32. Colchen, M.; Le Fort, P.; Pêcher, A. *Annapurna, Manaslu, Ganesh Himal*; Editions du Centre National de la Recherche Scientifique: Paris, France, 1986; 136p.
33. Carosi, R.; Montomoli, C.; Iaccarino, S. 20 years of geological mapping of the metamorphic core across Central and Eastern Himalayas. *Earth Sci. Rev.* **2017**, *177*, 124–138. [CrossRef]
34. Sakai, H. Rifting of the Gondwanaland and uplifting of the Himalayas recorded in Mesozoic and Tertiary fluvial sediments in the Nepal Himalayas. In *Sedimentary Facies in the Active Plate Margin*; Taira, A., Masuda, F., Eds.; Terra Scientific Publications: Tokyo, Japan, 1989; pp. 723–732.
35. DeCelles, P.G.; Robinson, D.M.; Quade, J.; Ojha, T.P.; Garzione, C.N.; Copeland, P.; Upreti, B.N. Stratigraphy, structure, and tectonic evolution of the Himalayan fold-thrust belt in western Nepal. *Tectonics* **2001**, *20*, 487–509. [CrossRef]
36. DeCelles, P.G.; Gehrels, G.E.; Quade, J.; Ojha, T.P.; Kapp, P.A.; Upreti, B.N. Neogene foreland basin deposits, erosional unroofing, and the kinematic history of the Himalayan fold-thrust belt, western Nepal Siwaliks. *Geol. Soc. Am. Bull.* **1998**, *110*, 2–21. [CrossRef]
37. Garzanti, E. The Himalayan foreland basin from collision onset to the present: A sedimentary-petrology perspective. In *Himalayan Tectonics: A Modern Synthesis*; Treloar, P., Searle, M.P., Eds.; Special Publication 483; Geological Society: London, UK, 2019; pp. 65–122. [CrossRef]
38. Zhu, D.C.; Zhao, Z.D.; Niu, Y.L.; Mo, X.X.; Chung, S.L.; Hou, Z.Q.; Wang, L.Q.; Wu, F.Y. The Lhasa terrane: Record of a microcontinent and its histories of drift and growth. *Earth Planet. Sci. Lett.* **2011**, *301*, 241–255. [CrossRef]
39. Wang, J.-G.; Hu, X.M.; Garzanti, E.; An, W.; Liu, X.-C. The birth of the Xigaze forearc basin in southern Tibet. *Earth Planet. Sci. Lett.* **2017**, *465*, 38–47. [CrossRef]
40. Hébert, R.; Bezard, R.; Guilmette, C.; Dostal, J.; Wang, C.S.; Liu, Z.F. The Indus–Yarlung Zangbo ophiolites from Nanga Parbat to Namche Barwa syntaxes, southern Tibet: First synthesis of petrology, geochemistry, and geochronology with incidences on geodynamic reconstructions of Neo-Tethys. *Gondwana Res.* **2012**, *22*, 377–397. [CrossRef]
41. Burg, J.P.; Nievergelt, P.; Oberli, F.; Seward, D.; Davy, P.; Maurin, J.C.; Diao, Z.; Meier, M. The Namche Barwa syntaxis: Evidence for exhumation related to compressional crustal folding. *J. Asian Earth Sci.* **1998**, *16*, 239–252. [CrossRef]
42. Goodbred, S.L.; Paolo, P.M.; Ullah, M.S.; Pate, R.D.; Khan, S.R.; Kuehl, S.A.; Singh, S.K.; Rahaman, W. Piecing together the Ganges-Brahmaputra-Meghna River delta: Use of sediment provenance to reconstruct the history and interaction of multiple fluvial systems during Holocene delta evolution. *Geol. Soc. Am. Bull.* **2014**, *126*, 1495–1510. [CrossRef]
43. Delft Hydraulics. River Survey Project, Flood Action Plan 24, Special Report NO. 18. In *Sediment Rating Curves and Balances*; Water Resources Planning Organization: Dhaka, Bangladesh, 1996.
44. France-Lanord, C.; Spiess, V.; Klaus, A.; Schwenk, T.; Adhikari, T.T.; Adhikari, S.K.; Bahk, J.J.; Baxter, A.T.; Cruz, J.W.; Das, S.K.; et al. Expedition 354 summary. In *International Ocean Discovery Program 354 Preliminary Report*; IODP: College Station, TX, USA, 2016; pp. 1–35.
45. Sinha, R.; Friend, P.F. River systems and their sediment flux, Indo-Gangetic Plains, northern Bihar, India. *Sedimentology* **1994**, *41*, 825–845. [CrossRef]
46. Milliman, J.D.; Syvitski, J.P. Geomorphic/tectonic control of sediment discharge to the ocean: The importance of small mountainous rivers. *J. Geol.* **1992**, *100*, 525–544. [CrossRef]
47. Summerfield, M.A.; Hulton, N.J. Natural controls of fluvial denudation rates in major world drainage basins. *J. Geophys. Res.* **1994**, *99*, 13871–13883. [CrossRef]

48. Goodbred, S.L.; Kuehl, S.A. Holocene and modern sediment budgets for the Ganga-Brahmaputra river system: Evidence for highstand dispersal to flood-plain, shelf, and deep-sea depocenters. *Geology* **1999**, *27*, 559–562. [CrossRef]
49. Milliman, J.D.; Meade, R.H. World delivery of river sediment to the oceans. *J. Geol.* **1983**, *91*, 1–21. [CrossRef]
50. Singh, I.B. The Ganga River. In *Large Rivers: Geomorphology and Management*; Gupta, A., Ed.; Wiley: New York, NY, USA, 2007; pp. 347–371.
51. Hübscher, C.; Spiess, V. Forced regression systems tracts on the Bengal Shelf. *Mar. Geol.* **2005**, *219*, 207–218. [CrossRef]
52. Palamenghi, L.; Schwenk, T.; Spiess, V.; Kudrass, H.R. Seismostratigraphic analysis with centennial to decadal time resolution of the sediment sink in the Ganges–Brahmaputra subaqueous delta. *Cont. Shelf Res.* **2011**, *31*, 712–730. [CrossRef]
53. Kudrass, H.R.; Machalett, B.; Palamenghi, L.; Meyer, I.; Zhang, W. Sediment transport by tropical cyclones recorded in a submarine canyon off Bangladesh. *Geo Mar. Lett.* **2018**, *38*, 481–496. [CrossRef]
54. Singh, S.K.; Sarin, M.M.; France-Lanord, C. Chemical erosion in the eastern Himalaya: Major ion composition of the Brahmaputra and $\delta^{13}C$ of dissolved inorganic carbon. *Geochim. Cosmochim. Acta* **2005**, *69*, 3573–3588. [CrossRef]
55. Garçon, M.; Chauvel, C. Where is basalt in river sediments, and why does it matter? *Earth Planet. Sci. Lett.* **2014**, *407*, 61–69. [CrossRef]
56. Sarin, M.M.; Krishnaswami, S.; Dilli, K.; Somayajulu, B.L.K.; Moore, W.S. Major ion chemistry of the Ganga-Brahmaputra river system: Weathering processes and fluxes to the Bay of Bengal. *Geochim. Cosmochim. Acta* **1989**, *53*, 997–1009. [CrossRef]
57. Datta, D.K.; Subramanian, V. Texture and mineralogy of sediments from the Ganges-Brahmaputra-Meghna river system in the Bengal Basin, Bangladesh and their environmental implications. *Environ. Geol.* **1997**, *30*, 181–188. [CrossRef]
58. Heroy, D.C.; Kuehl, S.A.; Goodbred, S.L. Mineralogy of the Ganges and Brahmaputra Rivers: Implications for river switching and Late Quaternary climate change. *Sediment. Geol.* **2003**, *155*, 343–359. [CrossRef]
59. Huyghe, P.; Guilbaud, R.; Bernet, M.; Galy, A.; Gajurel, A.P. Significance of the clay mineral distribution in fluvial sediments of the Neogene to recent Himalayan foreland basin (west-central Nepal). *Basin Res.* **2011**, *23*, 332–345. [CrossRef]
60. Khan, M.H.R.; Liu, J.; Liu, S.; Seddique, A.A.; Cao, L.; Rahman, A. Clay mineral compositions in surface sediments of the Ganges-Brahmaputra-Meghna river system of Bengal Basin, Bangladesh. *Mar. Geol.* **2019**, *412*, 27–36. [CrossRef]
61. Segall, M.P.; Kuehl, S.A. Sedimentary processes on the Bengal continental shelf as revealed by clay-size mineralogy. *Cont. Shelf Res.* **1992**, *12*, 517–541. [CrossRef]
62. Andò, S.; Vignola, P.; Garzanti, E. Raman counting: A new method to determine provenance of silt. *Rend. Lincei* **2011**, *22*, 327–347. [CrossRef]
63. Mange, M.A.; Maurer, H.F.W. *Heavy Minerals in Colour*; Chapman and Hall: London, UK, 1992; p. 147.
64. Andò, S.; Garzanti, E. Raman spectroscopy in heavy-mineral studies. *Geol. Soc. Lond. Spec. Publ.* **2014**, *386*, 395–412. [CrossRef]
65. Lünsdorf, N.K.; Kalies, J.; Ahlers, P.; Dunkl, I.; von Eynatten, H. Semi-Automated Heavy-Mineral Analysis by Raman Spectroscopy. *Minerals* **2019**, *9*, 385. [CrossRef]
66. Andò, S. Gravimetric separation of heavy-minerals in sediments. *Minerals* **2019**. in preparation.
67. Garzanti, E.; Andò, S. Heavy Minerals for Junior Woodchucks. *Minerals* **2019**, *9*, 148. [CrossRef]
68. Garzanti, E.; Andò, S. Heavy-mineral concentration in modern sands: Implications for provenance interpretation. In *Heavy Minerals in Use*; Mange, M.A., Wright, D.T., Eds.; Developments in Sedimentology Series 58; Elsevier: Amsterdam, The Netherlands, 2007; pp. 517–545.
69. Mernagh, T.P. Use of the laser Raman microprobe for discrimination amongst feldspar minerals. *J. Raman Spectrosc.* **1991**, *22*, 453–457. [CrossRef]
70. McKeown, D.A. Raman spectroscopy and vibrational analyses of albite: From 25 °C through the melting temperature. *Am. Min.* **2005**, *90*, 1506–1517. [CrossRef]
71. Freeman, J.J.; Wang, A.; Kuebler, K.E.; Jolliff, B.L.; Haskin, L.A. Characterization of natural feldspars by Raman spectroscopy for future planetary exploration. *Canad. Miner.* **2008**, *46*, 1477–1500. [CrossRef]

72. Bersani, D.; Aliatis, I.; Tribaudino, M.; Mantovani, L.; Benisek, A.; Carpenter, M.A.; Gatta, D.G.; Lottici, P.P. Plagioclase composition by Raman spectroscopy. *J Raman Spectrosc.* **2018**, *49*, 684–698. [CrossRef]
73. Kübler, K.; Wang, A.; Abbott, K.; Haskin, L.A. Can we detect carbonate and sulfate minerals on the surface of Mars by Raman spectroscopy? In Proceedings of the 32th Annual Lunar and Planetary Science Conference, Houston, TX, USA, 12–16 March 2001.
74. Borromeo, L.; Zimmermann, U.; Andò, S.; Coletti, G.; Bersani, D.; Basso, D.; Gentile, P.; Schulz, B.; Garzanti, E. Raman spectroscopy as a tool for magnesium estimation in Mg-calcite. *J Raman Spectrosc.* **2017**, *48*, 983–992. [CrossRef]
75. Rruff, database of Raman spectroscopy. Available online: www.rruff.info (accessed on 17 October 2019).
76. Laboratory of Photoinduced Effects, Vibrational and X-Ray Spectroscopies, University of Parma. Available online: www.fis.unipr.it/phevix/ramandb.php (accessed on 17 October 2019).
77. Raman Open Database. Available online: www.solsa.crystallography.net/rod/index.php (accessed on 17 October 2019).
78. Huang, E.; Chen, C.H.; Huang, T.; Lin, E.H.; Xu, J.A. Raman spectroscopic characteristics of Mg-Fe-Ca pyroxenes. *Am. Mineral.* **2000**, *85*, 473–479. [CrossRef]
79. Bersani, D.; Andò, S.; Vignola, P.; Moltifiori, G.; Marino, I.G.; Lottici, P.P.; Diella, V. Micro-Raman spectroscopy as a routine tool for garnet analysis. *Spectrochim. Acta Part A Mol. Biomol. Spectrosc.* **2009**, *73*, 484–491. [CrossRef] [PubMed]
80. Govindaraju, K.; Mevelle, G. Fully automated dissolution and separation methods for inductively coupled plasma atomic emission spectrometry rock analysis. Application to the determination of rare earth elements. Plenary lecture. *J. Anal. At. Spectrom.* **1987**, *2*, 615–621. [CrossRef]
81. Carignan, J.; Hild, P.; Mevelle, G.; Morel, J.; Yeghicheyan, D. Routine analyses of trace elements in geological samples using flow injection and low pressure on-line liquid chromatography coupled to ICP-MS: A study of geochemical reference materials BR, DR-N, UB-N, AN-G and GH. *Geostand. Newsl.* **2001**, *25*, 187–198. [CrossRef]
82. Köhn, M. Bemerkungen zur mechanischen Bodenanalyse. III. Ein neuer Pipettapparat. *Z. Für Pflanz. DüngungBodenkd.* **1928**, *11*, 50–54. [CrossRef]
83. Govin, A.; Holzwarth, U.; Heslop, D.; Ford Keeling, L.; Zabel, M.; Mulitza, S.; Collins, J.A.; Chiessi, C.M. Distribution of major elements in Atlantic surface sediments (36° N–49° S): Imprint of terrigenous input and continental weathering. *Geochem. Geophys. Geosyst.* **2012**, *13*, Q01013. [CrossRef]
84. Malvern Panalytical. Available online: www.malvernpanalytical.com (accessed on 17 October 2019).
85. MacDiff Software. Available online: www.uni-frankfurt.de/69620898/Petschick (accessed on 17 October 2019).
86. Petschick, R.; Kuhn, G.; Gingele, F. Clay mineral distribution in surface sediments of the South Atlantic: Sources, transport, and relation to oceanography. *Mar. Geol.* **1996**, *130*, 203–229. [CrossRef]
87. Vogt, C.; Lauterjung, J.; Fischer, R.X. Investigation of the clay fraction (<2 μm) of the clay mineral society reference clays. *Clays Clay Min.* **2002**, *50*, 388–400. [CrossRef]
88. Roberson, S.; Weltje, G.J. Inter-instrument comparison of particle-size analysers. *Sedimentology* **2014**, *61*, 1157–1174. [CrossRef]
89. Konert, M.; Vandenberghe, J. Comparison of laser grain size analysis with pipette and sieve analysis: A solution for the underestimation of the clay fraction. *Sedimentology* **1997**, *44*, 523–535. [CrossRef]
90. Beuselinck, L.; Govers, G.; Poesen, J.; Degraer, G.; Froyen, L. Grain-size analysis by laser diffractometry: Comparison with the sieve-pipette method. *Catena* **1998**, *32*, 193–208. [CrossRef]
91. Goossens, D. Techniques to measure grain-size distributions of loamy sediments: A comparative study of ten instruments for wet analysis. *Sedimentology* **2008**, *55*, 65–96. [CrossRef]
92. Murray, J.W. *Ecology and Palaeoecology of Benthic Foraminifera*; Routledge: Abingdon, UK, 2014; 408p.
93. Palamenghi, L. Tectonic and sea level control on the transport and depositional processes in a siliciclastic sedimentary basin. Insights from the Ganges-Brahmaputra Delta, Bengal Basin, Bangladesh. Ph.D. Thesis, Faculty of Geosciences, University of Bremen, Bremen, Germany, 2012; 147p.
94. Garzanti, E. Non-carbonate intrabasinal grains in arenites: Their recognition, significance and relationship to eustatic cycles and tectonic setting. *J. Sediment. Petrol.* **1991**, *61*, 959–975.
95. Opdyke, B.N.; Wilkinson, B.H. Paleolatitude distribution of Phanerozoic marine ooids and cements. *Palaeogeogr. Palaeoclimatol. Palaeoecol.* **1990**, *78*, 135–148. [CrossRef]

96. Gallagher, S.J.; Reuning, L.; Himmler, T.; Henderiks, J.; De Vleeschouwer, D.; Groeneveld, J.; Rastegar Lari, A.; Fulthorpe, C.S.; Bogus, K.; et al. The enigma of rare Quaternary oolites in the Indian and Pacific Oceans: A result of global oceanographic physicochemical conditions or a sampling bias? *Quat. Sci. Rev.* **2018**, *200*, 114–122. [CrossRef]
97. Wiedicke, M.; Kudrass, H.R.; Hübscher, C. Oolitic beach barriers of the last Glacial sea-level lowstand at the outer Bengal shelf. *Mar. Geol.* **1999**, *157*, 7–18. [CrossRef]
98. Garzanti, E. Petrographic classification of sand and sandstone. *Earth Sci. Rev.* **2019**, *192*, 545–563. [CrossRef]
99. Garzanti, E.; Andò, S. Plate tectonics and heavy-mineral suites of modern sands. In *Heavy Minerals in Use*; Mange, M.A., Wright, D.T., Eds.; Developments in Sedimentology Series 58; Elsevier: Amsterdam, The Netherlands, 2007; pp. 741–763.
100. Andò, S.; Garzanti, E.; Padoan, M.; Limonta, M. Corrosion of heavy minerals during weathering and diagenesis: A catalog for optical analysis. *Sediment. Geol.* **2012**, *280*, 165–178. [CrossRef]
101. Rubey, W.W. The size-distribution of heavy minerals within a water-laid sandstone. *J. Sediment. Pet.* **1933**, *3*, 3–29.
102. Garzanti, E.; Andò, S.; Vezzoli, G. Settling-equivalence of detrital minerals and grain-size dependence of sediment composition. *Earth Planet. Sci. Lett.* **2008**, *273*, 138–151. [CrossRef]
103. Gabriel, K.R. The biplot graphic display of matrices with application to principal component analysis. *Biometrika* **1971**, *58*, 453–467. [CrossRef]
104. Comas-Cufí, M.; Thió-Henestrosa, F.S. CoDaPack 2.0: A Stand-Alone, Multi-Platform Compositional Software. In Proceedings of the 4th International Workshop on Compositional Data Analysis, Girona, Spain, 10–13 May 2011.
105. Delmonte, B.; Paleari, C.I.; Andò, S.; Garzanti, E.; Andersson, P.S.; Petit, J.R.; Crosta, X.; Narcisi, B.; Baroni, C.; Salvatore, M.C.; et al. Causes of dust size variability in central East Antarctica (Dome B): Atmospheric transport from expanded South American sources during Marine Isotope Stage 2. *Quat. Sci. Rev.* **2017**, *168*, 55–68. [CrossRef]
106. Borromeo, L.; Egeland, N.; Minde, M.W.; Zimmermann, U.; Andò, S.; Madland, M.V.; Korsnes, R.I. Quick, easy, and economic mineralogical studies of flooded chalk for EOR experiments using Raman spectroscopy. *Minerals* **2018**, *8*, 221. [CrossRef]
107. Paleari, C.I.; Delmonte, B.; Andò, S.; Garzanti, E.; Petit, J.R.; Maggi, V. Aeolian dust provenance in central east Antarctica during the Holocene: Environmental constraints from single-grain Raman spectroscopy. *Geophys. Res. Lett.* **2019**, *46*, 9968–9979. [CrossRef]
108. Moore, D.; Reynolds, R. *X-Ray-Diffraction and the Identification and Analysis of Clay Minerals*; Oxford University Press: New York, NY, USA, 1997.
109. Galehouse, J.S. Point counting. In *Procedures in Sedimentary Petrology*; Carver, R.E., Ed.; Wiley: New York, NY, USA, 1971; pp. 385–407.
110. Flood, R.P.; Barra, I.D.; Weltje, G.J.; Roberson, S.; Russell, M.I.; Meneely, J.; Orford, J.D. Provenance and depositional variability of the Thin Mud Facies in the lower Ganges-Brahmaputra delta, West Bengal Sundarbans, India. *Mar. Geol.* **2018**, *395*, 198–218. [CrossRef]
111. Dickinson, W.R. Interpreting provenance relations from detrital modes of sandstones. In *Provenance of Arenites*; Zuffa, G.G., Ed.; NATO ASI Series 148; Reidel: Dordrecht, The Netherlands, 1985; pp. 333–361.
112. Odom, I.E.; Doe, T.W.; Dott, R.H. Nature of feldspar-grain size relations in some quartz-rich sandstones. *J. Sediment. Petrol.* **1976**, *46*, 862–870.
113. Dott, R.H. The importance of eolian abrasion in supermature quartz sandstones and the paradox of weathering on vegetation-free landscapes. *J. Geol.* **2003**, *111*, 387–405. [CrossRef]
114. Garzanti, E.; Resentini, A.; Vezzoli, G.; Andò, S.; Malusà, M.; Padoan, M. Forward compositional modelling of Alpine orogenic sediments. *Sediment. Geol.* **2012**, *280*, 149–164. [CrossRef]
115. Resentini, A.; Goren, L.; Castelltort, S.; Garzanti, E. Partitioning the sediment flux by provenance and tracing erosion patterns in Taiwan. *J. Geophys. Res. Earth Surf.* **2017**, *122*. [CrossRef]

minerals

MDPI

Article

Provenance of Bengal Shelf Sediments: 2. Petrology and Geochemistry of Sand

Eduardo Garzanti [1,*], Giovanni Vezzoli [1], Sergio Andò [1], Mara Limonta [1], Laura Borromeo [1] and Christian France-Lanord [2]

1 Laboratory for Provenance Studies, Department of Earth and Environmental Sciences, University of Milano-Bicocca, 20126 Milano, Italy; giovanni.vezzoli@unimib.it (G.V.); sergio.ando@unimib.it (S.A.); mara.limonta@unimib.it (M.L.); laura.borromeo@unimib.it (L.B.)

2 Centre de Recherches Pétrographiques et Géochimiques, Université de Lorraine-CNRS, BP 20, 54501 Vandoeuvre-lès-Nancy, France; cfl@crpg.cnrs-nancy.fr

* Correspondence: eduardo.garzanti@unimib.it; Tel.: +39-02-64482088

Received: 26 August 2019; Accepted: 12 October 2019; Published: 19 October 2019

Abstract: The Bangladesh lowlands are traversed by the largest sediment flux on the planet. Detritus generated mostly in Himalayan highlands and conveyed through the Ganga–Brahmaputra rivers and Meghna estuary reaches the Bay of Bengal, where it forms a composite deltaic system. This study integrates the vast existing database on Ganga–Brahmaputra sediments of all grain sizes from clay to sand with new petrographic, mineralogical, and geochemical data on estuarine and shallow-marine sands. A large spectrum of compositional signatures was used to: (i) assess the relative supply of the Ganga and Brahmaputra rivers to estuarine and shelfal sediments; (ii) define the compositional variability of estuarine sediments and the impact exerted by hydraulic sorting and climate-related chemical weathering on provenance signals; (iii) define the compositional variability of shelf sediments and the potential hydrodynamic segregation of fast-settling heavy minerals in coastal environments and of slow-settling platy micas on low-energy outer-shelf floors; (iv) consider the potential additional mud supply from the western subaerial part of the delta formerly built by the Ganga River; and (v) draw a preliminary mineralogical comparison between fluvio-deltaic sediments and turbidites of the Bengal–Nicobar deep-sea fan, thus tracing sediment dispersal across the huge sedimentary system extending from Tibet to the equatorial Indian Ocean. All investigated mineralogical and geochemical parameters, as well as Sr and Nd isotope ratios and clay–mineral assemblages, showed a clear prevalence in sediment supply from the Brahmaputra (60–70%) over the Ganga (30–40%). Heavy-mineral suites and Sr and Nd isotope fingerprints of Bengal shelf sediments are nearly identical to those of the Brahmaputra River and Meghna estuary, also because the Brahmaputra carries almost twice as many Ca-plagioclase grains and heavy minerals including epidote than the Ganga, and these minerals control the large majority of the Sr and Nd budgets. The experience gained in modern settings can be directly extrapolated only to the recent past, because sediments older than the late Pleistocene and buried more than a few hundred meters begin to lose less durable ferromagnesian minerals by selective chemical dissolution, which makes quantitative estimates progressively less robust in more deeply buried older strata.

Keywords: bulk-sediment petrography; bulk-sediment geochemistry; heavy minerals; selective entrainment; suspension sorting; chemical indices of weathering; sediment budgets; Brahmaputra River; Ganga River; Himalayan orogen

"Idly my mind follows the sinuous sweep of the Padma roaming under a distant sky. On the further side of hers stretches the sand-bank, insensitive to the living world, defiant in its sublime inutility."

Rabindranath Tagore, Poems: 94

1. Introduction

The Ganga and Brahmaputra rivers drain the Himalayan orogen and join in Bangladesh to form the Meghna estuary, which represents the largest single entry point of detritus in the world oceans ([1,2]). More than 1 billion tons of sediment have reached the Bengal Sea each year through most of the Neogene. Sediment partly accumulates in the delta and partly bypasses the shelf to generate turbiditic currents that feed the Bengal–Nicobar Fan, the largest deep-sea cone on our planet ([3,4]). The Bengal sediment system, which represents the focus of this study, is composed of an abandoned subaerial delta in the west, drained by tidal distributary channels and partially subject to erosion, and an actively prograding subaqueous delta that connects the estuary mouth with the Swatch of No Ground, the deep canyon through which detritus is conveyed toward the submarine fan ([5–7]).

This article is intended as complementary to previous studies carried out on both bedload and suspended-load sediments of the Ganga and Brahmaputra rivers in the proximal part of the sediment-routing system ([8–18]), and to the companion paper dedicated to the mineralogy of Bengal shelf mud ([19]). Our aim was to combine petrographic, heavy-mineral, and geochemical data on the entire range of size fractions from bedload sand to cohesive mud in order to investigate and wherever possible quantify how provenance signals are modified in transit from the subaerial fluvio-deltaic system to the subaqueous delta on the shelf and as far as the ocean floors. The thorough evaluation of the mineralogical variability associated with hydrodynamic sorting and other physical and chemical processes across deltaic and shallow-marine environments represents the fundamental pre-requisite to link the orogenic provenance signatures of Ganga and Brahmaputra sediments with the deep-sea sedimentary record ([20–23]).

A more accurate assessment of the relative amount of sediment supplied by the Ganga and Brahmaputra rivers to the Bay of Bengal is needed to unravel the interplay between tectonic and climatic processes that has controlled the evolution of the Himalayan orogen, and to circumscribe patterns of focused erosion associated with tectonic (e.g., rapid exhumation of the eastern Himalayan Syntaxis), climatic (e.g., impact of monsoon moisture from the Bay of Bengal on rain-drenched Himalayan foothills), or mixed tectonic/climatic processes (e.g., rain-shadow effect consequent to the recent rise of the Shillong Plateau in front of the Bhutan Himalaya).

2. The Ganga–Brahmaputra Rivers and the Bengal Shelf

The Ganga River is sourced from the Himalayan orogen and none of its tributaries reach into the Indus–Yarlung suture zone. The ~10^6 km^2-wide drainage basin includes the central Himalayan thrust belt from northern India to Nepal (~0.17×10^6 km^2) and the northern part of the Indian shield (~0.35×10^6 km^2), almost half of it being represented by foreland-basin lowlands ([24,25]). Detritus is mostly derived from the sub-parallel tectonic domains forming the Himalayan orogen, which comprise the Paleozoic to Eocene sedimentary succession of the Tethys Himalaya ([26,27]), amphibolite-facies metasedimentary rocks of the Greater Himalaya including Lower Paleozoic granitoids and Miocene leucogranites at the top ([28,29]), Proterozoic orthogneisses and metasedimentary rocks of the Lesser Himalaya covered by Paleozoic to Cenozoic strata ([30,31]), and Neogene foreland-basin siliciclastic rocks of the Sub-Himalaya ([32,33]). Major southern tributaries draining the Precambrian Indian shield and its sedimentary and Deccan Traps basaltic covers are the Chambal and Son rivers.

The ~0.63×10^6 km^2-wide Brahmaputra drainage basin includes in the upper catchment calc-alkaline batholiths and sedimentary rocks of the Lhasa block ([34]) and forearc-basin ophiolites, turbidites, and ophiolitic mélange exposed along the Yarlung Tsangpo suture ([35–37]) running from southern Tibet to Myanmar (~0.23×10^6 km^2; [38]). A large part of Brahmaputra sediment is derived from the eastern Himalayan syntaxis ([39]), where extreme erosion rates and huge sediment fluxes are produced by the combination of extraordinary river gradients with the large water discharge fueled by heavy summer-monsoon precipitation ([40,41]). Right-bank (northern) tributaries joining the trunk river in the Assam plains drain the eastern Himalayan belt in Arunachal Pradesh, Bhutan, and Sikkim (~0.13 × 10^6 km^2; [42–44]), whereas left-bank (eastern and southern) tributaries drain Cenozoic turbidites of the Indo-Burman Ranges and Precambrian granitoid gneisses of the Shillong Plateau (~0.04 × 10^6 km^2). Between 40% and 70% of suspended load carried by the Brahmaputra River is estimated to be trapped in foreland-basin floodplains ([45]).

Estimates of total water discharge range from 380 to 590 km^3/a for the Ganga and from 400 to 630 km^3/a for the Brahmaputra, whereas suspended load is estimated to range widely from 316 to 729 × 10^6 t/a for the Ganga and from 402 to 1157 × 10^6 t/a for the Brahmaputra ([16,18,46–48]). The most commonly cited figures range 380–480 × 10^6 t/a for the Ganga and 650–680 × 10^6 t/a for the Brahmaputra, which implies that ~60% of the sediment exported to the Bay of Bengal is contributed by the Brahmaputra and most of the remaining 40% by the Ganga ([17]). Bedload flux, however, is very difficult to estimate. Based on modeling of hydrodynamic conditions, a figure as low as 4 × 10^6 t/a (~1% of total load) was proposed for the Ganga ([14]). Conversely, estimates of Himalayan denudation based on ^{10}Be cosmogenic radionuclides point to an average total sediment export 50% to 80% higher than suspended-load fluxes (610 ± 230 × 10^6 t/a for the Ganga and 780–1430 × 10^6 t/a for the Brahmaputra; [16,18]), a surplus partitioned in unknown proportion between bedload flux and floodplain sequestration.

The high Himalayan range exerts a considerable influence on the climate of the region, and hence on sediment fluxes funneled via the Meghna estuary to the Bengal shelf. During the summer monsoon, the mountains are exposed to heavy rainfall, which reaches peak values (up to 5 m/a) in two major belts running along the foot of the mountains and in correspondence to the major topographic step ([49]). As much as 96% of the sediment load ([48,50]) and 82% of the annual water flow ([51]) are estimated to be delivered during the monsoon season, when water discharge reaches 83,000 m^3/s for the Ganga and 102,000 m^3/s for the Brahmaputra ([47,52,53]). Current velocity decreases with channel depth from ~3 to ~2 m/s in the Ganga ([8,14]), whereas depth-averaged velocity reaches 3.5 m/s for the Brahmaputra ([54]).

The Bengal Delta and the Shelf

The Ganga and Brahmaputra rivers join ~200 km upstream of the coastline forming a ~100 km-long river tract named Padma (the name of the Ganga in Bangladesh). After receiving the Meghna River, which drains the Shillong Plateau and the frontal part of the Indo-Burman Ranges, this huge river system debouches into the Bengal Sea in the ~100 km long, tide-dominated Lower Meghna estuary with four channels named, from west to east, Tetulia, Shahbazpur, Hatia, and Sandwip (Figure 1). To the west of the Meghna estuary lies the Sundarban mangrove forest, formed during progressive eastward migration of the Ganga during the late Holocene highstand ([55]).

Until the late 18th century, as shown in the map by Rennell [56] (Figure 19.4 in [54]), the Ganga discharged through the Tetulia channel, whereas the Brahmaputra flowed to the east of the Madhupur terrace and across the Sylhet basin to reach the sea joint with the Meghna River (Figure 1). This Old Brahmaputra (Burrampooter) course was progressively abandoned during catastrophic floods between the latest 18th century and the beginning of the 19th century, and the Ganga and Brahmaputra became eventually joined around 1830 ([57,58]).

The Ganga–Brahmaputra delta lies on an active plate-boundary and sits on the continent-ocean transition at the northern edge of the Indian plate, which is subducting under the Himalaya in the north and under the Tripura (Indo-Burman) fold-belt in the east ([5,59]). The delta has thus long been subjected to strong tectonic subsidence, which has determined an accumulation of ~16 km of orogenic sediment since Paleocene collision between India and Asia ([3,60,61]). Despite continuing subsidence ([62]; Figure 9 in [63]), huge sediment supply was sufficient to halt transgression during rapid sea-level rise in the early Holocene, and to initiate deltaic progradation at rates notably higher than most other deltas worldwide since ~11 ka, leading to seaward coastal advance by 100–300 km over a 250 km wide front after maximum sea level was reached at 8–7 ka ([64]). In the last 6–5 ka, the lower delta plain west of the modern estuary was deposited as a Ganga-dominated delta, whereas Brahmaputra influence remained confined to the eastern area of the Meghna estuary and to the supratidal part of the delta (Figure 13 in [65]; Figure 9 in [55]). Active sedimentation has continued until today in the estuary area, where on average 7 km^2 of new land have been added each year since 1792 (4.4 km^2/a since 1840) while the inactive western delta was undergoing tidal reworking and net erosion with as much as 3–4 km of shoreline retreat ([66]; Figure 2b in [6]).

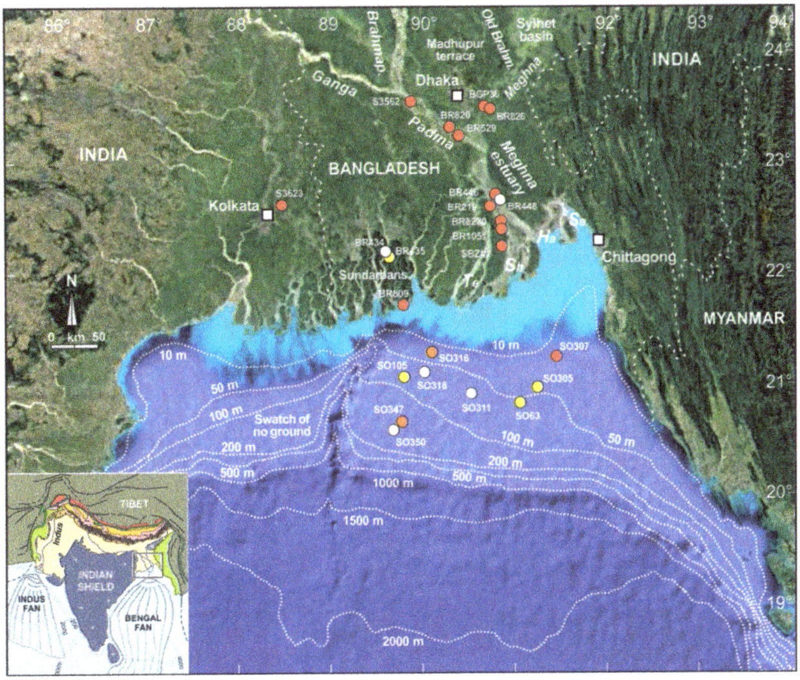

Figure 1. The subaerial and subaqueous delta formed by the Ganga and Brahmaputra rivers and prograding into the Bay of Bengal. Colored circles point at locations of the studied Padma, Meghna estuary, and shelfal sand (red), silty sand (orange), sandy silt (yellow), and silt (white) samples. Offshore sediments were collected during Sonne cruises SO93-3 and SO188-2 in February 1994 and July 2006. The four main channels of the estuary mouth are indicated (Te, Tetulia; Sh, Shahbazpur; Ha, Hatia; Sa, Sandwip).

Offshore, a large subaqueous delta is prograding seaward between the Hatia trough in the east and the Swatch of No Ground in the west, the gigantic submarine canyon that connected the Ganga–Brahmaputra mouth with the Bengal Fan during lowstands (Figure 1). The canyon deeply incised into the shelf, acts as a barrier for farther westward sediment transport and as a sink for approximately a third of fluvial sediment load ([5]). Beyond the estuary mouth, sandy topsets are wide and extend to ~100 km offshore between water depths of ~5 m and 15–30 m b.s.l. The slope steepens in delta-front foresets between water depths of 20–30 and ~80 m b.s.l., where grain size decreases from very fine sand to fine silt ([67]). The outer shelf, extending between 80 and 150 m b.s.l. at the shelf edge, includes muddy prodelta bottomsets passing seaward to a discontinuous layer of palimpsest sand and mud containing biogenic debris, originally deposited in tidal-flat and delta-plain environments during the last glacial period and reworked during the early Holocene transgression ([68,69]).

Tidal amplitude is ~2 m in the Sundarbans to the west, where saline waters may penetrate inland by more than 100 km, and then increases eastward in the Tetulia and Shahbazpur channels to reach ~5 m in the Hatia and Sandwip channels ([65]). Tidal range decreases in the inner shelf from 0.8 m on topset beds to 0.5 m on foreset and bottomset beds and becomes negligible in the outer shelf ([70]). Agents of sediment transport and resuspension in the inner shelf are not only tides but largely tropical cyclones, which generate high waves that sweep the shelf mobilizing large amounts of sediment redeposited as graded beds ([68]). In the November 1970 typhoon, winds of 220 km/h and a storm surge of 6–10 m recorded as far as 200 km inland resulted in 500,000 casualties in Bangladesh ([71,72]). Although the number and power of cyclones has seemingly decreased in the last decades, cyclones have continued to dominate sediment distribution across the shelf, with a main southwestward direction of transport toward the Swatch of No Ground ([7]).

Sedimentation rates have been assessed at 2–4 cm/a on the delta plain in the Holocene ([73]), increasing to a maximum of 5–10 cm/a on the delta front, and dropping to 0.3 cm/a on the prodelta farther offshore ([68,74]). Over the last centuries, the decline in monsoon precipitation contributed to a progressive decrease in the percentage of fluvial sediment load stored in the eastern part of the subaqueous delta, from ~22% in the 19th century, to ~18% in the first half of the 19th century, and to 11–17% thereafter. The depocentre is apparently shifting toward the western part of the subaqueous delta and to the Swatch of No Ground canyon, thus increasing the export as turbidity currents to the deep-sea fan ([67,75]).

3. Analytical Methods

To investigate provenance of Bengal shelf sand and quantify the compositional differences between fluvial and marine sediments we analyzed the petrographic, heavy-mineral, and geochemical compositions of samples SO188_307VC 0–20, SO188_316VC 0–100, and SO188_347VC 280–300 (named SO307, SO316, and SO347 for brevity throughout the article (Figure 1), collected by vibrocorer on the Bengal shelf at water depths of 16, 19, and 126 m b.s.l. during Sonne cruise 188-2 in July 2006 ([76]), and of bedload samples BR446, BR8230, and BR1051 collected from the active channel bed at water depths between 10 and 17 m in the Meghna estuary during the full monsoon season in July 2004, early September 2008, and July 2010.

These new results allowed comparison with mineralogical and geochemical data on Bengal shelf mud illustrated in the companion paper by [19], with the previously obtained extensive mineralogical dataset on fluvial bedload and suspended load of the Ganga–Brahmaputra system illustrated in [12,13], and with our own data on sediment samples from the Padma River (S3562, BR820, and vertical profile BR524–BR529), the Meghna River (BGP36 and BR826), the Meghna estuary (BR219 and vertical profiles BR439–BR448), the delta, and the shelf. Bengal delta sediments were cored on land ~20 km NNE of Kolkata (S3623; luminesce age ~7.764 years, Andy Carter written communication, 2006) and on Bhola Island between the Tetulia and Shahbazpur channels (SBZ#1), or collected from a tidal distributary channel (Pussur BR434, BR435) and from Katka beach in the Sundarbans (BR809). Samples SO93_63KL 3-7 and SO93_105KL 686–690 (named SO63 and SO105 for brevity throughout the article) were collected by piston corer on the Bengal shelf during Sonne cruise 93 in January 1994, at water depths of 66 and 80 m b.s.l. ([77]). Previously obtained results, including multiple-window petrographic and heavy-mineral data obtained separately for all significant size classes of two selected sand samples from the Padma River and Meghna estuary (S3562 collected during the dry season and BR446 collected during the wet monsoon season), and multiple-window heavy-mineral data on Meghna estuary shallow suspended load (sample BR448) and Bengal shelf sandy silt (sample SO63) allowed us to evaluate the compositional variability related to markedly seasonal transport, as well as various sources of error during sampling, sample treatment, and analysis.

Statistical techniques used to illustrate our petrographic and heavy-mineral datasets include multidimensional scaling, which produces a map of points in which samples with similar mineralogical signature cluster closely together and dissimilar samples plot far apart ([78,79]) and the compositional biplot ([80]). The biplot is a very efficient statistical/graphical tool that allows discrimination among multivariate observations (points) while shedding light on the mutual relationships among variables (rays). The length of each ray is proportional to the variance of the corresponding variable in the dataset. If the angle between two rays is close to 0°, 90° or 180°, then the corresponding variables are directly correlated, uncorrelated, or inversely correlated, respectively. Key compositional parameters for selected samples are summarized in Table 1. Full information on sampling locations (Table S1 and Google Earth™ file Bengalsand.kmz) and the complete petrographic (Table S2), heavy-mineral (Table S3), and geochemical datasets (Table S4) are provided in the Supplementary Materials.

3.1. Petrography

A quartered aliquot of each bulk sample was impregnated with araldite, cut into a standard thin section stained with alizarine red to distinguish dolomite and calcite, and analyzed by counting 400 or 450 points under the microscope (Gazzi–Dickinson method; [81]). Sand samples were classified according to the relative abundance of the three main groups of framework components (Q = quartz; F = feldspars; L = lithic fragments), considered where exceeding 10% QFL. According to standard use ([82,83]), the less abundant component goes first, the more abundant last (e.g., in a litho-feldspatho-quartzose sand Q > F > L> 10% QFL; classification scheme after [84,85]). The average rank of metamorphic rock fragments was expressed by the metamorphic indices MI or MI*, ranging, respectively, from 0 (detritus from sedimentary and volcanic rocks) or 100 (detritus from very low-grade metamorphic rocks) to 500 (detritus from high-grade metamorphic rocks; [86]). The median grain size of sand samples was determined both by sieving and in thin section by ranking and visual comparison with standards of $\phi/4$ classes prepared in our laboratory.

3.2. Heavy Minerals

From a split aliquot of the >15 μm size fraction obtained by wet sieving, heavy minerals were separated by centrifuging in Na-polytungstate (2.90 g/cm^3) and recovered by partial freezing with liquid nitrogen. In grain mounts, ≥200 transparent heavy minerals for each sample were point-counted at suitable regular spacing to obtain real volume percentages ([87]) and grains of uncertain identification were systematically checked by Raman spectroscopy ([88]). In previous years, samples were analyzed by grain-counting of the 32–500 μm or 63–250 μm size classes. Transparent heavy-mineral assemblages, called for brevity "tHM-suites" throughout the text, are defined as the spectrum of extrabasinal detrital minerals—except carbonates and slow-settling phyllosilicates—denser than 2.90 g/cm^3 and identifiable under a transmitted-light microscope. According to the transparent-heavy-mineral concentration in the sample (tHMC), tHM suites are defined as very poor (tHMC < 0.5), poor (0.5 ≤ tHMC < 1), moderately poor (1 ≤ tHMC < 2), moderately rich (2 ≤ tHMC < 5), rich (5 ≤ tHMC < 10), very rich (10 ≤ tHMC < 20), or extremely rich (20 ≤ tHMC < 50) ([89,90]).The ZTR index, expressing the "chemical durability" of the tHM suite ([91]), is the sum of zircon, tourmaline, and rutile over total transparent heavy minerals ([92]). Significant minerals are listed in order of abundance (high to low) throughout the text.

Table 1. Key petrographic, heavy-mineral, and geochemical parameters. GSZ = median grain size; Q = quartz; F = feldspar (K = K-feldspar, P = plagioclase); L = lithic grains (Lm = metamorphic: Lmb = metabasite, Lmv = low-rank metavolcanic; Ls = sedimentary: Lc = carbonate, Lh = chert, Lp = pelite; Lv = volcanic; Lu = ultramafic). HM = heavy-minerals; tHMC = transparent heavy-mineral concentration; ZTR = zircon + tourmaline + rutile; Ttn = titanite; Ap = apatite; Amp = amphibole; Px = pyroxene; Ep = epidote; Grt = garnet; CSKA = chloritoid + staurolite + andalusite + kyanite + sillimanite; and tHM = other transparent heavy minerals (anatase, brookite, monazite, olivine, Cr-spinel). Chemical indices explained in Section 3.3.

Sample	GSZ	Q	F	L	P/F	Lm	Lv	Ls	mica	tHMC	ZTR	Ttn	Ap	Amp	Px	Ep	Grt	CSKA	&tHM	Al/Si	Zr	CIA	CIX	WIP	$\alpha^{Al}Ca$	$\alpha^{Ti}Ca$	$\alpha^{Al}Na$	$\alpha^{Sm}Na$
PADMA	μm																											
S 3562	164	66	22	12	53%	66	22	12	4%	23	1	3	2	50	4	25	12	2	0	0.17	309	50	70	44	0.8	1.5	1.4	5.3
BR 820	130	67	17	15	51%	37	12	51	9%	12	3	2	1	37	5	34	13	3	1	0.18	2000	45	74	42	0.5	2.3	1.8	22.0
ESTUARY																												
BR 219	140	71	20	9	52%	60	13	28	6%	30	2	5	0	28	1	38	22	3	0	0.20	1345	47	75	43	0.5	2.6	1.9	17.9
BR 446	200	63	25	12	55%	68	12	21	4%	14	4	3	1	45	1	20	22	3	0	0.16	494	51	67	43	0.9	1.8	1.3	4.4
BR 1051	240	61	27	13	53%	63	11	26	8%	7	1	2	1	55	0	18	16	6	0	0.15	411	52	65	43	1.1	1.6	1.2	3.4
BR 8230	190	68	23	9	54%	70	12	18	2%	20	4	3	0	49	3	24	16	3	0	0.17	626	49	69	44	0.7	1.8	1.4	7.1
DELTA																												
S 3623	230	71	19	11	31%	32	17	52	12%	3	9	4	2	26	9	26	16	5	4	n.d.	n.d.	n.d.	n.d.	n.d.	n.d.	n.d.	n.d.	n.d.
BR 435	60	77	10	13	37%	32	0	68	23%	3	7	5	1	28	4	32	13	4	4	0.23	364	55	70	50	1.1	1.6	2.1	4.4
BR 809	140	72	16	12	43%	50	9	41	5%	13	7	4	1	28	4	18	31	8	1	0.13	1344	50	70	34	0.7	2.3	1.5	14.4
SHELF																												
SO 307	180	64	24	12	52%	69	4	27	10%	12	3	2	1	49	1	30	9	4	0	0.16	458	53	66	44	1.1	1.8	1.3	4.2
SO 316	65	72	21	8	47%	60	3	38	16%	8	5	2	2	50	3	26	9	2	0	0.17	361	53	66	46	1.1	1.7	1.4	3.4
SO 305	50	72	18	10	46%	42	5	53	38%	5	3	4	0	53	4	31	3	1	0	0.23	147	60	69	51	1.9	2.6	1.8	2.3
SO 105	55	72	7	22	47%	34	4	62	8%	2	2	2	0	64	2	26	2	2	0	n.d.	n.d.	n.d.	n.d.	n.d.	n.d.	n.d.	n.d.	n.d.
SO 347	80	72	18	10	58%	46	11	43	10%	4	18	0	0	28	0	38	8	6	0	0.11	581	38	66	38	0.4	1.0	1.4	5.2

3.3. Geochemistry

Bulk-sediment chemical analyses were carried out at Service d'Analyse des Roches et des Minéraux (SARM–CRPG, Nancy, France). Sediment samples were first powdered in an agate mortar. Element concentrations were measured by inductively-coupled-plasma optical emission spectrometry (ICP-OES) and mass spectrometry (ICP-MS) on bulk aliquots of ~100 mg of sediment after lithium metaborate fusion ([93,94]). The relative uncertainty for major element concentration is <5%, except for Si, Al, and Fe for which it is ~2% (Table S5). For full information on analytical procedures and geostandards used see [94]. Grain-size subclasses split by sieving at 0.50 ϕ were previously analyzed at ACME Laboratories Vancouver after lithium metaborate/tetraborate fusion and nitric acid digestion; major oxides and several minor elements were determined by ICP-ES, and trace elements by ICP-MS ([12]).

The Chemical Index of Alteration (CIA = $100Al_2O_3/(Al_2O_3 + CaO - 3.33P_2O_5 + Na_2O + K_2O)$; [95]) and the Weathering Index (WIP = $100 \cdot ((CaO - 3.33P_2O_5)/0.7 + 2Na_2O/0.35 + 2K_2O/0.25 + MgO/0.9)$; [96]) were calculated using molecular proportions of mobile alkali and alkaline earth metals corrected for CaO in apatite. Because correcting the CIA for CaO hosted in carbonates based on mineralogical data may result in significant error, a simplification of the CIA that does not consider CaO (CIX = $100Al_2O_3/(Al_2O_3 + Na_2O + K_2O)$; [97–100]) was also calculated. Enrichment or depletion of each single mobile element was evaluated by comparing its concentration to that of a non-mobile element of close magmatic compatibility (e.g., $\alpha Ca = [Ti/Ca]_{sed}/[Ti/Ca]_{UCC}$, $\alpha Na = [Sm/Na]_{sed}/[Sm/Na]_{UCC}$; [101]) and to non-mobile Al ($\alpha^{Al}E = [Al/E]_{sed}/[Al/E]_{UCC}$; [98,99]) in our samples and in the Upper Continental Crust standard (UCC; [102,103]). The comparison among these different chemical indices allowed us to assess how they differently respond to controls other than weathering, including source-rock mineralogy, hydraulic sorting, grain size, and quartz addition ([104]).

4. Results

In this section, we illustrate the petrographic, heavy-mineral, and geochemical data newly obtained on Meghna estuary and Bengal shelf sands.

4.1. Petrography and Heavy Minerals in Meghna Estuary Sand

Samples BR446, BR8230, and BR1051 (Figure 2A–C) are upper-fine-grained (2.5–2.0 ϕ) litho-feldspatho-quartzose sands, with plagioclase ≥ K-feldspar and mostly medium-to-high rank metamorphic rock fragments (MI* 254–309). Micas are common (biotite/muscovite up to 3.5). The tHM suites are rich to very rich and dominated by calcic amphibole, epidote, and garnet—the classic "triad" diagnostic of orogenic Himalayan provenance [105]. Other minerals are minor and include titanite, prismatic and fibrolitic sillimanite, schorlitic and dravitic tourmaline, diopsidic clinopyroxene, staurolite, zircon, apatite, kyanite, rutile, and rare hypersthene and andalusite (Table 1).

Figure 2. Petrography of litho-feldspatho-quartzose to feldspatho-quartzose sediments of the Meghna estuary, subaerial Bengal delta, and Bengal shelf. (**A–C**) Upper fine sands of the Meghna estuary. (**D–F**) Fine sand to silty sand of the Sundarbans. (**G**) Fine-grained micaceous sand in topsets of the subaqueous delta. (**H**) Mica-rich sandy silt in distal foresets of the subaqueous delta. (**I**) Very fine palimpsest sand in the outer-shelf; oversized ooids include benthic foraminifera with glaucony-filled chambers. All photos with crossed polars; all blue bars for scale are 100 microns.

4.2. Petrography and Heavy Minerals in Bengal Shelf Sand

The siliciclastic fraction of the studied samples cored in the subaqueous delta decreases seaward in grain size from fine sand on topsets (SO307; Figure 2G) to micaceous silty sand (SO316) and very coarse silt in upper foresets (SO105; Figure 2H), whereas it may be associated with ooids and bioclasts up to 300 µm in diameter in the outer shelf (SO347; Figure 2I). Bioclasts, commonly occurring at the core of ooids, include mostly foraminifera (miliolids and rotaliids with less common textularids, bolivinids, and rare globigerinids) and fragments of gastropods or echinoids ([19]). The chambers of foraminiferal tests are commonly filled with glaucony.

Sand is litho-feldspatho-quartzose to feldspatho-quartzose (Figure 3A) with plagioclase ≥ K-feldspar and mostly medium-rank metasedimentary rock fragments (MI* 182–264). Micas are common (biotite/muscovite 2.7). In samples SO307 and SO316, rich to very rich tHM suites include amphibole, epidote, garnet, and minor diopsidic clinopyroxene, zircon, titanite, tourmaline, apatite, sillimanite, kyanite, staurolite, and chloritoid. Sample SO347 yielded a moderately poor suite notably richer in zircon and other durable minerals, with a markedly higher epidote/amphibole ratio (0.7 versus 1.7–2.0), more common chloritoid, and rarer titanite (Table 1).

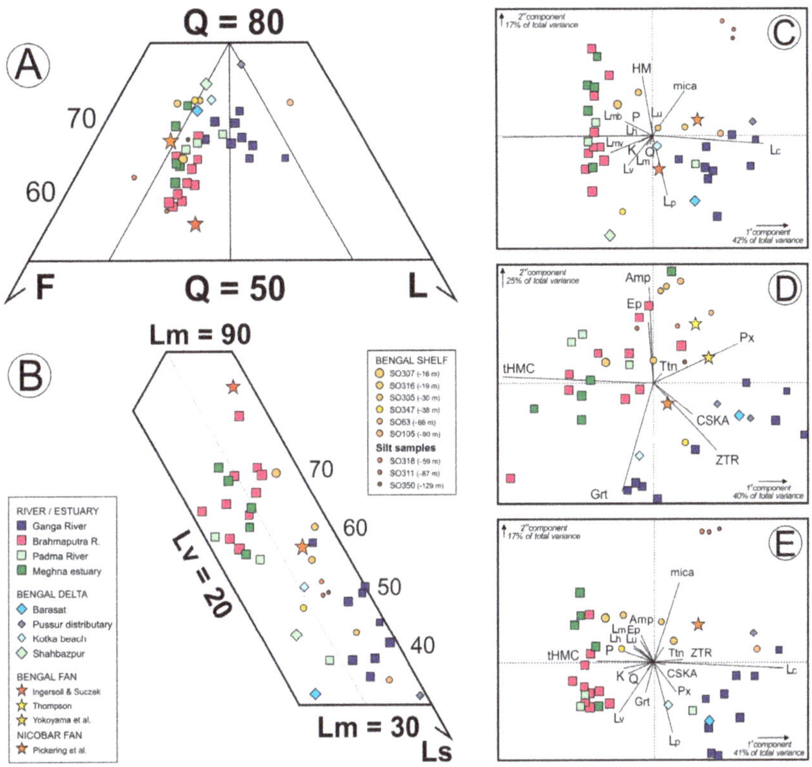

Figure 3. Petrography and heavy minerals in the Bengal sediment system (data for Ganga–Brahmaputra sand and Bengal shelf mud after [12,19]; data for Bengal–Nicobar Fan turbidites after [20–23]). Although detrital modes are controlled by several factors including seasonal and annual variability, hydraulic sorting, and grain size, litho-feldspatho-quartzose Brahmaputra sand is discriminated from mostly feldspatho-litho-quartzose Ganga sand in QFL (**A**) and LmLvLs (**B**) triangular diagrams, as well as in biplots ((**C**) petrography; (**D**) heavy minerals; (**E**) petrography and heavy minerals combined). Padma River and Meghna estuary sands range from almost pure Brahmaputra to mixed Ganga–Brahmaputra, whereas sediments of the subaerial delta are akin to the Ganga in the west and to the Brahmaputra in the east. Bengal shelf and fan sediments are either intermediate or close to the Brahmaputra. See Section 3 for an explanation of the biplot (drawn using CoDaPack software by [106]). Petrographic and mineralogical parameters are as in Table 1.

4.3. Geochemistry of Meghna Estuary and Bengal Shelf Sand

The estuary and shelf sands have quite similar major element concentrations (mostly SiO_2 73–76%; Al_2O_3 ~10%, Fe_2O_3 4.1–4.7%, MgO 1.3–1.6%, CaO 2.3–2.8%, Na_2O 1.7–1.8%, K_2O 1.9–2.3%, TiO_2 0.7–0.8%, P_2O_5 ~0.2%, MnO 0.07–0.11%; Table S4 and [17]). Estuary sand tends to be richer in Ti, Y, REEs (rare earth elements), Th, Nb, Ta, Cr, Mn, Cd, Sn, and Bi, reflecting concentration of ultra-dense minerals including garnet, monazite, and Fe–Cr–Ti oxides. Shelf sand, instead, displays higher Cs, Cu, As, and LOI; K and Rb also tend to be slightly higher, reflecting greater abundance of mica.

Among estuary sands, the coarsest-grained sample BR1051 is slightly enriched in Si and depleted in other major (Al, Fe, Mg, Ca, Ti, P) and minor elements (YREEs, Nb, Cr, Co, Ni). Conversely, sample BR8230 is depleted in Si and enriched in Fe, Mg, Ca, Ti, P, Mn, Sr, Ba, YREE, Th, U, Zr, Hf, V, Nb, Ta, Cr, Co, Ni, Zn, Cd, Sb, and Bi, which is largely explained by the much higher concentration in heavy minerals and less quartz and feldspar (Table 1). Among shelf sands, sample SO347 is rich in ooids and bioclasts (Figure 2I), which explains its anomalously high CaO and LOI (4.4% and 4.9%, respectively) and relatively high Sr. Instead, Al, Fe, Na, K, Rb, Ba, and Pb are low.

5. The Bengal Sediment System: Processes and Products

In this section, we integrate our new mineralogical and geochemical data on the Meghna estuary and Bengal shelf sand and mud ([19]) with previously obtained data from the Ganga and Brahmaputra rivers and Bengal delta ([8,9,11–15,17,107,108]). We also illustrate the strong variability of sediment composition observed in various parts of the Bengal sediment system and discuss specifically the control exerted by grain size and diverse hydrodynamic processes. These include suspension sorting (i.e., the partitioning of detrital grains at different depths in the water column according to their settling velocity, which depends in turn on their size, density, and shape; [109,110]) and selective entrainment (i.e., the preferential removal of lower density and, therefore, coarser settling-equivalent grains from deposits undergoing strong localized erosion; [111,112]).

As shown by previous studies, geochemical parameters may not be particularly helpful to discriminate sediment sources in provenance analysis ([113]), especially because of superposed grain-size and hydraulic-sorting control ([84,114]). This is particularly true for the Ganga–Brahmaputra sediment system, where powerful tractive currents locally promote a full order-of-magnitude increase in the concentration of dense minerals relative to "neutral" composition (i.e., the composition that sediment would ideally have everywhere in the absence of hydrodynamic control).

If, on the one hand, bulk-sediment chemical composition offers little help for provenance discrimination, on the other hand it provides essential information to identify and quantify hydraulic-sorting effects by the precise assessment of elements such as REE, Th, Zr, or Hf. These elements do not enter the crystal lattice of common silicates and are thus progressively concentrated in residual melts until the latest stages of magmatic crystallization, when they form ultra-dense accessory phases such as monazite or zircon. Chemical data also allow us to highlight the differences in weathering conditions found in different parts of the Bengal sediment system.

5.1. Ganga and Brahmaputra Rivers

Ganga and Brahmaputra sands transported from the central and eastern Himalaya to the Bengal Sea are feldspatho-litho-quartzose to litho-feldspatho-quartzose metamorphiclastic, with subequal amounts of K-feldspar and plagioclase, low-to-high rank metamorphic lithic grains, biotite, muscovite, amphibole, epidote, and garnet (Figure 3). This is the typical fingerprint of orogenic detritus produced by focused erosion of deeply exhumed neometamorphic axial belts at the core of high-relief collision orogens (Axial Belt Provenance; [115]).

Ganga sand is richer in quartz and contains limestone and dolostone grains from Himalayan sedimentary and metasedimentary rocks. Heavy minerals include common garnet, tourmaline, diopsidic clinopyroxene, kyanite, and sillimanite, documenting prominent contributions from amphibolite-facies metasedimentary rocks of the Greater Himalaya ([10]).

Brahmaputra sand is richer in feldspars (particularly plagioclase), lacks carbonate grains, and includes minor mafic and ultramafic detritus from ophiolitic units exposed along the Yarlung–Tidding suture zone. Heavy minerals are dominated by blue-green hornblende and epidote derived from amphibolite-facies Himalayan units exposed around the Namche–Barwa syntaxis as well as from Trans-Himalayan batholiths exposed in the Gangdese Range and Mishmi Hills. Hypersthene and chloritoid derived, respectively, from high-grade and low-grade rocks, also occur ([9]).

As it is the case for bedload sand, Ganga suspended load contains more quartz and significant calcite and dolomite, whereas Brahmaputra suspended load contains more Ca-plagioclase, very little dolomite, and no calcite. A moderately-rich epidote–amphibole–garnet tHM suite characterizes Ganga suspended load, whereas a rich amphibole–epidote suite characterizes Brahmaputra suspended load ([13]).

Hydraulic Sorting

Because of selective-entrainment effects, heavy-mineral concentration varies by one order of magnitude and more, from as low as 1.6 to 14.7 for Ganga bedload and from 3.0 to 22.6 for Brahmaputra bedload. Such a strong compositional variability is highlighted by geochemical data, showing that the concentration of zirconium, chiefly hosted in ultra-dense zircon grains, ranges from 116 to ~1470 ppm in Ganga sand and from 145 to ~2750 ppm in Brahmaputra sand ([12,107]). Even garnet placers may form in Brahmaputra bedload, reaching a heavy-mineral concentration of 73.5, a grain density of ~3.65 g/cm^3 (i.e., much higher than eclogite), and Th and Zr concentrations of ~350 ppm and ~6500 ppm (i.e., ~33 times higher than the UCC standard; [12]).

Because of suspension sorting, quartz, feldspar, and heavy minerals steadily increase with depth in the river channel relative to slow-settling platy micas and clay. Among heavy minerals, fast-settling zircon and garnet tend to increase with channel depth relatively to less dense amphibole and epidote. Steady mineralogical trends are faithfully reflected by geochemical trends, with higher concentration in chemical elements preferentially hosted in micas or associated with clay and oxy-hydroxides (i.e., Al, K, Rb, Cs, Ba, and heavy metals) close to the surface, and progressive increase in elements preferentially hosted in denser minerals (e.g., Zr, Hf) and more equant tectosilicates (Si, Na, Ca) toward the bed (Table 2).

5.2. *Meghna Estuary and Subaerial Bengal Delta*

Similar hydrodynamic processes characterize the Padma River and the Meghna estuary downstream, where fluvial and tidal currents locally cause strong erosion of river bars and redeposition in lower-energy estuarine tracts (Figure 6b in [6]). In litho-feldspatho-quartzose sand, tHM suites range from rich to extremely rich and contain abundant amphibole, epidote, and common garnet (12–22%; Table 1). Zr concentration varies from 218 to 2000 ppm, and also Y, REE, Th, Hf, Ta, Mn, and P—all preferentially hosted in dense and ultra-dense minerals (monazite, zircon, rutile, garnet, allanite, apatite)—may be enriched by factors between three and nine (Figure 4A). In the shallow suspended load of the Meghna estuary, instead, the rich amphibole–epidote tHM suite contains < 3% garnet and only 163–169 Zr ppm.

Table 2. Comparison among key mineralogical and geochemical parameters. Note the close similarity between tHM suites of Bengal shelf silt/sand and shallow and deep suspended load in the Brahmaputra River and Meghna estuary. In the subaerial delta, instead, tidal mud and sand have tHM suites closest to Ganga suspended load and bedload. Data sources cited in text; n° = number of samples; Cc = calcite; Do = dolomite. Chemical indices explained in Section 3.3. Other parameters are as in Table 1.

Sample	n°	Q	K	P	Cc	Do	mica	clay	L	HM		tHMC	ZTR	Ttn	Ap	Amp	Px	Ep	Grt	CSKA	&tHM		Al/Si	Zr	CIA	CIX	WIP	$\alpha^{Al}Ca$	$\alpha^{Al}Na$
																							mol						
GANGA RIVER																													
Shallow susp. load	4	27	2	6	1	2	33	26	—	3	100.0	3	11	3	3	22	5	36	9	5	6	100.0	0.38	167	75	76	46	1.5	4.2
Deep susp. load	4	48	6	10	3	3	17	10	—	3	100.0	3	10	4	2	28	6	28	13	5	4	100.0	0.22	228	64	70	41	0.9	2.4
Bedload sand	8	57	7	7	3	2	9	—	10	6	100.0	8	8	4	2	28	6	17	28	7	1	100.0	0.15	445	58	68	34	0.6	1.7
BRAHMAPUTRA RIVER																													
Shallow susp. load	4	35	4	10	0.1	0.6	36	9	—	5	100.0	4	3	3	1	45	4	36	5	2	2	100.0	0.32	186	64	73	54	2.2	2.4
Deep susp.load	2	39	7	19	0	0.4	23	5	—	7	100.0	7	4	4	2	46	5	31	4	3	2	100.0	0.23	253	59	68	51	1.6	1.7
Bedload sand	6	51	10	12	0	0	6	0	10	11	100.0	12	2	3	1	45	4	26	14	3	1	100.0	0.18	328	53	66	45	1.3	1.3
PADMA/MEGHNA ESTUARY																													
Shallow susp.load	1	—	—	—	—	—	—	—	—	—		6	1	3	1	50	5	34	3	3	0	100.0	0.34	164	66	74	53	2.4	2.8
Bedload sand	6	53	9	10	0.5	0.2	5	—	8	14	100.0	14	3	3	1	50	1	21	18	4	0	100.0	0.16	453	51	67	43	0.9	1.3
BENGAL DELTA																													
Very coarse silt	2	56	4	3	3	0.6	23	—	5	4	100.0	3	8	5	2	29	6	28	13	6	4	100.0	0.27	307	59	72	51	1.5	2.6
Sand	3	63	9	6	0.3	0.5	7	—	9	5	100.0	8	8	4	1	27	7	22	23	6	2	100.0	0.13	—	50	70	34	0.7	1.5
BENGAL SHELF																													
Fine silt	3	24	3	8	0.9	0.9	30	25	2	6	100.0	5	5	4	2	51	2	28	4	4	1	100.0	0.36	129	71	76	50	4.6	3.7
Very coarse silt	2	49	5	3	2	2	23	—	7	8	100.0	4	3	3	1	58	3	29	3	2	0	100.0	0.23	147	60	69	51	1.9	1.8
Sand	3	57	8	9	0	0	12	—	8	5	100.0	7	9	2	1	42	2	32	8	4	0	100.0	0.16	467	53	66	45	1.1	1.4

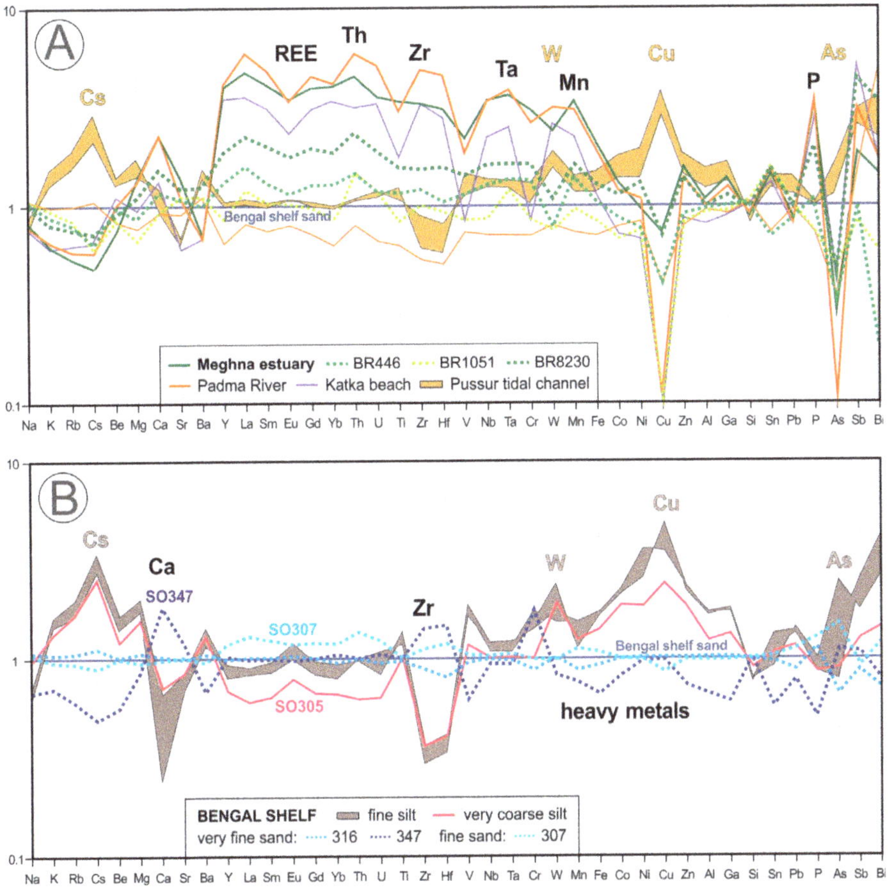

Figure 4. Sedimentary geochemistry (elements arranged following the periodic table group by group; data normalized to the median composition of shelf sand samples SO307, SO316, and SO347; Table S4). (**A**) Fluvial, estuary, and subaerial delta sediments compared to Bengal shelf sand. The Katka beach and some Padma and Meghna estuary samples are markedly enriched in elements preferentially hosted in heavy minerals (e.g., Y, REE, Th, Zr, Ta, Mn, and P), indicating that they are semi-placer lags generated by selective entrainment of lower density grains. Shelf sand has more Cs, W, Cu, and As than fluvial and estuary sands, elements all markedly enriched in tidal mud of the Pussur distributary. (**B**) Bengal shelf mud compared to Bengal shelf sand. Geochemical signatures are chiefly controlled by grain size and clay content. Besides abundant Ca associated with allochems (Figure 2I), sample SO347 displays low concentration in heavy metals, relatively high Zr, and low Al, suggesting effective reworking and winnowing of palimpsest sand on the outer shelf.

Subaerial Bengal delta samples range from feldspatho-litho-quartzose to litho-feldspatho-quartzose and tend to be richer in quartz and much poorer in plagioclase than Meghna estuary sand. The tHM suites are only moderately rich, including 13–16% garnet, <3% kyanite, and ≤0.5% staurolite; Zr is <400 ppm. The Katka beach sand, instead, has a very rich tHM suite including 32% garnet, 4% kyanite, and 2.5% staurolite; Zr is 1344 ppm. This sample can thus be considered as a semi-placer lag formed by selective removal of low-density grains by storm waves during beach erosion.

5.3. Bengal Shelf

The same mineralogical trend observed from bedload to deep and shallow suspended load in fluvial and estuary sediments is reproduced from sandy topsets to silty foresets and bottomsets on the subaqueous delta, with progressive enrichment in platy phyllosilicates in distal settings at the expense of tectosilicates and heavy minerals (Table 2). Among heavy minerals, slower-settling amphibole tends to increase in fine silt at the expense of fast-settling garnet, which is more common in sand. The notable enrichment in durable ZTR minerals, dearth of pyroxene, and the high epidote/amphibole ratio in sample SO347 hint at early diagenetic dissolution of palimpsest sediment originally deposited by the Ganga River.

The analyzed Bengal shelf sands display Zr concentrations between 361 and 581 ppm, higher than the UCC values by factors between approximately two and three but not as high as the Padma River, Meghna estuary, and Katka beach semi-placers (Zr 1344–2000; Table 1). This reflects prevailing depositional processes on the subaqueous delta and only mild concentration of fast-settling heavy minerals by winnowing of slow-settling grains such as phyllosilicates rather than by strong erosional processes.

The joint inspection of Figure 4A,B reveals another aspect of hydraulic control on the compositional variability of Bengal Shelf sediments associated with either selective winnowing or settling of clay. The analyzed shelf sands display higher concentrations than fluvial and estuarine bedload in heavy metals and elements such as Cs and As, largely adsorbed on clay or associated with oxy-hydroxides and organic matter (e.g., [116–118]). The concentration of these elements is even markedly higher in tidal muds of the Pussur distributary channel, and highest in fine silt of distal foresets and bottomsets of the subaqueous delta.

Besides abundant Ca associated with ooids and bioclasts (Figure 2I), the outer shelf sample SO347 displays the lowest concentration in heavy metals, relatively high Zr, and low Al (Figure 4B), suggesting effective winnowing of clay by storm-generated bottom currents. Palimpsest sand including ooids and bioclasts represents drowned beach barriers deposited during the last lowstand stage around 20 ka ([119,120]), followed by reworking and sediment starvation during the Holocene transgression, as also indicated by growth of glaucony within foraminiferal chambers and lack of reported sites where ooids are forming at present in the Indian Ocean ([121,122]). Protracted sediment starvation just in front of the largest deltaic sediment system on Earth indicates that throughout the Holocene sea-level rise sediment was systematically swept by tropical cyclones toward the Swatch of No Ground, and from there funneled toward the Bengal Fan ([7]), thus systematically bypassing the outer shelf.

5.4. Chemical Weathering

Besides physical effects such as selective mechanical breakdown, which can be considered as negligible for the purpose of this study ([123,124]), and hydraulic-sorting, agents capable of compositional modification include chemical weathering. Previous studies have highlighted the difference between the essentially physical Brahmaputra River system—characterized by a cold and arid climate in Tibet, ultra-rapid erosion across the eastern Himalayan syntaxis, and very high runoff in the narrow floodplain—and the Ganga river system, where storage time in the floodplain is longer and weathering notably more intense ([15,125]). This is reflected by clay–mineral assemblages ([126–130]), which are dominated by illite (70–80%) with subequal amounts of chlorite and kaolinite (~10% each) and negligible smectite (≤5%) in Brahmaputra suspended load, whereas smectite/montmorillonite is much more common (20–40%) in Ganga suspended load.

The analysis presented here adds to the thorough study by [17], who used mobile to immobile element ratios including K/Si and H_2O^+/Si as well as the abundance of detrital calcite to document an increase of chemical weathering in the Himalayan system since the Last Glacial Maximum. The use of diverse chemical indices allows us not only to clarify their different response to different controls (Table 1) but also to perceive the different qualitative character of weathering processes in the Brahmaputra and Ganga catchments (Table 2).

5.4.1. The Meaning of Chemical Indices

The broadly homogeneous composition of Meghna estuary and Bengal shelf sediments reflects the continuous supply of enormous volumes of detritus dominantly derived from the Himalayan orogen via the Ganga and Brahmaputra rivers. Because, in this huge system, sediment provenance can be safely considered as substantially unvaried in the short term, modern sediment samples offer a way to test whether the so-called chemical indices of weathering are indeed mainly controlled by climate-related weathering, which at the very short time scale of a few years can be held as roughly invariant, or they are largely or even overwhelmingly controlled by diverse factors including source-rock lithologies, hydraulic sorting, quartz addition, and grain size. Such a critical assessment is needed to prevent that these indices are used indiscriminately as an act of faith.

Close inspection of Table 1 is instructive in this regard. Al/Si, a widely used proxy for grain size (e.g., [8]), correlates quite well with the WIP ($r = 0.83$; significance level = 0.1%), which confirms that the WIP is strongly dependent on grain size and quartz content and, therefore, can hardly be used as a measure of weathering in sediments ([98,99,131]). The CIA also correlates with Al/Si (i.e., with grain size; $r = 0.64$) and best with α^{Al}Ca ($r = 0.89$), a correlation that disappears if the CIX is calculated instead. The addition of carbonate grains, also formed as allochems within the sedimentary basin (e.g., Figure 2I), may lead to the underestimation of weathering, even markedly if not properly identified and corrected for.

An almost perfect correlation exists between α^{Sm}Na and Zr ($r = 0.98$; significance level = 0.1%), which demonstrates the overwhelming effect of concomitant hydraulic concentration of ultra-dense minerals such as REE-rich monazite and zircon. Dominant hydraulic-sorting control on trace-element concentration confirms that α indices using as a reference non-mobile elements such as Ti, Sm, Nd or Th ([101]), which are preferentially hosted in ultra-dense minerals (e.g., monazite, allanite, ilmenite, rutile), cannot be used to evaluate weathering in sediments deposited by tractive currents ([98,99]). The hydraulic-sorting effect is efficiently sterilized by using, instead, the non-mobile element Al as a reference (Table 1).

The study of African sediments has shown that α^{Al}Na is the chemical index most sensitive to weathering, largely because of the high emigration rate of Na^+ from plagioclase under active leaching ([132–135]). The use of α^{Al}Na—and in general of α^{Al}E indices instead of α^{Sm}Na, α^{Ti}Ca, α^{Th}K, α^{Nd}Sr, and α^{Th}Ba indices—avoids undesired perturbations up to even more than one order of magnitude associated with the locally strong hydraulic concentration or depletion of heavy minerals preferentially hosting Ti or REE (Table 1), and is thus recommended.

No chemical index, however, can be claimed to be a truthful proxy for weathering if taken alone. The use of the CIX instead of the CIA circumvents the problems associated with correction for CaO not hosted in silicates, although the CIX is more affected than the CIA by selective-entrainment processes and reaches extreme values even in the absence of weathering in placer deposits strongly depleted in low-density alkali feldspars. Being strongly affected by quartz addition, the WIP is best used in conjunction with the CIA or the CIX to evaluate the extent of recycling rather than the intensity of weathering ([98,131]).

5.4.2. Weathering Processes

The inability of the WIP to measure weathering is clearly exposed by inspection of Table 2, which shows that all indices increase steadily with decreasing grain size from bedload to deep and shallow suspended load, being minimum in sand and maximum in clay in all tracts of the Bengal sediment system ([8,13,15,108]). Higher values of the WIP, however, indicate lower and not higher weathering intensity. The reason why the WIP is unduly lower for sand than for clay is that the WIP is affected principally by quartz content—which is markedly higher in bedload sand than in phyllosilicate-rich suspended load—rather than by chemical weathering.

All indices but α^{Al}Ca concur to confirm more intense weathering in the Ganga catchment than in the Brahmaputra catchment. The main reasons why α^{Al}Ca and α^{Al}Na give opposite indications for the two rivers do not only reflect lithological differences in source terranes but also the distinct character of weathering in the two river systems. Carbonate rocks provide a much larger share of detritus to the Ganga River, largely via the Gandak tributary that drains Tethys Himalayan and Greater Himalayan carbonate and metacarbonate rocks in the upper course and some Lesser Himalayan metacarbonates in the middle course ([10]). The Gandak River is estimated to contribute possibly even as much as half of total Ganga load ([100]). Consequently, despite conspicuous chemical weathering in the floodplain, the Ca/Al ratio is higher in Ganga bedload and deep suspended load than in the average upper continental crust. In Brahmaputra sediments, instead, α^{Al}Ca displays the same values and trend as α^{Al}Na from bedload to deep and shallow suspended load (Table 2) because in this case both elements are chiefly controlled by plagioclase weathering. The Brahmaputra River carries a significant amount of carbonate grains as far as the Himalayan mountain front (~6%, 20% of which calcite and 80% dolomite), reduced to ~1% (100% dolomite) within the first 100 km of transport across the foreland basin, and progressively reduced to zero farther downstream ([9]). On the same samples studied for mineralogy, manometrical measurement of CO_2 after 3 h for calcite and 1 week for dolomite gave comparable results: 2.2% calcite and 2.2% dolomite at the mountain front, 0.2% calcite and 0.4% dolomite after 100 km of floodplain transport, and 0.003–0.2% calcite and 0.004–0.5% dolomite at the entry point in Bangladesh, where calcite is, however, still 0.15% and dolomite 1.2% in suspended load ([107). Near complete dissolution of carbonate grains in the Assam foreland-basin lowlands is a weathering effect ascribed to huge Brahmaputra discharge with high pCO_2 levels in river waters undersaturated with respect to carbonates ([136,137]).

In the Meghna estuary, weathering indices are either similar to Brahmaputra sediments or intermediate between Brahmaputra and Ganga sediments. In the subaerial Bengal delta, tidal mud of the Pussur distributary results as weathered as Ganga–Brahmaputra suspended load. In the subaqueous delta, sediments are markedly depleted in Ca and Na as grain size decreases and clay—the product of chemical weathering—progressively increases from sandy topsets to muddy distal foresets and bottomsets.

6. Provenance

In this section, we draw inferences about provenance and sediment mixing along the routing system that includes the Meghna estuary and the Bengal shelf, and terminates in the Bengal–Nicobar Fan.

6.1. Meghna Estuary

The Ganga and Brahmaputra rivers drain different geological domains besides the Himalayan thrust belt and, thus, carry sediments with distinguishable mineralogical and isotopic fingerprints (Figure 3). Ganga sediments, largely derived from Greater and Lesser Himalayan rocks, are, for instance, characterized by more negative ε_{Nd} and higher $^{87}Sr/^{86}Sr$ than Brahmaputra sediments, which are partly derived from juvenile Trans-Himalayan magmatic rocks in south Tibet ([100,107]). The relative supply from the two big rivers to the Padma River, the Meghna estuary, and the Bengal shelf is, however, difficult to calculate precisely because of the strong variability of sediment composition, largely caused by hydraulic-sorting effects.

In the Meghna estuary, detrital modes are closer to Brahmaputra sand than to Ganga sand (Figure 3), and tHM suites are virtually indistinguishable from Brahmaputra tHM suites (Table 2). The close similarity of tHM suites in the Brahmaputra, Padma River, and Meghna estuary (Figure 6) is, however, partly an effect of heavy-mineral concentration, which is 50% to 100% higher in Brahmaputra bedload and suspended load than in Ganga bedload and suspended load ([12,13]). The same holds largely true also for Nd and Sr isotopic ratios, which in the Padma River, Meghna estuary, and Bengal shelf are virtually the same as in Brahmaputra sediments (Figures 7 and 8 in [17,55]). As highlighted in [138], most Nd and much Sr are hosted in heavy minerals, the abundance of which is markedly higher in Brahmaputra than in Ganga sediments, as discussed further in the following subsection.

The mineralogical and geochemical signatures of suspended load are intermediate between the Ganga and Brahmaputra rivers, but notably closer to the Brahmaputra (e.g., relatively low quartz/feldspar ratio, heavy-mineral assemblages dominated by hornblende and epidote with minor garnet, diopside, sillimanite, and kyanite). A few carbonate grains occur in Padma bedload, and suspended load has more CaO than in the Brahmaputra river at any water depth but more so in shallow suspended load, suggesting that contributions from the Ganga River, although invariably subordinate, is more significant for finer-grained fractions (Figure 5). Greater Ganga contribution for cohesive mud fractions is also indicated by the anomalous increase in the Q/F ratio with decreasing grain size of Bengal shelf muds ([19]).

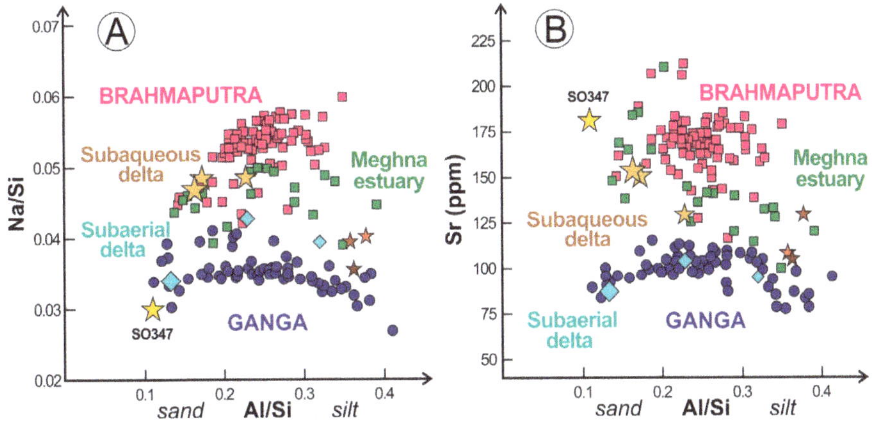

Figure 5. Geochemical discrimination between Ganga and Brahmaputra provenance. Brahmaputra sediments contain more plagioclase and epidote and, therefore, more Na (**A**) and Sr (**B**). Subaerial-delta sediments are akin to the Ganga, whereas subaqueous-delta sediments are akin to the Meghna estuary and to the Brahmaputra for sand but closer to the Ganga for fine silt. Sample SO347 is as low in Na as Ganga sand, Al poor because of the winnowing of clay, and Sr rich because of common calcareous ooids and bioclasts. The size of symbols is roughly proportional to the grain size for subaerial and subaqueous delta samples.

Forward mixing calculations based on integrated petrographic and heavy-mineral data ([139,140]) indicate that sand supply from the Brahmaputra river is at least twice that of the Ganga River, despite the drainage basin of the Ganga being notably greater than that of the Brahmaputra. Similar calculations based on clay–mineral assemblages (as reported in Table 1 in [130]) indicate that the Ganga River contributes ≤ 40% and the Brahmaputra River ~60% of clay minerals fed into the Bay of Bengal ([19]). Because the drainage area of the Brahmaputra is ~60% that of the Ganga, such a broad agreement among estimates based on different compositional parameters and grain-size fractions would imply that average denudation rates are about three times higher in the Brahmaputra catchment than in the Ganga catchment. However, a larger part of the Ganga sediment flux is sequestered in the foreland basin and part of it is diverted southward before reaching the Brahmaputra confluence.

Tracing of Sr concentration throughout the Holocene succession of the Bengal delta indicates that the western part of the subaerial delta consists almost entirely of low-Sr Ganga sediments, whereas high-Sr Brahmaputra sediments have remained dominant in the eastern region including the modern estuary ([55]). This is consistent with petrographic and heavy-mineral modes of sand cored in the western part of the subaerial Bengal delta (Figure 2D) and collected at Katka beach (Figure 2E), which are closer to Ganga sand than to Brahmaputra sand (Figure 3), and with the tHM suite of tidal mud in the Pussur distributary (Figure 2F), which is indistinguishable from deep suspended load in the Ganga River (Figure 6).

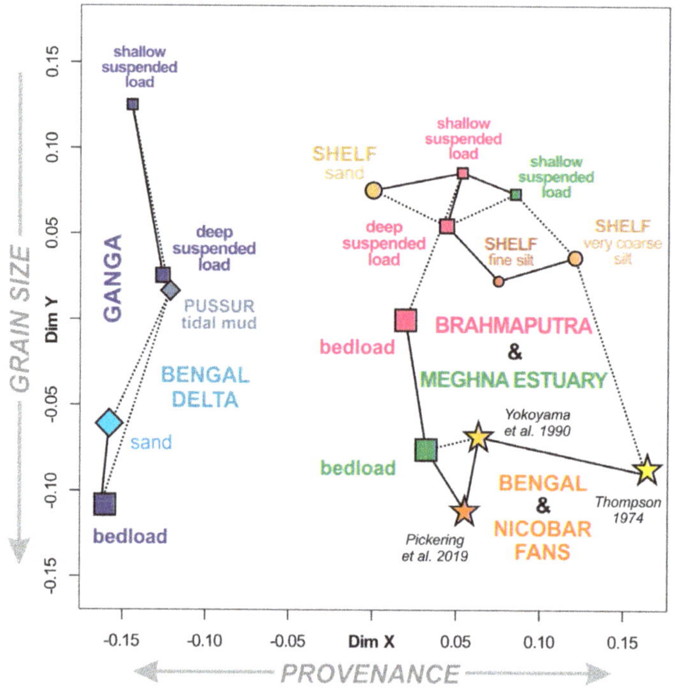

Figure 6. Multidimensional scaling map based on heavy-mineral data (plotted using the "provenance" package of [141]). The distance among samples is approximately proportional to the Kolmogorov–Smirnov dissimilarity of their tHM suites. Solid and dashed lines link closest and second-closest neighbors, respectively; the "stress" value of the configuration is 5.4, indicating a "good" fit ([78]). Samples are separated chiefly by grain size along the *y*-dimension and by provenance along the *x*-dimension, which highlights the Ganga affinity of subaerial Bengal delta sediments, and the Brahmaputra affinity of Meghna estuary, Bengal shelf, and Bengal Fan sediments. As in Figure 3D, Brahmaputra affinity of estuary, shelf, and fan sediments partly reflects the higher heavy-mineral concentration in Brahmaputra than in Ganga sediments.

Minerals Controlling the Sr and Nd Budgets

In shallow to deep suspended load, heavy minerals are estimated to contribute 17% to 21% of total Sr (13% to 16% from epidote) for the Ganga, and 25% (~20% from epidote) for the Brahmaputra. Heavy minerals are estimated to contribute 53% to 76% of total Nd (43–46% to 60–65% from allanite + monazite + titanite) in both Ganga and Brahmaputra shallow to deep suspended load. In bedload, heavy minerals may contribute up to 28% of total Sr (20% from epidote) and up to 96% of total Nd (88% from allanite + monazite + titanite) for the Ganga, and even up to 84% of total Sr (66% from epidote) and almost all of Nd (93% from allanite + monazite + titanite) for the Brahmaputra ([12,13]).

Most of the remaining Sr is hosted in Ca-plagioclase, which is at least twice as abundant in Brahmaputra than in Ganga bedload and suspended load, which explains why Brahmaputra sediments contain 50–60% more Sr than Ganga sediments in any size fraction (Figure 5). In summary, the Sr and Nd budgets are chiefly dependent on Ca-plagioclase and heavy minerals (between 56% and 70% for the Ganga and between 70% and 96% for the Brahmaputra), which in any size fraction are more abundant in Brahmaputra than in Ganga sediments by factors between 1.5 and ≥2 (Table 2).

6.2. Bengal Shelf Sediments

The petrographic and chemical composition of shelf sand is broadly intermediate between Brahmaputra and Ganga bedload (Figure 5), indicating that both rivers contribute large amounts of sediment. Partly because of the reasons discussed above, the tHM suite of shelf sediments (Figure 6) and their Sr and Nd isotopic fingerprints (Figures 7 and 8 in [17]) are instead barely distinguishable from those of Brahmaputra and Meghna estuary sediments.

Silt deposited on the Bengal shelf has quartz/feldspar ratio and proportions among feldspar minerals very close to Brahmaputra silt, and a tHM suite virtually indistinguishable from suspended load in both the Brahmaputra River and Meghna estuary (Figure 6). Fine cohesive mud deposited on the distal foresets and bottomsets of the subaqueous delta may be interpreted as suspended load largely delivered by the Meghna estuary mixed with clay largely supplied by tidal distributaries draining the Sundarban swamps ([19]). Illite-dominated clay-mineral assemblages include subordinate chlorite, smectite, and kaolinite in very similar proportions as clay minerals in Meghna estuary mud ([19,130]).

These estimates confirm that sediment of all size fractions deposited on the subaqueous Bengal delta is mostly derived from the Meghna estuary—with possibly significant contributions of mud delivered by tidal distributaries draining the western part of the inactive subaerial delta—and ultimately supplied between 30% and 40% by the Ganga River and between 60% and 70% by the Brahmaputra River. Sediment flux across the Bay of Bengal and subsequently funneled via the Swatch of No Ground to the Bengal Fan is thus presently Brahmaputra-dominated (Figure 7).

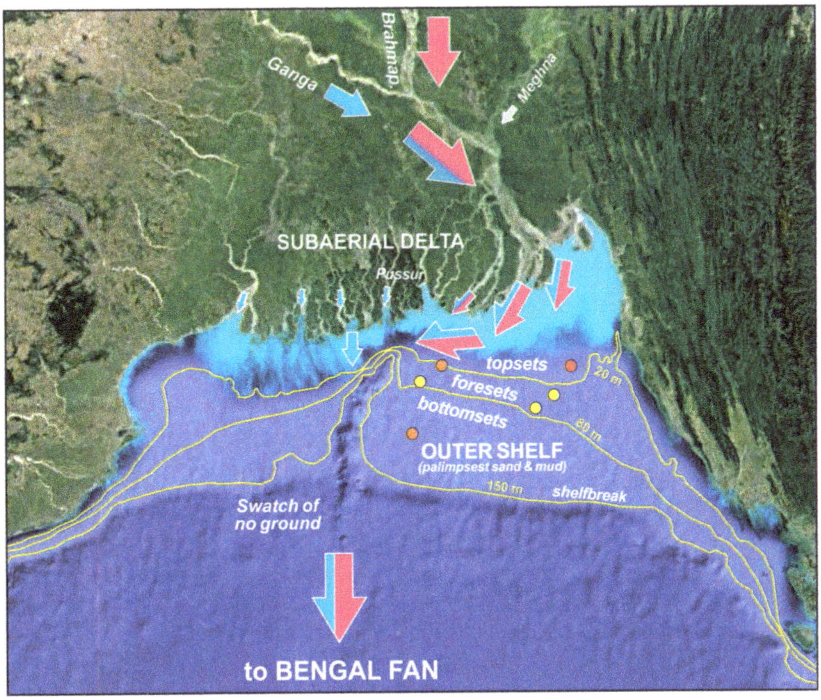

Figure 7. Sediment in transit across the Bengal system ([5–7]). Note the predominance of Brahmaputra (magenta) over Ganga (cyan) supply to the Meghna estuary and subaqueous delta. The subaerial delta, built by the Ganga River and progressively abandoned during the late Holocene highstand is now undergoing tidal reworking and locally net erosion, thus contributing mostly mud originally deposited by the Ganga River to the western part of the subaqueous delta. The studied sand and sandy silt samples are color-coded according to grain size as in Figure 1.

6.3. From the Bengal Shelf to the Bengal–Nicobar Fan

Neogene Bengal Fan turbidites display litho-feldspatho-quartzose metamorphiclastic composition with hornblende-dominated heavy-mineral assemblages ([20,21]), comparing closely with modern Brahmaputra and Meghna estuary sediments and, thus, indicating continuous supply from the Ganga–Brahmaputra sediment system and dominant Himalayan provenance since the Miocene ([4,142,143]). The mineralogy of upper Quaternary silty turbidites cored at ODP Sites 717–719 ([22]) is very close to that of Meghna estuary sediments (Figure 6), documenting a remarkable compositional homogeneity of the fluvio-deltaic to deep-sea-fan depositional system over ~3000 km, from Bangladesh to the equatorial Indian Ocean.

Additional detritus from the Indian subcontinent is, however, indicated in clay-rich intervals deposited during periods when turbidite channels shifted eastward ([22]). This is confirmed by geochemical data on uppermost Quaternary sediments of the western Bay of Bengal ([144]) and by the occurrence of olivine—which is rare in Meghna estuary sediments—in upper Pleistocene distal-fan turbidites, suggesting supply from Deccan Trap basalts via the Godavari and Krishna rivers ([145]).

Nicobar Fan turbidites also display feldspatho-quartzose to litho-feldspatho-quartzose metamorphiclastic detrital modes and mostly moderately rich amphibole-epidote-garnet tHM suites that compare closely with those of Meghna estuary and Bengal shelf sediments (Figure 3), confirming that they also belong to the colossal Bengal sediment system ([23]). The abundance of feldspar and the very close similarity of amphibole-epidote tHM suites of Bengal–Nicobar Fan turbidites with Brahmaputra River and Meghna estuary bedload (Figure 6) indicates a clear prevalence of Brahmaputra contribution lasting since the late Miocene and, thus, since the onset of rapid unroofing of the eastern Himalayan syntaxis ([146,147]).

A robust quantitative comparison among mineralogical assemblages, however, can be drawn only for sediments buried less than a few hundreds of meters, because the tHM suites of ancient strata are exposed to selective intrastratal dissolution of less durable minerals progressing with increasing age and burial depth ([148–150]). Heavy-mineral studies of Bengal Fan turbidites have documented the common occurrence of unstable pyroxene and olivine only in the upper part of the studied cores (i.e., Zone I of [22], dated as ~0.5 Ma, burial depth ≤ 160 m b.s.f.). Ferromagnesian minerals including amphibole progressively decrease in the Pliocene and older sediments buried more than 250–300 m at DSDP Sites 211 and 218 ([20,151]), and the relative abundance of more durable zircon, tourmaline, rutile, and garnet consequently increases (correlation coefficients with core depth 0.54 to 0.72, all significant at the 1% level). Amphibole prevails over epidote + garnet in Quaternary strata (Zones I and IIa, burial depth ≤ 200 m b.s.f.), is less than both epidote and garnet in middle Miocene strata (Zone V, burial depth 650–750 m b.s.f.), to eventually become negligible in lower Miocene strata (Zone VI, burial depth > 750 m b.s.f.; [22]).

7. Conclusions

This study adds new data and observations that, combined with extensive previous work, contributes to a better understanding and quantification of the enormous sediment flux that throughout the Neogene has transited across the Bengal basin and accumulated on Bengal Sea floors. Detritus mostly generated from rapid erosion of the active Himalayan orogen and entrained by the Ganga and Brahmaputra rivers, which joined two centuries ago to form the Meghna estuary, is partly stored in the subaqueous delta prograding onto the Bengal shelf, as in the subaerial delta formed by the Ganga River to the west and progressively abandoned during the late Holocene highstand.

From the Meghna estuary and the Bengal shelf all the way to the Bengal and Nicobar deep-sea fans, detritus is closer in composition to Brahmaputra sediments than to Ganga sediments, as testified by all mineralogical and geochemical parameters. Transparent heavy-mineral suites in sand and silt of the Meghna estuary, Bengal shelf, and Bengal–Nicobar Fans, as well as Sr and Nd isotope fingerprints ([17]), are virtually indistinguishable from those of Brahmaputra bedload and suspended load. Such a predominance over the Ganga is overemphasized because the Brahmaputra carries between 50% and ≥100% more Ca-plagioclase, epidote and heavy minerals in general, which together control most of the Sr and Nd budget in the sediment ([138]). A more balanced contribution from the two rivers is suggested by the most commonly cited estimates of their suspended loads as well as by bulk-sediment mineralogy and geochemistry, which converge to indicate that the Ganga may supply as much as 40% of the total sediment flux. We conclude that, although the Brahmaputra River supplies no less than two-thirds of the sand, Ganga contribution may be more significant for finer silt fractions and clay. This is because the Ganga River, having lower flow velocity and competence, entrains finer particles than the Brahmaputra at all channel depths ([8,14]), and also because the subaerial delta actively formed by the Ganga until a few ka ago is presently undergoing tidal reworking and erosion ([6,66]), thus contributing fine material to the western part of the active subaqueous delta ([19]). Although denser minerals tend to be segregated in coastal and shallow-marine environments, and slow-settling platy micas and clay are effectively winnowed and accumulated offshore, the studied shelf sediments never display the strong hydraulic-sorting effects observed in fluvial, estuarine, and beach sediments, reflecting the prevalence of depositional processes in the prograding subaqueous delta.

Although the Ganga and Brahmaputra drain also different geological domains, the Indian shield and the Trans-Himalayan batholiths and suture zone, respectively, the two rivers receive a large amount of detritus from the same Himalayan rock units. As a consequence, their sediments have mineralogical and, therefore, geochemical signatures that, although distinct, are not different enough to allow very precise estimates of their relative detrital supply. The task is made particularly difficult by the superposed effects of physical and chemical processes, which can produce a marked distortion of provenance signals. Hydraulic-sorting bias tends to efface any potential for distinction based on the geochemistry of bedload. Together with clay–mineral assemblages, geochemical data instead provide precious information on weathering conditions, highlighting not only the different intensity but also the different processes of weathering in Ganga and Brahmaputra lowlands. Ganga sediments undergo stronger alteration during more prolonged storage in the wider foreland basin. The Brahmaputra, instead, is chiefly a physical system, and yet the huge runoff leads to virtually complete dissolution of carbonate grains in bedload, whereas very little calcite and some dolomite is preserved in suspended load. Chemical dissolution becomes a much harder obstacle in the quantitative provenance assessment of ancient sediments, because unstable ferromagnesian minerals are selectively dissolved during burial diagenesis and the original mineralogical and geochemical fingerprints are thus progressively blurred in older strata. Nonetheless, the experience obtained from modern settings remains a fundamental aid to investigate the relative role played by climate and rainfall versus tectonically sustained relief in controlling the intensity of erosive processes across an active orogenic belt such as the Himalaya.

Supplementary Materials: The following are available online at http://www.mdpi.com/2075-163X/9/10/642/s1, Table S1: Information on the location of studied samples, Table S2: Petrographic dataset, Table S3: Heavy-mineral dataset, Table S4: Geochemical dataset, Table S5: Analytical uncertainties. The Google-Earth™ map of the sampling sites file, Bengalsand.kmz, is also provided.

Author Contributions: Project design and writing: E.G.; Petrographic analyses: G.V.; Heavy mineral analyses: S.A., M.L., and L.B.; Geochemical analyses: C.F.-L.

Funding: Funding was provided by Projects MIUR-PRIN 2015EC9PJ5 "The Subduction and Exhumation of the Continental Lithosphere: Their Effects on the Structure and Evolution of the Orogens" and MIUR-Dipartimenti di Eccellenza 2018–2022, Department of Earth and Environmental Sciences, University of Milano-Bicocca.

Acknowledgments: The article benefited from critical comments by two anonymous reviewers. Alberto Resentini carried out multidimensional scaling statistical analysis. We warmly thank Hermann Kudrass (BGR Hannover) for providing samples from Sonne campaigns SO93 and SO188. Yani Najman, Ruth Allen, and Andy Carter kindly shared samples S3562 and S3623.

Conflicts of Interest: The authors declare no conflict of interest.

References

1. Milliman, J.D.; Syvitski, J.P.M. Geomorphic/tectonic control of sediment discharge to the ocean: The importance of small mountainous rivers. *J. Geol.* **1992**, *100*, 525–544. [CrossRef]
2. Goodbred, S.L.; Kuehl, S.A. Enormous Ganges-Brahmaputra sediment discharge during strengthened early Holocene monsoon. *Geology* **2000**, *28*, 1083–1086. [CrossRef]
3. Curray, J.R.; Emmel, F.J.; Moore, D.G. The Bengal Fan: Morphology, geometry, stratigraphy, history and processes. *Marine Petrol. Geol.* **2003**, *19*, 1191–1223. [CrossRef]
4. France-Lanord, C.; Spiess, V.; Klaus, A.; Schwenk, T.; Expedition 354 Scientists. Expedition 354 summary. In Proceedings of the International Ocean Discovery Program 354, Colombo, Sri Lanka, 29 January–31 March 2015; pp. 1–35.
5. Kuehl, S.A.; Allison, M.A.; Goodbred, S.L.; Kudrass, H. The Ganges-Brahmaputra Delta. In *River Deltas—Concepts*, Models, and Examples; Giosan, L., Bhattacharya, J.P., Eds.; SEPM Special Publication 83; SEPM: Tulsa, OK, USA, 2005; pp. 413–434.
6. Wilson, C.A.; Goodbred, S.L. Construction and maintenance of the Ganges-Brahmaputra-Meghna Delta: Linking process, morphology, and stratigraphy. *Annu. Rev. Mar. Sci.* **2015**, *7*, 67–88. [CrossRef] [PubMed]
7. Kudrass, H.R.; Machalett, B.; Palamenghi, L.; Meyer, I.; Zhang, W. Sediment transport by tropical cyclones recorded in a submarine canyon off Bangladesh. *Geo Mar. Lett.* **2018**, *38*, 481–496. [CrossRef]
8. Galy, V.; France-Lanord, C.; Beyssac, O.; Faure, P.; Kudrass, H.; Palhol, F. Efficient organic carbon burial in the Bengal fan sustained by the Himalayan erosional system. *Nature* **2007**, *450*, 407. [CrossRef]
9. Garzanti, E.; Vezzoli, G.; Andò, S.; France-Lanord, C.; Singh, S.K.; Foster, G. Sand petrology and focused erosion in collision orogens: The Brahmaputra case. *Earth Planet. Sci. Lett.* **2004**, *220*, 157–174. [CrossRef]
10. Garzanti, E.; Vezzoli, G.; Andò, S.; Lavé, J.; Attal, M.; France-Lanord, C.; DeCelles, P. Quantifying sand provenance and erosion (Marsyandi River, Nepal Himalaya). *Earth Planet. Sci. Lett.* **2007**, *258*, 500–515. [CrossRef]
11. Garzanti, E.; Andò, S.; Vezzoli, G. Settling-equivalence of detrital minerals and grain-size dependence of sediment composition. *Earth Planet. Sci. Lett.* **2008**, *273*, 138–151. [CrossRef]
12. Garzanti, E.; Andò, S.; France-Lanord, C.; Vezzoli, G.; Censi, P.; Galy, V.; Najman, Y. Mineralogical and chemical variability of fluvial sediments: 1. Bedload sand (Ganga–Brahmaputra, Bangladesh). *Earth Planet. Sci. Lett.* **2010**, *299*, 368–381. [CrossRef]
13. Garzanti, E.; Andò, S.; France-Lanord, C.; Censi, P.; Vignola, P.; Galy, V.; Lupker, M. Mineralogical and chemical variability of fluvial sediments. 2. Suspended-load silt (Ganga-Brahmaputra, Bangladesh). *Earth Planet. Sci. Lett.* **2011**, *302*, 107–120. [CrossRef]
14. Lupker, M.; France-Lanord, C.; Lavé, J.; Bouchez, J.; Galy, V.; Métivier, F.; Gaillardet, J.; Lartiges, B.; Mugnier, J.L. A Rouse-based method to integrate the chemical composition of river sediments: Application to the Ganga basin. *J. Geophys. Res. Earth Surf.* **2011**, *116*, 1–24. [CrossRef]
15. Lupker, M.; France-Lanord, C.; Galy, V.; Lavé, J.; Gaillardet, J.; Gajurel, A.P.; Guilmette, C.; Rahman, M.; Singh, S.K.; Sinha, R. Predominant floodplain over mountain weathering of Himalayan sediments (Ganga basin). *Geochim. Cosmochim. Acta* **2012**, *84*, 410–432. [CrossRef]
16. Lupker, M.; Blard, P.H.; Lavé, J.; France-Lanord, C.; Leanni, L.; Puchol, N.; Charreau, J.; Bourlès, D. ^{10}Be-derived Himalayan denudation rates and sediment budgets in the Ganga basin. *Earth Planet. Sci. Lett.* **2012**, *333–334*, 146–156. [CrossRef]
17. Lupker, M.; France-Lanord, C.; Galy, V.; Lavé, J.; Kudrass, H. Increasing chemical weathering in the Himalayan system since the Last Glacial Maximum. *Earth Planet. Sci. Lett.* **2013**, *365*, 243–252. [CrossRef]
18. Lupker, M.; Lavé, J.; France-Lanord, C.; Christl, M.; Bourlès, D.; Carcaillet, J.; Maden, C.; Wieler, R.; Rahman, M.; Bezbaruah, D.; et al. ^{10}Be systematics in the Tsangpo-Brahmaputra catchment: The cosmogenic nuclide legacy of the eastern Himalayan syntaxis. *Earth Surf. Dyn.* **2017**, *5*, 429–449. [CrossRef]
19. Borromeo, L.; Andò, S.; France-Lanord, C.; Hahn, A.; Coletti, G.; Garzanti, E. Provenance of Bengal Shelf Sediments: 1. Mineralogy of Silt. *Minerals* **2019**, *9*, 640. [CrossRef]
20. Thompson, R.W. *Mineralogy of sands from the Bengal and Nicobar fans, Sites 218 and 211, Eastern Indian Ocean*; Initial Reports DSDP 22; U.S. Government Printing Office: Washington, DC, USA, 1974; pp. 711–713.
21. Ingersoll, R.V.; Suczek, C.A. Petrology and provenance of Neogene sand from Nicobar and Bengal Fans, DSDP Sites 211 and 218. *J. Sediment. Petrol.* **1979**, *49*, 1217–1228.

22. Yokoyama, K.; Amano, K.; Taira, A.; Saito, Y. *Mineralogy of silts from the Bengal Fan*; Proc. ODP Sci. Results; Texas A&M University: College Station, TX, USA, 1990; Volume 116, pp. 59–73.
23. Pickering, K.T.; Carter, A.; Andò, S.; Garzanti, E.; Limonta, M.; Vezzoli, G.; Milliken, K.L.; Chemale, F.; Pouderoux, H.; Kutterolfn, S.; et al. Sediment provenance of the Nicobar Submarine Fan, Indian Ocean. *Earth Planet. Sci. Lett.* **2019**, in press.
24. Galy, A.; France-Lanord, C. Weathering processes in the Ganges–Brahmaputra basin and the riverine alkalinity budget. *Chem. Geol.* **1999**, *159*, 31–60. [CrossRef]
25. Galy, A.; France-Lanord, C.; Derry, L.A. The strontium isotopic budget of Himalayan rivers in Nepal and Bangladesh. *Geochim. Cosmochim. Acta* **1999**, *63*, 1905–1925. [CrossRef]
26. Garzanti, E. Stratigraphy and sedimentary history of the Nepal Tethys Himalayan passive margin. *J. Asian Earth Sci.* **1999**, *17*, 805–827. [CrossRef]
27. Sciunnach, D.; Garzanti, E. Subsidence history of the Tethys Himalaya. *Earth Sci. Rev.* **2012**, *25*, 179–198. [CrossRef]
28. Colchen, M.; Le Fort, P.; Pêcher, A. *Annapurna, Manaslu, Ganesh Himal*; Editions du C.N.R.S.: Paris, France, 1986; 136p.
29. Carosi, R.; Montomoli, C.; Iaccarino, S. 20 years of geological mapping of the metamorphic core across Central and Eastern Himalayas. *Earth Sci. Rev.* **2017**, *177*, 124–138. [CrossRef]
30. Sakai, H. Rifting of the Gondwanaland and uplifting of the Himalayas recorded in Mesozoic and Tertiary fluvial sediments in the Nepal Himalayas. In *Sedimentary Facies in the Active Plate Margin*; Taira, A., Masuda, F., Eds.; Terra Scientific Publications: Tokyo, Japan, 1989; pp. 723–732.
31. DeCelles, P.G.; Robinson, D.M.; Quade, J.; Ojha, T.P.; Garzione, C.N.; Copeland, P.; Upreti, B.N. Stratigraphy, structure, and tectonic evolution of the Himalayan fold-thrust belt in western Nepal. *Tectonics* **2001**, *20*, 487–509. [CrossRef]
32. DeCelles, P.G.; Gehrels, G.E.; Quade, J.; Ojha, T.P.; Kapp, P.A.; Upreti, B.N. Neogene foreland basin deposits, erosional unroofing, and the kinematic history of the Himalayan fold-thrust belt, western Nepal Siwaliks. *Geol. Soc. Am. Bull.* **1998**, *110*, 2–21. [CrossRef]
33. Garzanti, E. The Himalayan foreland basin from collision onset to the present: A sedimentary-petrology perspective. In *Himalayan Tectonics: A Modern Synthesis*; Treloar, P., Searle, M.P., Eds.; Geological Society: London, UK, 2019; Special Publication 483; pp. 65–122. [CrossRef]
34. Zhu, D.C.; Zhao, Z.D.; Niu, Y.L.; Mo, X.X.; Chung, S.L.; Hou, Z.Q.; Wang, L.Q.; Wu, F.Y. The Lhasa terrane: Record of a microcontinent and its histories of drift and growth. *Earth Planet. Sci. Lett.* **2011**, *301*, 241–255. [CrossRef]
35. Hébert, R.; Bezard, R.; Guilmette, C.; Dostal, J.; Wang, C.S.; Liu, Z.F. The Indus–Yarlung Zangbo ophiolites from Nanga Parbat to Namche Barwa syntaxes, southern Tibet: First synthesis of petrology, geochemistry, and geochronology with incidences on geodynamic reconstructions of Neo-Tethys. *Gondwana Res.* **2012**, *22*, 377–397. [CrossRef]
36. An, W.; Hu, X.; Garzanti, E. Sandstone provenance and tectonic evolution of the Xiukang Mélange from Neotethyan subduction to India-Asia collision (Yarlung-Zangbo suture). *Gondwana Res.* **2017**, *41*, 222–234. [CrossRef]
37. Wang, J.-G.; Hu, X.M.; Garzanti, E.; An, W.; Liu, X.-C. The birth of the Xigaze forearc basin in southern Tibet. *Earth Planet. Sci. Lett.* **2017**, *465*, 38–47. [CrossRef]
38. Gansser, A. The Peri-Indian suture zone. In *Géologie des Châines Alpines Issues de la Téthys*; Auboin, J., Debelmas, J., Latreille, M., Eds.; Mémoires B.R.G.M.: Paris, France, 1980; Volume 115, pp. 140–148.
39. Burg, J.P.; Nievergelt, P.; Oberli, F.; Seward, D.; Davy, P.; Maurin, J.C.; Diao, Z.; Meier, M. The Namche Barwa syntaxis: Evidence for exhumation related to compressional crustal folding. *J. Asian Earth Sci.* **1998**, *16*, 239–252. [CrossRef]
40. Zeitler, P.K.; Meltzer, A.S.; Koons, P.O.; Craw, D.; Hallet, B.; Chamberlain, C.P.; Kidd, W.S.F.; Park, S.K.; Seeber, L.; Bishop, M.; et al. Erosion, Himalayan geodynamics, and the geomorphology of metamorphism. *GSA Today* **2001**, *11*, 4–9. [CrossRef]
41. Gemignani, L.; van der Beek, P.; Braun, J.; Najman, Y.; Bernet, M.; Garzanti, E.; Wijbrans, J.R. Downstream evolution of the thermochronologic age signal in the Brahmaputra catchment (eastern Himalaya): Implications for the detrital record of erosion. *Earth Planet. Sci. Lett.* **2018**, *499*, 48–61. [CrossRef]

42. Gansser, A. *Geology of the Bhutan Himalaya*; Denkschriften der Schweizerischen Naturforschenden Geselschaft; Birkhäuser Verlag: Basel, Switzerland, 1983; 181p.
43. Kumar, G. *Geology of Arunachal Pradesh*; Geological Society of India: Bengaluru, India, 1997; 217p.
44. Long, S.P.; McQuarrie, N.; Tobgay, T.; Grujic, D.; Hollister, L. Geologic map of Bhutan. *J. Maps* **2011**, 7. [CrossRef]
45. Allison, M.A.; Kuehl, S.A.; Martin, T.C.; Hassan, A. Importance of flood-plain sedimentation for river sediment budgets and terrigenous input to the oceans: Insights from the Brahmaputra-Jamuna River. *Geology* **1998**, *26*, 175–178. [CrossRef]
46. Delft Hydraulics. River Survey Project, Flood Action Plan 24, Special Report n°18. In *Sediment Rating Curves and Balances*; Water Resources Planning Organization: Dhaka, Bangladesh, 1996.
47. Hay, W.W. Detrital sediment fluxes from continents to oceans. *Chem. Geol.* **1998**, *145*, 287–323. [CrossRef]
48. Islam, M.R.; Begum, S.F.; Yamaguchi, Y.; Ogawa, K. The Ganges and Brahmaputra rivers in Bangladesh: Basin denudation and sedimentation. *Hydrol. Process.* **1999**, *13*, 2907–2923. [CrossRef]
49. Bookhagen, B.; Burbank, D.W. Topography, relief, and TRMM-derived rainfall variations along the Himalaya. *Geophys. Res. Lett.* **2006**, *33*, L08405. [CrossRef]
50. Coleman, J.M. Brahmaputra River: Channel processes and sedimentation. *Sediment. Geol.* **1969**, *3*, 129–239. [CrossRef]
51. Whitaker, D.W.; Wasimi, S.A.; Islam, S. The El Niño–southern oscillation and long-range forecasting of flows in the Ganges. *Int. J. Climatol.* **2001**, *21*, 77–87. [CrossRef]
52. Singh, I.B. The Ganga River. In *Large Rivers: Geomorphology and Management*; Gupta, A., Ed.; Wiley: Chichester, UK, 2007; pp. 347–371.
53. Singh, S.K. Erosion and weathering in the Brahmaputra River system. In *Large Rivers: Geomorphology and Management*; Wiley: Chichester, UK, 2007; pp. 373–393.
54. Best, J.L.; Ashworth, P.J.; Sarker, M.H.; Roden, J.E. The Brahmaputra-Jamuna River, Bangladesh. In *Large Rivers: Geomorphology and Management*; Gupta, A., Ed.; Wiley: New York, NY, USA, 2007; pp. 395–433.
55. Goodbred, S.L.; Paolo, P.M.; Ullah, M.S.; Pate, R.D.; Khan, S.R.; Kuehl, S.A.; Singh, S.K.; Rahaman, W. Piecing together the Ganges-Brahmaputra-Meghna River delta: Use of sediment provenance to reconstruct the history and interaction of multiple fluvial systems during Holocene delta evolution. *Geol. Soc. Am. Bull.* **2014**, *126*, 1495–1510. [CrossRef]
56. Rennell, J.J. *An Actual Survey of the Provinces of BENGAL, BAHAR etc. by Major General James Rennell, Surveyor to the Honourable East India Company in Bengal, 1776*; Faden, W., Ed.; Geographer to the King, Charing Cross; East India Company: London, UK, 1786.
57. Bristow, C.S. Avulsion, river metamorphosis and reworking by underfit streams: A modern example from the Brahmaputra River in Bangladesh and a possible ancient example in the Spanish Pyrenees. In *Fluvial Sedimentology 6*; Smith, N.D., Rogers, J., Eds.; Special Publication of the International Association of Sedimentologists: Devon, UK, 1999; Volume 28, pp. 221–230.
58. Pickering, J.L.; Goodbred, S.L.; Reitz, M.D.; Hartzog, T.H.; Mondal, D.R.; Hossain, M.S. Late Quaternary sedimentary record and Holocene channel avulsions of the Jamuna and Old Brahmaputra River valleys in the upper Bengal delta plain. *Geomorphology* **2014**, *227*, 123–136. [CrossRef]
59. Alam, M.; Alam, M.M.; Curray, J.R.; Chowdhury, M.L.R.; Gani, M.R. An overview of the sedimentary geology of the Bengal Basin in relation to the regional tectonic framework and basin-fill history. *Sediment. Geol.* **2003**, *155*, 179–208. [CrossRef]
60. Najman, Y.; Bickle, M.; BouDagher-Fadel, M.; Carter, A.; Garzanti, E.; Paul, M.; Wijbrans, J.; Willett, E.; Oliver, G.; Parrish, R.; et al. The Paleogene record of Himalayan erosion: Bengal Basin, Bangladesh. *Earth Planet. Sci. Lett.* **2008**, *273*, 1–14. [CrossRef]
61. Najman, Y.; Allen, R.; Willett, E.A.F.; Carter, A.; Barfod, D.; Garzanti, E.; Wijbrans, J.; Bickle, M.J.; Vezzoli, G.; Andò, S.; et al. The record of Himalayan erosion preserved in the sedimentary rocks of the Hatia Trough of the Bengal Basin and the Chittagong Hill Tracts, Bangladesh. *Basin Res.* **2012**, *24*, 1–21. [CrossRef]
62. Brown, S.; Nicholls, R.J. Subsidence and human influences in mega deltas: The case of the Ganges–Brahmaputra–Meghna. *Sci. Total Environ.* **2015**, *527–528*, 362–374. [CrossRef]
63. Grall, C.; Steckler, M.S.; Pickering, J.L.; Goodbred, S.; Sincavage, R.; Paola, C.; Akhter, S.H.; Spiess, V. A base-level stratigraphic approach to determining Holocene subsidence of the Ganges–Meghna–Brahmaputra Delta plain. *Earth Planet. Sci. Lett.* **2018**, *499*, 23–36. [CrossRef]

64. Goodbred, S.L.; Kuehl, S.A.; Steckler, M.S.; Sarker, M.H. Controls on facies distribution and stratigraphic preservation in the Ganges–Brahmaputra delta sequence. *Sediment. Geol.* **2003**, *155*, 301–316. [CrossRef]
65. Allison, M.A.; Khan, S.R.; Goodbred, S.L.; Kuehl, S.A. Stratigraphic evolution of the late Holocene Ganges–Brahmaputra lower delta plain. *Sediment. Geol.* **2003**, *155*, 317–342. [CrossRef]
66. Allison, M.A. Historical changes in the Ganges-Brahmaputra delta front. *J. Coast. Res.* **1998**, *14*, 1269–1275.
67. Palamenghi, L.; Schwenk, T.; Spies, V.; Kudrass, H.R. Seismostratigraphic analysis with centennial to decadal time resolution of the sediment sink in the Ganges–Brahmaputra subaqueous delta. *Cont. Shelf Res.* **2011**, *31*, 712–730. [CrossRef]
68. Michels, K.H.; Kudrass, H.R.; Hubscher, C.; Suckow, A.; Wiedicke, M. The submarine delta of the Ganges–Brahmaputra: Cyclone-dominated sedimentation patterns. *Mar. Geol.* **1998**, *149*, 133–154. [CrossRef]
69. Hübscher, C.; Spiess, V. Forced regression systems tracts on the Bengal Shelf. *Mar. Geol.* **2005**, *219*, 207–218. [CrossRef]
70. Murty, T.S.; Henry, R.F. Tides in the Bay of Bengal. *J. Geophys. Res.* **1983**, *88*, 6069–6076. [CrossRef]
71. Islam, T.; Peterson, R.E. Climatology of landfalling tropical cyclones in Bangladesh 1877–2003. *Nat. Hazards* **2009**, *48*, 115–135. [CrossRef]
72. Hossain, M.D.; Aucoin, J. George Harrison and the Concert for Bangladesh: When rock music forever fused with politics on a World stage. In *Music as a Platform for Political Communication*; IGI Global: Pennsylvania, PA, USA, 2017; pp. 149–166.
73. Goodbred, S.L.; Kuehl, S.A. The significance of large sediment supply, active tectonism, and eustasy on margin sequence development: Late Quaternary stratigraphy and evolution of the Ganges—Brahmaputra delta. *Sediment. Geol.* **2000**, *133*, 227–248. [CrossRef]
74. Kuehl, S.A.; Levy, B.M.; Moore, W.S.; Allison, M.A. Subaqueous delta of the Ganges–Brahmaputra river system. *Mar. Geol.* **1997**, *144*, 81–96. [CrossRef]
75. Rogers, K.G.; Goodbred, S.L.; Khan, S.R. Shelf-to-canyon connections: Transport-related morphology and mass balance at the shallow-headed, rapidly aggrading Swatch of No Ground (Bay of Bengal). *Mar. Geol.* **2015**, *369*, 288–299. [CrossRef]
76. Kudrass, H.R.; Spiess, V.; Bruns, A.; Ding, F.; Fekete, N.; Fenner, J.; France-Lanord, C.; Palamenghi, L.; Reinhardt, L.; Rühlemann, C.; et al. *Cruise Report SO 188-2 Chittagong (Bangladesh)—Penang (Malaysia) 06.07.2006–31.07.2006*; BGR: Hannover, Germany, 2007; pp. 1–100.
77. Kudrass, H.R.; Scientific Shipboard Party. *Cruise Report SO93/1-3 Bengal Fan*; Bundesanstalt für Geowissenschaften und Rohstoffe: Hannover, Germany, 1994; 135p.
78. Vermeesch, P. Multi-sample comparison of detrital age distributions. *Chem. Geol.* **2013**, *341*, 140–146. [CrossRef]
79. Vermeesch, P.; Garzanti, E. Making geological sense of 'Big Data' in sedimentary provenance analysis. *Chem. Geol.* **2015**, *409*, 20–27. [CrossRef]
80. Gabriel, K.R. The biplot graphic display of matrices with application to principal component analysis. *Biometrika* **1971**, *58*, 453–467. [CrossRef]
81. Ingersoll, R.V.; Bullard, T.F.; Ford, R.L.; Grimm, J.P.; Pickle, J.D.; Sares, S.W. The effect of grain size on detrital modes: A test of the Gazzi-Dickinson point-counting method. *J. Sediment. Petrol.* **1984**, *54*, 103–116.
82. Crook, K.A.W. Classification of arenites. *Am. J. Sci.* **1960**, *258*, 419–428. [CrossRef]
83. Dickinson, W.R. Interpreting detrital modes of graywacke and arkose. *J. Sediment. Petrol.* **1970**, *40*, 695–707.
84. Garzanti, E. From static to dynamic provenance analysis—Sedimentary petrology upgraded. *Sediment. Geol.* **2016**, *336*, 3–13. [CrossRef]
85. Garzanti, E. Petrographic classification of sand and sandstone. *Earth Sci. Rev.* **2019**, *192*, 545–563. [CrossRef]
86. Garzanti, E.; Vezzoli, G. A classification of metamorphic grains in sands based on their composition and grade. *J. Sediment. Res.* **2003**, *73*, 830–837. [CrossRef]
87. Galehouse, J.S. Point counting. In *Procedures in Sedimentary Petrology*; Carver, R.E., Ed.; Wiley: New York, NY, USA, 1971; pp. 385–407.
88. Andò, S.; Garzanti, E. Raman spectroscopy in heavy-mineral studies. *Geol. Soc. Lond. Spec. Publ.* **2014**, *386*, 395–412. [CrossRef]
89. Garzanti, E.; Andò, S. Heavy-mineral concentration in modern sands: Implications for provenance interpretation. In *Heavy Minerals in Use*; Mange, M.A., Wright, D.T., Eds.; Developments in Sedimentology Series 58; Elsevier: Amsterdam, The Netherlands, 2007; pp. 517–545.

90. Garzanti, E.; Andò, S. Heavy Minerals for Junior Woodchucks. *Minerals* **2019**, *9*, 148. [CrossRef]
91. Garzanti, E. The maturity myth in sedimentology and provenance analysis. *J. Sediment. Res.* **2017**, *87*, 353–365. [CrossRef]
92. Hubert, J.F. A zircon–tourmaline–rutile maturity index and the interdependence of the composition of heavy mineral assemblages with the gross composition and texture of sandstones. *J. Sediment. Petrol.* **1962**, *32*, 440–450.
93. Govindaraju, K.; Mevelle, G. Fully automated dissolution and separation methods for inductively coupled plasma atomic emission spectrometry rock analysis. Application to the determination of rare earth elements. Plenary lecture. *J. Anal. Atomic Spectrom.* **1987**, *2*, 615–621. [CrossRef]
94. Carignan, J.; Hild, P.; Mevelle, G.; Morel, J.; Yeghicheyan, D. Routine analyses of trace elements in geological samples using flow injection and low pressure on-line liquid chromatography coupled to ICP-MS: A study of geochemical reference materials BR, DR-N, UB-N, AN-G and GH. *Geostand. Newsl.* **2001**, *25*, 187–198. [CrossRef]
95. Nesbitt, H.W.; Young, G.M. Early Proterozoic climates and plate motions inferred from major element chemistry of lutites. *Nature* **1982**, *299*, 715–717. [CrossRef]
96. Parker, A. An index of weathering for silicate rocks. *Geol. Mag.* **1970**, *107*, 501–504. [CrossRef]
97. Colin, C.; Turpin, L.; Bertaux, J.; Desprairies, A.; Kissel, C. Erosional history of the Himalayan and Burman ranges during the last two glacial–interglacial cycles. *Earth Planet. Sci. Lett.* **1999**, *171*, 647–660. [CrossRef]
98. Garzanti, E.; Padoan, M.; Andò, S.; Resentini, A.; Vezzoli, G.; Lustrino, M. Weathering and relative durability of detrital minerals in equatorial climate: Sand petrology and geochemistry in the East African Rift. *J. Geol.* **2013**, *121*, 547–580. [CrossRef]
99. Garzanti, E.; Padoan, M.; Setti, M.; Peruta, L.; Najman, Y.; Villa, I.M. Weathering geochemistry and Sr-Nd fingerprints of equatorial upper Nile and Congo muds. *Geochem. Geophys. Geosyst.* **2013**, *14*, 292–316. [CrossRef]
100. Singh, S.K.; Rai, S.K.; Krishnaswami, S. Sr and Nd isotopes in river sediments from the Ganga Basin: Sediment provenance and spatial variability in physical erosion. *J. Geophys. Res. Earth Surf.* **2008**, *113*, F03006. [CrossRef]
101. Gaillardet, J.; Dupré, B.; Allègre, C.J. Geochemistry of large river suspended sediments: Silicate weathering or recycling tracer? *Geochim. Cosmochim. Acta* **1999**, *63*, 4037–4051. [CrossRef]
102. Taylor, S.R.; McLennan, S.M. The geochemical evolution of the continental crust. *Rev. Geophys.* **1995**, *33*, 241–265. [CrossRef]
103. Rudnick, R.L.; Gao, S. Composition of the continental crust. In *Treatise on Geochemistry, The Crust*; Rudnick, R.L., Holland, H.D., Turekian, K.K., Eds.; Elsevier Pergamon: Oxford, UK, 2003; Volume 3, pp. 1–64.
104. Garzanti, E.; Resentini, A. Provenance control on chemical indices of weathering (Taiwan river sands). *Sediment. Geol.* **2016**, *336*, 81–95. [CrossRef]
105. Garzanti, E.; Andò, S. Plate tectonics and heavy-mineral suites of modern sands. In *Heavy Minerals in Use*; Mange, M.A., Wright, D.T., Eds.; Developments in Sedimentology Series 58; Elsevier: Amsterdam, The Netherlands, 2007; pp. 741–763.
106. Comas-Cufí, M.; Thió-Henestrosa, F.S. CoDaPack 2.0: A stand-alone, multi-platform compositional software. In Proceedings of the 4th International Workshopon Compositional Data Analysis, Sant Feliu de Guíxols, Spain, 10–13 May 2011.
107. Singh, S.K.; France-Lanord, C. Tracing the distribution of erosion in the Brahmaputra watershed from isotopic compositions of stream sediments. *Earth Planet. Sci. Lett.* **2002**, *202*, 645–662. [CrossRef]
108. Galy, V.; France-Lanord, C.; Lartiges, B. Loading and fate of riverine particulate organic carbon from the Himalaya to the Ganga–Brahmaputra delta. *Geochim. Cosmochim. Acta* **2008**, *72*, 1767–1787. [CrossRef]
109. Rouse, H. Modern conceptions of the mechanics of fluid turbulence. *Trans. Am. Soc. Civ. Eng.* **1937**, *102*, 463–543.
110. Vanoni, V.A. *Sedimentation Engineering*; Manuals and Reports in Engineering Practice; American Society of Civil Engineers: Reston, VA, USA, 2006; Volume 54, 418p.
111. Komar, P.D. The entrainment, transport and sorting of heavy minerals by waves and currents. In *Heavy Minerals in Use*; Mange, M.A., Wright, D.T., Eds.; Developments in Sedimentology Series; Elsevier: Amsterdam, The Netherlands, 2007; Volume 58, pp. 3–48.

112. Garzanti, E.; Andò, S.; Vezzoli, G. Grain-size dependence of sediment composition and environmental bias in provenance studies. *Earth Planet. Sci. Lett.* **2009**, *277*, 422–432. [CrossRef]
113. Basu, A.; Bickford, M.E.; Deasy, R. Inferring tectonic provenance of siliciclastic rocks from their chemical compositions: A dissent. *Sediment. Geol.* **2016**, *336*, 26–35. [CrossRef]
114. Von Eynatten, H.; Tolosana-Delgado, R.; Karius, V. Sediment generation in modern glacial settings: Grain-size and source-rock control on sediment composition. *Sediment. Geol.* **2012**, *280*, 80–92. [CrossRef]
115. Garzanti, E.; Resentini, A.; Vezzoli, G.; Andò, S.; Malusà, M.G.; Padoan, M.; Paparella, P. Detrital fingerprints of fossil continental-subduction zones (Axial Belt Provenance, European Alps). *J. Geol.* **2010**, *118*, 341–362. [CrossRef]
116. McBride, M.B. Reactions controlling heavy metal solubility in soils. In *Advances in Soil Science*; Springer: New York, NY, USA, 1989; Volume 10, pp. 1–56.
117. Bostick, B.C.; Vairavamurthy, M.A.; Karthikeyan, K.G.; Chorover, J. Cesium adsorption on clay minerals: An EXAFS spectroscopic investigation. *Environ. Sci. Technol.* **2002**, *36*, 2670–2676. [CrossRef]
118. Shamsudduha, M.; Uddin, A.; Saunders, J.A.; Lee, M.K. Quaternary stratigraphy, sediment characteristics and geochemistry of arsenic-contaminated alluvial aquifers in the Ganges–Brahmaputra floodplain in central Bangladesh. *J. Contam. Hydrol.* **2008**, *99*, 112–136. [CrossRef] [PubMed]
119. Sengupta, R.; Basu, P.; Bandyopadhyay, R.; Bandyopadhyay, A.; Rakshit, S.; Sharma, B. *Sediments in the Continental Shelf in and Around the Swatch of No Ground*; Special Publications; Geological Survey of India: Kolkata, India, 1992; Volume 29, pp. 201–207.
120. Wiedicke, M.; Kudrass, H.R.; Hübscher, C. Oolitic beach barriers of the last Glacial sea-level lowstand at the outer Bengal shelf. *Mar. Geol.* **1999**, *157*, 7–18. [CrossRef]
121. Opdyke, B.N.; Wilkinson, B.H. Paleolatitude distribution of Phanerozoic marine ooids and cements. *Palaeogeogr. Palaeoclimatol. Palaeoecol.* **1990**, *78*, 135–148. [CrossRef]
122. Gallagher, S.J.; Reuning, L.; Himmler, T.; Henderiks, J.; De Vleeschouwer, D.; Groeneveld, J.; Rastegar Lari, A.; Fulthorpe, C.S.; Bogus, K.; Expedition 356 Shipboard Scientists. The enigma of rare Quaternary oolites in the Indian and Pacific Oceans: A result of global oceanographic physico-chemical conditions or a sampling bias? *Quat. Sci. Rev.* **2018**, *200*, 114–122. [CrossRef]
123. Garzanti, E.; Resentini, A.; Andò, S.; Vezzoli, G.; Vermeesch, P. Physical controls on sand composition and relative durability of detrital minerals during long-distance littoral and eolian transport (coastal Namibia). *Sedimentology* **2015**, *62*, 971–996. [CrossRef]
124. Resentini, A.; Andò, S.; Garzanti, E. Quantifying roundness of detrital minerals by image analysis: Sediment transport, shape effects, and provenance implications. *J. Sediment. Res.* **2018**, *88*, 276–289. [CrossRef]
125. Singh, S.K.; Sarin, M.M.; France-Lanord, C. Chemical erosion in the eastern Himalaya: Major ion composition of the Brahmaputra and $\delta^{13}C$ of dissolved inorganic carbon. *Geochim. Cosmochim. Acta* **2005**, *69*, 3573–3588. [CrossRef]
126. Sarin, M.M.; Krishnaswami, S.; Dilli, K.; Somayajulu, B.L.K.; Moore, W.S. Major ion chemistry of the Ganga-Brahmaputra river system: Weathering processes and fluxes to the Bay of Bengal. *Geochim. Cosmochim. Acta* **1989**, *53*, 997–1009. [CrossRef]
127. Datta, D.K.; Subramanian, V. Texture and mineralogy of sediments from the Ganges-Brahmaputra-Meghna river system in the Bengal Basin, Bangladesh and their environmental implications. *Environ. Geol.* **1997**, *30*, 181–188. [CrossRef]
128. Heroy, D.C.; Kuehl, S.A.; Goodbred, S.L. Mineralogy of the Ganges and Brahmaputra Rivers: Implications for river switching and Late Quaternary climate change. *Sediment. Geol.* **2003**, *155*, 343–359. [CrossRef]
129. Huyghe, P.; Guilbaud, R.; Bernet, M.; Galy, A.; Gajurel, A.P. Significance of the clay mineral distribution in fluvial sediments of the Neogene to recent Himalayan foreland basin (west-central Nepal). *Basin Res.* **2011**, *23*, 332–345. [CrossRef]
130. Khan, M.H.R.; Liu, J.; Liu, S.; Seddique, A.A.; Cao, L.; Rahman, A. Clay mineral compositions in surface sediments of the Ganges-Brahmaputra-Meghna river system of Bengal Basin, Bangladesh. *Mar. Geol.* **2019**, *412*, 27–36. [CrossRef]
131. Garzanti, E.; Vermeesch, P.; Andò, S.; Botti, E.; Limonta, M.; Vezzoli, G.; Dinis, P.; Hahn, A.; Baudet, D.; De Grave, J.; et al. Congo River Sand and the Equatorial Quartz Factory. *Earth Sci. Rev.* **2019**, *197*, 102918. [CrossRef]

132. Garzanti, E.; Vermeesch, P.; Padoan, M.; Resentini, A.; Vezzoli, G.; Andò, S. Provenance of passive-margin sand (southern Africa). *J. Geol.* **2014**, *122*, 17–42. [CrossRef]
133. Garzanti, E.; Padoan, M.; Setti, M.; López-Galindo, A.; Villa, I.M. Provenance versus weathering control on the composition of tropical river mud (southern Africa). *Chem. Geol.* **2014**, *366*, 61–74. [CrossRef]
134. Dinis, P.; Garzanti, E.; Vermeesch, P.; Huvi, J. Climatic zonation and weathering control on sediment composition (Angola). *Chem. Geol.* **2017**, *467*, 110–121. [CrossRef]
135. Dinis, P.A.; Garzanti, E.; Hahn, A.; Vermeesch, P.; Pinto, M.C. Weathering indices as climate proxies. A step forward based on Congo and SW African river muds. *Earth Sci. Rev.* **2019**, in press.
136. Stumm, W.; Morgan, J.J. *Aquatic Chemistry*; John Wiley & Sons: New York, NY, USA, 1996; 1022p.
137. Karim, A.; Veizer, J. Weathering processes in the Indus River Basin: Implications from riverine carbon, sulfur, oxygen, and strontium isotopes. *Chem. Geol.* **2000**, *170*, 153–177. [CrossRef]
138. Garçon, M.; Chauvel, C.; France-Lanord, C.; Limonta, M.; Garzanti, E. Which minerals control the Nd–Hf–Sr–Pb isotopic compositions of river sediments? *Chem. Geol.* **2014**, *364*, 42–55. [CrossRef]
139. Garzanti, E.; Resentini, A.; Vezzoli, G.; Andò, S.; Malusà, M.; Padoan, M. Forward compositional modelling of Alpine orogenic sediments. *Sediment. Geol.* **2012**, *280*, 149–164. [CrossRef]
140. Resentini, A.; Goren, L.; Castelltort, S.; Garzanti, E. Partitioning the sediment flux by provenance and tracing erosion patterns in Taiwan. *J. Geophys. Res. Earth Surf.* **2017**, *122*. [CrossRef]
141. Vermeesch, P.; Resentini, A.; Garzanti, E. An R package for statistical provenance analysis. *Sediment. Geol.* **2016**, *336*, 14–25.
142. France-Lanord, C.; Derry, L.; Michard, A. Evolution of the Himalaya since Miocene time: Isotopic and sedimentological evidence from the Bengal Fan. *Geol. Soc. Lond. Spec. Publ.* **1993**, *74*, 603–621. [CrossRef]
143. Yoshida, K.; Cruz, J.W.; Osaki, A.; Masuda, A.; Manoj, M.C.; France-Lanord, C. Detrital heavy mineral record in the Bengal Fan constrains the evolution of the Himalaya and the linkage of the Ganga and Brahmaputra Rivers. In Proceedings of the American Geophysical Union, Fall Meeting 2018, Washington, DC, USA, 9–14 December 2018.
144. Tripathy, G.R.; Singh, S.K.; Ramaswamy, V. Major and trace element geochemistry of Bay of Bengal sediments: Implications to provenances and their controlling factors. *Palaeogeogr. Palaeoclimatol. Palaeoecol.* **2014**, *397*, 20–30. [CrossRef]
145. Subramanian, V.; Van't Dack, L.; Van Grieken, R. Chemical composition of river sediments from the Indian sub-continent. *Chem. Geol.* **1985**, *48*, 271–279. [CrossRef]
146. Zeitler, P.K.; Meltzer, A.S.; Brown, L.; Kidd, W.S.F.; Lim, C.; Enkelmann, E. Tectonics and topographic evolution of Namche Barwa and the easternmost Lhasa block, Tibet. In *Toward an Improved Understanding of Uplift Mechanisms and the Elevation History of the Tibetan Plateau*; Nie, J., Hoke, G.D., Horton, B.K., Eds.; Geological Society of America Special Papers; Geological Society of America: Bould, CO, USA, 2014; Volume 507, pp. 23–58.
147. Govin, G.; Najman, Y.; Dupont-Nivet, G.; Millar, I.; van der Beek, P.; Huyghe, P.; O'Sullivan, P.; Mark, C.; Vögeli, N. The tectonics and paleo-drainage of the easternmost Himalaya (Arunachal Pradesh, India) recorded in the Siwalik rocks of the foreland basin. *Am. J. Sci.* **2018**, *318*, 764–798. [CrossRef]
148. Morton, A.C.; Hallsworth, C. Stability of detrital heavy minerals during burial diagenesis. In *Heavy Minerals in Use*; Developments in Sedimentology Series; Mange, M.A., Wright, D.T., Eds.; Elsevier: Amsterdam, The Netherlands, 2007; Volume 58, pp. 215–245.
149. Andò, S.; Garzanti, E.; Padoan, M.; Limonta, M. Corrosion of heavy minerals during weathering and diagenesis: A catalog for optical analysis. *Sediment. Geol.* **2012**, *280*, 165–178. [CrossRef]
150. Garzanti, E.; Andò, S.; Limonta, M.; Fielding, L.; Najman, Y. Diagenetic control on mineralogical suites in sand, silt, and mud (Cenozoic Nile Delta): Implications for provenance reconstructions. *Earth Sci. Rev.* **2018**, *185*, 122–139. [CrossRef]
151. Mallik, T.K. Mineralogy of deep-sea sands of the Indian Ocean. *Mar. Geol.* **1978**, *27*, 161–176. [CrossRef]

minerals

MDPI

Article

The Chemical Composition and Surface Texture of Transparent Heavy Minerals from Core LQ24 in the Changjiang Delta

Wei Yue [1,2,*], Xiyuan Yue [3], Sugandha Panwar [2], Lingmin Zhang [2] and Bingfu Jin [4]

1. School of Geography, Geomatics and Planning, Jiangsu Normal University, Xuzhou 221116, China
2. State Key Laboratory of Marine Geology, Tongji University, Shanghai 200092, China
3. Shandong Key Laboratory of Eco-environmental Science for Yellow River Delta, Binzhou University, Binzhou 256603, China
4. School of Resources and Environmental Engineering, Ludong University, Yantai 264025, China

* Correspondence: yuewei@jsnu.edu.cn

Received: 12 May 2019; Accepted: 19 July 2019; Published: 22 July 2019

Abstract: The assessment of textural and compositional modifications of detrital sediments is required to reconstruct past source to sink dynamics. The Changjiang Delta is an ideal location to study the sedimentary environment from the Pliocene to Quaternary transition. In the present study, we aim to decipher the response of heavy minerals to mechanical wear and chemical weathering since the Pliocene. With the application of a scanning electron microscope and an electron probe, the geochemistry and surface texture of different heavy minerals (amphibole, epidote, and tourmaline groups) with grain-size fractions of 32–63 μm and 63–125 μm were studied. The result shows that the surface texture of unstable minerals (amphibole, epidote) changed under strong chemical weathering in the Pliocene sediments. By contrast, unstable minerals of the Pleistocene sediments are relatively fresh and similar to those of the modern Changjiang sediment. The stable mineral tourmaline does not exhibit morphology changes in different chemical weathering conditions. No effect of grain size on geochemical composition is noticed. The single minerals of very fine sand and coarse silt show similar geochemical and morphological features. The integration of mineralogy, geochemical data, and grain size parameters yield a more precise understanding of the physical and chemical response of heavy minerals to different weathering conditions. The outcome of the study is also helpful in deciphering sediment provenance changes and environmental changes in the Changjiang basin.

Keywords: heavy mineral; Pliocene; the Changjiang Delta; amphibole; surface texture

1. Introduction

The study of source to sink dynamics of large river systems such as Changjiang are vital to understand the hydrological, geomorphological, ecological, and anthropogenic modifications in the ever-changing climatic conditions [1–3]. The pre-dam (Three Gorges Dam) sediment yield of the Changjiang River was >300 × 10^6 tons per year, whereas, at present, it is ~120 × 10^6 tons [4]. The reconstruction of the source and distribution patterns of the delta sediments can help to unravel the history of erosion processes, source area characteristics, and the factors controlling and determining the production (source), transport, dispersal and accumulation (sink), and reworking (number of burial-erosion cycles during sediment transport) at different temporal and spatial scales [5–7].

The identification and quantification of heavy minerals is one of the most sensitive approaches to classify the various provenance terrains and depositional environment conditions [8,9]. However, heavy mineral assemblages are affected by hydraulic sorting during transportation, chemical weathering during deposition, and dissolution during diagenesis [10–13]. Moreover, physical and chemical etching and corrosion processes intensify with burial depth due to intrastratal dissolution, resulting in the depletion of unstable minerals (like amphibole and epidote) and the enrichment of stable minerals (ZTR) in core sediments [8,14]. To overcome this bias, several provenance-sensitive heavy mineral ratios (e.g., the garnet-zircon (GZi), ZTR, and stability indexes) are widely used [11,15]. Moreover, in situ geochemical analysis of heavy minerals is also widely applied in geoscience and adds a considerable complexity of interpretation [16–18]. Chemical analysis on single-grain minerals (e.g., zircon, tourmaline, apatite, rutile, monazite, amphibole, and garnet) has promoted the study of sediment source to sink [19–21]. However, there are only a few studies that deal with the relationship between the physical and chemical parameters of single heavy minerals and address these significant research lacunae: (i) How do different minerals respond to different sedimentary environments in the deltaic area? (ii) Do the physical and chemical properties of stable and unstable heavy minerals change depending on depth?

Thicknesses of 150–400 m sediments have been deposited in the Changjiang Delta since the Pliocene [22]. This area is a natural laboratory that can be used to carry out studies on sediment provenance, tectonic evolution, sedimentary environment, and river evolution [23–25]. With the application of clay minerals, geochemistry, and heavy mineral analyses, researchers have documented different sedimentary environments during the Pliocene to Pleistocene transition [26,27]. Heavy mineral analysis evinces the presence of high contents of zircon and extremely low contents of amphibole (5%) in Pliocene sediments of the Changjiang Delta, which is related to strong chemical weathering or diagenesis during this time (Figure 1). In contrast, high proportions of unstable mineral (amphibole 40%) in the Pleistocene strata indicate relatively weak chemical weathering or minor diagenetic effects [3,27]. The high heavy mineral concentration (HMC) values in some of the Pleistocene samples define a placer trend, probably due to hydraulic sorting [13,20]. The different chemical weathering and sedimentary conditions at the Pliocene to Quaternary transition make the Changjiang Delta an ideal location to document the physical and geochemical response of single minerals to different sedimentary environments [23,27].

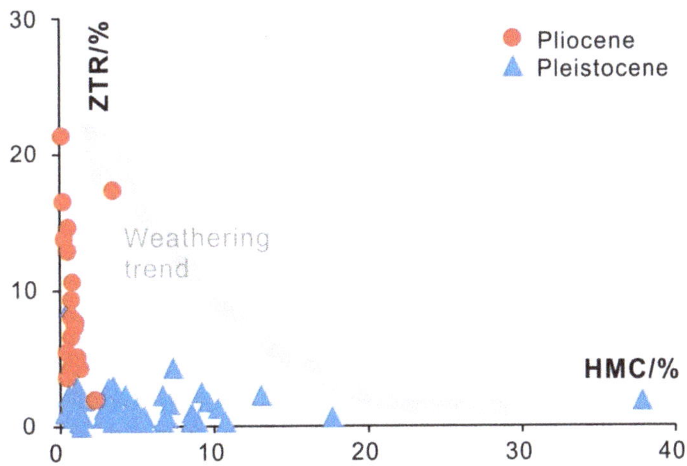

Figure 1. Stable minerals (ZTR) vs. heavy mineral concentration of Pliocene and Pleistocene sediments from the Core LQ24 of the Changjiang Delta. HMC = Heavy mineral concentration; the ZTR index [11] is the percentage of relatively durable zircon, tourmaline, and rutile among the transparent heavy minerals. The high HMC values in some of the Pleistocene samples define a placer trend [13], and diagenetic dissolution in the Pliocene sand produces a decrease in HMC and an increase in ZTR.

In this study, unstable (amphibole, epidote) and stable (tourmaline) heavy mineral in grain size fractions of 32–63 μm and 63–125 μm were analyzed by using a scanning electron microscope (SEM) and an electron probe (EP). Through the assessment of the surface texture and chemical composition of different transparent heavy minerals, we aim to study the physical and chemical response of different single minerals and discuss their influence on provenance discrimination.

2. Materials and Methods

One continuous Late Cenozoic borehole (LQ24, 400 m in length) was extracted from Chongming Island in the Changjiang Delta. The bottom (400–337.2 m) of LQ24 is constrained to the Pliocene according to the paleomagnetic analysis results, which disclosed that the boundary of Gauss/Matuyama (G/M) is located at a core depth of 337.2 m (Figure 2). The Pliocene strata above the bedrocks consisted of stiff clay. There were many calcareous and iron-manganese nodules (Figure 2). Moreover, a block of grey calcareous sand was found in the bottom strata (Figure 2). No microfossils were found in the Pliocene sediments. In addition, poorly sorted gravels can be found, which indicate the origin of sedimentary facies as an alluvial fan or a meandering river [22]. The Quaternary strata comprised several sedimentary sequences, which were composed of sand at the bottom and clayey silt at the top, suggesting a fluvial environment (Figure 2). In some mid–late Quaternary sediments, foraminifera and shell debris were easily found, implying marine transgression since then [3,23,27]. In-depth detailed stratigraphic information, including magnetostratigraphy, grain size, sedimentary facies, magnetic susceptibility, and foraminifera have been documented by [3]. In the present study, two Pliocene sediment samples (core depth of 378 m and 355 m), and three Pleistocene samples (core depth of 320, 259, and 178 m) were selected (Figure 2).

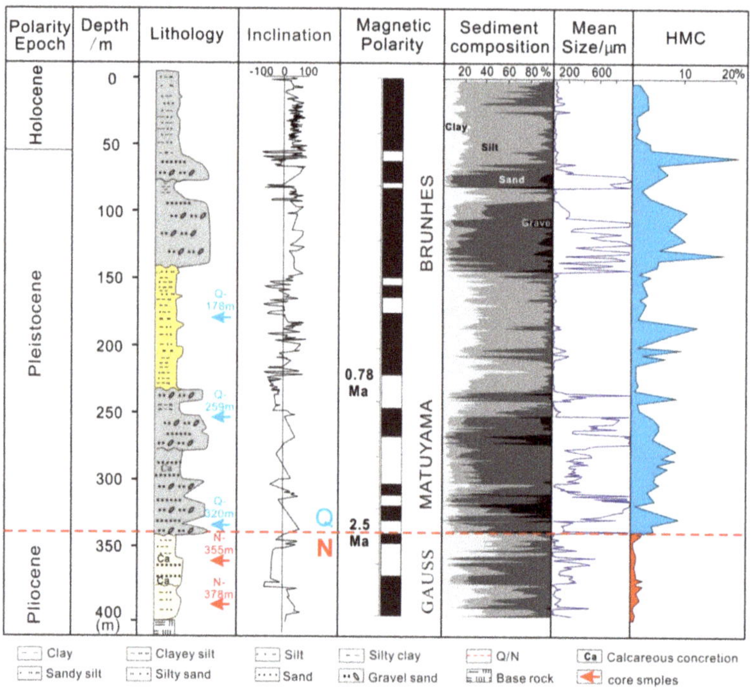

Figure 2. The Plio-Pleistocene magneto-stratigraphy of Core LQ24 from the Changjiang Delta. Paleo-magnetic dating, lithology, and grain size parameters were from [3]. N = Pliocene, Q = Pleistocene.

In order to make a comparison with the Changjiang Delta sediment, one modern sediment sample was collected from its estuary levee (Figure 3).

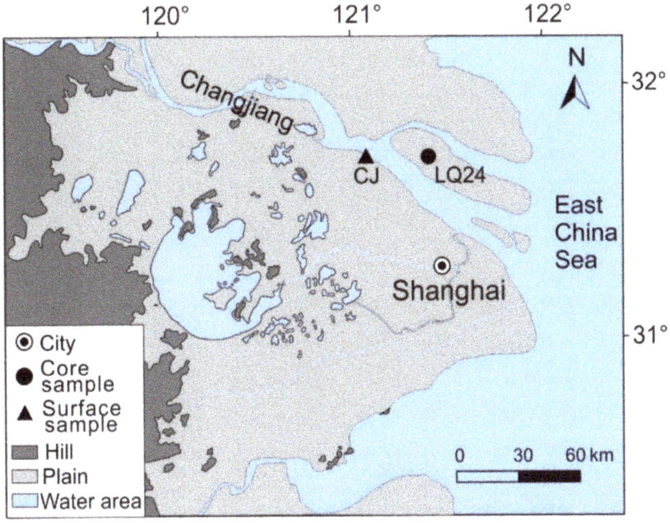

Figure 3. Sampling locations of the core and modern Changjiang River sediment.

For each sample (1 kg each), the very fine sand (63–125 μm) and coarse silt (32–63 μm) fractions were separated and selected for single mineral analyses. These size fractions were wet-sieved through 125, 63, and 32 μm nylon meshes, respectively. After drying in the oven with a temperature of <40 °C, the very fine sand and coarse silt fractions were then put into a sodium polytungstate solution (2.90 kg/dm^3 at 20 °C; e.g., [5,21,28]) and heavy minerals were separated from the light fraction by the gravity separation method (e.g., [14,29]). The content of the heavy minerals (HMC) is calculated as the total mass of the heavy mineral fraction contained in the analyzed fraction [30].

After heavy liquid separation, >100 mineral grains (amphibole, epidote, and tourmaline) were handpicked randomly under a stereoscopic microscope. Further, all the handpicked minerals were observed and checked under a polarizing microscope. All of the amphibole grains of the Pliocene samples were selected due to their low concentrations. At least 30 single mineral grains were measured randomly by SEM analysis. After SEM analysis, single mineral grains were then embedded in epoxy, polished, and coated with carbon to carry out electron probe analysis. A uniform flat surface of each grain was selected as a test spot. Single mineral grains were photographed by a scanning electron microscope (XL-30 ESEM, Philips Company, Eindhoven, The Netherlands), and the mineral chemical composition was measured by an electron probe microanalysis (JXA-8230 electron probe system, JEOL, Kyoto, Japan) at the State Key Laboratory of Marine Geology, Tongji University. The measurement conditions were a 15 kV accelerating voltage, a 10 nA probe current, and a 5 μm beam diameter. The analysis time of each element was 30 s for Si, Mg, Al, Fe, Ca, and Mn; 10 s for Na; 12 s for K; and 60 s for Ti. Natural and synthetic mineral standards (SPI) were used to calibrate all quantitative analyses, and a ZAF correction was used for data reduction.

3. Results

3.1. Geochemical Characteristics of Different Single Minerals in Core LQ24

3.1.1. Geochemical Characteristics of Unstable Minerals in Core LQ24

No change in the composition content (TFeO, CaO, MgO, Al_2O_3, K_2O, TiO_2, MnO, Cr_2O_3, and V_2O_3) of the amphibole derived from very fine sand and the coarse silt fractions was noticed (Figure 4). The detailed EPMA results of this study are presented in Table 1.

Table 1. Average element compositions (%) of amphibole, epidote, tourmaline of Core LQ24 and modern Changjiang River. s = Very fine sand fraction (63–125 μm), cs = Coarse silt fraction (32–63 μm). CJ = Changjiang River.

Samples (Number)	SiO_2	Al_2O_3	Na_2O	MgO	K_2O	TiO_2	CaO	FeO	MnO	Cr_2O_3	V_2O_3
LQ24-355m-cs-Amp(17)	46.83	6.55	0.86	12.75	0.46	0.31	11.39	13.15	0.38	0.02	0.05
LQ24-320m-cs-Amp(31)	45.43	8.74	1.18	11.31	0.76	0.50	11.33	15.34	0.45	0.04	0.06
LQ24-259m-cs-Amp(25)	46.79	7.62	0.91	11.33	0.55	0.34	11.29	15.65	0.52	0.07	0.05
LQ24-178m-cs-Amp(15)	46.32	8.05	0.92	11.41	0.37	0.23	12.32	13.56	0.35	0.03	0.05
CJ-Amp-cs(29)	46.19	7.95	1.00	11.27	0.69	0.74	11.86	15.74	0.38	0.08	0.06
LQ24-378m-cs-Ep(12)	37.14	22.12	0.00	0.02	0.01	0.01	21.64	11.31	0.62	0.01	0.03
LQ24-355m-cs-Ep(44)	36.80	21.9	0.03	0.17	0.01	0.05	22.06	11.34	0.18	0.03	0.05
LQ24-320m-cs-Ep(27)	37.66	22.61	0.01	0.03	0.00	0.06	22.49	11.71	0.21	0.02	0.08
LQ24-259m-cs-Ep(26)	37.03	21.98	0.01	0.05	0.00	0.06	22.25	11.94	0.19	0.02	0.06
LQ24-178m-cs-Ep(25)	37.55	21.40	0.05	0.02	0.00	0.04	22.74	13.25	0.15	0.04	0.1
CJ-Ep-cs(27)	36.92	22.67	0.01	0.03	0.00	0.06	22.5	11.5	0.17	0.04	0.07
LQ24-378m-cs-Tur(29)	35.92	30.43	1.93	6.03	0.03	0.34	0.83	8.40	0.05	0.05	0.04
LQ24-355m-cs-Tur(23)	35.78	31.17	1.91	5.03	0.03	0.35	0.44	8.50	0.05	0.04	0.06
LQ24-178m-cs-Tur(26)	36.42	31.88	1.84	5.52	0.03	0.3	0.56	7.72	0.05	0.07	0.04
CJ-Tur-cs(20)	34.65	31.25	1.92	6.18	0.03	0.81	0.95	8.39	0.04	0.05	0.07
LQ24-378m-s-Amp(7)	46.62	8.24	1.05	11.13	0.76	0.45	11.69	15.81	0.34	0.03	0.07
LQ24-355m-s-Amp(12)	46.58	9.40	0.92	11.48	1.26	0.35	10.36	15.19	0.33	0.03	0.07
LQ24-320m-s-Amp(27)	46.55	7.88	1.17	12.26	0.63	0.40	11.42	14.38	0.46	0.05	0.04
LQ24-259m-s-Amp(24)	46.59	7.41	1.08	12.38	0.54	0.38	12.22	13.48	0.42	0.07	0.06
LQ24-178m-s-Amp(15)	47.32	7.30	0.76	12.41	0.34	0.37	12.49	12.66	0.25	0.10	0.06
CJ-Amp-s(25)	46.18	8.80	1.07	11.28	0.53	0.65	11.44	15.45	0.39	0.04	0.06
LQ24-378m-s-Ep(24)	38.08	22.93	0.08	0.19	0.01	0.16	22.58	11.00	0.16	0.01	0.08
LQ24-355m-s-Ep(52)	37.31	22.3	0.02	0.08	0.01	0.06	23.05	11.97	0.21	0.03	0.05
LQ24-320m-s-Ep(33)	37.25	22.86	0.01	0.25	0.01	0.06	22.53	10.32	0.20	0.01	0.07
LQ24-259m-s-Ep(28)	37.66	22.29	0.07	0.03	0.12	0.06	22.23	11.65	0.22	0.02	0.04
LQ24-178m-s-Ep(25)	37.39	21.87	0.01	0.15	0.01	0.04	22.45	11.92	0.19	0.04	0.08
CJ-Ep-s(22)	36.41	21.98	0.01	0.07	0.01	0.22	22.22	11.7	0.22	0.04	0.09
LQ24-378m-s-Tur(47)	35.95	32.55	1.86	4.41	0.04	0.31	0.57	9.40	0.07	0.03	0.03
LQ24-355m-s-Tur(21)	35.88	33.02	1.92	5.62	0.04	0.32	0.55	7.28	0.06	0.05	0.04
LQ24-259m-s-Tur(6)	37.51	31.45	1.92	8.38	0.05	0.28	1.25	5.38	0.05	0.07	0.06
LQ24-178m-s-Tur(25)	35.36	32.29	1.71	5.35	0.03	0.36	0.85	7.75	0.04	0.04	0.03
CJ-Tur-s(16)	34.98	31.74	2.11	6.40	0.03	0.75	0.70	7.49	0.04	0.05	0.06

Further classification of the amphiboles was made in Figure 5. Results show that members of the amphibole group in all of the Changjiang sediment samples belong to the calcic–amphibole group [31], in which ten different kinds of hornblendes were included. These hornblendes were dominated by magnesio-hornblende, followed by actinolite, tschermakite, and edenite (Figure 5). Other amphibole species, including ferrotschermakite, ferroedenite, magnesio hastingsite, and ferropargasite, were low in content. Moreover, amphibole compositions in very fine sand and coarse silt fractions were very similar (Figure 5).

Similar to the characteristics of the amphibole in Core LQ24, the epidote geochemistry of the Pliocene is almost the same with the Pleistocene and modern Changjiang samples (Figure 6). No difference in the composition (main element content) of the epidote between 63–125 μm and 32–63 μm size fractions was found.

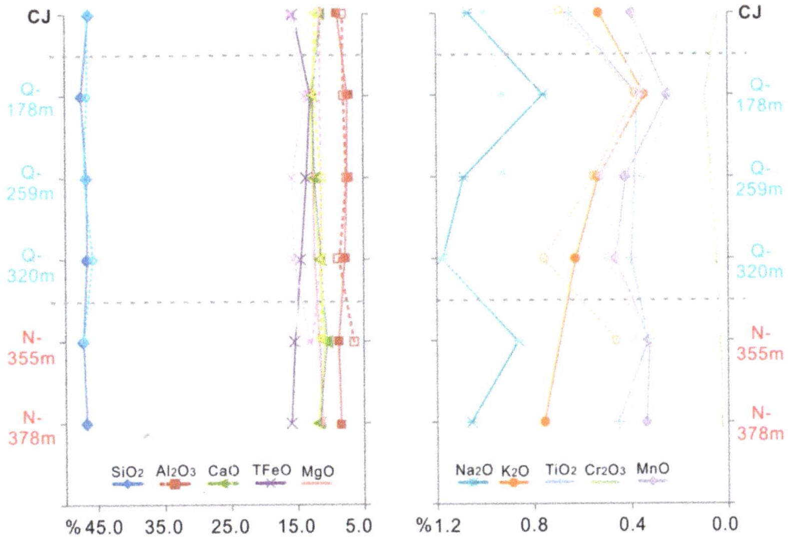

Figure 4. The mean element compositions of the amphibole of Core LQ24 and modern Changjiang River. A very fine sand fraction (63–125 μm) is indicated by full lines and coarse silt (32–63 μm) is indicated by dashed lines. Core depth is set in the vertical axis. CJ = Changjiang River. Core samples that were sampled from different depths of burial are marked in a red and light blue color. The red one is representative of the Pliocene; the light blue is representative of Pleistocene; N = Pliocene, Q = Pleistocene. Please see Table 1 for the numbers of grains analysed per sample.

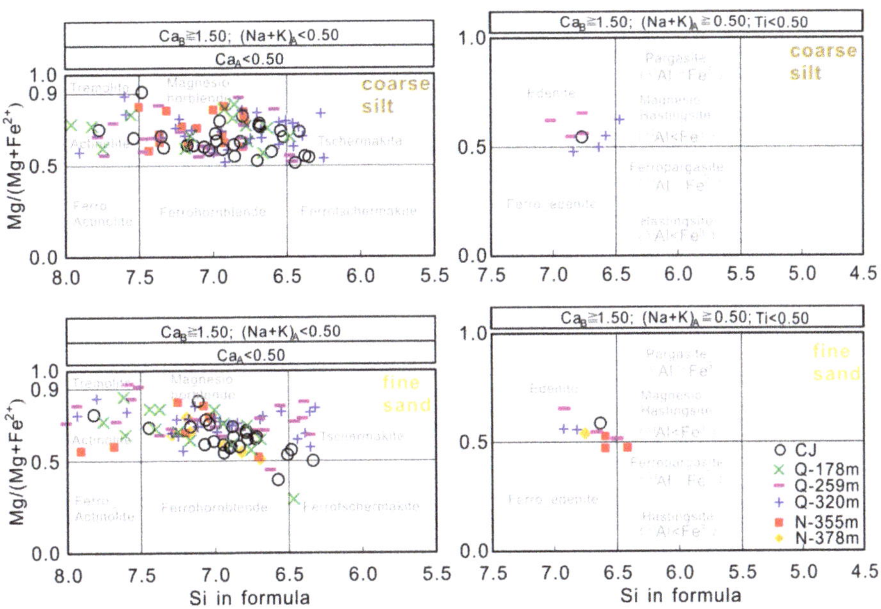

Figure 5. Amphibole classification diagrams for Core LQ24 and the modern Changjing River. CJ = Changjiang, N = Pliocene, Q = Pleistocene. Ca-amphibole classification was performed according to [31].

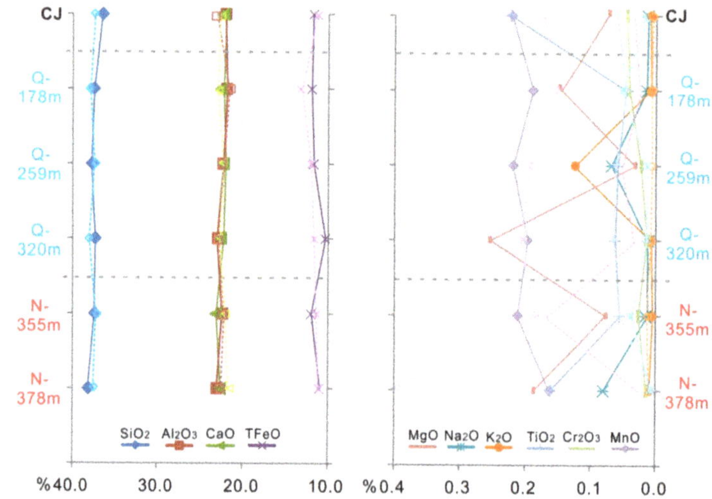

Figure 6. Mean element compositions of the epidote of Core LQ24 and the modern Changjiang River. A very fine sand fraction (63–125 μm) is indicated by full lines and coarse silt (32–63 μm) is indicated by dashed lines. Core depth is set in the vertical axis. CJ = Changjiang River. Core samples that were sampled from different depths of burial are marked in a red and light blue color. The red one is representative of the Pliocene; the light blue is representative of the Pleistocene; N = Pliocene, Q = Pleistocene. Please see Table 1 for the numbers of grains analysed per sample.

3.1.2. Geochemical Characteristics of Stable Mineral in Core LQ24

The element composition (Al_2O_3, TFeO, MgO, Na_2O, CaO, K_2O, TiO_2, MnO, Cr_2O_3, and V_2O_3) of stable tourmaline is shown in Figure 7. Similar to the geochemical characteristics of amphibole and epidote in Core LQ24, the element characteristics of tourmaline are also the same at different core depths and the same as the modern Changjiang river samples (Figure 7).

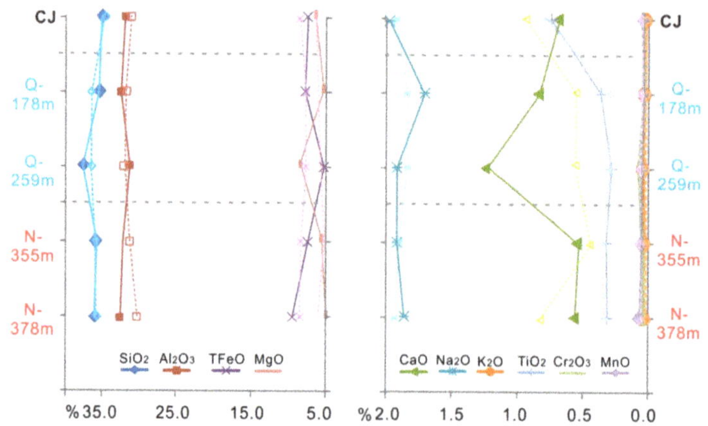

Figure 7. The mean element compositions of the tourmaline of Core LQ24 and the modern Changjiang River. A very fine sand fraction (63–125 μm) is indicated by full lines and coarse silt (32–63 μm) is indicated by dashed lines. The core depth is set in the vertical axis. CJ = Changjiang River. Core samples that were sampled from different depths of burial are marked in a red and light blue color. The red one is representative of the Pliocene; the light blue is representative of the Pleistocene; N = Pliocene, Q = Pleistocene. Please see Table 1 for the numbers of grains analysed per sample.

3.2. *Surface Texture of Different Transparent Heavy Minerals in Core Sediment of the Changjiang Delta*

The surface texture characteristic of unstable heavy minerals are different for the Pliocene and Pleistocene sediment samples (Figure 8). The corroded unstable mineral grains were abundant in the Pliocene sediment compared to the Pleistocene samples. The corroded amphibole accounted for about 30% in the Pliocene sediment and 15% in the Pleistocene samples. On the other hand, corroded epidote grains accounted for about 50% in Pliocene sediments, which was approximately twice that of the Pleistocene samples. In contrast to the unstable minerals, the surface texture of the stable minerals did not exhibit an obvious change in the depth of the core. The corroded tourmaline grains were about 10% on average, which was less than that of unstable minerals. For the same single mineral, the surface texture characteristic of the very fine sand was similar to that of the coarse silt (Figure 8).

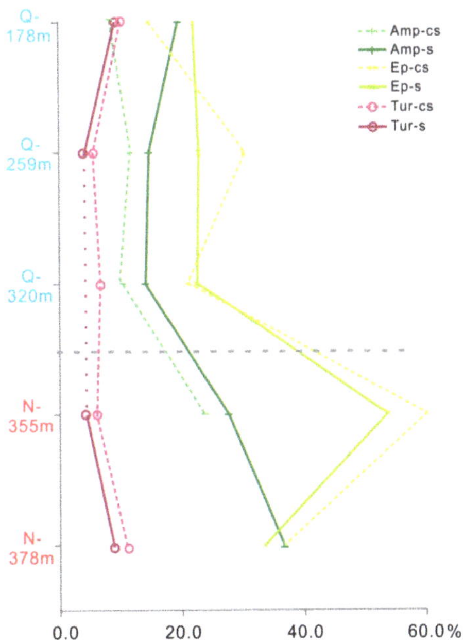

Figure 8. Percentage of corroded single mineral grains in the Core LQ24 of Changjiang Delta. A very fine sand fraction (63–125 μm) is indicated by full lines and coarse silt (32–63 μm) is indicated by dashed lines. N = Pliocene, Q = Pleistocene.

3.2.1. Crystal Morphology of Unstable Heavy Minerals in Core LQ24

SEM results showed that the surface texture of the unstable heavy minerals was different in the Pliocene and Pleistocene samples. For the Pliocene sediment of Core LQ24, both the amphibole and epidote displayed evidence of chemical etching (Figures 9 and 10). Etch pits on the crystal surfaces of these single mineral grains were visible. The crystal corrosion along the cleavage plane and fuzzy edges were readily visible. Microscope observation found that the transparency of some mineral grains was significantly reduced. Due to strong dissolution, some amphibole grains were found to be deeply etched (Figure 9).

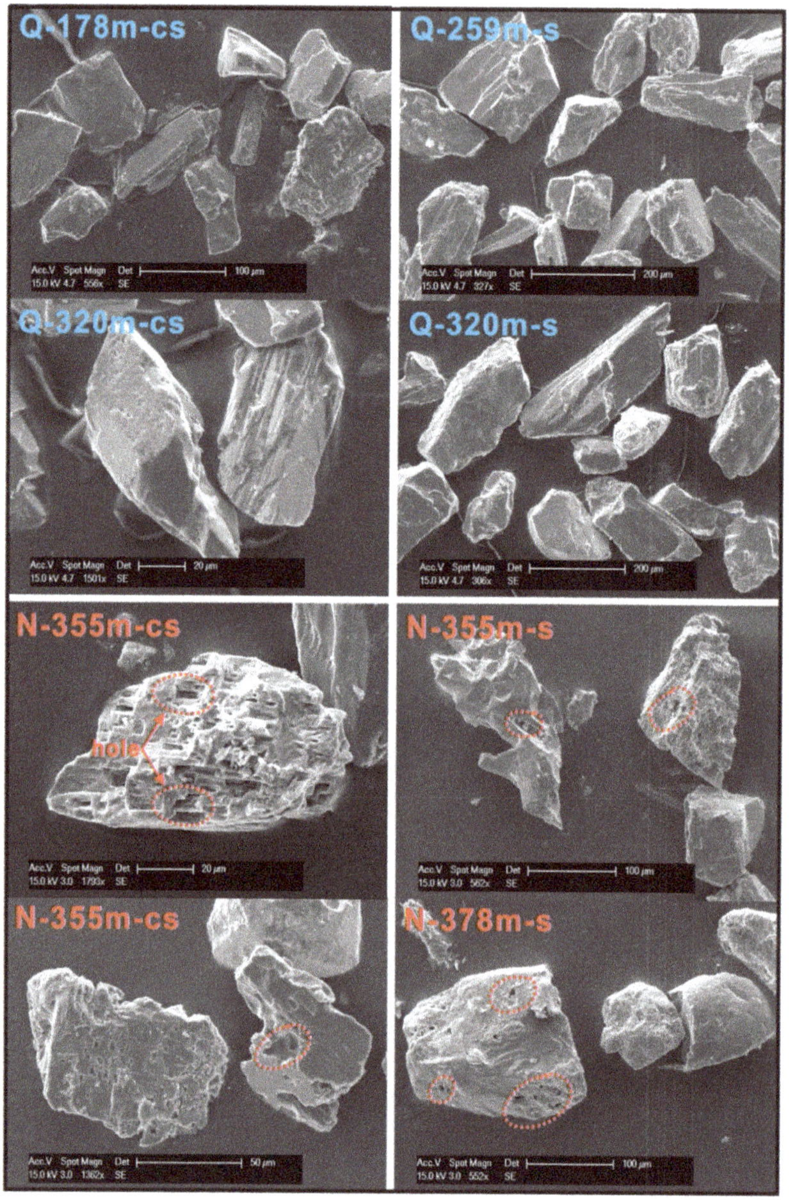

Figure 9. Surface texture of the amphibole in the Core LQ24 of the Changjiang Delta. s = Very fine sand fraction (63–125 µm), cs = coarse silt fraction (32–63 µm). N = Pliocene, Q = Pleistocene.

Unlike the sample of the Pliocene strata, the surface texture of the unstable heavy minerals in the Pleistocene sample displayed different characteristics. Amphibole and epidote grains generally had an irregular-angular shape, and their crystal surfaces were relatively fresh (Figures 9 and 10). Two systems of cleavage were obvious at a specific angle. Only a few particles had a fuzzy crystal surface and plane due to dissolution and abrasion (Figures 9 and 10). The crystal morphology of the different grain size displayed little surface texture difference between the very fine sand and coarse silt.

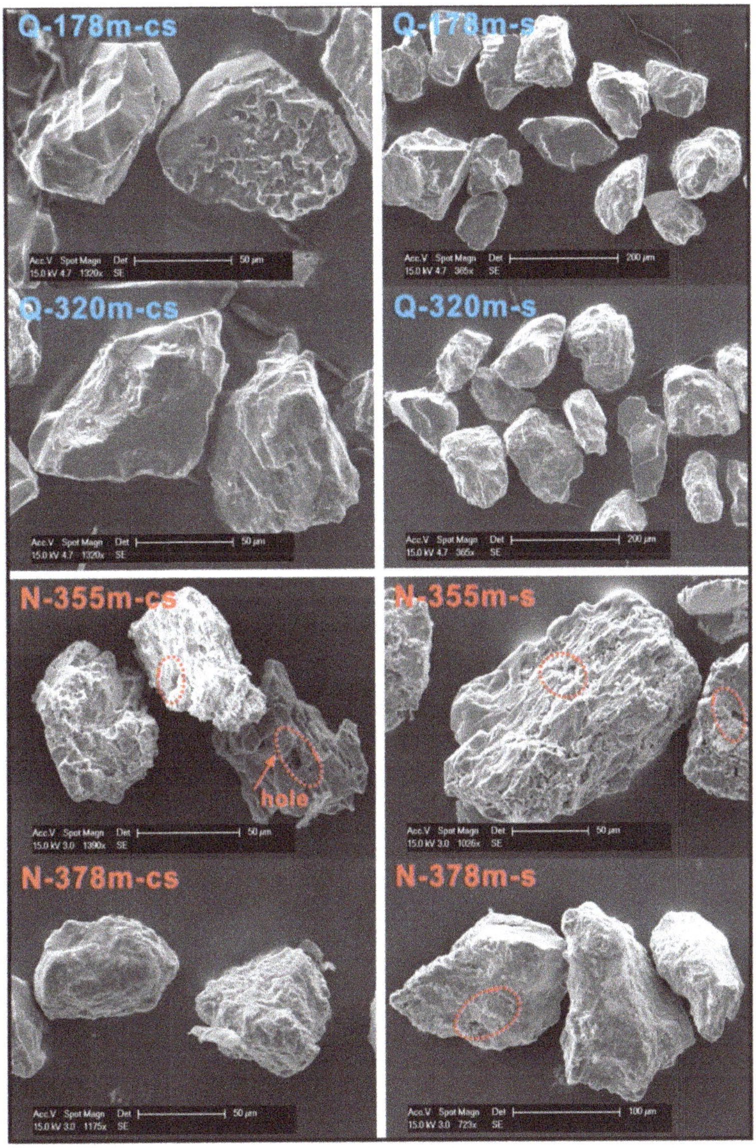

Figure 10. Surface texture of the epidote in the Core LQ24 of the Changjiang Delta. s = Very fine sand fraction (63–125 μm), cs = coarse silt fraction (32–63 μm). N = Pliocene, Q = Pleistocene.

3.2.2. Surface Texture of Stable Heavy Mineral

The SEM images of the stable heavy mineral of tourmaline showed that crystal particles in the Pliocene and Pleistocene sediments had short prismatic or granular morphologies, and were mostly sub-angular or sub-rounded in shape (Figure 11). The edges of the tourmaline crystal were clear with generally flat crystal surfaces. Etch pits were noticed in a few particles (Figure 11). Similar crystal morphologies were observed in the Pleistocene and Pliocene samples of tourmaline. The tourmaline grain size in the very fine sand and coarse silt has a similar surface texture. In general, the unstable heavy minerals of the Pliocene samples show signs of weathering, while these minerals in the Quaternary strata are relatively fresh. A slight alteration or dissolution was found in the crystal morphology of the stable mineral from the Changjiang Delta, and no variations were seen in the Pliocene and Pleistocene samples.

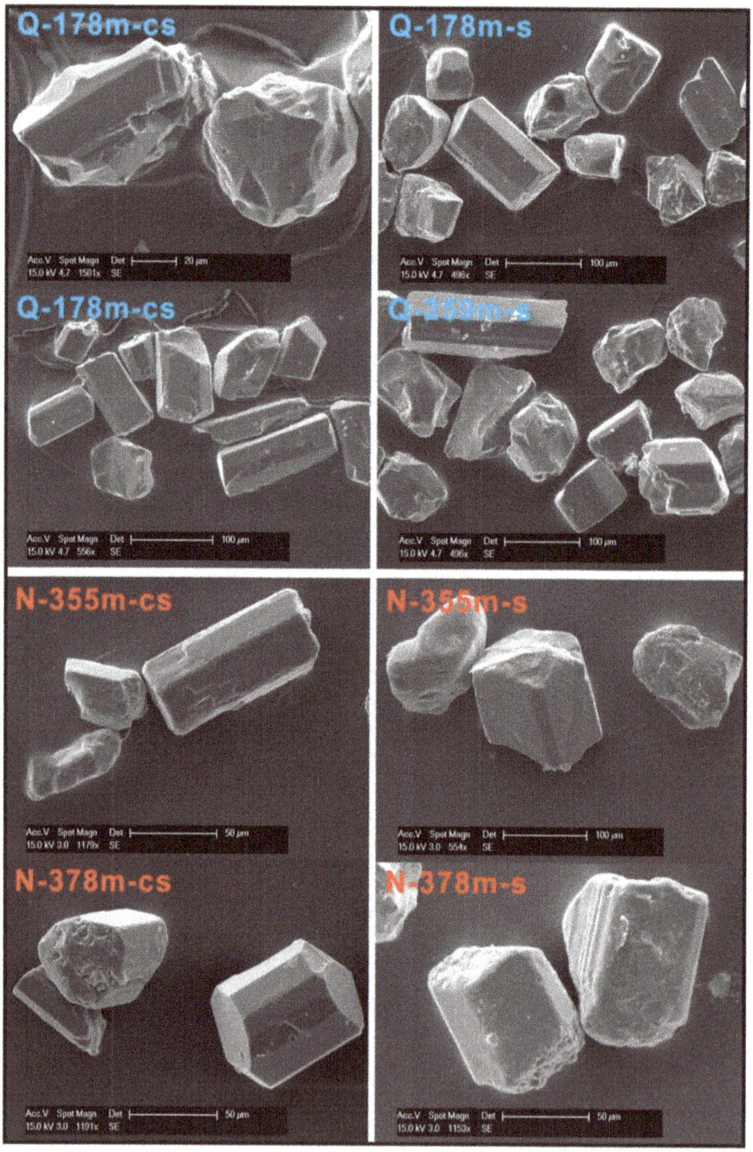

Figure 11. Surface texture of the tourmaline in the Core LQ24 of the Changjiang Delta. s = Very fine sand fraction (63–125 μm), cs = coarse silt fraction (32–63 μm). N = Pliocene, Q = Pleistocene.

4. Discussion

4.1. The Factors Which may Cause Variations in Surface Texture of Different Single Minerals

The corrosion signature of heavy minerals indicates different transportation, sedimentation, and burial conditions [32–34]. The SEM images of Core LQ24 revealed that the amphibole and epidote in the Pliocene strata were more corroded than the tourmaline mineral present in both the Pliocene and Pleistocene sediments (Figures 9–11). In our interpretation, weathering is the main factor influencing the dissolution of unstable minerals in the analyzed sediment samples. In the warm-humid climate, chemical weathering proceeds at a faster rate and can result in the progressive corrosion and depletion of unstable minerals [35,36]. Previous studies have documented a drastic environment change in the Changjiang Delta during the Pliocene to Pleistocene transition [23,27]. Then, warm and wet climate conditions were prevalent in East China, along with a low deposition rate in the Changjiang Delta [24,37,38]. Before deposition, weathering may have been responsible for the surface texture variation in amphibole, epidote, and tourmaline. In such strong weathering conditions, unstable minerals are progressively corroded and finally depleted.

The abundance of apatite is a reliable indicator of surface weathering. This mineral is susceptible to dissolution during weathering but remains stable in deep burial [8,13,15]. In order to reflect the weathering intensity, we used the ATi (Apatite-Tourmaline index), whose value is calculated as (100 × Apatite%)/(Apatite% + tourmaline%) [39]. In this parameter, two minerals (Apatite and tourmaline) possess identical densities but show contrasting behaviours in weathering, thereby reducing the effect of hydrodynamics. Furthermore, apatite fertility and erosion rates may have an impact on the content of apatite [20,21]. In the present study, low ATi and apatite content in the Pliocene imply stronger weathering scenarios (Figure 12). Moreover, a strong enrichment of ZTR and a significant loss of unstable minerals (amphibole <5%) also suggest that weathering was much stronger during the Pliocene than during the Quaternary time (Figure 12).

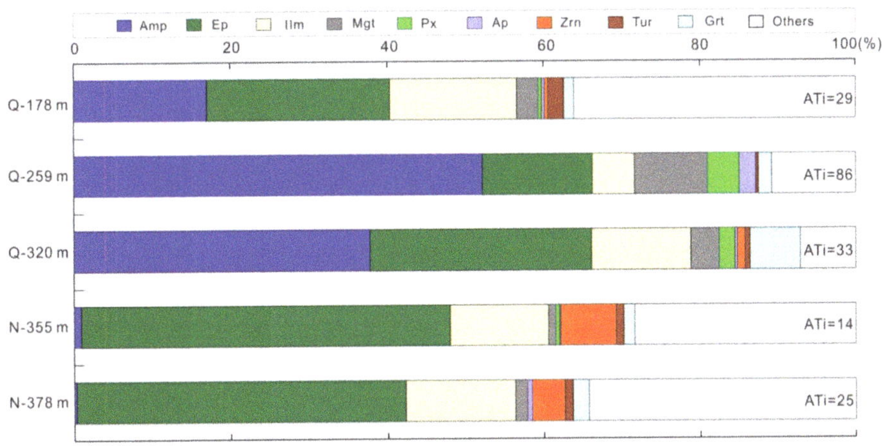

Figure 12. Heavy mineral distribution for Core LQ24 of the Changjiang Delta. Heavy mineral data are from [3]. The analyzed size fraction is very fine sand (63–125 μm). N = Pliocene, Q = Pleistocene. Systematic mineral abbreviation list: ATi (Apatite-Tourmaline index) = (100 × Apatite%)/(Apatite% + tourmaline%). Amp = Amphibole, Ep = Epidote, Ilm = Ilmenite, Mgt = Magnetite, Px = Pyroxene, Ap = Apatite, Zrn = Zircon; Tur = Tourmaline, Grt = Garnet, Others = Other heavy minerals and mineral aggregate, including limonite, fluorite, titanite, rutile, anatase, kyanite, andalusite, un-identified grains, rock fragments, etc.

Strong weathering conditions resulted in a decline in the concentration of unstable heavy minerals [8,27], and amphibole and epidote developed etched surfaces, as shown in Figures 9 and 10.

As an unstable heavy mineral, amphibole is dissolved in strong weathering conditions [10,27]. In such environments, unstable minerals were easily etched during the Pliocene. SEM images show that amphibole and epidote were corroded along the direction of the crystal cleavage and left tiny holes on the crystal face after dissolution (Figures 9 and 10). At the same time, the crystal's transparency reduced.

Amphibole and epidote etching were found to be comparable to previously published studies [9,10,32]. In modern sediments of subequatorial regions, surface textures of detrital minerals ranged from incipient corrosion to deep etching, reflecting a progressive increase in the degree of alteration [9]. In addition, some studies show that etching developed from mamillations, through facetted grains, to skeletal grains [32]. Therefore, we infer that the etching of amphibole and epidote grains would have occurred during the sediment transport process before deposition in the sink area of the Changjiang Delta. In this process, the least stable mineral, amphibole, dissolved faster and decreased in the bottom of Core LQ24. On the other hand, the more stable minerals of ZTR dissolved at slower rates and thus were relatively enriched.

The intrastratal dissolution during the diagenetic process may also be listed as a major factor that promotes unstable mineral dissolution in a core sediment. For ancient sediments, temperatures and pressures that increase with burial depth cause the intrastratal dissolution of unstable heavy minerals [14]. Being a depocenter since the Pliocene, the Changjiang Delta has a deposition of a ~300 m thick siliciclastic sedimentary sequence. Affected by compaction and diagenesis, a block of calcite-cemented sand formed at the core bottom of the Changjiang Delta [3,27]. However, the burial depths involved in this study area are modest (the deepest thickness = ca. 400 m), and the sediments of the Changjiang Delta are very young (Neogene) [22,38]. Calcite-cemented sands in the bottom strata are not fully diagenized. Environmental magnetic parameters have shown that diagenesis has not erased the detrital magnetic signal in the Changjiang Delta. This observation suggests weak post-depositional diagenesis [38]. Compared to similar studies in other parts of the world, the process of dissolution during burial diagenesis needs a deeper burial and higher temperatures. Studies in older (Paleocene) sandstones of the North Sea show that amphibole and epidote dissolutions are not complete until they reach a 600 m and 1100 m burial depth, respectively [40]. Walderhaug and Porten suggest that amphibole and epidote dissolution requires pore water temperatures of 40 and 95 °C, respectively [41]. These studies suggest that the process of burial diagenesis did not play a significant role in the single mineral dissolution in the Changjiang Delta.

4.2. Chemical Responses of Different Single Minerals to Different Weathering Environments

Irrespective of the corroded surface texture of susceptible heavy minerals in the Pliocene strata (as a result of strong weathering conditions), no change in the major element composition of single minerals was observed (Figures 4–7). The EPMA results show the prevalence of a similar geochemical character in single minerals extracted from the Pliocene, Pleistocene, and modern Changjiang sediments. The result of amphibole classification showed that all amphibole grains in the boreholes and river samples were mainly composed of magnesio-hornblende, actinolite, tschermakite, and edenite (Table 2, Figure 5). For different samples, the percentage of the same amphibole species varied slightly (Table 2). This bias might have been introduced by natural processes, such as hydraulic sorting, mechanical abrasion, and weathering, or by inaccurate procedures during sampling, laboratory treatment, and analysis [20,28]. Furthermore, for the same type of heavy mineral, the geochemical characteristics are similar in the size fractions of 63–125 μm and 32–63 μm.

Table 2. Amphibole species (%) for Core LQ24 and the modern Changjing River. s = Very fine sand fraction (63–125 μm), cs = coarse silt fraction (32–63 μm). CJ = Changjiang River. Others = other amphibole species, including Ferrotschermakite, Ferroedenite, Magnesio Hastingsite, and Ferropargasite.

Samples	Tremolite	Actinolite	Magnesio-Hornblende	Ferro-Hornblende	Tschermakite	Edenite	Others
LQ24-N-s	0.00	11.11	66.67	0.00	0.00	11.11	11.10
LQ24-Q-s	3.03	15.15	57.58	1.52	13.64	7.58	1.50
CJ-s	0.00	4.00	80.00	4.00	8.00	4.00	0.00
LQ24-N-cs	0.00	5.88	94.12	0.00	0.00	0.00	0.00
LQ24-Q-cs	1.45	13.04	59.42	0.00	13.04	7.25	5.80
CJ-cs	0.00	6.90	72.41	0.00	13.79	3.45	3.50

Under a strong weathering environment, unstable minerals are easily corroded due to their crystal lattice feature. As chemical weathering proceeds, unstable minerals gradually disappear, thereby resulting in an extremely low (about 5%) content of amphibole in the Pliocene strata. Unstable mineral dissolution further affected the HMC. Studies show that the HMC of most Pliocene samples is only 0.5% (Figure 2), which is much lower than that of the modern river sediment samples [27,42].

5. Conclusions

(1) The major element (Si, Fe, Mn, Al, Mg, Ti, and Na) contents of single minerals at different core depths of LQ24 are similar to those of the modern Changjiang sample. There is no distinct depth control of elemental richness or loss of single minerals in the Plio-Pleistocene sediment.

(2) The surface texture of unstable minerals (amphibole, epidote) in the Pliocene strata of Changjiang Delta shows corrosion. By contrast, unstable minerals from the Pleistocene sediment are relatively fresh, like those of the modern Changjiang samples. The stable mineral tourmaline does not show a morphological difference in different strata.

(3) Different minerals within grains of 63–125 μm and 32–63 μm size fractions do not show physical and chemical variations at different core depths in the Changjiang Delta.

Author Contributions: Conceptualization, W.Y.; Methodology, L.Z. and B.J.; Writing—original draft preparation, W.Y; Writing—review and editing, W.Y., S.P. and X.Y.

Funding: This research was funded by a Project Funded by the Priority Academic Program Development of Jiangsu Higher Education Institutions, the National Science Foundation of China (Grant No. 41706048; 41730531; 41576057), and the Natural Science Foundation of Shandong Province, China (No: ZR2019MD009).

Acknowledgments: Special thanks to Sergio Andò and three anonymous reviewers for their constructive comments that helped improve the manuscript.

Conflicts of Interest: The authors declare no conflict of interest.

References

1. Milliman, J.D.; Shen, H.T.; Yang, Z.S.; Mead, R.H. Transport and deposition of river sediment in the Changjiang estuary and adjacent continental shelf. *Cont. Shelf Res.* **1985**, *4*, 37–45. [CrossRef]
2. Yang, S.L.; Xu, K.H.; Milliman, J.D.; Yang, H.F.; Wu, C.S. Decline of Yangtze River water and sediment discharge: Impact from natural and anthropogenic changes. *Sci. Rep.* **2015**, *5*, 12581. [CrossRef] [PubMed]
3. Yue, W.; Jin, B.F.; Zhao, B.C. Transparent heavy minerals and magnetite geochemical composition of the Yangtze River sediments: Implication for provenance evolution of the Yangtze Delta. *Sediment. Geol.* **2018**, *364*, 42–52. [CrossRef]
4. Changjiang Water Resources Commission. *Bulletin of the Changjiang Sediment in 1951–2016*; Changjiang Press: Wuhan, China, 2016. (In Chinese)
5. Vezzoli, G.; Garzanti, E.; Limonta, M.; Andò, S.; Yang, S.Y. Erosion patterns in the Changjiang (Yangtze River) catchment revealed by bulk-sample versus single-mineral provenance budgets. *Geomorphology* **2016**, *261*, 177–192. [CrossRef]
6. Hori, K.; Saito, Y.; Zhao, Q.; Cheng, X.; Wang, P.; Sato, Y.; Li, C. Sedimentary facies and Holocene progradation rates of the Changjiang (Yangtze) delta, China. *Geomorphology* **2001**, *41*, 233–248. [CrossRef]

7. Wang, Z.H.; Saito, Y.; Zhan, Q.; Nian, X.M.; Pan, D.D.; Wang, L.; Chen, T.; Xie, J.L.; Li, X.; Jiang, X.Z. Three-dimensional evolution of the Yangtze River mouth, China during the Holocene: Impacts of sea level, climate and human activity. *Earth Sci. Rev.* **2018**, *185*, 938–955. [CrossRef]
8. Morton, A.C.; Hallsworth, C.R. Processes controlling the composition of heavy mineral assemblages in sandstones. *Sediment. Geol.* **1999**, *124*, 3–29. [CrossRef]
9. Andò, S.; Garzanti, E.; Padoan, M.; Limonta, M. Corrosion of heavy minerals during weathering and diagenesis: a catalog for optical analysis. *Sediment. Geol.* **2012**, *280*, 165–178. [CrossRef]
10. Velbel, M.A. Surface textures and dissolution processes of heavy minerals in the sedimentary cycle: Examples from pyroxenes and amphiboles. *Dev. Sedimentol.* **2007**, *58*, 113–150.
11. Hubert, J.F.A. Zircon-tourmaline-rutile maturity index and the interdependence of the composition of heavy mineral assemblages with the gross composition and texture of sandstones. *J. Sediment. Res.* **1962**, *32*, 440–450.
12. Garzanti, E.; Andò, S.; Vezzoli, G. Grain-size dependence of sediment composition and environmental bias in provenance studies. *Earth Planet. Sci. Lett.* **2009**, *277*, 422–432. [CrossRef]
13. Malusà, M.G.; Garzanti, E. The Sedimentology of Detrital Thermochronology. In *Fission-Track Thermochronology and Its Application to Geology*; Springer: Cham, Switzerland, 2019; pp. 123–143.
14. Morton, A.C. Depth control of intrastratal solution of heavy minerals from the Palaeocene of the North Sea. *J. Sedim. Petrol.* **1979**, *49*, 281–286.
15. Garzanti, E.; Andò, S. Heavy mineral concentration in modern sands: Implications for provenance interpretation. *Dev. Sedimentol.* **2007**, *58*, 517–545.
16. Von Eynatten, H.; Gaupp, R. Provenance of Cretaceous synorogenic sandstones in the Eastern Alps: Constraints from framework petrography, heavy mineral analysis and mineral chemistry. *Sediment. Geol.* **1999**, *124*, 81–111. [CrossRef]
17. Garzanti, E.; Vermeesch, P.; Rittner, M.; Simmons, M. The zircon story of the Nile: Time-structure maps of source rocks and discontinuous propagation of detrital signals. *Basin Res.* **2018**, *30*, 1098–1117. [CrossRef]
18. Morton, A.C. A new approach to provenance studies: Electron microprobe analysis of detrital garnets from Middle Jurassic sandstones of the northern North Sea. *Sedimentology* **1985**, *32*, 553–566. [CrossRef]
19. Von Eynatten, H.; Dunkl, I. Assessing the sediment factory: The role of single grain analysis. *Earth Sci. Rev.* **2012**, *115*, 97–120. [CrossRef]
20. Malusà, M.G.; Resentini, A.; Garzanti, E. Hydraulic sorting and mineral fertility bias in detrital geochronology. *Gondwana Res.* **2016**, *31*, 1–19. [CrossRef]
21. Malusà, M.G.; Wang, J.; Garzanti, E.; Liu, Z.C.; Villa, I.M.; Wittmann, H. Trace-element and Nd-isotope systematics in detrital apatite of the Po river catchment: Implications for provenance discrimination and the lag-time approach to detrital thermochronology. *Lithos* **2017**, *290–291*, 48–59. [CrossRef]
22. Yue, W.; Yue, X.Y.; Zhang, L.M.; Liu, X.B.; Song, J. Morphology of Detrital Zircon as a Fingerprint to Trace Sediment Provenance: Case Study of the Yangtze Delta. *Minerals* **2019**, *9*, 438. [CrossRef]
23. Yang, S.Y.; Li, C.X.; Cai, J.G. Geochemical compositions of core sediments in eastern China: Implication for Late Cenozoic palaeoenvironmental changes. *Palaeogeogr. Palaeoclimatol. Palaeoecol.* **2006**, *229*, 287–302. [CrossRef]
24. Zhang, P.; Miao, Y.F.; Zhang, Z.Y.; Lu, S.M.; Zhang, Y.J.; Chen, H.G.; Li, X.Q.; Miao, Q.Y.; Feng, W.L.; Ou, J.; et al. Late Cenozoic sporopollen records in the Yangtze River Delta, East China and implications for East Asian summer monsoon evolution. *Palaeogeogr. Palaeoclimatol. Palaeoecol.* **2013**, *388*, 153–165. [CrossRef]
25. Yue, W.; Liu, J.T.; Zhang, D.; Wang, Z.H.; Zhao, B.C.; Chen, Z.Y.; Chen, J. Magnetite with anomalously high Cr_2O_3 as a fingerprint to trace upper Yangtze sediments to the sea. *Geomorphology* **2016**, *268*, 14–20. [CrossRef]
26. Yang, S.Y.; Li, C.X.; Yokoyama, K. Elemental compositions and monazite age patterns of core sediments in the Changjiang Delta: Implications for sediment provenance and development history of the Changjiang River. *Earth Planet. Sci. Lett.* **2006**, *245*, 762–776. [CrossRef]
27. Yue, W.; Yang, S.Y.; Zhao, B.C.; Chen, Z.Y.; Yu, J.J.; Liu, X.B.; Huang, X.T.; Jin, B.F.; Chen, J. Changes in environment and provenance within the Changjiang (Yangtze River) Delta during Pliocene to Pleistocene transition. *Mar. Geol.* **2019**, *416*. [CrossRef]
28. Malusà, M.G.; Carter, A.; Limoncelli, M.; Villa, I.M.; Garzanti, E. Bias in detrital zircon geochronology and thermochronometry. *Chem. Geol.* **2013**, *359*, 90–107. [CrossRef]
29. Mange, M.A.; Maurer, H.F.W. *Heavy Minerals in Colour*; Chapman and Hall: London, UK, 1992; pp. 135–151.

30. Jin, B.; Wang, M.; Yue, W.; Zhang, L.; Wang, Y. Heavy Mineral Variability in the Yellow River Sediments as Determined by the Multiple-Window Strategy. *Minerals* **2019**, *9*, 85. [CrossRef]
31. Leake, B.E.; Woolley, A.R.; Birch, W.D.; Burke, E.A.J.; Ferraris, G.; Grice, J.D.; Hawthorne, F.C.; Kisch, H.J.; Krivovichev, V.G.; Schumacher, J.C.; et al. Nomenclature of amphiboles: Additions and revisions to the International Mineralogical Association's amphibole nomenclature. *Am. Mineral.* **2004**, *89*, 883–887.
32. Morton, A.C. Surface features of heavy mineral grains from Palaeocene sands of the central North Sea. *Scot. J. Geol.* **1979**, *15*, 293–300. [CrossRef]
33. Berner, R.A.; Sjöberg, E.L.; Velbel, M.A.; Krom, M.D. Dissolution of pyroxenes and amphiboles during weathering. *Science* **1980**, *207*, 1205–1206. [CrossRef]
34. Lång, L.O. Heavy mineral weathering under acidic soil conditions. *Appl. Geochem.* **2000**, *15*, 415–423. [CrossRef]
35. White, A.F.; Blum, A.E. Effects of climate on chemical weathering in watersheds. *Geochim. Cosmochim. Acta.* **1995**, *59*, 1729–1747. [CrossRef]
36. Brantley, S.L. Reaction Kinetics of Primary Rock-Forming Minerals Under Ambient Conditions. In *Treatise on Geochemistry*; Drever, J.I., Holland, H.D., Turekian, K.K., Eds.; Elsevier: Amsterdam, The Netherland, 2003; pp. 73–117.
37. Han, J.; Fyfe, W.S.; Longstaffe, F.J.; Palmer, H.C.; Yan, F.H.; Mai, X.S. Pliocene-Pleistocene climatic change recorded in fluviolacustrine sediments in central China. *Palaeogeogr. Palaeoclimatol. Palaeoecol.* **1997**, *135*, 27–39. [CrossRef]
38. Liu, X.; Chen, J.; Maher, B.A.; Zhao, B.; Yue, W.; Sun, Q.; Chen, Z. Connection of the proto-Yangtze River to the East China Sea traced by sediment magnetic properties. *Geomorphology* **2018**, *303*, 162–171. [CrossRef]
39. Morton, A.C.; Hallsworth, C.R. Identifying provenance-specific features of detrital heavy mineral assemblages in sandstones. *Sediment. Geol.* **1994**, *90*, 241–256. [CrossRef]
40. Morton, A.C. Stability of detrital heavy minerals in Tertiary sandstones of the North Sea Basin. *Clay Miner.* **1984**, *19*, 287–308. [CrossRef]
41. Walderhaug, O.; Porten, K.W. Stability of detrital heavy minerals on the Norwegian continental shelf as a function of depth and temperature. *J. Sediment. Res.* **2007**, *77*, 992–1002. [CrossRef]
42. Yang, S.Y.; Wang, Z.; Guo, Y.; Li, C.X.; Cai, J.G. Heavy mineral compositions of the Changjiang (Yangtze River) sediments and their provenance-tracing implication. *J. Asian Earth Sci.* **2009**, *35*, 56–65. [CrossRef]

minerals

MDPI

Article

Evolution of the Upper Yellow River as Revealed by Changes in Heavy-Mineral and Geochemical (REE) Signatures of Fluvial Terraces (Lanzhou, China)

Zhao Wang [1,2], Haobo Zhang [1], Eduardo Garzanti [2], Junsheng Nie [1,3,*], Wenbin Peng [4], Sergio Andò [2], Xiaofei Hu [1], Baotian Pan [1] and Katharina Pfaff [5]

1 Key Laboratory of Western China's Environment Systems (Ministry of Education), College of Earth and Environmental Sciences, Lanzhou University, Lanzhou 730000, China; zhaowang14@lzu.edu.cn (Z.W.); hbzhang17@lzu.edu.cn (H.Z.); feixhu@lzu.edu.cn (X.H.); panbt@lzu.edu.cn (B.P.)

2 Laboratory for Provenance Studies, Department of Earth and Environmental Sciences, University of Milano-Bicocca, 20126 Milano, Italy; eduardo.garzanti@unimib.it (E.G.); sergio.ando@unimib.it (S.A.)

3 CAS Center for Excellence in Tibetan Plateau Earth Sciences, Chinese Academy of Sciences (CAS), Beijing 100101, China

4 School of Tourism and Resource Environment, Zaozhuang University, Zaozhuang 277160, China; pengwb14@163.com

5 Department of Geology and Geological Engineering, Colorado School of Mines, 1516 Illinois Street, Golden, CO 80401, USA; kpfaff@mines.edu

* Correspondence: jnie@lzu.edu.cn

Received: 5 September 2019; Accepted: 28 September 2019; Published: 30 September 2019

Abstract: Despite decades of study, the factors that controlled the formation and evolution of the upper reaches of the Yellow River, including uplift of the northeastern Tibetan Plateau, Pliocene-Pleistocene climate change, and autogenetic processes are still poorly constrained. The stratigraphic record of such paleogeographic evolution is recorded in the sequence of nine terraces formed during progressive incision of the Yellow River in the last 1.7 Ma. This article investigates in detail for sediment provenance in terraces of the Lanzhou area, based on heavy-mineral and geochemical (REE) signatures. Two main provenance changes are identified, pointing each to a major paleogeographic reorganization coupled with expansion of the upper Yellow River catchment and enhanced sediment fluxes. The first change took place between the deposition of terrace T9 (formed around 1.7 Ma) and terrace T8 (formed around 1.5 Ma), when rapid fluvial incision point to tectonic control and active uplift of northeastern Tibetan Plateau. The second change took place between deposition of terrace T4 (formed around 0.86 Ma) and terrace T3 (formed around 0.14 Ma), during a period of low incision rates and notably enhanced sediment fluxes as a response to enhanced East Asian Summer Monsoon and consequently increased precipitations, pointing instead chiefly to climatic control.

Keywords: provenance analysis; tectonic versus climatic control; early-middle Pleistocene transition; Yellow River terraces; Lanzhou (northern China)

1. Introduction

The Yellow River is the sixth longest river and carries the highest sediment load in the world [1,2]. The upper course of the Yellow River drains the central and northeastern Tibetan Plateau, and reconstructing its origin and early history is key to understanding the relationship between fluvial incision, climate variation, and basement uplift [3–9]. In the northeastern Tibetan Plateau, the Yellow River is deeply incised into a series of intermontane basins and tectonic ranges. The valley contains a series of fluvial terraces, which preserve the record of successive stages of plateau uplift and climatic change [10]. Based on the chronological sequence of the terraces, the Yellow River system was interpreted to have been formed since the middle Pleistocene as a response either to the accelerated tectonic uplift of the northeastern Tibetan Plateau [3,4,6,7,10,11], or to climatically-driven expansion of lake systems breaching topographic barriers [12–14].

Previous provenance studies of Yellow River sediments emphasized mainly the spatial differentiation among upper, middle, and lower modern fluvial reaches using zircon and heavy-mineral data [9,15–20]. However, there are few studies using heavy-mineral data to study paleo-drainage changes of the upper Yellow River on time scale. Zircon U-Pb geochronology was used to infer provenance of the upper Yellow River terraces in the Lanzhou area [18,21], but based on a limited number of grains (~100) per sample. In the complex northeastern Tibetan Plateau area, a considerably higher number of valid zircon ages need to be obtained (≥250 per sample) [22,23] in order to reliably identify subtle provenance differences in the dataset.

To improve our understanding of the evolution of the upper Yellow River catchment and to explain the role of several interconnected factors—including regional topography, tectonic deformation, and climatic change—we have carried out detailed provenance analysis of the Yellow River terraces well exposed in the Lanzhou area based on heavy-mineral and geochemical data. Heavy mineral compositions are often used to infer sedimentary provenance. Many heavy mineral species are found in sediment or sedimentary rocks, many of which are diagnostic sedimentary source indicators [24]. Differences in heavy mineral assemblages can often be used to differentiate different sediment transport routes [25]. Rare earth elements (REE) are generally regarded as useful in identifying the origin and formation mechanism of source rocks [26] and have been increasingly used in recent years as provenance tracers, because REE are less fractionated during the progress of weathering, transport, and sedimentation [27–31]. In order to increase throughput, we relied on the QEMSCAN (Quantitative Evaluation of Minerals by Scanning Electron Microscopy) method for heavy-mineral analyses, whereas geochemical signatures were determined by LA-ICP-MS (laser ablation and Inductively coupled plasma mass spectrometry).

2. Yellow River Terraces in the Lanzhou Area

The Lanzhou Basin, located at the northeastern edge of the actively uplifting Tibetan Plateau, is part of the larger Longzhong Basin crossed by the upper Yellow River [32,33] (Figure 1) and including deposits of Late Cretaceous to Cenozoic age [34]. The Lanzhou basin, a graben-like structure 5–10 km wide and extending over a length of 40 km from Hutouya in the west-northwest to Sangyuanxia in the east-southeast, and it is divided into a western sub-basin delimited to the north by the Jinchengguan fault and an eastern sub-basin delimited by the Baitashan and Leitanhe fault in the west and by the Gaolan Mountains in the south [35] (Figure 1b).

Along the northern margin of the eastern Lanzhou sub-basin, the stepwise incision of the Yellow River has formed nine terraces (from the oldest T9 at the top to the youngest T1 at the base) (Figure 1a,b). Terraces T9 to T2 are strath terraces covered by loess-paleosol of different thickness and beveled across either Cretaceous-Cenozoic sandstone or Precambrian basement [10,11] (Figure 1c). The highest terrace T9, studied in the Yaogou section, lies 333 m above the modern river, is based on red Cenozoic sandstone, and consists of 10 m of fluvial gravel overlain by ~22 m of fluvial sand and clay [36–38]. Terrace T8, studied in the Jiuzhoutai section, lies 100 m below T9, is also based on red Cenozoic sandstone, and consists of 2–3 m of fluvial gravel [11]. Terrace T7, studied in the Dunwashan section, lies 140 m above the modern river and consists of 6 m of fluvial gravel overlain by 12 m of overbank deposits. Terrace T6, exposed along the northern margin of the eastern Lanzhou basin and studied in the Dalanggou section, lies 127 m above the modern river and consists of 3–4 m of fluvial gravel overlain by 18 m of fluvial sand. Terrace T5, exposed in a narrow strip between the Xiaoshagou and Xiaogou sections, lies 108 m above the modern river consists of 3–4 m of fluvial gravel overlain by ~11 m of interchannel sand. Terrace T4, exposed almost along the entire length of the Lanzhou basin and studied in the Zaoshugou section, lies 99 m above the modern river and consists of 4–6 m of fluvial gravel overlain by 6–10 m of fluvial sand. The most widely exposed and preserved best terrace T3 lies

60 m above the modern river and comprises of 5–7 m of fluvial gravel overlain by 10 m of fluvial sand in the Baitashan section. The lowest strath terrace T2 lies 23 m above the modern river and consists of 5 m of fluvial gravel and ~10 m of fluvial sand in the Luoguogou section. Terrace T1, consisting of fluvial gravel overlain by fluvial silt, is not studied here because it is poorly exposed and mostly occupied by buildings of Lanzhou city. Based on paleomagnetic and optically-stimulated-luminescence (OSL) dating of the base of loess-paleosol sequences, the ages of these terraces have been determined as 0.05 Ma (T2), 0.14 Ma (T3), 0.86 Ma (T4), 0.96 Ma (T5), 1.05 Ma (T6), 1.24 Ma (T7), 1.5 Ma (T8), and 1.7 Ma (T9) [10,11,38].

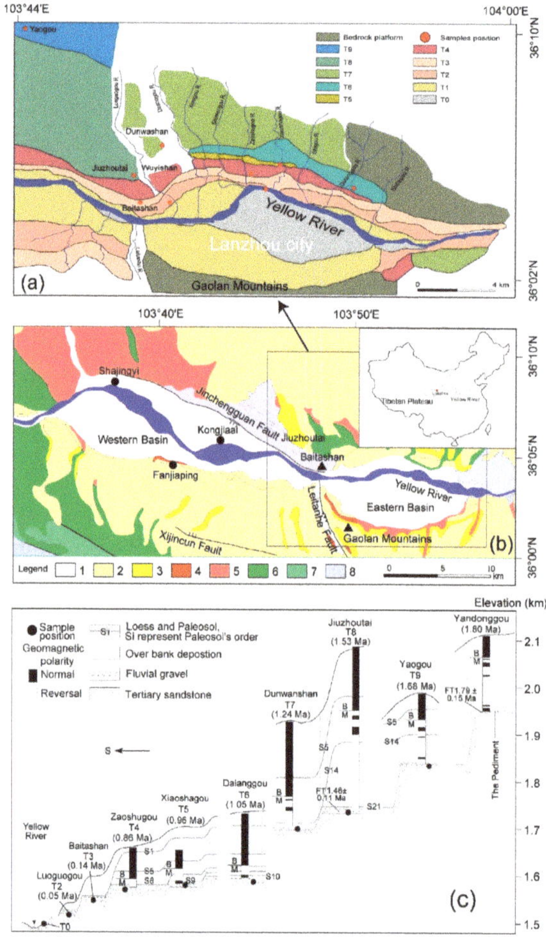

Figure 1. Yellow River terraces in the Lanzhou area. (**a**) Terraces distribution and sample location (modified from [10,38]). (**b**) Geological map of the Lanzhou area (modified from [19,21]). Geological units: 1: Holocene fluvial sediments; 2–3: Pleistocene loess-paleosol sequence; 4: Wuquan fluvial gravel formation; 5: Tertiary sandstone and glutenite; 6: Cretaceous purplish-red sandstone and greyish green siltstone; 7: Ordovician greyish green andesite and meta-andesite; 8: Precambrian basement. (**c**) Syntheses of stratigraphic and chronological data from loess-paleosol sequences covering terrace surfaces. The terraces' ages were obtained by measuring the magnetostratigraphy (T9–T4) and optically stimulated luminescence (OSL, T3–T2) dating of loess-paleosol sequences covering these terraces (modified from [10,38]).

3. Sampling and Methods

Provenance analysis was carried out on eight sand samples collected in the eastern Lanzhou basin from alluvial terraces T9 to T2, and on one modern sand sample (T0) collected from an active bar on the Yellow River bed at Lanzhou (Figure 1a).

From each sample, heavy minerals were separated using the dense liquid tribromomethane (density 2.89 g/cm3). Heavy-mineral separates were analyzed by QEMSCAN at Colorado School of Mines (Golden, CO, USA), following the procedure described in detail in [39–41]. From 290 to 398 heavy-mineral grains were counted in each sample. For composite grains including more than one mineral, the QEMSCAN determines the surface area percentage of each (Figure 2), thus providing full information on all components.

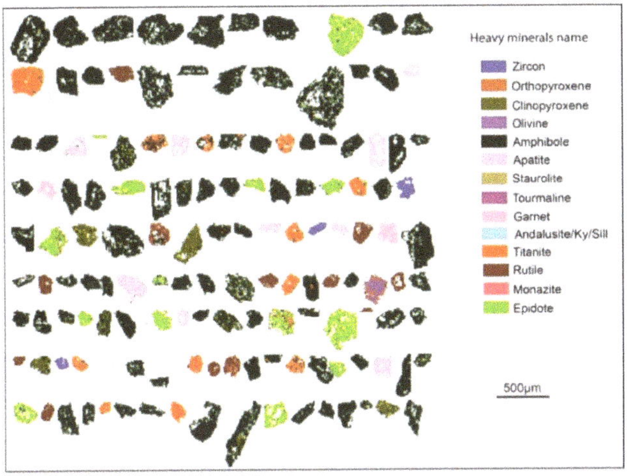

Figure 2. QEMSCAN image of heavy minerals in the modern Yellow River sample T0. Ky = Kyanite, Sill = Sillimanite.

The concentration of major and trace elements, including REE (rare earth elements) are widely considered as useful tracers of provenance [42–46]. Geochemical analyses were carried out on a 30–40 mg aliquot of each bulk sample crushed in an agate mortar, weighed into an airtight Teflon decomposer, and moistened with a little extra-pure water. Next, we added 1 mL HNO_3 and 1 mL HF in Teflon containers left in ultrasonic bath for over 20 min, and the samples were steamed nearly dry on a heating plate at 150 °C After repeating the operation three times, the residues placed in a stainless-steel pot with 2 mL HNO_3 and 3 mL extra-pure water were heated in oven at ~150 °C for over 24 h. The concentration of chemical elements was determined by LA-ICP-MS using Agilent 7700X and Photon Machine Analyte 193nm at Key Laboratory of Mineral Resources in Western China, Lanzhou University (Lanzhou, China).

Grain-size analyses on bulk samples were carried out by Laser Diffraction Particle Size Analyzer at the Key Laboratory of Western China's Environment Systems (Ministry of Education) of Lanzhou University. All samples were air-dried and analyzed by a Mastersizer 2000 produced by Malvern Instruments Ltd. (Malvern, UK) with Hydro2000G automatic sampler. The measuring range is from 0.02 to 2000 μm, and the median particle size D50 has an accuracy of ±1%.

4. Results

Different heavy-mineral assemblages in superposed Yellow River terraces of the Lanzhou area, together with varying REE patterns, indicate major provenance changes through time.

4.1. Heavy-Mineral Assemblages

Transparent heavy-mineral suites in modern Yellow River sand and in Pleistocene to Holocene terraces are invariably dominated by amphibole (46–58%) and epidote (22–29%) (Figure 3a, Table 1). The abundance of amphibole indicates prominent contributions from basement rocks [24,47], with the amphibole series forming in a wide range of temperature and pressure conditions in igneous and metamorphic rocks [48]. A ratio of amphibole versus total heavy minerals has been used to reflect past changes in the origin of loess from the central Chinese Loess Plateau [49]. Because more than 73% of the heavy-mineral fraction is composed of amphibole plus epidote in Yellow River terraces, we here use the ratio between these two dominant minerals as a tracer of provenance change. Amphibole accounts for less than 50% of the assemblage in terraces T8 and T3–T2 as well as in T0, whereas epidote displays peak abundance in terraces T3 and T2 where the amphibole/epidote ratio is close to 1 (1.01–1.08 versus 1.5–2.7 in older terraces).

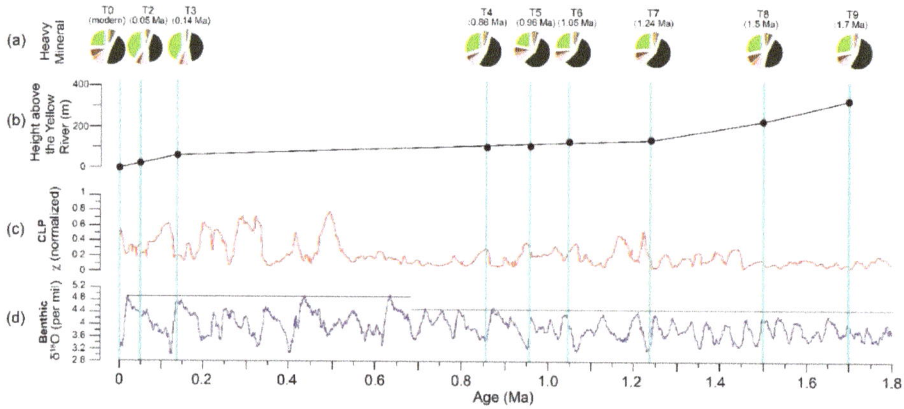

Figure 3. Comparison between heavy-mineral spectra from Yellow River terraces and the paleoclimatic record. (**a**) Heavy-mineral data from Yellow River terraces in the Lanzhou area. (**b**) Correlation between formation age of Yellow River terraces and height above the river (mod. from [10,38]). The less steep slope between T7 and T3 indicates lower rates of rock uplift and river incision at that time. (**c**) Variations of the East Asian summer monsoon during the past 1.8 Ma, based on the stacked magnetic susceptibility (χ) records on the Chinese Loess Plateau (CLP) [50]. (**d**) Variations of global ice volume during the past 1.8 Ma, based on the benthic oxygen-isotope record [51]. Mineral color-code as in Figure 2.

Table 1. Heavy-mineral assemblages in samples from the Yellow River terraces in the Lanzhou area in volume% (data from [52])

Sample Number	Age	Zircon	Opx #	Cpx #	Olivine	Amphibole	Apatite	Andalusite /Ky/Sill#	Staurolite	Tourmaline	Garnet	Titanite	Rutile	Monazite	Epidote
T9 *	1.70 Ma	0	2	1	0	55	3	2	0	0	2	2	5	0	28
T8	1.50 Ma	0	2	5	0	46	4	0	0	0	2	3	8	0	29
T7	1.24 Ma	0	2	3	1	57	1	1	0	0	2	2	6	0	26
T6	1.05 Ma	1	2	3	1	58	1	0	0	0	4	2	4	0	25
T5	0.96 Ma	0	1	3	1	58	2	0	0	0	3	2	6	0	22
T4	0.86 Ma	0	2	3	0	54	3	1	0	0	1	1	4	0	30
T3	0.14 Ma	0	0	2	0	44	2	0	0	0	3	1	3	0	44
T2	0.05 Ma	0	1	3	0	44	1	0	0	0	3	1	4	0	41
T0	modern	1	1	6	0	47	3	0	0	0	4	4	8	0	26

* The name Tx stands for Yellow River terrace sample where x is a terrace number, T0 is the modern Yellow River bed sample. # Opx = Orthopyroxene; Cpx = Clinopyroxene; Ky = Kyanite; Sill = Sillimanite.

4.2. Rare Earth Elements

The total REE (∑REE) content in Yellow River terraces varies widely from 46 to 116 ppm. In chondrite-normalized REE diagrams (Figure 4, Table 2), all samples show enrichment in light rare earth elements (LREE, 43–108 ppm) and depletion in heavy rare earth elements (HREE, 3.4–8.2 ppm), with negative Eu (Eu/Eu* = 0.67–0.84) and Ce (Ce/Ce* = 0.86–0.90) anomalies.

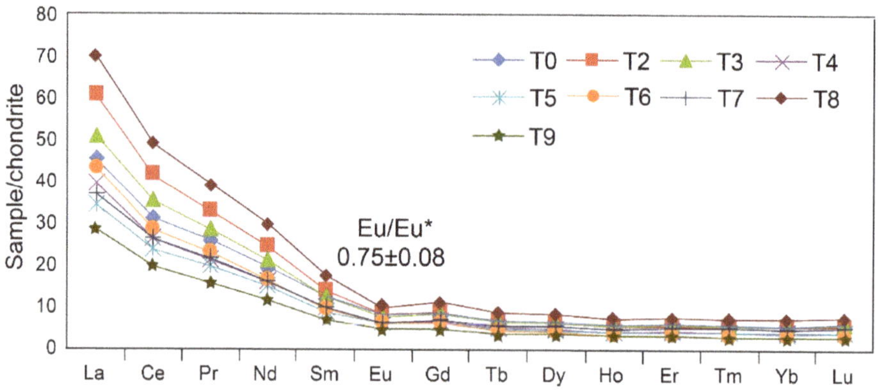

Figure 4. Chondrite-normalized REE patterns for Yellow River terraces of the Lanzhou area.

Table 2. Rare earth elements concentrations (ppm) in sediments from the Yellow River terraces in the Lanzhou area

Sample Number	Age	La	Ce	Pr	Nd	Sm	Eu	Gd	Tb	Dy	Ho	Er	Tm	Yb	Lu
T9 *	1.70 Ma	10.470	18.676	2.140	8.260	1.580	0.383	1.402	0.210	1.306	0.250	0.759	0.100	0.677	0.105
T8	1.50 Ma	25.715	47.243	5.410	21.115	3.985	0.832	3.412	0.497	3.068	0.597	1.839	0.254	1.744	0.280
T7	1.24 Ma	13.480	24.928	2.934	11.420	2.219	0.533	2.031	0.312	2.024	0.396	1.211	0.172	1.175	0.194
T6	1.05 Ma	15.838	27.250	3.112	11.699	2.141	0.519	1.789	0.238	1.389	0.257	0.744	0.103	0.687	0.105
T5	0.96 Ma	12.582	22.681	2.666	10.430	2.025	0.515	1.843	0.269	1.664	0.316	0.966	0.132	0.895	0.144
T4	0.86 Ma	14.512	24.976	2.892	11.228	2.202	0.533	1.939	0.281	1.733	0.331	1.016	0.138	0.937	0.150
T3	0.14 Ma	18.592	33.793	3.882	14.859	2.853	0.643	2.471	0.371	2.252	0.442	1.386	0.189	1.318	0.213
T2	0.05 Ma	22.269	39.731	4.508	17.364	3.184	0.731	2.630	0.375	2.314	0.450	1.314	0.189	1.268	0.203
T0	modern	16.664	29.551	3.470	13.511	2.640	0.622	2.320	0.334	2.048	0.393	1.193	0.164	1.081	0.188
chondrite		0.367	0.957	0.137	0.711	0.231	0.087	0.306	0.058	0.381	0.0851	0.249	0.0356	0.248	0.0381

* The name Tx stands for Yellow River terrace sample where x is a terrace number, T0 is the modern Yellow River bed sample.

REE curves are all similar, with minimum values for the oldest terrace T9 and maximum values for terrace T8. REE contents are relatively high for both terraces T3 and T2 (Figure 4). For the interpretation of REE concentration; however, it must be kept in mind that REE are preferentially hosted in ultra-dense minerals such as monazite, and their abundance in sediments is consequently strongly affected by hydraulic-sorting processes [53].

5. Discussion

The mineralogical and geochemical fingerprints of siliciclastic sediments are mainly controlled by the lithology of source rocks, with superposed effects associated with climatic conditions in the catchment and hydraulic sorting during erosion, transport, and deposition [9,54].

Most of the river sediments have similar and uniform REE patterns displaying enrichment in light REE [55,56] which is inherited from source rocks, confirming that the sediments are not anthropogenically polluted [57,58]. Heavy-mineral assemblages in Yellow River terraces and geochemical data suggest that changes in sediment composition took place between terraces T9 and T8, and between terraces T4 and T3. The provenance changes cannot be discounted as a grain-size effect because T3 and T2 have similar composition despite their markedly different grain-size distributions (Figure 5). This is confirmed by the principal-component diagram shown in Figure 6, where terraces T3 and T2 plot on the right side of the PC1 component, accounting for ~82% of data variance. Principal-component analysis also suggest that terrace T8 and modern fluvial sand T0 are distinguished from other samples on PC2, accounting for ~15% of data variance. This reflects at least in part the slightly lower content in hornblende relative to samples T9 and T7–T4. Some studies in Europe have shown that higher uplift rates may decrease the preservation of climate-controlled fluvial terraces, whereas moderate to low uplift rates will enable their preservation [59–61]. Accurate field investigations have failed to find any Yellow River terrace in the Lanzhou area formed between 0.86 Ma and 0.14 Ma [10]. However, it is noteworthy that this provenance change took place between 0.86 and 0.14 Ma, a long period of decreased tectonic activity when river terraces were not formed— or were formed but were not preserved—in the Lanzhou basin [10] (Figure 3b). For this reason, we could not more precisely constrain the age of such a compositional change.

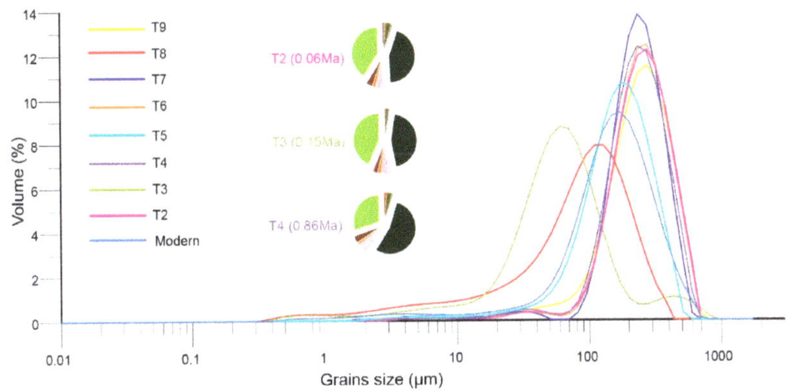

Figure 5. Grain-size curves of sediment samples from Yellow River terraces of the Lanzhou area, and heavy-mineral spectra for terraces T2, T3, and T4. Mineral color-code as in Figure 2.

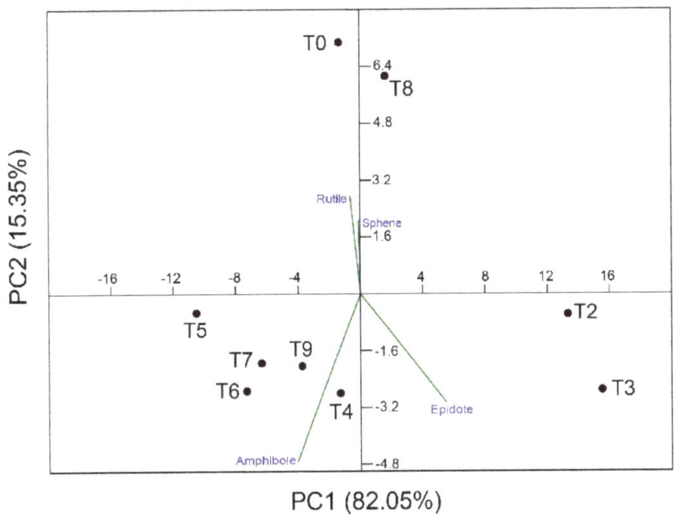

Figure 6. Principal component analysis of heavy-mineral assemblages in Yellow River terraces of the Lanzhou area.

Before 1.45 Ma (terraces T8 and T9), the East Asian summer monsoon was relatively stable, as documented by magnetic-susceptibility data on Chinese loess [50] (Figure 3c), and there is no evidence of large variations in ice volume [51] (Figure 3d). This speaks against climate change as a main cause for the change in heavy-mineral assemblages and geochemical signatures observed between terraces T9 and T8. In this time window—i.e., between 1.7 and 1.5 Ma—the Yellow River rapidly incised its talweg by as much as 100 m. If this was not triggered by climatic change, then active tectonic uplift remains as the most plausible cause [62].

Between 1.24 Ma (terrace T7) and 0.86 Ma (terrace T4), the Yellow River course was shorter, its drainage basin smaller, and its sediment load notably less than today [12,13,63]. The timing of major paleogeographic change took place during, and in the final part of the so-called "mid-Pleistocene climatic revolution" (or early-middle Pleistocene transition; 1.4-0.424 Ma), characterized by a progressive increase in the amplitude of climatic oscillations [64,65]. At that time, the East Asian summer monsoon intensified and the amplitude of ice-volume variation between glacial and interglacial stages increased, as documented by magnetic-susceptibility data on Chinese loess [50,51] (Figure 3c,d). Increased glacial erosion may have enhanced sediment fluxes [66–68] at that time, when a phase of decreased rock uplift and deformation in the Lanzhou area is suggested by the decrease in incision rates as highlighted in Figure 3b. Tectonic control is unlikely for this time period [62], during which heavy-mineral and geochemical signatures have remained remarkably constant.

After 0.86 Ma, the change in heavy-mineral assemblages and geochemical fingerprints observed between terraces T4 (0.86 Ma) and T3 (0.14 Ma) may be explained with the stepwise addition of large areas to the upper reaches of the Yellow River by headward erosion [12,63]. This scenario is consistent with the increase in precipitations as a consequence of enhanced East Asian Summer Monsoon [50] and with the increased amplitude in ice-volume variation [51]. Between ~0.5 and 0.1 Ma, Yellow River sediment fluxes markedly increased, while the Gonghe basin, the Tongde basin, and the Jungong basin were successively incorporated in the Yellow River drainage system [4,12,13].

6. Conclusions

The study of heavy-mineral assemblages and REE patterns characterizing Yellow River terraces in the Lanzhou area reveals two notable provenance changes, from ~1.7 Ma (terrace T9) to ~1.5 Ma (terrace T8) and from ~0.86 Ma (terrace T4) to ~0.14 Ma (terrace T3). Two paleogeographic reorganizations took place during those time intervals, both associated with phases of expansion of the upper Yellow River catchment and consequently increased sediment fluxes.

The height difference of ~100 m between terraces T9 and T8 indicates rapid incision during the early Pleistocene (early Calabrian), suggesting that the first shift was triggered tectonically by a phase of rapid uplift of the northeastern Tibetan plateau. The second reorganization took place during the final part of the middle Pleistocene transition (late Calabrian), when the East Asian summer monsoon intensified and the amplitude of ice-volume variation between glacial and interglacial stages increased. River incision was much slower at this time, indicating that this second, major reorganization was chiefly climatically controlled.

Author Contributions: Writing—Original Draft Preparation by Z.W. and H.Z.; Writing—Review and Editing by E.G., J.N., and S.A.; Data Analysis and Figures by Z.W., H.Z., and W.P.; Sampling by W.P., X.H., B.P., and J.N.; Experiments by Z.W., W.P., H.Z., and K.P.

Funding: This work was funded by the Second Tibetan Plateau Scientific Expedition (grant no. 2019QZKK0704), the National Natural Science Foundation of China (grant no. 41761144063, 41672157, 41807427, 41907073, 41422204), the Open Foundation of MOE Key Laboratory of Western China's Environmental System, Lanzhou University (grant no. lzujbky-2018-kb01, lzujbky-2019-kb01), the Open Project of Key Laboratory for Magnetism and Magnetic Materials of the Ministry of Education, Lanzhou University (grant no. LZUMMM2018007), the State Scholarship Fund organized by China Scholarship Council (grant no. 201806180066), and by MIUR–Dipartimenti di Eccellenza 2018-2022, Department of Earth and Environmental Sciences, University of Milano-Bicocca.

Acknowledgments: Heartfelt thanks to Assistant Editor and three reviewers, whose comments helped us to improve on the paper significantly.

Conflicts of Interest: The authors declare no conflict of interest.

References

1. Milliman, J.D.; Meade, R.H. World-wide delivery of river sediment to the oceans. *J. Geol.* **1983**, *91*, 1–21. [CrossRef]
2. Ren, M.E. Sediment discharge of the Yellow River, China: Past, present and future-A synthesis. *Acta Oceanol. Sin.* **2015**, *34*, 1–8. [CrossRef]
3. Li, J.J. The environmental effects of the uplift of the Qinghai-Xizang Plateau. *Quat. Sci. Rev.* **1991**, *10*, 479–483. [CrossRef]
4. Harkins, N.; Kirby, E.; Heimsath, A.; Robinson, R.; Reiser, U. Transient fluvial incision in the headwaters of the Yellow River, northeastern Tibet, China. *J. Geophys. Res.* **2007**, *112*, F03S04. [CrossRef]
5. Lease, R.O.; Burbank, D.W.; Gehrels, G.E.; Wang, Z.C.; Yuan, D.Y. Signatures of mountain building: Detrital zircon U/Pb ages from northeastern Tibet. *Geology* **2007**, *35*, 239–242. [CrossRef]
6. Hu, Z.B.; Pan, B.T.; Guo, L.Y.; Vandenbergheb, J.; Liu, X.P.; Wang, J.P.; Fan, Y.L.; Mao, J.W.; Gao, H.S. Rapid fluvial incision and headward erosion by the Yellow River along the Jinshaan gorge during the past 1.2 Ma as a result of tectonic extension. *Quat. Sci. Rev.* **2016**, *133*, 1–14. [CrossRef]
7. Hu, Z.B.; Pan, B.T.; Bridgland, D.; Vandenberghe, J.; Guo, L.Y.; Fan, Y.L.; Westaway, R. The linking of the upper-middle and lower reaches of the Yellow River as a result of fluvial entrenchment. *Quat. Sci. Rev.* **2017**, *166*, 324–338. [CrossRef]
8. Wang, W.T.; Zhang, P.Z.; Liu, C.K.; Zheng, D.W.; Yu, J.X.; Zheng, W.J.; Wang, Y.Z.; Zhang, H.P.; Chen, X.Y. Pulsed growth of the West Qinling at ~30 Ma in northeastern Tibet: Evidence from Lanzhou Basin magneostratigraphy and provenance. *J. Geophys. Res.-Solid Earth* **2016**, *121*, 7754–7774. [CrossRef]
9. Pang, H.L.; Pan, B.T.; Garzanti, E.; Gao, H.S.; Zhao, X.; Chen, D.B. Mineralogy and geochemistry of modern Yellow River sediments: Implications for wreathing and provenance. *Chem. Geol.* **2018**, *488*, 76–86. [CrossRef]
10. Pan, B.T.; Su, H.; Hu, Z.B.; Hu, X.F.; Gao, H.S.; Li, J.J.; Kirby, E. Evaluating the role of climate and tectonics during non-steady incision of the Yellow River: Evidence from a 1.24 Ma terrace record near Lanzhou, China. *Quat. Sci. Rev.* **2009**, *28*, 3281–3290. [CrossRef]
11. Pan, B.T. Problems on the Development of the Yellow River and Uplift of Qinghai-Xizang Plateau. Ph.D. Thesis, Lanzhou University, Lanzhou, China, 1991. (In Chinese).
12. Craddock, W.H.; Kirby, E.; Harkins, W.N.; Zhang, H.P.; Shi, X.H.; Liu, J.H. Rapid fluvial incision along the Yellow River during headward basin integration. *Nat. Geosci.* **2010**, *3*, 209–213. [CrossRef]
13. Perrineau, A.; Van Der Woerd, J.; Gaudemer, Y.; Zeng-Liu, J.; Pik, P.; Tapponnier, P.; Thuizat, R.; Zheng, R.Z. Incision rate of the Yellow River in Northeastern Tibet constrained by ^{10}Be and ^{26}Al cosmogenic isotope dating of fluvial terraces: Implications for catchment evolution and plateau building. *Geol. Soc. Lond. Spec. Publ.* **2011**, *353*, 189–219. [CrossRef]
14. Kong, P.; Jia, J.; Zheng, Y. Time constraints for the Yellow River traversing the Sanmen Gorge. *Geochem. Geophy. Geosyst.* **2014**, *15*, 395–407. [CrossRef]
15. Zhang, J.; Huang, W.W.; Liu, M.G.; Zhou, Q. Drainage basin weathering and major element transport of two larger Chinese river (Huanghe and Changjiang). *J. Geophy. Res.* **1990**, *95*, 13277–13288. [CrossRef]
16. Yang, S.Y.; Li, C.X. Research progress in REE trace for sediment source. *Adv. Earth Sci.* **1999**, *14*, 164–167. (In Chinese)
17. Yang, S.Y.; Li, C.X.; Jung, H.S.; Lee, H.J. Discrimination of geochemical compositions between the Changjiang and the Huanghe sediments and its application for the identification of sediment source in the Jiangsu coastal plain, China. *Mar. Geol.* **2002**, *186*, 229–241. [CrossRef]
18. Nie, J.S.; Stevens, T.; Rittner, M.; Stockli, D.; Garzanti, E.; Limonta, M.; Bird, A.; Andò, S.; Vermeesch, P.; Saylor, J.; et al. Loess Plateau storage of northeastern Tibetan Plateau-derived Yellow River sediment. *Nat. Commun.* **2015**, *6*, 8511. [CrossRef] [PubMed]
19. Pan, B.T.; Pang, H.L.; Gao, H.S.; Garzanti, E.; Zou, Y.; Liu, X.P.; Li, F.Q.; Jia, Y.X. Heavy-mineral analysis and provenance of Yellow River sediments around the China Loess Plateau. *J. Asian Earth Sci.* **2016**, *127*, 1–11. [CrossRef]
20. Jin, B.F.; Wang, M.Y.; Yue, W.; Zhang, L.N.; Wang, Y.J. Heavy Mineral Variability in the Yellow River Sediments as Determined by the Multiple-Window Strategy. *Minerals* **2019**, *9*, 85. [CrossRef]

21. Guo, B.H.; Liu, S.P.; Peng, T.J.; Ma, Z.H.; Feng, Z.T.; Li, M.; Li, X.M.; Li, J.J.; Song, C.H.; Zhao, Z.J.; et al. Late Pliocene establishment of exorheic drainage in the northeastern Tibetan Plateau as evidenced by the Wuquan Formation in the Lanzhou Basin. *Geomorphology* **2018**, *303*, 271–283. [CrossRef]
22. Che, X.D.; Li, G.J. Binary sources of loess on the Chinese Loess Plateau revealed by U-Pb ages of zircon. *Quat. Res.* **2013**, *80*, 545–551. [CrossRef]
23. Licht, A.; Pullen, A.; Kapp, P.; Abell, J.; Giesler, N. Eolian cannibalism: Reworked loess and fluvial sediment as the main sources of the Chinese Loess Plateau. *Geol. Soc. Am. Bull.* **2016**, *128*, 944–956. [CrossRef]
24. Garzanti, E.; Andò, S. Plate tectonics and heavy-mineral suites of modern sands. *Dev. Sedimentol.* **2007**, *58*, 741–763. [CrossRef]
25. Morton, A.C.; Hallsworth, C. Stability of detrital heavy minerals during burial diagenesis. *Dev. Sedimentol.* **2007**, *58*, 215–245. [CrossRef]
26. Taylor, S.R.; McLennan, S.M. *The Continental Crust: Its Composition and Evolution*; Blackwell Scientific Publications: Oxford, UK, 1985; p. 312.
27. Cullers, R.L. Implication of elemental concentrations for provenance, redox conditions, and metamorphic studies of shales and limestones near Pueblo, CO, USA. *Chem. Geol.* **2002**, *191*, 305–327. [CrossRef]
28. Moreno, T.; Querol, X.; Castillo, S.; Alastuey, A.; Cuevas, E.; Herrmann, L.; Mounkaila, M.; Elvira, J.; Gibbons, W. Geochemical variations in aeolian mineral particles from the Sahara-Sahel Dust Corridor. *Chemosphere* **2006**, *65*, 261–270. [CrossRef] [PubMed]
29. Roy, P.D.; Smykatz-Kloss, W. REE geochemistry of the recent playa sediments from the Thar desert, India: An implication to play a sediment provenance. *Chemie der Erde-Geochemistry* **2007**, *67*, 55–68. [CrossRef]
30. Castillo, S.; Moreno, T.; Querol, S.; Alastuey, A.; Cuevas, E.; Herrmann, L.; Mounkaila, M.; Gibbons, W. Trace element variation in size-fractionated African desert dusts. *J. Arid Environ.* **2008**, *72*, 1034–1045. [CrossRef]
31. Kasper-Zublillaga, J.J.; Acevedo-Vargas, B.; Bermea, O.M.; Zamora, G.O. Rare earth elements of the Altar Desert dune and coastal sands, Northwestern Mexico. *Chemie der Erde-Geochemistry* **2008**, *68*, 45–59. [CrossRef]
32. Wang, P.; Jiang, H.C.; Yuan, D.Y.; Liu, X.W.; Zhang, B. Optically stimulated luminescence dating of sediments from the Yellow River terraces in Lanzhou: Tectonic and climatic implications. *Quat. Geochronol.* **2010**, *5*, 181–186. [CrossRef]
33. Miao, Y.F.; Wu, F.L.; Herrmann, M.; Yan, X.L.; Meng, Q.Q. Late early Oligocene East Asian summer monsoon in the NE Tibetan Plateau: Evidence from a palynological record from the Lanzhou Basin China. *J. Asian Earth Sci.* **2013**, *75*, 46–57. [CrossRef]
34. Yue, L.P.; Heller, F.; Qiu, Z.X.; Zhang, L.; Jie, G.P.; Qiu, Z.D.; Zhang, Y.X. Magnetostratigraphy and paeleo-environmental record of Tertiary deposits of Lanzhou Basin. *Chin. Sci. Bull.* **2001**, *46*, 770–773. [CrossRef]
35. Zhang, L.Y. Preliminary study on Quaternary geology of Lanzhou valley basin. *J. Lanzhou Univ.* **1962**, *2*, 89–98. (In Chinese)
36. Chen, J.F. Study on Landform Development Model at Longzhong Basin since Late Pliocene. Master's Thesis, Lanzhou University, Lanzhou, China, 2010. (In Chinese).
37. Zhu, J.J.; Zhong, W.; Li, J.J.; Cao, J.X.; Wang, J.L. The oldest eolian loess deposition in the Longxi Basin-Yandonggou profile in Lanzhou. *Sci. Geogr. Sin.* **1996**, *16*, 365–369. (In Chinese)
38. Hu, X.F.; Kirby, E.; Pan, B.T.; Granger, D.E.; Su, H. Cosmogenic burial ages reveal sediment reservoir dynamics along the Yellow River, China. *Geology* **2011**, *39*, 839–842. [CrossRef]
39. Nie, J.S.; Peng, W.B.; Pfaff, K.; Möller, A.; Garzanti, E.; Andò, S.; Stevens, T.; Bird, A.; Chang, H.; Song, Y.G.; et al. Controlling factors on heavy mineral assemblages in Chinese loess and Red Clay. *Palaeogeogr. Palaeoclimatol. Palaeoecol.* **2013**, *381–382*, 110–118. [CrossRef]
40. Nie, J.S.; Peng, W.B. Automated SEM-EDS heavy mineral analysis reveals no provenance shift between glacial loess and interglacial paleosol on the Chinese Loess Plateau. *Aeolian Res.* **2014**, *13*, 71–75. [CrossRef]
41. Peng, W.B.; Wang, Z.; Song, Y.G.; Pfaff, K.; Luo, Z.; Nie, J.S.; Chen, W.H. A comparison of heavy mineral assemblage between the loess and the Red Clay sequences on the Chinese Loess Plateau. *Aeolian Res.* **2016**, *21*, 87–91. [CrossRef]
42. McLennan, S.M. Relationships between the trace element composition of sedimentary rocks and upper continental crust. *Geochem. Geophy. Geosyst.* **2001**, *2*, 2000GC000109. [CrossRef]
43. Yang, S.; Li, C.; Lee, C.B.; Na, T.K. REE geochemistry of suspended sediments from the rivers around the Yellow Sea and provenance indicators. *Chin. Sci. Bull.* **2003**, *48*, 1135–1139. [CrossRef]

44. Singh, P. Major, trace and REE geochemistry of the Ganga River sediments: Influence of provenance and sedimentary processes. *Chem. Geol.* **2009**, *266*, 251–264. [CrossRef]
45. Fleet, A.J. Aqueous and sedimentary geochemistry of the rare earth elements. In *Developments in Geochemistry*; Elsevier: Amsterdam, The Netherlands, 1984; Volume 2, pp. 343–374.
46. Garzanti, E.; Padoan, M.; Setti, M.; López-Galindo, A.; Villa, I.M. Provenance versus weathering control on the composition of tropical river mud (southern Africa). *Chem. Geol.* **2014**, *366*, 61–74. [CrossRef]
47. Garzanti, E.; Vezzoli, G.; Andò, S.; Castiglioni, G. Petrology of rifted-margin sand (Red Sea and Gulf of Aden, Yemen). *J. Geol.* **2001**, *109*, 277–297. [CrossRef]
48. Andò, S.; Morton, A.; Garzanti, E. Metamorphic grade of source rocks revealed by chemical fingerprints of detrital amphibole and garnet. *Geol. Soc. Lond. Spec. Publ.* **2014**, *386*, 351–371. [CrossRef]
49. He, T.; Liu, L.W.; Chen, Y.; Sheng, X.F.; Ji, J.F. A seven-million-year hornblende mineral record from the central Chinese Loess Plateau. *Sci. Rep.* **2017**, *7*, 2382. [CrossRef]
50. Sun, Y.B.; Clemens, S.C.; An, Z.S.; Yu, Z.W. Astronomical timescale and palaeoclimatic implication of stacked 3.6-Myr monsoon records from the Chinese Loess Plateau. *Quat. Sci. Rev.* **2006**, *25*, 33–48. [CrossRef]
51. Lisiecki, L.E.; Raymo, M.E. A Plio-Pleistocene stack of 57 globally distributed benthic $\delta^{18}O$ records. *Paleoceanography* **2005**, *20*, PA1003. [CrossRef]
52. Wang, Z.; Nie, J.S.; Wang, J.P.; Zhang, H.B.; Peng, W.B.; Garzanti, E.; Hu, X.F.; Stevens, T.; Pfaff, K.; Pan, B.T. Testing contrasting models of the formation of the upper Yellow River using heavy-mineral data from the Yinchuan Basin drill cores. *Geophys. Res. Lett.* **2019**, *46*. [CrossRef]
53. Garzanti, E.; Andò, S.; France-Lanord, C.; Vezzoli, G.; Censi, P.; Galy, V.; Najman, Y. Mineralogical and chemical variability of fluvial sediments: 1. Bedload sand (Ganga–Brahmaputra, Bangladesh). *Earth Planet. Sci. Lett.* **2010**, *299*, 368–381. [CrossRef]
54. Garzanti, E.; Andò, S. Heavy Minerals for Junior Woodchucks. *Minerals* **2019**, *9*, 148. [CrossRef]
55. Goldstein, S.J.; Jacobsen, S.B. Rare earth elements in river waters. *Earth Planet. Sci. Lett.* **1988**, *89*, 35–47. [CrossRef]
56. Sholkovitz, E.R. Chemical evolution of rare earth elements: Fractionation between colloidal and solution phases of filtered river water. *Earth Planet. Sci. Lett.* **1992**, *114*, 77–84. [CrossRef]
57. Yang, S.Y.; Jung, H.S.; Choi, M.S.; Li, C.X. The rare earth element compositions of the Changjiang (Yangtze) and Huanghe (Yellow) river sediments. *Earth Planet. Sci. Lett.* **2002**, *201*, 407–419. [CrossRef]
58. Romero-Freire, A.; Minguez, L.; Pelletier, M.; Cayer, A.; Caillet, C.; Devin, S.; Gross, E.M.; Guérold, F.; Pain-Devin, S.; Vignati, D.A.L.; et al. Assessment of baseline ecotoxicity of sediments from a prospective mining area enriched in light rare earth elements. *Sci. Total Environ.* **2018**, *612*, 831–839. [CrossRef] [PubMed]
59. Veldkamp, A.; Van Dijke, J.J. Simulation internal and external controls on fluvial terrace stratigraphy: A qualitative comparison with the Maas record. *Geomorphology* **2000**, *33*, 225–236. [CrossRef]
60. Gibbard, P.L.; Lewin, J. River incision and terrace formation in the late Cenozoic of Europe. *Tectonophysics* **2009**, *474*, 41–55. [CrossRef]
61. Winsemann, J.; Lang, J.; Böhner, U.; Polom, U.; Brandes, C.; Roskosch, J.; Glotzbach, C.; Frechen, M. Terrace styles and timing of terrace formation in the Weser and Leine valleys, northern Germany: Response of a fluvial system to climate change and glaciations. *Quat. Sci. Rev.* **2015**, *123*, 31–57. [CrossRef]
62. Li, J.J.; Fnag, X.M.; Van der Voo, R.; Zhu, J.J.; Nicocaill, C.M.; Ono, Y.; Pan, B.T.; Zhong, W.; Wang, J.L.; Sasaki, T.; et al. Magnetostratigraphic dating of river terraces: Rapid and intermittent incision by the Yellow River of the northeastern margin of the Tibetan Plateau during the Quaternary. *J. Geophys. Res.-Solid Earth* **1997**, *102*, 10121–10132. [CrossRef]
63. Zhang, H.P.; Zhang, P.Z.; Champagnac, J.D.; Molnar, P.; Anderson, R.S.; Kirby, E.; Craddock, W.H.; Liu, S.F. Pleistocene drainage reorganization driven by the isostatic response to deep incision into the northeastern Tibetan Plateau. *Geology* **2014**, *42*, 303–306. [CrossRef]
64. Ruddiman, W.F.; Raymo, M.; McIntyre, A. Matuyama 41,000-year cycles: North Atlantic Ocean and northern hemisphere ice sheets. *Earth Planet. Sci. Lett.* **1986**, *80*, 117–129. [CrossRef]
65. Head, M.J.; Gibbard, P.L. Early-Middle Pleistocene transitions: Linking terrestrial and marine realms. *Quat. Int.* **2015**, *389*, 7–46. [CrossRef]
66. McLennan, S.M. Weathering and global denudation. *J. Geol.* **1993**, *101*, 295–303. [CrossRef]

67. Zhou, S.Z.; Jijun, L.; Zhang, S.Q.; Zhao, J.D.; Cui, J.X. Quaternary glaciations in China. In *Developments in Quaternary Sciences*; Ehlers, J., Gibbard, P.L., Eds.; Elsevier: Amsterdam, The Netherlands, 2004; Volume 2, pp. 105–113.
68. Sun, J.M.; Zhu, X.K. Temporal variations in Pb isotopes and trace element concentrations within Chinese eolian deposits during the past 8 Ma: Implications for provenance change. *Earth Planet. Sci. Lett.* **2010**, *290*, 438–447. [CrossRef]

minerals

MDPI

Article

Provenance of Heavy Minerals: A Case Study from the WNW Portuguese Continental Margin

João Cascalho

Instituto D. Luiz and Departamento de Geologia, Faculdade de Ciências (Universidade de Lisboa), Edifício C6, Campo Grande, 1749-016 Lisboa, Portugal; jpcascalho@ciencias.ulisboa.pt

Received: 27 February 2019; Accepted: 10 June 2019; Published: 12 June 2019

Abstract: This work describes and interprets the presence of heavy minerals in the WNW Portuguese continental margin using a set of 78 bottom samples collected from three distinct areas of this margin: the Porto, Aveiro, and Nazaré canyon head areas. The main transparent heavy mineral assemblage (mineral grains with frequencies ≥1% identified under a petrographic microscope) is composed of amphibole, andalusite, tourmaline, biotite, garnet, staurolite, pyroxene, zircon, and apatite. The felsic igneous and metamorphic rock outcrops in the main Northern Portuguese river basins and the relict sedimentary continental shelf deposits explained the presence of most of these mineral grains (both considered as distal sources). However, the presence of pargasite, augite, diopside-hedenbergite, enstatite-ferrosilite, and forsterite in the Porto and Aveiro areas (minerals identified by electronic microprobe analysis) is probably related to the presence of an igneous basic source next to dolomitic limestones affected by thermal metamorphism. These geological formations are considered as local sources. The high concentration of biotite observed in the Nazaré area is the result of the selective transport of the most lamellar sand particles of this mineral.

Keywords: mineral grains composition; surface textures; sources; WNW Portuguese Continental Margin

---

1. Introduction

"Thus, we live in a universe primed for complexification: hydrogen atoms form stars, stars form the elements of the periodic table, those elements form planets, which in turn form minerals abundantly. Minerals catalyze the formation of biomolecules, which on Earth led to life. In this sweeping scenario, minerals represent but one inexorable step in the evolution of a cosmos that is learning to know itself." [1] (p.58)

The presence of heavy minerals in sedimentary deposits represents a detrital occurrence of either constituent or accessory rock-forming minerals. Consequently, they have been used as an important tool to understand sedimentary processes and to identify operating factors that control depositional environments [2]. For example, on continental margins and on littoral environments, heavy mineral analysis is often used to diagnose sediment sources, to trace sand transport paths, to understand grain sorting processes, to deduce energy levels of transporting agents, and to interpret global sediment provenance in terms of sedimentary cycles [3–6]. Normally the presence of these minerals is easier to interpret on terrestrial and coastal environments than on continental shelf areas. This is because the direct relationship between sourcing/distributor processes and sedimentary deposits is more straightforward in the accessible terrestrial and coastal environments than on the underwater continental shelf areas that usually contain a mixture of relict and modern sedimentary particles [7,8]. The presence of heavy minerals on the Northern Portuguese continental shelf and corresponding coastal areas is mentioned in several works published over the last decades, from where it is possible to know their distribution patterns [9–11]. Biotite, andalusite, zircon, tourmaline, apatite, amphibole, garnet, and staurolite are referred on those works as the main transparent mineral species. A first attempt to interpret the provenance of these minerals pointed to rocky outcrops from the watersheds of the Northern Portuguese rivers as the most probable sources. Hence, the main sources for these minerals were attributed to metamorphic and felsic igneous rocks of the Old Iberian Massif [12]. However, the presence of a mafic mineral assemblage composed by "brown hornblende", pyroxene, and olivine initially detected south of Porto canyon in the early 1990s, it was explained at that time by the existence of a nearby source made up of some kind of basic igneous rocks [9]. Later, based on the acquisition of the first mineral chemical composition data (microprobe analysis) it was discovered that this mafic assemblage was composed by pyroxenes of diopside-hedenbergite, enstatite-ferrosilite, and augite compositions, and by olivine with dominant forsterite composition [10]. It should be noted that this mineralogical assemblage was only found on the south of Porto's canyon and at depths exceeding 100 m, which led [12] to state the probable existence of an igneous basic source located "in the outer shelf/upper slope south of Porto canyon" [12] (p. 99). It turns out that the presence of these minerals was described through chemical composition analysis, expressed by the percentage of oxides that were present. Therefore, the relationship between these minerals and their specific sources has never been properly interpreted using suitable mineralogical diagrams. In addition to these findings, seismic data and remote operated vehicle (ROV) images of the seafloor near the Porto upper canyon head area obtained by [13] showed the existence of a geological structure of possible volcanic origin. This structure was described as a rock relief at more than 15 m in height, standing out from neighboring geological formations being recognized in a seismic profile by a very distinctive diffractive hyperbola [13]. Its presence was attributed to a hard rock body (of dolomitic nature) that stands out from the nesting sedimentary rocks (detrital sediments with evidence of carbonate cement). The heavy mineral assemblage of the Nazaré canyon area stands out from the assemblages of Porto and Aveiro areas because of the absence of pyroxene and olivine and by the high frequency of biotite mineral grains [11]. Due to its localization, Nazaré's canyon head remains active in capturing sedimentary particles transported by littoral drift and along the inner continental shelf [14–16]. Despite the existence of all this relevant information, it happens that in the surrounding areas of the Porto, Aveiro and Nazaré canyons heads (Figure 1), the relationship of the referred mineral species with specific sources has never been properly investigated. Thus, to study this relationship in detail, the present work uses 78 sea bottom samples collected from Porto, Aveiro, and Nazaré canyon upper head areas, from which the identification of transparent heavy minerals was made. This identification is based on two different datasets. One of them concerns the optical identification of the transparent heavy minerals. The other set uses the chemical composition of heavy mineral grains obtained by microprobe analysis. This second dataset includes old mineral chemical composition results covering the Northern Portuguese continental shelf [10], and new mineral chemical composition results from four samples collected from the Porto canyon head area. Thus, the central issue that is addressed in this work is the search of the specific sources for the main transparent heavy minerals identified in the three target areas, Porto, Aveiro, and Nazaré canyon head areas, considering the supply and distribution of sedimentary particles in the Western Portuguese continental margin.

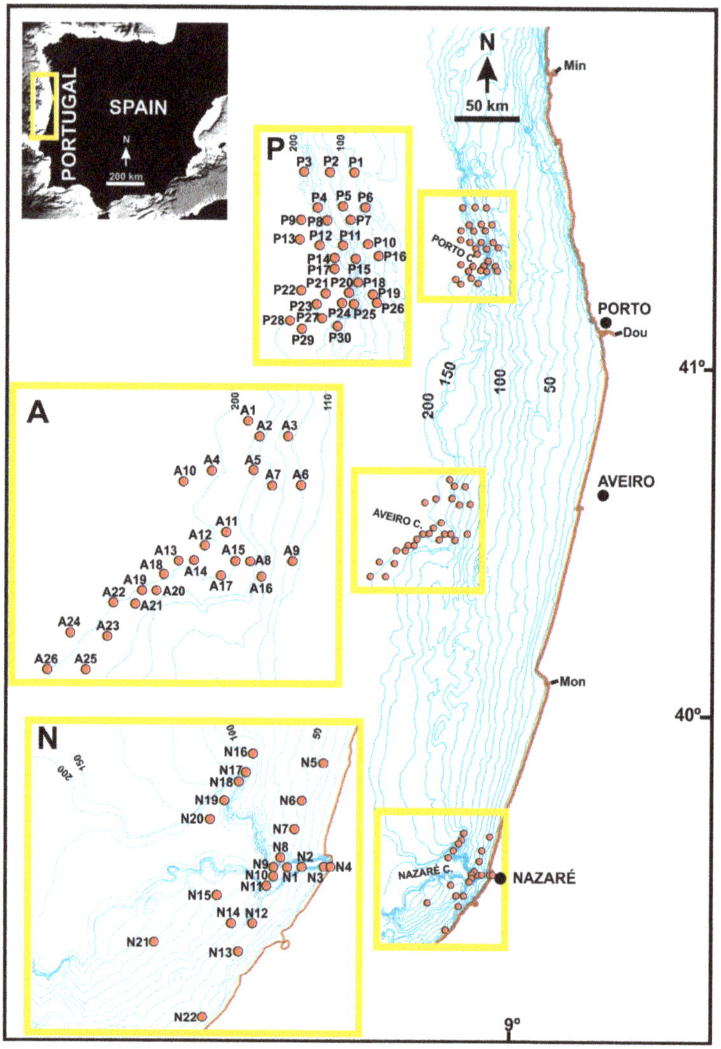

Figure 1. Location of the samples according the three studies areas: Porto (P), Aveiro (A), and Nazaré (N) canyon head areas. Min, Minho River; Dou, Douro River; Mon, Mondego River. Bathymetric contour lines are in meters.

2. Geological Setting

Porto, Aveiro, and Nazaré canyons are important geomorphologic features of the WNW Portuguese continental margin. However, while the Porto and Aveiro canyons are considered as minor submarine valleys because they weakly indent the shelf [17], the Nazaré canyon is one of the largest canyons of the European Margin (170 km) and it cuts the entire width of the Portuguese Margin, from the Iberia Abyssal Plain (at a 5000 m depth) to the infralittoral zone off the Nazaré beach [18]. The geological nature of the canyon heads surrounding the area is also different: The Porto canyon is carved in carbonated to detrital rocks that are highly dolomitized, of Paleocene age; the Aveiro canyon is carved in biogenic and detrital limestones rocks of Neogenic and Eocenic ages; and the Nazaré canyon is carved on Mesozoic rocks (essentially limestones) [13,18].

The sedimentary cover of the referred canyon head's surrounding areas is mainly composed of sand with the presence of some other deposits enriched in gravel or silt particles [19]. The Porto area reveals a higher grain size variability, ranging between sand and gravel at shallower depths (less than 100 m) up to fine sediment particles (silt and clay), which is well represented at the middle shelf (Douro muddy deposit) and upper slope where some isolated spots of these finer particles reach up to 70% of the sediment total weight [19]. The Aveiro area reveals a more homogenous sedimentary cover where sand is the dominant textural type, always representing more than 60% of the total sediment. In some small areas between 100 and 150 m in depth, the gravel particles represent up to $^1/_3$ of the total sediment. Finer sediments are only important in some small areas of the upper continental slope with almost 30% of the total sediment weight [19]. The shelf sedimentary cover near the Nazaré canyon is dominated by coarse-grained particles (sandy gravel) in some locations, namely at 40–80 m in depth. At these depths, these particles constitute a sedimentary deposit with a geometry sub-parallel to the coast line orientation (paleo littorals). Fine and very fine sands have been recorded in the inner shelf north of the canyon and close to its head. Additionally, two important muddy deposit areas are present in the middle shelf north and south of this canyon, at approximately 100 m in depth [14,20]. Most of the Norwest of Iberian Peninsula in an area corresponding to the main river basins is characterized by the presence of Precambrian and Paleozoic igneous (mainly granites) and metamorphic rocks (mainly schists, gneisses, and graywackes). These old rocks are covered by more recent terrains (of Mesozoic and Cenozoic ages) composed by detrital and carbonate sedimentary rocks correspondent to the Douro basin and to the Portuguese Occidental sedimentary basin (known as the Lusitanian basin) (Figure 2).

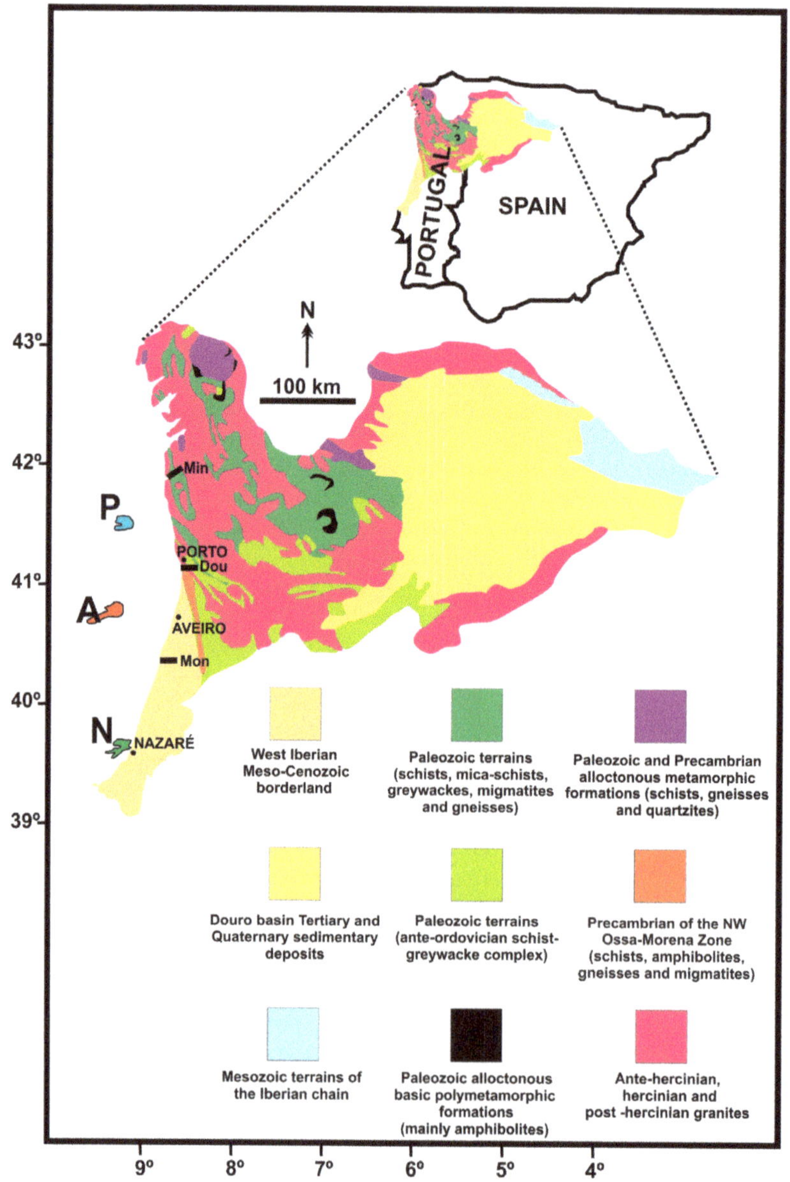

Figure 2. Simplified geological map considering the areas of the main northern Iberian Peninsula and Portuguese river basins: Minho (Min), Douro (Dou) and Mondego (Mon). P, A, and N represent the location of the sea bottom areas with samples collected respectively from Porto, Aveiro, and Nazaré upper canyon areas (adapted from [21]).

The referred igneous and metamorphic rocks can be considered as primary sources and the more recent terrains (of West Iberian Meso-Cenozoic borderland) can be considered as sedimentary sources of the heavy minerals found in the three studied areas. According to [22–32] these geological formations have a diverse heavy mineral composition that is summarized in Table 1.

Table 1. Heavy mineral composition of the main igneous and metamorphic rocks and from the sedimentary rocks from the West Iberian Meso-Cenozoic borderland. These geological formations are present in the Portuguese north and western river basins (based on the information contained in works [22–31].

Source Rocks	Location (River Basin)	Heavy Minerals
Granites	Outcropping in the six Portuguese northern river basins (from Minho to Mondego). Main outcrops are present in Minho, Lima, Ave and Cávado river basins.	Biotite, tourmaline, apatite, zircon, rutile, amphibole and iron, titanium oxides, and occasionally garnet.
Micaschists, gneisses and migmatites	Outcropping in the SW limit of Douro basin (Douro River mouth) and near the coastal zone south of the Douro river.	Biotite, garnet, sillimanite, apatite, and zircon.
Amphibolites and amphibolitic schists	Outcropping in the SW limit of Douro basin (Douro River mouth) and near the coastal zone south of the Douro river.	Amphibole (abundant), Apatite (accessory).
Porphyroblastic schists	Outcropping in the SW limit of Douro basin and in the littoral south of the Douro river.	Garnet, staurolite and biotite (abundant), zircon, tourmaline, apatite, sillimanite and magnetite (accessories).
Schist-greywacke complex	Outcropping mainly in the Douro and Mondego River basins.	Andalusite, garnet and staurolite (abundant in some schists and greywackes. Kyanite occasionally present.
Schists, greywacke, quartzites, hornfels and meatasediments	Outcropping mainly north of the Douro river, and present in all river basins. There are several important outcrops which are crossed by the Minho, Lima, and Mondego rivers	Biotite, andalusite in hornfels. Garnet and andalusite in schists. Apatite, tourmaline, silimanite, amphibole, pyrite, ilmenite and zircon are also present.
West Iberian Meso-Cenozoic borderland	Outcropping mainly on the Mondego, Vouga, Lis, Alcoa, and Tornada river basins.	Tourmaline, zircon and andalusite in Cretaceous formations. Andalusite, tourmaline, biotite, staurolite and zircon in Pliocene/Pleistocene deposits.

The presence of heavy minerals in the Northern Portuguese continental shelf, main river sediments and other continental sedimentary deposits is well known from several published works ([9–12,33,34]. In the Northern Portuguese rivers, from Minho to Douro rivers, and in the sedimentary filling of Minho and Douro estuaries biotite is the main transparent mineral, followed by andalusite and tourmaline. Further south in the Vouga, Lis, Alcoa, and Tornada rivers, tourmaline, andalusite, garnet, and staurolite became the dominant minerals, while in the Mondego basin sediments the most frequent minerals are tourmaline and andalusite. In the continental shelf the presence of heavy minerals is described in 5 sectors using samples collected between 10 and 120 m below mean sea level [10]. In the northernmost one (S1) the biotite is the main mineral followed by andalusite, tourmaline, zircon and garnet. In the second sector (S2) the major difference from the precedent one is the high frequency of amphibole. In the third one (S3) the principal difference from the precedent sector is the relative low frequency of biotite. The fourth one (S4) shows a similar heavy mineral distribution pattern from the precedent sector. Finally, in the fifth sector (S5) the frequency of tourmaline and andalusite represent more than 50% of the main heavy mineral assemblage (Figure 3A,B).

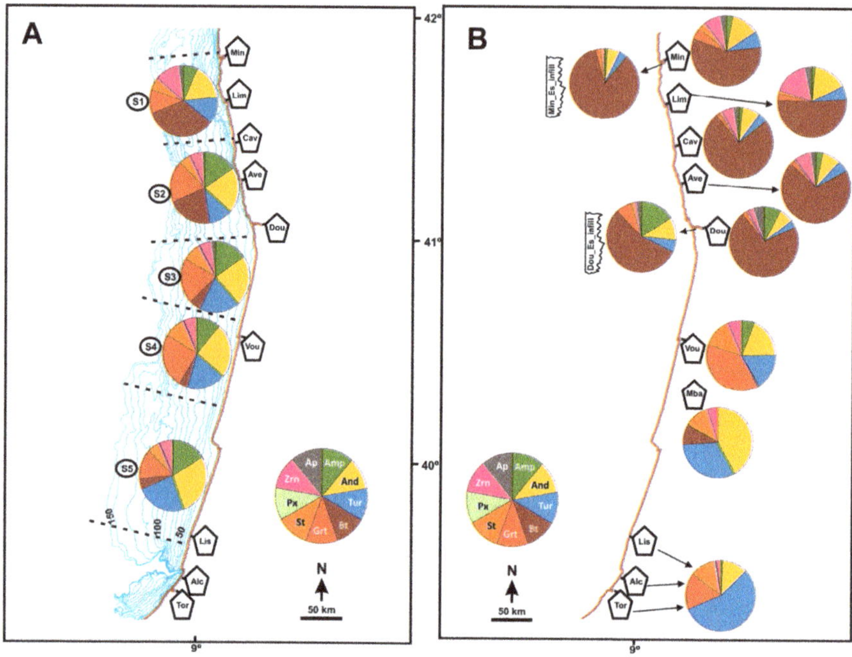

Figure 3. Heavy mineral previous data from diverse locations. (**A**) NW Portuguese continental shelf, five sectors (S1–S5). (**B**) River sediments: Minho (Min), Lima (Lim), Cávado (Cav), Ave (Ave), Douro (Dou), Vouga (Vou), Lis (Lis), Alcoa (Alc), and Tornada (Tor). Estuaries sedimentary infill: Minho (Min_Es_infill) and Douro (Dou_Es_infill). Continental sedimentary deposits: Mondego basin sediments (Mba). The heavy mineral relative frequencies were extracted from [9–11,31–34].

3. Materials and Methods

The present work is based on a set of 78 samples that were collected from three different areas of the Northern Portuguese continental margin, which match the Porto (30 samples), Aveiro (26 samples), and Nazaré (22 samples) submarine canyon upper heads (Figure 1). Samples from Porto and Aveiro areas were used for the first time in the study of heavy minerals and did not coincide with those used in previously published works that have focused on the northern part of the Portuguese continental margin [10,12]

These samples were collected during several cruises in the periods of 1990/99 and 2000/09 (Table S1) using the Smith–McIntyre grab (that collect a sediment sample with a maximum thick of 20 cm) on board hydrographical vessels (*Almeida Carvalho, NRP D. Carlos I, Andrómeda*, and *Auriga*) within the scope of the Portuguese Instituto Hidrográfico program of cartography of the continental shelf sediments (SEPLAT), Sedimentary Dynamics of the Northern Portuguese Continental Shelf project (DISEPLA II), Hotspot Ecossystem Research on the Margins of European Seas project (HERMES), and the Sedimentary Conduits of the West-Iberian Margin project (DEEPCO).

All of the samples were first washed using hydrogen peroxide and distilled water to eliminate the organic matter and marine salts. Grain-size analysis was done using the classic sieving method for sediments coarser than 4 ϕ and the settling method for finer fractions < 4 ϕ. The textural statistical parameters (mean and sorting) were computed using the method of moment [35]. Heavy minerals were separated from fine (2 ϕ) to very fine sand (4 ϕ) using bromoform in a safety laboratory equipped with an air extraction system. After this, the heavy fraction was sieved in two grain size classes: fine sand (from 2 to 3 ϕ) and very fine sand (from 3 to 4 ϕ). Each of these fractions was mounted in Canada balsam on glass slides and the required amount of heavy minerals to fill an area of 25 ×

30 mm on each slide was obtained using a micro-splitter. An average of more than 300 transparent heavy minerals per sample (considering simultaneously the two grain sizes) were counted under the petrographic microscope according to the ribbon counting method [36]. The results of the most frequent minerals (minerals with mean frequency ≥ 1%) are analyzed using the principal component analysis (PCA) based on the extraction of the correlation matrix from the initial data matrix [37]. The grain surface morphology of the most frequent heavy minerals was qualitatively evaluated under the optical microscope. For this purpose two fundamental classes were considered: angular to sub-angular and rounded to sub-rounded. When the mineral grains show little or no evidence of wear and have frequent edges, corners sharp or faces virtually untouched, they belong to the angular to sub-angular class. When the mineral grains show considerable wear, less or no original faces, edges or corners, they belong to the rounded to sub-rounded class. Using these characteristics, a visual estimative of the predominant grain morphological class in each sample was made. Additionally, to confirm the identities of heavy minerals that could be indicative of specific igneous and metamorphic sources (namely, garnet, amphibole, pyroxene, and olivine), some specific grain mounts were made using epoxy resin polished with silicon carbide (sic) and diamond polishing in polishing cloths. These grain mounts were than analyzed by an electron microprobe (JEOL Superprobe 733 at Lisbon University, Lisbon, Portugal).

The heavy mineral percentage weighted in the total sample sediment (HMwt%) was computed according to the Equations (1) and (2):

$$\mathrm{HM\%} = \left(\frac{\mathrm{HMw}(2-4\Phi)}{\mathrm{Sedw}(2-4\Phi)}\right) \times 100 \tag{1}$$

$$\mathrm{HMwt\ \%} = \mathrm{HM\%} \times (\mathrm{Sed\%}(2-4\Phi)/100) \tag{2}$$

where HM% is the weight percentage of heavy minerals in the 2–4 Φ fraction, HMw (2–4 Φ) is the heavy mineral weight (in grams) in the 2–4 Φ fraction, Sedw is the sediment weight (in grams) in the 2–4 Φ fraction, HMwt % is the heavy mineral percentage weighted in total sediment and Sed% (2–4 Φ) is the sediment weight percentage in the 2–4 Φ fraction.

4. Results

4.1. Sediment Texture (Mean and Sorting)

The mean grain size of the sampled sediments (corresponding to the 78 samples) is between 2.08 ϕ (fine sand; Aveiro canyon) and 3.96 ϕ (very fine sand; Nazaré canyon) while the sorting is between 1.58 ϕ (poorly sorted; Nazaré canyon) and 2.01 ϕ (very poorly sorted; Porto canyon) (for more details see Table S1). Additionally, the textural data reveal that the medium and fine sand are the dominant classes of the Porto and Aveiro canyon head areas, while the Nazaré canyon head area denote the presence of more heterometric sediments from very coarse sand to medium silt (Table 2).

Table 2. Average values, maximum, and minimum of mean and sorting. Values in ϕ units for the three canyon head areas (# = number of samples).

Canyon	#	Mean (Average)	Mean (Maximum)	Mean (Minimum)	Sorting (Average)	Sorting (Maximum)	Sorting (Minimum)
Porto	30	2.26	5.24	1.12	2.01	3.24	0.82
Aveiro	26	2.08	3.65	0.70	1.88	2.76	0.81
Nazaré	22	3.96	6.28	−0.25	1.58	3.03	0.43

4.2. Heavy Minerals Analysis

Under the petrographic microscope, it was possible to identify the presence of 18 transparent species that can be mentioned in descending order of their mean frequency in the 78 analyzed samples, considering the sum of the two grain size classes (2–4 ϕ): amphibole, biotite, andalusite, tourmaline, garnets, pyroxene, staurolite, zircon, apatite, rutile, olivine, monazite, kyanite, epidote, titanite, anatase, silimanite, and brookite. The relative frequencies of these mineral grains are represented in Table 3. The results (in count values) are displayed in Table S2 as a data matrix with 78 rows (samples) and 18 columns (minerals).

Table 3. Results of the heavy mineral relative frequency for all the analyzed samples (Porto, Aveiro and Nazaré areas) considering the grain size interval between 2 and 4 ϕ. Mean: mean frequency for each transparent heavy mineral (values in% referred to the total transparent heavy minerals); Max: higher frequency of each transparent heavy mineral; and Min: lower frequency of each transparent heavy mineral. Heavy Minerals (HM): amphibole (Amp), andalusite (And), tourmaline (Tur), biotite (Bt), garnet (Grt), staurolite (St), pyroxene (Px), zircon (Zrn), apatite (Ap), rutile (Rt), kyanite (Ky), olivine (Ol), monazite (Mnz), epidote (Ep), titanite (Ttn), anatase (Ant), silimanite (Sil), and brookite (Brk). *HMwt*: heavy mineral percentage weighted in the total sample sediment.

	Porto			Aveiro			Nazaré		
HM	**Mean**	**Max.**	**Min.**	**Mean**	**Max.**	**Min.**	**Mean**	**Max.**	**Min.**
Amp	15.8	24.9	9.5	24.6	40.1	9.1	20.3	50.5	4.8
And	23.1	46.7	13.7	15.5	29.2	1.0	12.2	28.2	1.9
Tur	16.0	24.4	7.9	14.5	38.5	1.0	13.1	42.3	1.1
Bt	4.6	31.3	0.0	0.8	5.2	0.0	43.0	87.3	0.5
Grt	16.6	26.9	1.1	18.1	27.9	9.0	5.6	27.4	0.0
St	7.5	12.4	4.1	5.3	10.7	0.9	3.0	10.7	0.0
Px	2.1	10.3	0.0	14.0	37.8	0.2	0.0	0.0	0.0
Zrn	5.9	13.0	0.9	1.7	5.6	0.0	1.5	10.0	0.0
Ap	2.7	8.8	0.0	0.0	0.3	0.0	0.1	1.1	0.0
Rt	1.4	4.1	0.0	0.7	2.3	0.0	0.2	1.7	0.0
Ky	1.1	2.4	0.0	0.5	2.4	0.0	0.3	0.9	0.0
Ol	0.4	3.0	0.0	1.2	3.8	0.0	0.0	0.5	0.0
Mnz	0.9	3.8	0.0	0.5	1.9	0.0	0.2	0.9	0.0
Ep	0.5	1.5	0.0	0.7	3.5	0.0	0.1	0.9	0.0
Ttn	0.7	2.5	0.0	0.2	1.0	0.0	0.2	1.1	0.0
Ant	0.3	1.8	0.0	0.7	2.7	0.0	0.1	0.8	0.0
Sil	0.3	1.9	0.0	0.3	1.4	0.0	0.2	1.1	0.0
Brk	0.1	0.8	0.0	0.5	1.7	0.0	0.0	0.5	0.0
HMwt	*0.7*	*6.7*	*0.0*	*0.1*	*0.3*	*0.0*	*0.5*	*2.8*	*0.0*

4.3. Main Heavy Minerals (Porto, Aveiro, and Nazaré Areas)

The main heavy mineral suite (mean frequencies > 1%) is composed of amphibole, andalusite, tourmaline, biotite, garnet, staurolite, pyroxene, zircon, and apatite (Table 2). This mineral suite can be represented according to each studied area (Figure 4). From this figure, it is possible to observe that each area had a specific heavy mineral signature. The Porto canyon head is characterized by a main mineral suite composed, in decreasing order, of mean frequency by andalusite (23.1%), garnet (16.6%), tourmaline (16.0%), amphibole (15.8%), staurolite (7.5%), zircon (5.9%), biotite (4.6%), apatite (2.7%), and pyroxene (2.1%) (Figure 4, chart P). Aveiro contains the most representative heavy suite made up by amphibole (24.7%), garnet (18.1%), andalusite (15.5%), tourmaline (14.5%), pyroxene (14.0%), staurolite (5.3%), zircon (1.7%), biotite (0.8%), and apatite (<0.1%) (Figure 4, chart A). The Nazaré area is dominated by the presence of biotite (43.0%), amphibole (20.3%), tourmaline (13.1%), andalusite (12.2%), garnet (5.1%), staurolite (3.0%), zircon (1.5%), and apatite (0.1%). In this area, the mineral pyroxene was not identified (Figure 4, chart N).

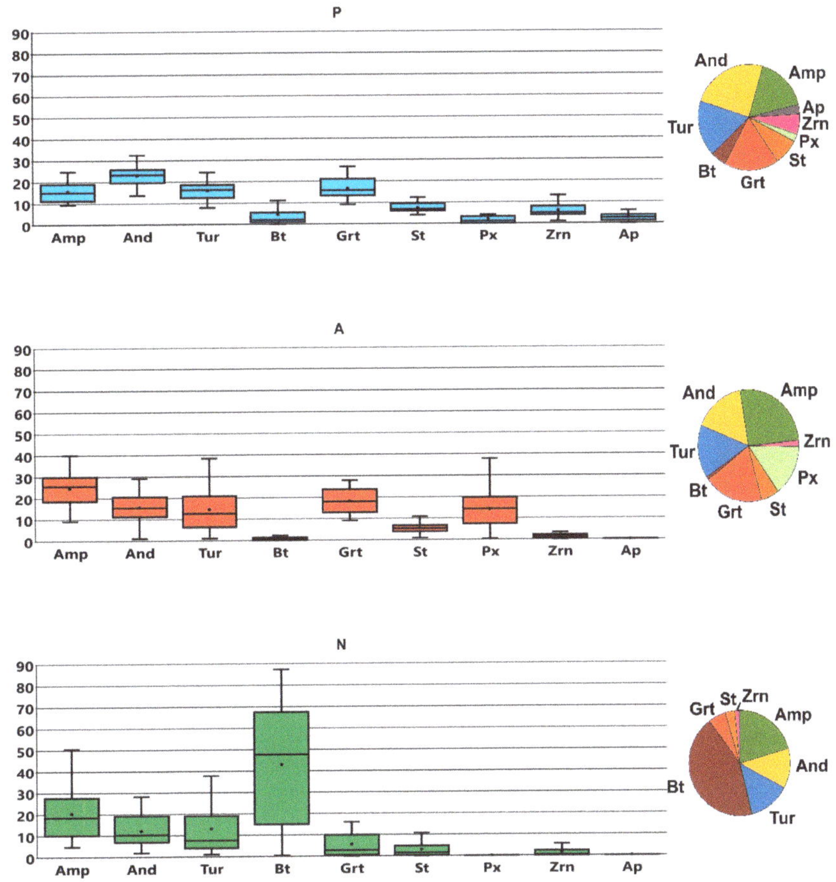

Figure 4. The heavy mineral transparent frequencies for each studied area (bar chart P—Porto; bar chart A—Aveiro; bar chart N—Nazaré) of specimens with more than 1% of mean frequency (Table 2). The box-plots represent the 25th and 75th quartiles. The horizontal line represents the 50th quartile (median). The small dots represent the mean value for each mineral. The extremes of each box vertical line represent the maximum and minimum values. Each pie chart next to the right of each bar chart represent the average frequency values of each mineral.

The distribution pattern of the main heavy mineral assemblage can be represented in detail for each studied area. In the case of the Porto area, it is verified that amphibole, andalusite, tourmaline, and garnet together represented more than 50% of the main mineral suite in all the samples. Biotite showed a high variable frequency pattern that is well represented in some samples (e.g., P22, P24, or P28) but is also absent in many samples (e.g., in most of the northernmost samples). Additionally, pyroxene tends to be more frequent in the southernmost samples, particularly in P21, P22, P23, P27, P28, P29, and P30. In a different way, zircon tend to be more frequent in the samples collected at lower depths, particularly in P1, P5, P6, P10, P16, P18, P25, P26, and P30. Staurolite and apatite show undefined distribution patterns (Figure 5).

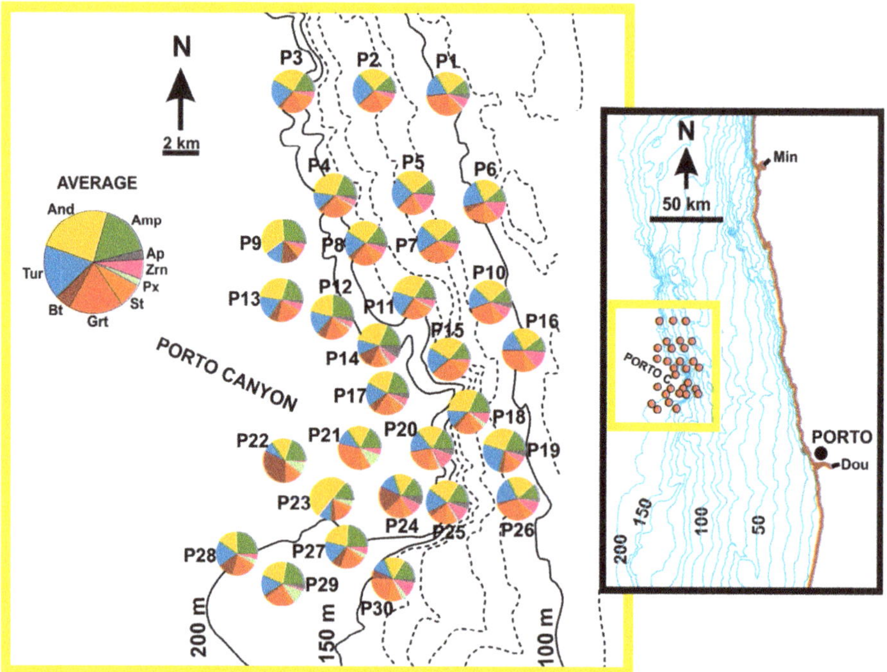

Figure 5. Main heavy mineral suite distribution according to the samples collected from the Porto upper canyon area. The average values are represented on the larger pie chart.

In the case of the Aveiro area, amphibole has a regular distribution, revealing a slight tendency to be more frequent in the southernmost samples, particularly in the A18–A24 samples. In turn, in the case of andalusite distribution, it is observed that the higher frequency values are reached in the northernmost samples (A1–A7, excluding A5), and in the ones collected along the canyon's main axis, particularly in A12, A14, and A17. Garnet is more frequent in the southernmost samples and, in contrast, tourmaline appear to be more concentrated in the northernmost samples. Pyroxene is more concentrated in the southernmost samples, representing in some cases more than 25% of the main spectrum (A20 and A21). Biotite, staurolite, and zircon are poorly represented in all the samples collected from this area (Figure 6).

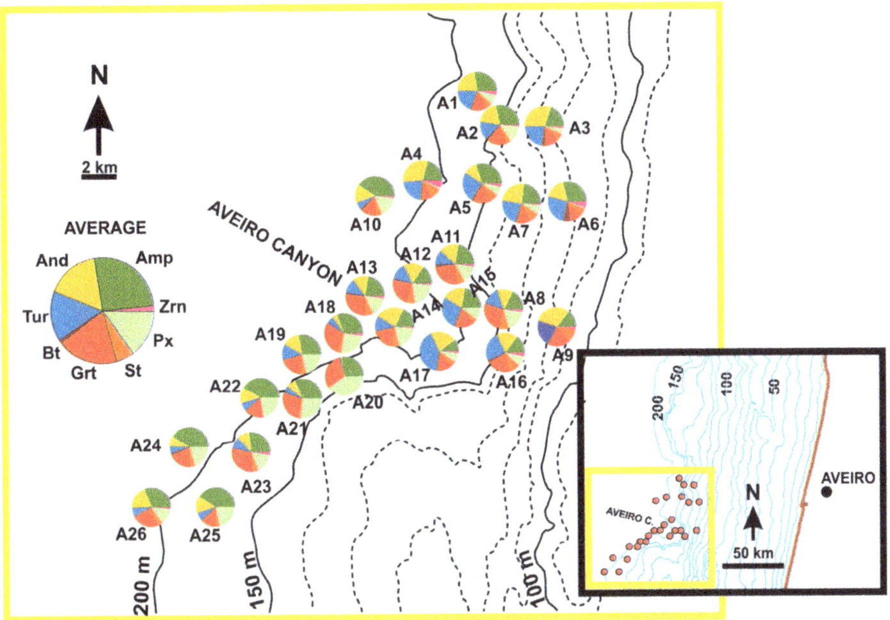

Figure 6. Main heavy mineral suite distribution according to the samples collected from the Aveiro upper canyon area. The average values are represented on the larger pie chart.

In the case of the Nazaré area, the presence of biotite is clearly dominant in most of the samples, particularly in those collected near the canyon valley where this mineral grain can represent more than two thirds of the main mineral assemblage (e.g., N9, N10, N17, N18, and N20). Amphibole seemed to be more frequent in the northernmost samples, where in some cases it can represent more than one third of the main mineral spectrum (samples N4, N5, and N6). Andalusite, tourmaline, garnet, staurolite, and zircon are more frequent in samples collected at lower depths, particularly in samples N5, N6, N7, N12, N13, N14, and N22 (Figure 7).

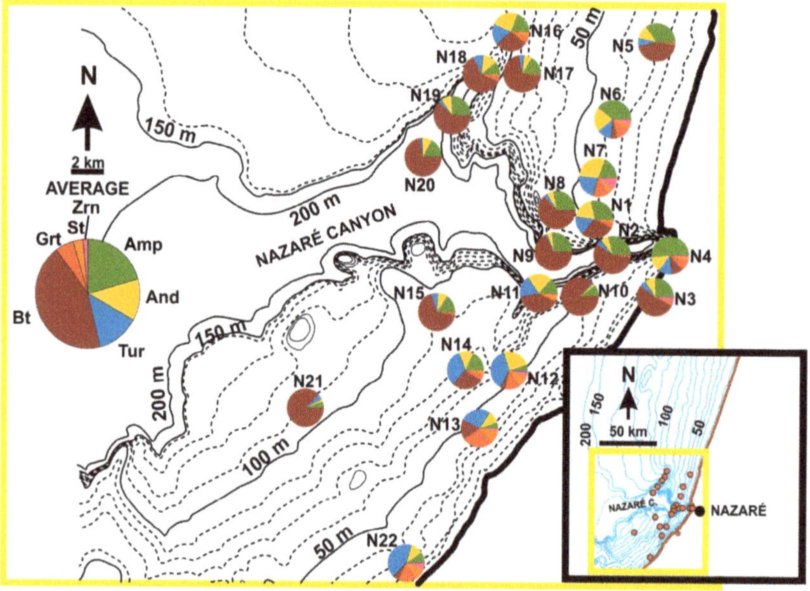

Figure 7. Main heavy mineral suite distribution according to the samples collected from the Nazaré upper canyon area. The average values are represented on the larger pie chart.

4.4. Principal Component Analysis

By the application of the principal component analysis (PCA) to the data matrix composed of nine heavy minerals (from amphibole to apatite) and by 78 samples it is possible to extract two components with eigenvalues higher than 1 that together explain about 65% of variance (Table 4).

Table 4. Results of the extracted principal components. Only the first two have eigenvalues higher than 1. The sum of the variance explained by the first two components is about 65%.

Component	Eigenvalue	Variance (%)
1	3.98	44.25
2	1.84	20.39
3	0.97	10.74
4	0.76	8.46
5	0.49	5.40
6	0.38	4.24
7	0.28	3.14
8	0.19	2.09
9	0.12	1.29

The first component accounts for about 44% of variance and the mineral loadings show an opposition between biotite and the mineral set composed by andalusite, tourmaline, garnet, staurolite, zircon, and apatite (Figure 8). The second component accounts for about 20% of variance and the mineral loadings on this component show an opposition between biotite and the mineral set composed by garnet and pyroxene (Figure 8).

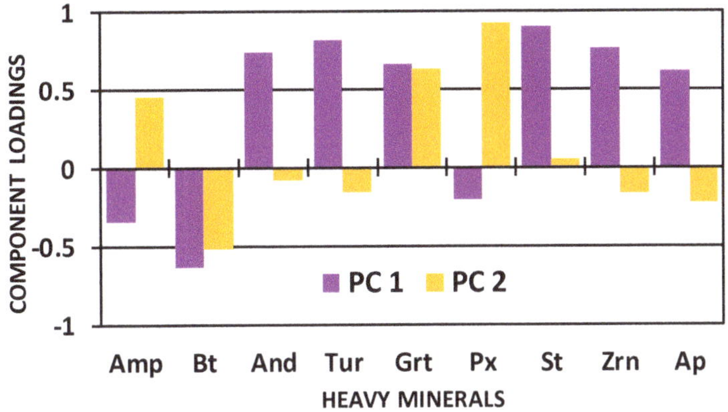

Figure 8. Mineral loadings according to the first (PC1) and second (PC2) components accounting for about 65% of variance. Amphibole (Amp), biotite (Bt), andalusite (And), tourmaline (Tur), garnet (Grt), pyroxene (Px), staurolite (St), zircon (Zrn), and apatite (Ap).

The plot of the scores of each sample considering these two principal components give rise to the scatter diagram of Figure 9. The separation of the samples according to their respective area is very clear on this diagram where the first component separates Porto samples from Nazaré samples while the second one separates the Aveiro samples from the other two sample groups (Figure 9).

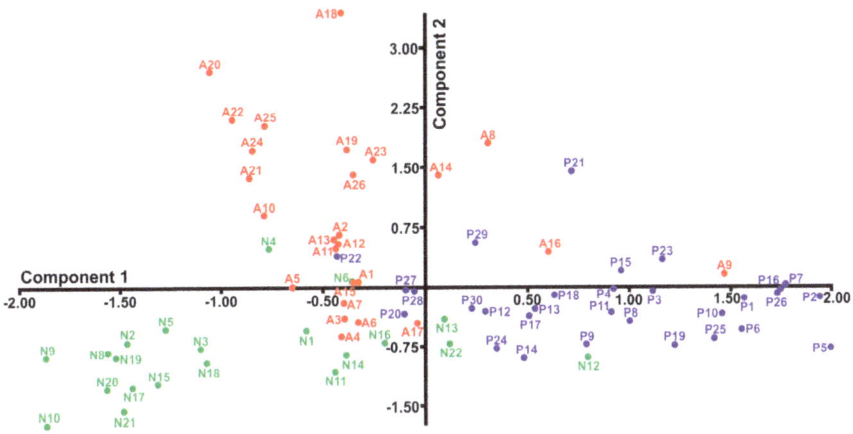

Figure 9. Scatter diagram considering the first two principal components: C1 accounts for about 44% of variance and C2 accounts for about 20% of variance.

4.5. Microprobe Analysis (Garnet, Amphibole, Pyroxene, and Olivine Mineral Groups)

The first set of heavy mineral chemical composition data it was obtained by [10]. These data correspond to samples collected from the Northern Portuguese continental shelf (Figure 10).

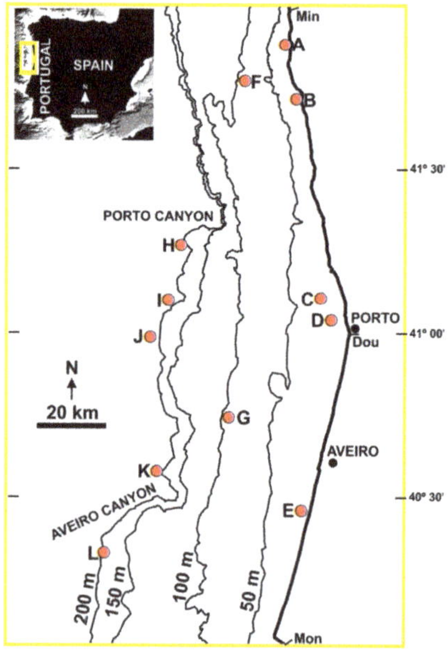

Figure 10. Location of the samples subjected to electron microprobe analysis (red dots from A–L). Min: Minho River; Dou: Douro River; Mon: Mondego River (adapted from [10]).

The second set of heavy mineral chemical composition data correspond to four samples collected from the Porto canyon head area (Figure 11).

Figure 11. Location of the samples subjected to electron microprobe analysis (new data) in the Porto canyon area (red dots corresponding to samples P3, P16, P17, and P20).

4.5.1. Garnet Group

The geochemical data from detrital garnets have been used by several researchers with the aim of interpreting the sedimentary provenance [38–42]. This mineral group is known for its potential in the analysis of the sedimentary provenance of detrital sediments because "it has a wide compositional variation that may be specific to certain lithologies and, therefore, source areas, it is mechanically resistant during transport, and it is resistant to chemical modification during transport, diagenesis and low-grade metamorphism" [42] (p.373). All the available data corresponding to the microprobe analysis of detrital grains of garnet are shown in Figure 12A,B). This figure was built using the excel spreadsheets made available by [42,43] and following the recommendations of [43].

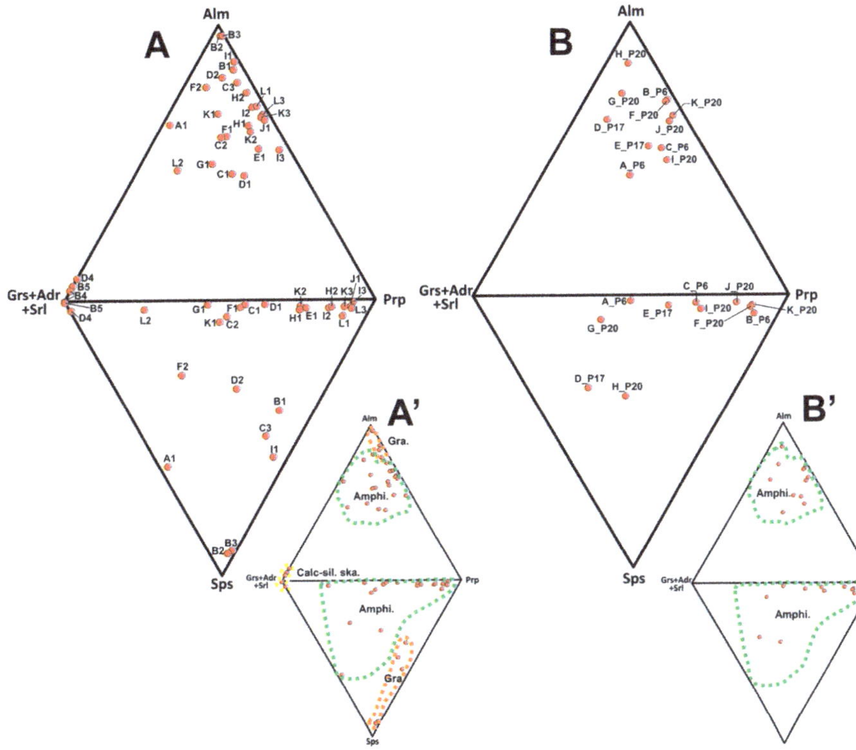

Figure 12. Ternary plots using end-members grossular (Grs) + andradite (Adr) + schorlomite (Srl), almandine (Alm), pyrope (Prp), and spessartine (Sps). **A** shows the results corresponding to several samples collected from the Northern Portuguese continental shelf (first set of chemical composition data obtained by [10]). The capital letter next to each red dot indicates the location of the sample on the map of the Figure 10. The number next to the capital letter is the reference to the analyzed mineral grain. **B** shows the results corresponding to the second set of heavy mineral composition data (samples collected from the Porto canyon head area). In this case the capital letters represent each analyzed mineral grain. Each sample is referenced by the letter P followed by a number. The location of the samples is shown in Figure 11. **A′** and **B′** are the ternary plots showing sub-areas characteristic of garnets with different protoliths: Gra—granites, Amphi—amphibolites, Calc-sil. ska—calc-silicate skarns.

The results of the ternary plot of Figure 12A show the predominance of garnets with the dominant presence of the almandine end-member. However, a small number of samples (B4, B5, and D5) show the predominance of the grossular (Grs) + andradite (Adr) + schorlomite (Srl) end-member. The protoliths corresponding to the analyzed garnets can be observed in Figure 12A′. Figure 12B shows only the predominance of the almandine end-member and the correspondent protoliths can be observed in Figure 12B′ (see Table S3 for details).

4.5.2. Amphibole Group

The classification of these amphiboles is made according to the spreadsheet of [44] that follows the nomenclature recommended by the International Mineralogical Association [45]. The chemical composition of the analyzed amphiboles is plotted according to the diagrams defined by [46]. For the samples collected from the Northern Portuguese continental shelf [10] (Figure 10) the presence of magnesio-hornblende (Mhb) is predominant (it is detected in 14 samples). Other types of calcic amphiboles are tschermakite (Ts), edenite (Ed), pargasite (Prg) (all detected in three samples), ferro-hornblende (Fhb) detected in two samples, and finally actinolite (Act), magnesio-hastingsite (Mhst), and hastingsite (Hst) (all detected only in one sample) (Figure 13A,B); see Table S4 for details).

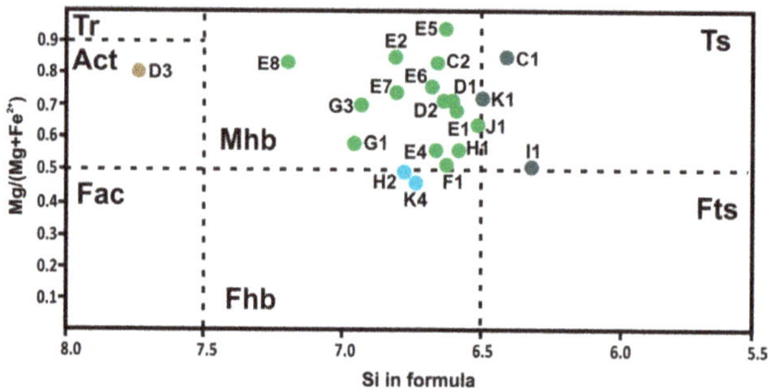

A

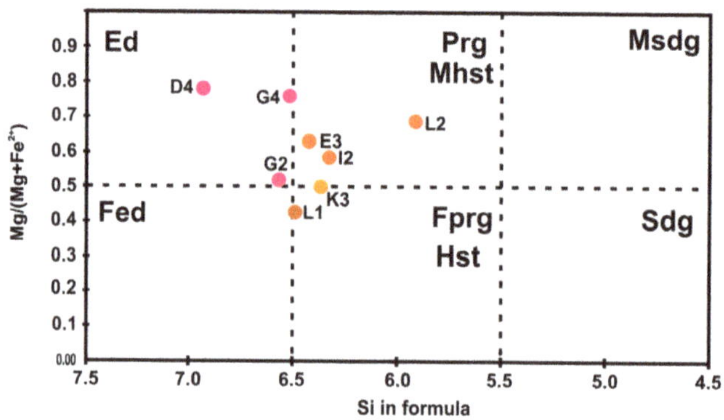

B

Figure 13. A (upper diagram) shows the presence of magnesio-hornblend (Mhb), tschermakite (Ts), ferro-hornblende (Fhb) and actinolite (Act). In this diagram tremolite (Tr), ferro-actinolite (Fac) and ferro-tschermakite (Fts) were not found. **B (lower diagram)** shows the presence of pargasite (Prg), edenite (Ed), magnesio-hastingsite (Mhst), and hastingsite. In this diagram ferro-edenite (Fed), magnesio-sadanagaite (Msdg) and sadanagaite (Sdg) were not found. The capital letters next to each dot indicates the location of the sample on the map of the Figure 10. Each number next to each capital letter is the reference of the analyzed mineral grain. These diagrams project chemical analysis data of calcic amphiboles from samples collected from the Northern Portuguese continental shelf.

For the samples collected from the Porto canyon area (Figure 11) the analyzed amphiboles belong to the group of calcic amphiboles. The presence of magnesio-hornblende (Mhb) is predominant (detected in 11 samples), followed by tchermakite (Ts) (in three samples), pargasite (Prg) (in three samples), and by ferro-tschermakite (Fts) and actinolite (Act) both detected in one sample (Figure 14A,B; see Table S4 for details).

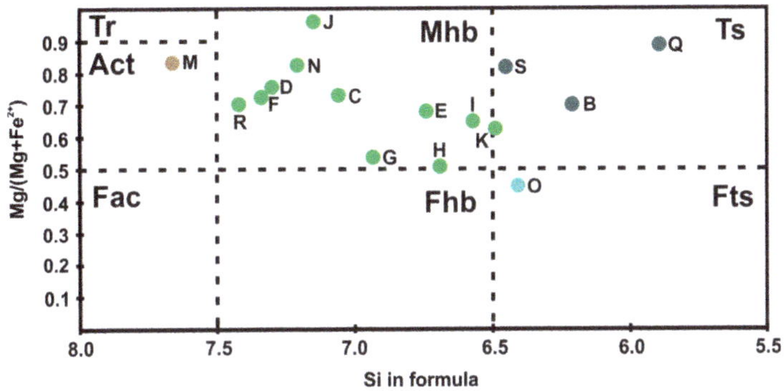

A

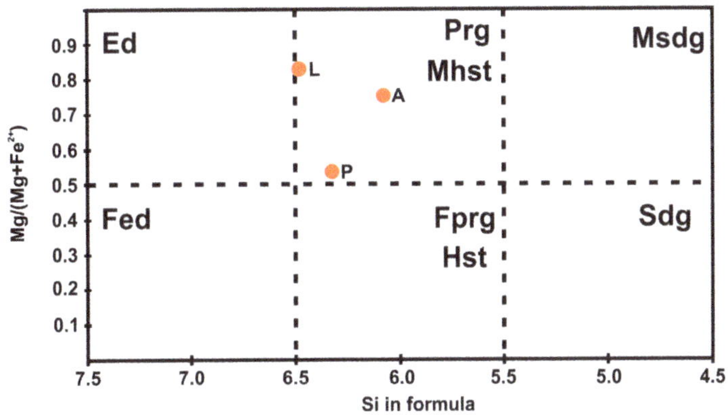

B

Figure 14. A (upper diagram) shows the presence of magnesio-horneblend (Mhb), tschermakite (Ts), ferro-tschermakite (Fts) and actinolite (Act). In this diagram tremolite (Tr), ferro-actinolite (Fac), ferro-hornblend (Fhb) were not found. **B (lower diagram)** shows the presence of pargasite (Prg). In this diagram edenite (Ed), magnesio-hastingsite (Mhst), hastingsite (Hst) ferro-edenite (Fed), magnesio-sadanagaite (Msdg), and sadanagaite (Sdg) were not found. Mineral grains D–N are from sample P6; A–C are from sample P3; P–S are from sample P20; O is from P17. The location of the samples can be seen on Figure 11. These diagrams project chemical analysis data of calcic amphiboles from samples collected from the Porto canyon area.

4.5.3. Pyroxene Group

The classification of the pyroxenes is made using the available spreadsheet by [47] and the results were plotted according to the ternary diagrams defined by [48]. For the samples collected from the Northern Portuguese continental shelf [10] and considering the group of clinopyroxenes it is possible to detect the presence of diopside (Di) in five samples, of hedenbergite (Hd) in seven samples, and of augite (Aug) in three samples. For the orthopyroxene group it is possible to detect the presence of enstatite (En) in seven samples, and ferrosilite (Fs) in two samples (Figure 15; see Table S5 for details).

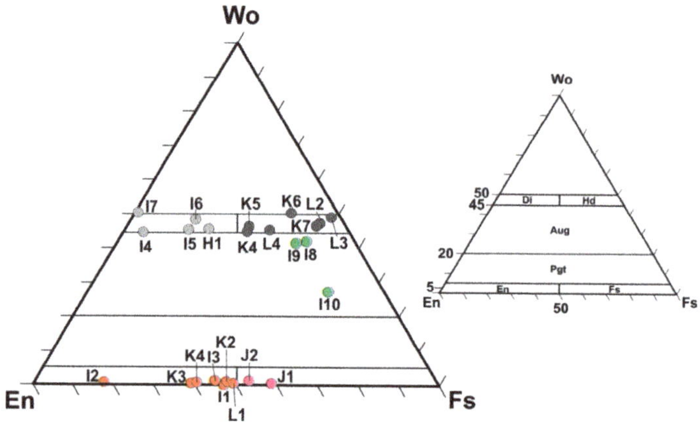

Figure 15. Classification diagram for identified pyroxenes in samples collected from the Northern Portuguese continental shelf by combining the classification diagrams of clinopyroxenes and orthopyroxenes proposed by [48]. The classification shows the presence of diopside (Di), hedenbergite (Hd), augite (Aug), enstatite (En), and ferrosilite (Fs). Pigeonite (Pgt) was not found. The capital letters next to each dot indicates the location of the sample on the map of the Figure 10. Each number next to each capital letter is the reference of the analyzed mineral grain.

For the samples collected from the Porto canyon area and considering the group of clinopyroxenes it is possible to detect the presence of diopside (Di) in five samples, of hedenbergite (Hd) in one sample, and of augite (Aug) in six samples. For the orthopyroxene group it is possible to detect the presence of enstatite (En) in four samples, and ferrosilite (Fs) in two samples (Figure 16; see Table S5 for details).

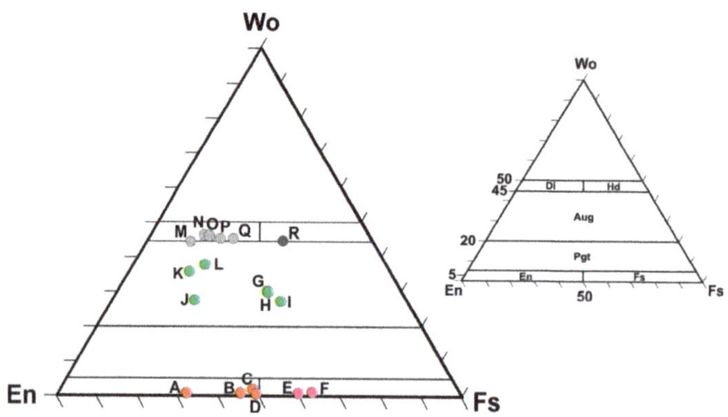

Figure 16. Classification diagram for identified pyroxenes in samples collected from the Northern Portuguese continental shelf by combining the classification diagrams of clinopyroxenes and orthopyroxenes proposed by [48]. The classification shows the presence of diopside (Di), hedenbergite (Hd), augite (Aug), enstatite (En), and ferrosilite (Fs). Pigeonite (Pgt) was not found. A–D are En mineral grains from sample P20; E and F are Fs mineral grains from sample P20; G–I are Aug mineral grains from sample P6; J and K are Aug mineral grains from sample P17; L is a Aug mineral grain from sample P20. The location of the samples can be seen on Figure 11.

4.5.4. Olivine Group

The classification of the olivine is made using the available spreadsheet by [47] and the results are plotted using a diagram expressing the compositional variation between the two extreme members:

forsterite (Fo) and fayalite (Fa). For the samples collected from the Northern Portuguese continental shelf it is possible to detect two olivine mineral grains with strong composition in Fo end-member (Figure 17A,A'; see Table S6 for details).

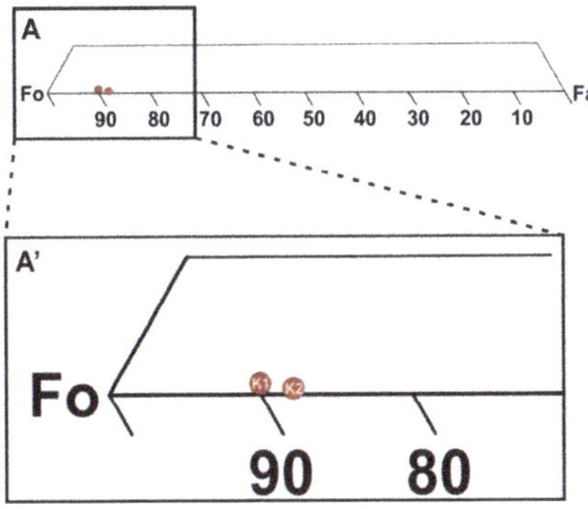

Figure 17. Classification diagram of identified olivine's in samples collected from the Northern Portuguese continental shelf. The capital letters inside to each dot indicates the location of the sample on the map of Figure 10. Each number next to each capital letter is the reference of the analyzed mineral grain. In A', it is possible to observe that the Fo end-member has a value greater than 85%.

For the samples collected from the Porto canyon area the strong presence of the Fo end-member is also found (Figure 18A,A').

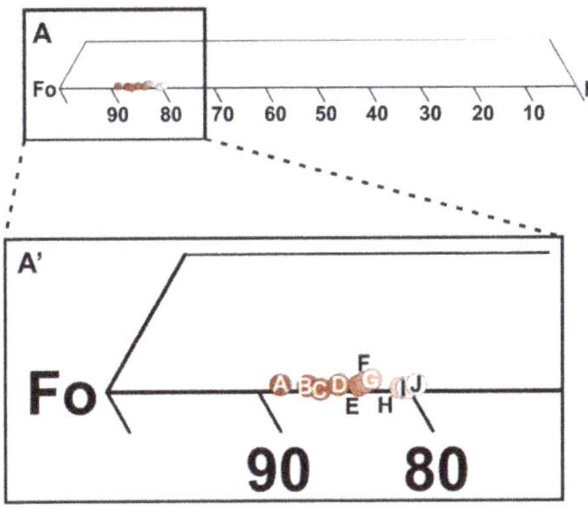

Figure 18. Classification diagram of identified olivines in samples collected from the Porto canyon area. A, D, and G are Fo mineral grains from sample P20; B, C and J are Fo mineral grains from sample P3; F and H are Fo mineral grains from sample P17; E is a Fo mineral grain from sample P3. In A', it is possible to observe that the end member Fo ranged between 80 and 90%.

4.6. Mineral Grain Surface Morphologies

The most common transparent heavy minerals identified under the optical microscope appear with contrasting morphological surface morphologies. Two fundamental classes of surface morphologies are considered: angular to sub-angular and rounded to sub-rounded. The first class includes heavy minerals correspondent to "first-cycle" particles independently from their source. These minerals can be found in all the samples analyzed and they are dominant (>50%) in samples collected from the three studied areas at depths exceeding 120 m. The second class includes heavy minerals that show a long evolution in the sedimentary environment and for that reason they can be considered as "multi-cycle" particles. These minerals are dominant (>50%) in samples collected from the three studied areas at depths of less than 120 m (Figure 19). In this figure the heavy-mineral classification is based on the description of the minerals surficial textures referred by [49,50]

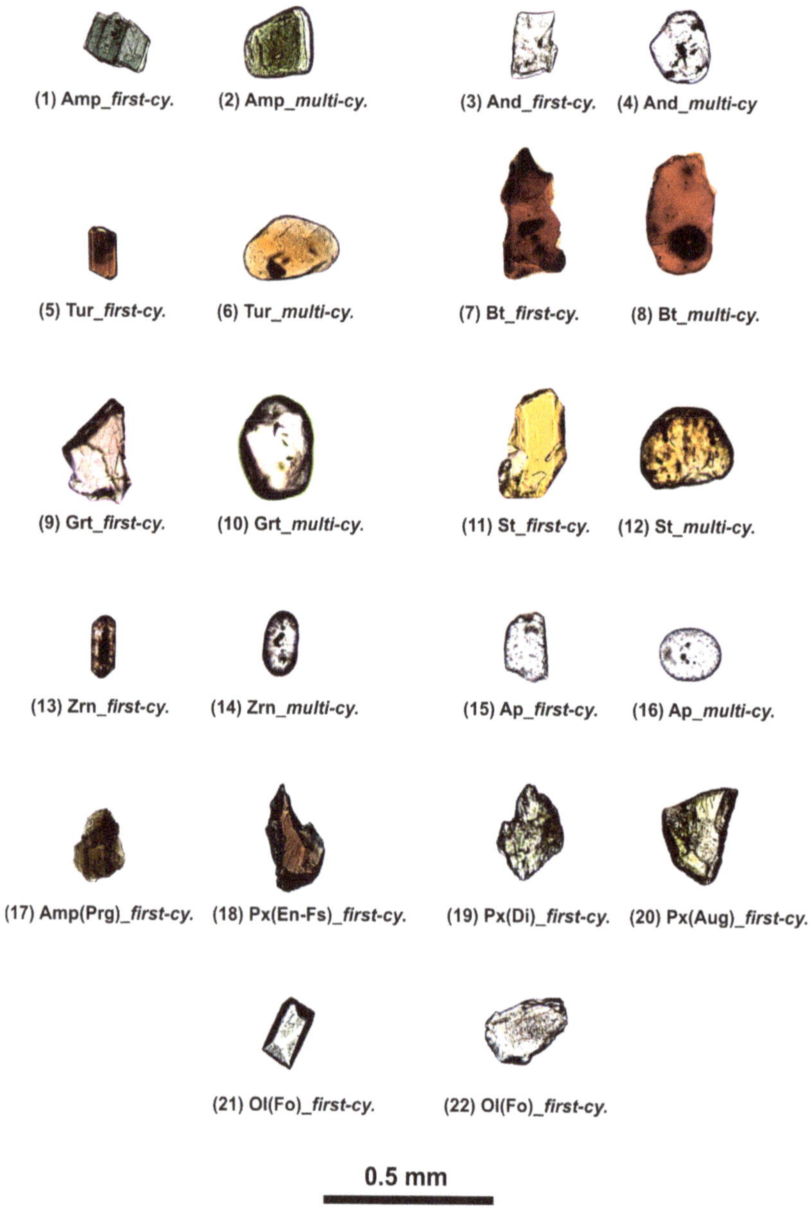

Figure 19. Typical visual aspects of the main heavy mineral transparent suite according to their dominant surface texture. Mineral grains classified with *first-cy.* are "first-cycle" particles that have angular to sub-angular surface textures and are catalogued with odd numbers from 1–15 and continuously from 17–22. In this class it was detected the presence of several euhedral Fo minerals (grain 21). Mineral grains classified with *multi-cy.* (catalogued with odd numbers from 2–16) correspond to the "multi-cycle" ones.

5. Discussion

5.1. Heavy Mineral Sources

One can verify that a correlation exists between the presence of the main heavy mineral assemblage made of amphibole, andalusite, tourmaline, biotite, garnet, staurolite, zircon, and apatite with the fluvial heavy mineral spectrum formerly identified in the Northern Portuguese river basins (Figures 4 and 6–8) [31–34]. Furthermore, this mineral assemblage is also compatible with the igneous and metamorphic rocks that appear in these river basins, which can be considered as primary sources of these heavy minerals (Figure 3 and Table 1). Thus, it is possible to cite the examples of andalusite-rich metamorphic rocks as important sources of andalusite, of porphyroblastic schists known as main sources of garnet, staurolite, and biotite, of micaschists, gneisses, and granites identified as important sources of biotite, tourmaline, apatite, and zircon, as well as amphibolites of the Douro metamorphic complex known as main sources of amphibole [30].

The two sets of microprobe analysis results (Figures 12–18) make a strong mineral-chemical tool to determine heavy mineral provenance. In this context, the garnets from the Northern Portuguese continental shelf that were analyzed show a source compatibility with granites, with intermediate to high-grade metamorphic rocks (of amphibolitic and granulitic facies), and with calc-silicate skarns. This interpretation is based on the relationship between the sub-area's characteristic of garnets with the different protoliths referred by [42] (Figures 12A and 20A). The garnets compatible with granites are mostly found in samples collected at low depths north of the Douro River (samples B and C, Figure 10). This means that the main outcrops of granites present at Minho, Lima, Ave, and Cávado river basins (Figure 2 and Table 1) are the most likely primary sources of these garnets. However, in a sample collected from the upper continental slope, between Porto and Aveiro canyon areas (sample I, Figure 10), the presence of one garnet compatible with this granitic source was detected. The garnets sourced from metamorphic rocks (of amphibolitic and granulitic facies) are found throughout the Northern Portuguese continental shelf (samples A to L, Figure 10). The occurrence pattern of these garnets certainly reflects the widespread presence of metamorphic rocks (of amphibolitic and granulitic facies) all over the Northern Portuguese river basins (Figure 2 and Table 1). The presence of grossular end-member rich garnets (with more than 80% in grossular composition; Table S3) is confined to two samples collected at low depths (samples B and D, Figure 10). According to the parameters of [42] these garnets can be sourced from "calc-silicate skarns and rodingites". The most important Portuguese calc-silicate skarn deposits are found in the central-northern part of the country (Central Iberian Zone). These deposits are within the Dúrico-Beirão Supergroup lithostratigraphic sequences [51]. Of these occurrences it is possible to highlight the presence of skarns at the mining districts of Covas (Minho region) and Tabuaço (Douro river basin) where the presence of grossular is known [52,53]. Thus, it is possible to conclude that these kinds of garnets (rich in grossular end-member) have, for their main primary sources, these skarn deposits present in the Douro, Lima, and Minho river basins (Tabuaço and Covas areas). In regard of the chemical composition of garnets concerning the second sample set (samples collected from the Porto canyon area), the results only show a source compatibility with metamorphic rocks of amphibolitic and granulitic facies (Figures 12B and 20B), based on the parameters defined by [42]. Since these rocks are well represented in the Douro river basin (as the most important group of metamorphic rocks; Figure 2 and Table 1) it is possible to conclude that these garnets were essentially sourced from the Douro river basin.

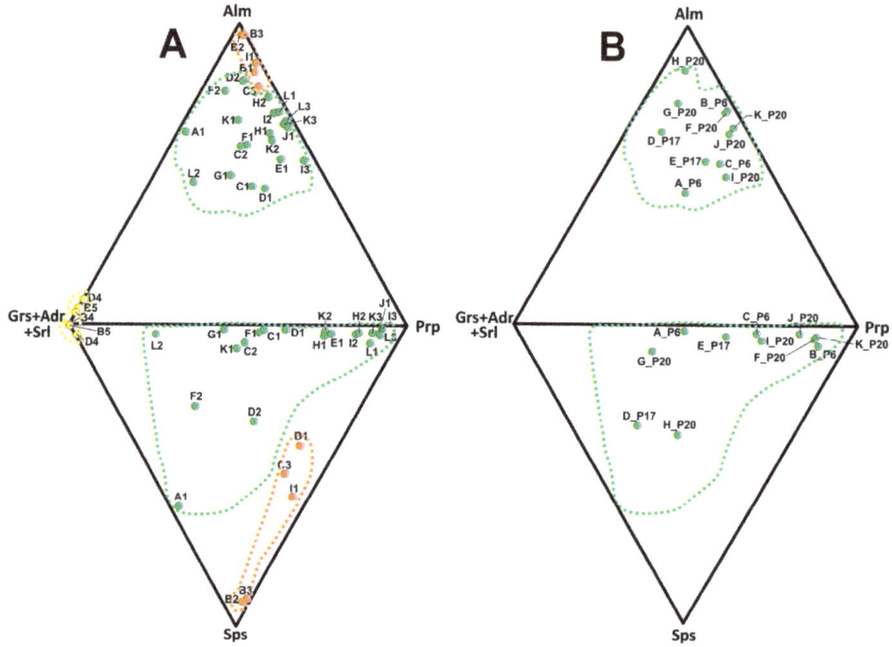

Figure 20. (**A**). Northern Portuguese continental shelf. Orange dots and correspondent surrounding area indicate garnets sourced from granites. Green dots and correspondent surrounding area indicate garnets sourced from metamorphic rocks of amphibolitic and granulitic facies. Yellow dots and correspondent surrounding area represent garnets sourced from calc-silicate skarns. (**B**) Porto canyon area. Green dots indicate garnets sourced from metamorphic rocks of amphibolitic and granulitic facies.

For the first sample set (samples collected from the continental shelf), the available data on the amphibole chemical composition show the predominance of magnesio-hornblende, followed by other calcic amphiboles, such as tschermakite, edenite, pargasite, ferro-hornblende, actinolite, magnesio-hastingsite, and hastingsite (Figure 13A,B). The most likely sources for these amphiboles are the amphibolitic rocks of the Douro metamorphic complex (Figure 2 and Table 1). However, among the mineral grains with pargasite composition, two grains with a relatively high content in TiO_2 (I2 with 2.0% and L2 with 3.4%; Table S4) were detected. Under optical microscope observation, in plane polarized light, these amphiboles appear with shades of brown which makes them easily distinguishable from other amphiboles (Figure 19, mineral #17). These cases of pargasite may be indicative of a provenance related to basic igneous rocks, such as gabbro [54]. Similar source interpretation can be considered for the presence of the magnesio-hastingsite (mineral grain K3; Figure 13A) detected near the Aveiro canyon area (Figure 10). As these amphiboles were only found in the deepest areas of the continental shelf, south of Porto canyon (samples I, K and L, Figure 10), and knowing that this igneous basic source is not represented in the Northern Portuguese river basins, then it will be necessary to admit the existence of a compatible source located elsewhere in the outershelf/upper slope south of Porto canyon. For the second sample set (samples collected from the Porto canyon area), the available data of the amphibole chemical composition show the dominant presence of magnesio-hornblende (Figure 14A). The provenance of these amphiboles is compatible with the amphibolitic rocks of the Douro metamorphic complex (Figure 2 and Table 1). In this sample set it is also identified the presence of pargasite in three samples with relatively high values of TiO_2 (between 0.9 and 1.4%, Figure 14B and Table S4). Moreover, these pargasite mineral grains appear with shades of brown under the microscope observation in plane polarized light (Figure 19, mineral #17). Therefore, these data point to the fact that these amphiboles have an origin related with basic igneous rocks as discussed above. The source of the other identified amphiboles (actinolite, tschermakite, and ferro-tschermakite) is compatible with the metamorphic rocks outcropping in the referred river basins (Figure 2 and Table 1). For a better understanding the relationship between the identified amphiboles and their most typical sources, Table 5 makes the synthesis of the results referent to this mineral group.

Table 5. Correspondence between the identified amphiboles, most typical primary sources and sample sets. 1st—first set of heavy mineral chemical compositional data (Northern Portuguese continental shelf). 2nd—second set of heavy mineral compositional data (Porto canyon head area). Mhb—magnesio-hornblende, Act—actinolite, Ts—tschermakite, Fts—ferro-tschermakite, Fhb—ferro-hornblende, Ed—edenite, Prg—pargasite, Mhst—magnesio-hastingsite, Hst—hastingsite.

Amphibole	Typical Primary Sources	Sample Set
Mhb	amphibolite, schist	1st; 2nd
Act	metamorphized carbonate rocks	1st; 2nd
Ts	amphibolite	1st; 2nd
Fts	amphibolite, schist, gneiss	2nd
Fhb	amphibolite, schist	1st; 2nd
Ed	amphibolite	1st; 2nd
Prg	gabbro; amphibolite, schist, calc-silicate skarns	1st; 2nd
Mhst	alkali basalts	1st
Hst	amphibolite, schist, granite, gneiss	1st

The presence of pyroxene and olivine mineral grains in the deeper areas of continental shelf around Porto and Aveiro canyons areas is not compatible with the felsic igneous and metamorphic rock outcroppings in the Northern Portuguese river basins, since these mineral grains are genetically linked to basic igneous rocks such as basalt, gabbro, or dolerite [55], whose presence is not known in these river basin areas (Figure 2 and Table 1). The first reference to these mineral grains in the Northern Portuguese continental shelf describes the occurrence of "augite" and "hypersthene" in a restricted area located south of the Porto canyon, at depths greater than 100 m [56]. In this work, a source interpretation was outlined which emphasized the existence of basic igneous rocks near this canyon head area, given the fact that these mineral grains have always fresh and angular surface textures which demonstrates their incompatibility with a multi-cycle source as for example from an ancient shoreline, that is, from a relict or palimpsest continental shelf deposits. The optical identification of pyroxenes (mineral grains #18 to #20; Figure 19) is confirmed by the microprobe analysis of several mineral grains belonging to the two samples sets. The chemical composition of pyroxenes shows the presence of some mineral grains compatible with the diopside-hedenbergite and enstatite-ferrosilite series composition, and with augite composition (Figures 15 and 16). It turns out that the optical identification of olivine (mineral grains #21 and #22; Figure 19) is also confirmed by the microprobe analysis of several mineral grains belonging to samples collected from the continental shelf and from the Porto canyon areas, revealing a high content in forsterite end-member (Figures 17 and 18). At the same time, the existence of a seismic reflection profile complemented by bathymetric data and images of the sea floor captured by a remote operated vehicle (ROV), detected the presence of a geological structure near the Porto canyon head area (Figure 21A) with a probable volcanic origin [13,57]. This structure was described as a rock relief more than 15 m height, standing out from neighboring geological formations and it was recognized in a seismic profile by a very distinctive diffractive hyperbola (Figure 21B,C). Its presence was attributed to a hard rock body (of limestone to dolomitic nature) that stood out from the nesting sedimentary rocks (detrital sediments with evidence of carbonate cement). This rock body seems to be embedded in a fault zone that shows evidence of relative movement between the two adjacent blocks; the WSW block has lowered about 2.5 m relative to the ENE block (Figure 21D). During a ROV dive, it was possible to confirm that the referred rock body was made of dolomitic rocks showing signs of karstic erosion [13]. Indirect evidence for the existence of volcanic rocks was found inside this dolomitic structure due to the presence of an elongated depression (with an approximate N–S direction) that could be a match for a volcanic dyke that is now completely eroded [13].

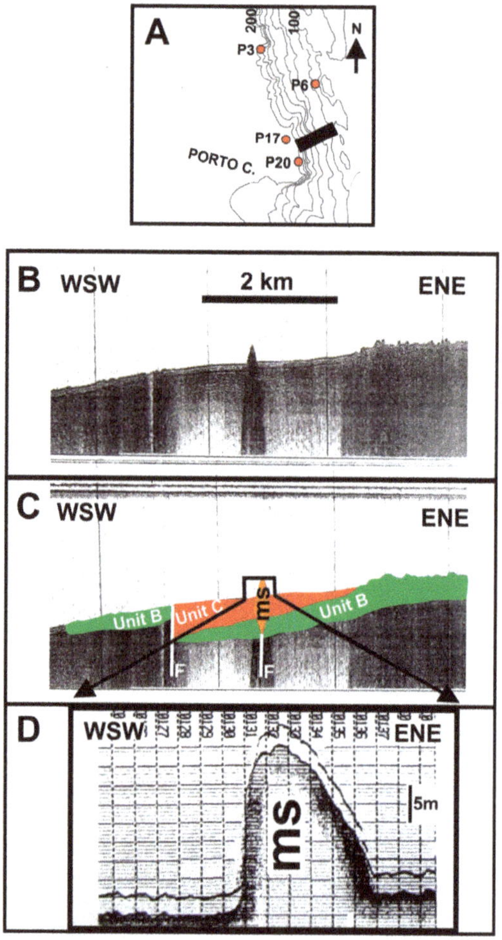

Figure 21. (A–D). Seismic reflection profile obtained at the Porto canyon head latitude. (**A**) Localization of the profile; (**B**) Seismic profile without interpretation; (**C**) interpreted seismic profile with localization of the metamorphic structure (ms); and (**D**) Detail of the bathymetric profile of the metamorphic structure. F corresponds to the interpreted faults. Unit B corresponds to a Mesozoic rock unit. Unit C corresponds to a Cenozoic rock unit (interpreted as dolomitized detrital limestones). Adapted from [13].

All the available data concerning to the optical characteristics and the chemical composition of diopside-hedenbergite, augite, enstatite-ferrosilite, fosrterite, pargasite rich in Ti, and magnesio-hastingsite mineral grains allow to sustain that this mineralogical assemblage is exclusively represented in the areas around the Porto and Aveiro canyons (Figures 13B, 14B and 15, Figures 16–18). Within this mineral assemblage it is possible to distinguish the influence of two different main sources. While augite and enstatite-ferrosilite could be sourced from igneous basic rocks, the presence of diopside-hedenbergite and forsterite may be derived from metasomatic processes resulting from chemical reactions between an igneous basic rock with the nesting sedimentary rocks (limestones and dolomites) [58]. During these processes, the circulation of fluids in the limestone-dolomitic formations and the high temperature inherent to the installation of the volcanic body could explain the formation of these specific minerals, as their presence is referred in other geological contexts associated with the formation of magnesian and/or calcic skarns [59,60]. References to the presence of detrital forsterite with euhedral shapes as it is detected in some of the studied samples (mineral grain #21; Figure 19) are not easy to find in the literature. However, in a different geological context, the presence of euhedral Fo (among other minerals) can be interpreted as a result of re-crystallization processes during contact metamorphism between igneous basic rocks (gabbro) and dolomitic limestones [61]. Therefore, with all due precautions and based on the existing data, it is possible to consider that the presence of this specific mineral assemblage (diopside-hedenbergite and forsterite) is probably related with metasomatic process derived from the thermal contact between basic igneous rocks and dolomitic limestones (possible formation of calcic and magnesian skarn rocks). Together, these rocks (basic igneous and dolomitic limestones affected by metasomatic processes) can be considered as local sources.

5.2. The Nazaré Canyon Area

For the specific case of the Nazaré area, the dominant presence of biotite, followed by amphibole, tourmaline, andalusite, and garnet (Figures 4N and 7) calls for a specific interpretation from the point of view of mineral source and physical grain sorting. According to available data, the high frequency of biotite (>40%—Figure 7) only occurs when the sediment has a mean grain size higher than 2.5 ϕ, that is, when the presence of fine sand to very coarse silt is dominant (Table S1). Moreover, when the biotite frequency is extremely high (biotite > 50%) the sediment mean grain size is equal or higher than 2.5 ϕ (Figure 22). This is consistent with the fact that mica flakes are preferentially concentrated in the coarser part of the sediment tail because of their lamellar shape, that is, they are hydraulically equivalent to finer-grained sediments [62]. Considering that the most proximal sources of heavy minerals are depleted in biotite as it can be observed in the continental shelf sector S5 and in Lis, Alcoa and Tornada river sediments (Figure 3), this high concentration of biotite could result from hydrodynamic fractionation (mineral grain sorting). This can only be understood in the context of a long transport path from a distal source (Northern Portuguese river basins) to the main depocenters located on this canyon head area (Nazaré). This interpretation is supported by the heavy mineral composition of the Minho to Douro river sediments where biotite is, by far, the most important heavy mineral (Figure 3). Thus, the biotite sorting may occur in several steps. The first sorting affects the original source (the Portuguese northern river input) when sand particles are selectively transported from the river into the inner shelf domain [12]. The second step occurs during the inner shelf southward transport of fine sand [63] and, finally, the third step happens when only the finer (and lamellar) sand particles are captured and temporary deposited on the canyon upper head valleys according to the complex oceanographic processes that take place in this area [15,16,63,64]. Given that biotite flakes are hydraulically equivalent to fine grained sediments, their resuspension could happen during the typical oceanographic regimes that affected the Nazaré canyon area [16]. During summer sediments are laterally transported in suspension into the canyon during the upwelling regime, and the resuspension of fine sediments present in the mid-shelf deposits happens due to the internal wave's activity. Additionally, during this regime, the sediments transported through the north-south littoral drift are captured in the head of the canyon. During the winter regime fine sediments captured by the canyon are essentially sourced from southern rivers and from southern continental shelf. When these sediments are sourced from the continental shelf they are resuspended and transported by the combined effect of the waves and the poleward current that is established during the downwelling regime [14–16]. Thus, the high concentration of this mineral found in most of the Nazaré samples is related to how easy it is for biotite to be transported in suspension due to its lamellar shape, which is a characteristic that makes it hydraulically equivalent to finer sedimentary particles. This interpretation agrees with the knowledge of the hydraulic behavior of mica flakes in sand sediments known since the 1960s [65–69].

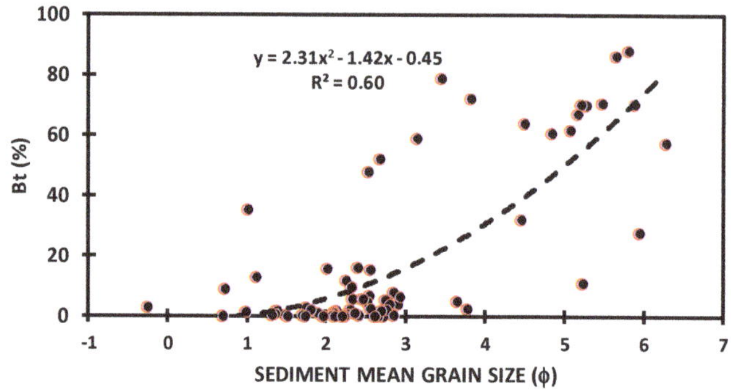

Figure 22. Correlation between biotite frequency and sediment mean grain size using a polynomial function of degree 2.

5.3. The Meaning of the PCA Results

It is possible to observe in the scatter diagram of Figure 9 regarding the application of PCA that there is a clear separation of the samples according to the area from where they were collected. The comparation of the samples position in the scatter diagram (Figure 9) with their respective geographical location visible in Figures 5–7 is quite straightforward. For example, concerning the Aveiro area, samples that are further away from the center of the diagram of Figure 9 (A18–A26) are present in the extreme SW of the sampled area (Figure 6). They also have in common the high frequency of pyroxene (Figures 6 and 9). For Porto samples, no correlation between the way the samples are placed on the scatter diagram (Figure 9) and their respective geographical position exist (Figure 5). However, samples P21, P22, and P29 can be considered exceptions as they were collected from the SW part of the sampled area (Figure 5). These samples together with the correspondent ones collected from the Aveiro area have in common the relative high frequency of pyroxene. Thus, the influence of the igneous basic local source seems to be more distinct on both most southwestern samples collected from Porto and Aveiro canyon areas. Regarding Nazaré, the samples that are further away from the center of the Figure 9 diagram (N2, N3, N5, N8, N9, N10, N15, N17, N18, N19, N20, and N21) have in common the high frequency of biotite. The position of these samples will then correspond to the sites where sediment resuspension phenomena is more frequent.

5.4. The Interpretation of the Heavy Mineral Grain Surface Morphologies

The presence of mineral grains with contrasting surface morphologies, from the most angular to the most rounded ones (Figure 19), could be indicative of potential sources diversity, transport pathways, and sediment deposit's nature. Previous studies dealing with the presence of heavy minerals in the Northern Portuguese continental shelf have considered that the mineral grains with "predominantly rounded to sub-rounded forms" suggest a "polycyclic origin or a long exposure to dynamic processes prior to deposition" [12]. These mineral grains make a contrast with the presence of more angular mineral ones that are believed to be delivered more directly from primary sources (felsic igneous and metamorphic rocks) [12]. The presence of rounded mineral grains is more common in the outer shelf at depths less than 120 m in the three studied areas, and their presence suggests a "multi-cycle" sedimentary origin compatible with long exposure to dynamic processes, which are typical of high energy environments such as, for example, beach environments where the intense grain abrasion is frequent [70]. Although most of these rounded mineral grains were found at the referred depths, some of them are found at the shelf break and in the upper slope (depths >140 m), which could be happening due to of some particle remobilization and transportation into deeper areas. This means that the presence of such mineral grains could be suggestive of a source corresponding related with reworked sedimentary shelf deposits, namely the medium to coarse sand deposits present on this continental margin at depths between 60 and 100 m genetically linked to ancient littorals (relict sediments) [63]. The presence of angular mineral grains among the main mineral suite is very frequent at the outer shelf, shelf break and upper continental slope (depths > 120 m). The presence of such particles could result from the selective transport of river borne terrigenous particles into deeper areas of the continental shelf/upper slope domains without a long residence in higher energy environments (such as the inner shelf or littoral zones). Thus, the sedimentary history of these angular mineral grains may coincide with the one deduced from the quartz immature grains found at the shelf break [63]. It is also possible to consider the hypothesis that some angular heavy minerals were sourced and transported into the continental shelf deeper areas/upper slope during periods of lower sea levels. During these periods, they would have been directly transported into these areas without a long exposure to the intense dynamic processes that characterize shallow water environments (inner shelf and littoral zones). This is likely to have happened during the Last Glacial Maximum period (18,000 BP) when the environmental conditions allowed the transportation of large numbers of terrigenous particles by rivers into the continental slope [71,72]. This interpretation could explain the morphological similarities between the mineral grains identified in the Porto, Aveiro, and Nazaré areas and the ones observed either on the inner continental shelf or in the Northern Portuguese river sediments [12,33,34]. As such, it can be said that the main heavy mineral assemblage (amphibole, biotite, andalusite, tourmaline, garnet, staurolite, zircon, and apatite) includes the presence of "first-cycle" specimens recognized by their angular to very angular grain surface textures. Additionally, this heavy mineral assemblage includes the presence of "multi-cycle" mineral grains identified by their rounded to very rounded surface textures (Figure 19). All these mineral grains have a distal source either represented by the felsic igneous and metamorphic rocks of the NW Iberian Massif (angular ones) or by the reworked relict sediments from the continental shelf (rounded ones). The influence of the local source is recognized by the presence of pyroxene (diopside-hefenbergite, augite, enstatite-ferrosilite), amphibole (pargasite), and olivine (forsterite), mineral grains which always appear with angular surface textures and, in some sporadic cases, with euhedral forms. They can also be considered as "first-cycle" mineral grains although delivered from a local source (Figure 19, mineral grains #17–#22).

5.5. Heavy Mineral Source Synthesis

Figure 23 represents the synthesis of the main mineralogical sources. This synthesis considers the influence of distal and local sources. The distal ones are responsible for supplying the most representative mineralogical species: amphibole, andalusite, tourmaline, biotite, garnet, staurolite, zircon, and apatite. These minerals exhibit angular ("first-cycle" mineral grains) or rounded shapes ("multi-cycle" mineral grains). Local sources, in turn, are responsible for supplying pyroxene (diopside-hedenbergite, augite, enstatite-ferrosilite), amphibole (pargasite), and olivine (forsterite) mineral grains. These minerals have always angular or, in more sporadic situations, euhedral forms ("first-cycle" mineral grains).

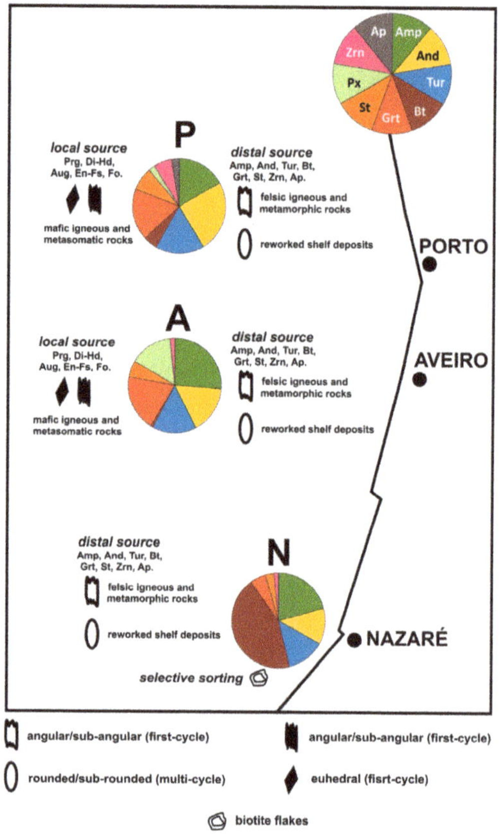

Figure 23. Heavy mineral source synthesis considering the 78 collected samples from the Porto (P), Aveiro (A), and Nazaré (N) areas. Distal and local sources are responsible for two different kinds of supplied minerals. The distal sources are responsible for the presence of angular mineral grains with fresh surface textures, considered as "first-cycle" detrital particles. They are also responsible for the presence of rounded mineral grains that are considered as "multicycle" detrital particles. The felsic igneous and metamorphic rocks of the Iberian Massif and derived terrigenous sediments are the main source of the angular mineral grains and the reworked relict-sedimentary deposits of the continental shelf are the main source of the rounded mineral grains. The mineral grains compatible with these distal sources are: amphiboles (Amp), andalusite (And), tourmaline (Tur), biotite (Bt), garnet (Grt), staurolite (St), zircon (Zrn), and apatite (Ap). Basic igneous rocks and thermal metamorphized dolomitic limestones are the most likely sources of the mineral assemblage derived from the local sources. This mineral assemblage is composed of diopside-hedenbergite (Dp-Hd), augite (Aug), enstatite-ferrosilite (En-Fs), pargasite (Prg) and forsterite (Fo). In the Nazaré area, a high concentration in biotite is observed due to the physical grain sorting of the lamellar mineral grains of this specimen.

6. Conclusions

This study identified the fundamental processes that control the presence of heavy minerals in three distinct areas of the Western Portuguese continental margin: the Porto, Aveiro, and Nazaré canyon head areas. In a broad view, the main heavy mineral assemblage identified in each area is composed by amphibole, andalusite, tourmaline, biotite, garnet, staurolite, pyroxene, zircon, and apatite. However, each studied area has a specific mineral signature that is controlled mainly by the source influence and, in a secondary plan, by the physical mineral grain sorting. In the Porto area, the high frequency of andalusite, amphibole, tourmaline and garnet marked the specificity of the heavy mineral signature. In the Aveiro area, the high frequency of amphibole and pyroxene stood out as a distinctive mineral assemblage. In the Nazaré area, the extreme high frequency of biotite showed the peculiarity of the heavy mineral suite. Together, these results point to the influence of distal sources (erosion of the Iberian Massif rocks) as a fundamental factor in controlling heavy mineral variability. The specific influence of some geological formations of the Iberian Massif can be recognized by the chemical composition of garnets and amphiboles identified all over the Northern Portuguese continental shelf and in the Porto canyon head area. Granites, metamorphic rocks of amphibolitic and granulitic facies, are the most important sources for the identified garnets. However, in the Porto area the source of garnets seems to be limited to the referred metamorphic rocks. Most of the chemical composition of amphiboles is compatible with magnesio-hornblende and for that reason they were sourced from several types of metamorphic rocks (amphibolite and schist). The high frequency of biotite detected at the Nazaré area reflects the peculiar oceanographic setting of this canyon head area and, simultaneously, illustrates what is known as "hydraulic sorting", where the lamellar fine sand-sized biotite particles are concentrated in finer sediments that are preferentially transported together as a suspended load. The peculiar presence of pyroxene and olivine mineral grains at the Porto and Aveiro areas indicates the influence of specific local sources corresponding to basic igneous rocks and dolomitic limestone rocks affected by thermal metamorphism. This hypothesis is supported by the seismic data collected near the Porto canyon head area and by the chemical composition of several mineral grains of the referred species, confirming the presence of diospside-hedenbergite, augite, enstatite-ferrosilite and forsterite. Additionally, the presence of pargasite also supports the existence of an igneous basic source.

Supplementary Materials: The following are available online at http://www.mdpi.com/2075-163X/9/6/355/s1, Table S1. Sediment texture. Table S2. Heavy mineral suite (counts and relative frequencies). Table S3. Garnet chemical composition. Table S4. Amphibole chemical. Table S5. Pyroxene chemical composition. Table S6. Olivine chemical composition.

Funding: Publication supported by FCT—project UID/GEO/50019/2019—Instituto Dom Luiz.

Acknowledgments: The ideas exposed in this manuscript benefit from the fruitful collaboration with Rui Taborda (Faculdade de Ciências, Lisbon University), Aurora Rodrigues, Anabela Oliveira, and Mónica Ribeiro (Portuguese Instituto Hidrográfico). The author is also grateful to Álvaro Pinto (University of Lisbon) to produce the heavy mineral preparations suitable for electron microprobe analysis. The author is also grateful to Joana Reis (Museu Nacional de História Natural e da Ciência, University of Lisbon) for the English revision. Finally, special thanks are due to the Portuguese Instituto Hidrográfico in the scope of the Sedimentos da Plataforma Program (SEPLAT) which provide some of the samples used in this work.

Conflicts of Interest: The author declares no conflict of interest.

References

1. Hazen, R.M. Evolution of minerals. *Sci. Am.* **2010**, *302*, 58–65. [CrossRef] [PubMed]
2. Morton, A.C.; Hallsworth, C.R. Processes controlling the composition of heavy minerals assemblages in sandstones. *Sediment. Geol.* **1999**, *124*, 3–29. [CrossRef]
3. Mange-Rajetzky, M.A. Sediment dispersal from source to shelf on an active continental margin, S. Turkey. *Mar. Geol.* **1983**, *52*, 1–26. [CrossRef]
4. Clemens, K.E.; Komar, P.D. Tracers of Sand Movement on the Oregon Coast. In *Coastal Engineering*; M.ASCE: Reston, VA, USA, 1998; pp. 1338–1351.
5. Komar, P.D. The Entrainment, Transport and Sorting of Heavy Minerals by Waves and Currents. In *Developments in Sedimentology*, 1st ed.; Mange, M.A., Wright, D.T., Eds.; Elsevier: Amsterdam, The Netherlands, 2007; Volume 58, pp. 3–48.
6. Garzanti, E.; Vermeesch, P.; Andò, S.; Lustrino, M.; Padoan, M.; Vezzoli, G. Ultra-long distance littoral transport of Orange sand and provenance of the Skeleton Coast Erg (Namibia). *Mar. Geol.* **2014**, *357*, 25–36. [CrossRef]

7. Emery, K.O. Continental shelf sediments of southern California. *Geol. Soc. Am. Bull.* **1952**, *63*, 1105–1108. [CrossRef]
8. Swift, D.J.; Stanley, D.J.; Curray, J.R. Relict sediments on continental shelves: A reconsideration. *J. Geol.* **1971**, *79*, 322–346. [CrossRef]
9. Cascalho, J.P. Minerais Pesados: Aplicação ao Estudo da Dinâmica Sedimentar da Plataforma Continental Setentrional Portuguesa. Master's Thesis, Lisbon University, Lisbon, Portugal, 1993; 140p.
10. Cascalho, J. Mineralogia dos Sedimentos Arenosos da Margem Continental Setentrional Portuguesa. Ph.D. Thesis, Lisbon University, Lisbon, Portugal, 2000; 400p.
11. Ribeiro, M. *Dinâmica Sedimentar da Cabeceira do Canhão Submarino da Nazaré*; Scientific Report; Lisbon University: Lisbon, Portugal, 2008; 77p.
12. Cascalho, J.; Fradique, C. The Sources and Hydraulic Sorting of Heavy Minerals on the Northern Portuguese Continental Margin. In *Developments in Sedimentology*, 1st ed.; Mange, M.A., Wright, D.T., Eds.; Elsevier: Amsterdam, The Netherlands, 2007; Volume 58, pp. 75–110.
13. Rodrigues, A. Tectono-Estartigrafia da Plataforma Continental Setentrional Portuguesa. Ph.D. Thesis, Lisbon University, Lisbon, Portugal, 2001.
14. Duarte, J.F. Distribuição Espacial dos Sedimentos do Canhão da Nazaré e Plataforma Adjacente. In Proceedings of the Assembleia Luso-Espanhola de Geodesia e Geofísica, Valência, Spain, 4–8 February 2002; Garcia, F.G., Valero, J.L.B., Eds.; Volume Tomo III, pp. 1162–1666.
15. De Stigter, H.C.; Boer, W.; de Jesus Mendes, P.A.; Jesus, C.C.; Thomsen, L.; van den Bergh, G.D.; van Weering, T.C. Recent sediment transport and deposition in the Nazaré Canyon, Portuguese continental margin. *Mar. Geol.* **2007**, *246*, 144–164. [CrossRef]
16. Oliveira, A.; Santos, A.I.; Rodrigues, A.; Vitorino, J. Sedimentary particle distribution and dynamics on the Nazaré canyon system and adjacent shelf (Portugal). *Mar. Geol.* **2007**, *246*, 105–122. [CrossRef]
17. Guerreiro, C.; Oliveira, A.; Rodrigues, A. Shelf-break canyons versus "Gouf" canyons: A comparative study based on the silt-clay mineralogy of bottom sediments from Oporto, Aveiro and Nazaré submarine canyons (NW off Portugal). *J. Coast. Res.* **2009**, *SI 56*, 722–726.
18. Vanney, J.R.; Mougenot, D. La Plate-Forme Continentale du Portugal et les Provinces Adjacentes: Analyse Geomorphologique. *Mem. Serv. Geo. Portugal.* **1981**, *28*, 86.
19. Magalhães, F. A Cobertura Sedimentar da Plataforma Continental Portuguesa. Distribuição Espacial. Contrastes Temporais. Potencialidades Económicas. Ph.D. Thesis, Lisbon University, Lisbon, Portugal, 1999.
20. Pombo, J. Sedimentos Superficiais da Plataforma Continental Portuguesa Entre o Cabo Mondego e S. Martinho do Porto. Master's Thesis, Coimbra University, Coimbra, Portugal, 2004.
21. Rodríguez Fernández, L.R.; López Olmedo, F.; Oliveira, J.T.; Medialdea, T.; Terrinha, P.; Matas, J.; Martín-Serrano, A.; Martín Parra, L.M.; Rubio, F.; Marín, C.; et al. *Mapa Geológico de la Península Ibérica, Baleares y Canarias a escala 1: 1.000. 000*; IGME: Madrid, Spain, 2015.
22. Carvalho, G. *Les Minéraux Lourds et la Paléogéographie des Dépôts Pliocènes du Portugal (Région Entre Vouga et Mondego)*; Museu e Laboratório Mineralógico e Geológico Centro de Estudos Geológicos: Coimbra, Portugal, 1954; Volume 37, pp. 61–64. Available online: http://hdl.handle.net/10316.2/38022 (accessed on 28 May 2019).
23. Teixeira, C. *Notícia explicativa da folha 1-A (Valença) da Carta Geológica de Portugal na escala 1:50 000*; Serviços Geológicos de Portugal: Lisboa, Portugal, 1956; 16p.
24. Costa, C.J.; Teixeira, C. *Notícia explicativa da folha 5-A (Viana do Castelo) da Carta Geológica de Portugal na escala 1:50 000*; Serviços Geológicos de Portugal: Lisboa, Portugal, 1957; 38p.
25. Teixeira, C.; Assunção, C.T. *Notícia explicativa da folha 1-C (Caminha) da Carta Geológica de Portugal na escala 1:50 000*; Serviços Geológicos de Portugal: Lisboa, Portugal, 1961; 41p.
26. Teixeira, C.; Medeiros, A.C.; Assunção, C.T. *Notícia explicativa da folha 9-A (Póvoa de Varzim) da Carta Geológica de Portugal na escala 1:50 000*; Serviços Geológicos de Portugal: Lisboa, Portugal, 1965; 50p.
27. Teixeira, C.; Medeiros, A.C.; Alves, C.A.M.; Moreira, M.M. *Notícia explicativa da folha 5-C (Barcelos) da Carta Geológica de Portugal na escala 1:50 000*; Serviços Geológicos de Portugal: Lisboa, Portugal, 1969; 49p.
28. Teixeira, C.; Medeiros, A.C.; Coelho, A.P. *Notícia explicativa da folha 5-A (Viana do Castelo) da Carta Geológica de Portugal na escala 1:50 000*; Serviços Geológicos de Portugal: Lisboa, Portugal, 1972; 43p.
29. Ribeiro, A.; Antunes, M.T.; Ferreira, M.P.; Rocha, R.B.; Soares, A.F.; Zbyszewski, G.; Almeida, F.M.; Carvalho, D.; Monteiro, J.H. *Introduction à la Géologie Générale du Portugal*; Serviços Geológicos de Portugal: Lisboa, Portugal, 1979.

30. Noronha, F.; Leterrier, J. *Complexo Metamórfico da Foz do Douro. Geoquímica e Geocronologia. Resultados Preliminares. Mem. n. 4*; Museu e Laboratório Mineralógico e Geológico da Faculdade de Ciências da Universidade do Porto: Porto, Portugal, 1995; pp. 769–774.
31. Dinis, P.A.; Soares, A.F. Stable and ultrastable heavy minerals of alluvial to nearshore marine sediments from Central Portugal: Facies related trends. *Sediment. Geol.* **2007**, *201*, 1–20. [CrossRef]
32. Fradique, C.; Cascalho, J. Sedimentary processes in Douro Estuary (Portugal). A heavy mineral study. *Thalassas* **2004**, *20*, 61–68.
33. Fradique, C.; Cascalho, J.; Drago, T.; Rocha, F.; Silveira, T. The meaning of heavy minerals in the recent sedimentary record of the Douro estuary (Portugal). *J. Coast. Res.* **2006**, *SI 39*, 165–169.
34. Fradique, C.; Cascalho, J.; Drago, T. Translucent heavy minerals from the Minho Estuary sedimentary record. In Proceedings of the Costal Hope Conference, Lisbon, Portugal, 24–29 July 2005; Freitas, M.D., Drago, T., Eds.; Faculdade de Ciências da Universidade de Lisboa: Lisboa, Portugal, 2005; pp. 62–63.
35. McManus, J. Grain size determination and interpretation. In *Techniques in Sedimentology*, 1st ed.; Tucker, M., Ed.; Blackell Science: Oxford, UK, 1988; Volume 3, pp. 63–85. ISBN 0-632-01361-3.
36. Galehouse, J.S. Counting grain mounts; number percentage vs. number frequency. *J. Sediment. Res.* **1969**, *39*, 812–815. [CrossRef]
37. Davis, J.C. *Statistics and Data Analysis in Geology*; John Wiley & Sons: New York, NY, USA, 1986; 646p.
38. Morton, A.C. A new approach to provenance studies: Electron microprobe analysis of detrital garnets from Middle Jurassic sandstones of the northern North Sea. *Sedimentology* **1985**, *32*, 553–566. [CrossRef]
39. Haughton, P.D.W.; Farrow, C.M. Compositional variation in Lower Old Red Sandstone detrital garnets from the Midland valley of Scotland and the Anglo-Welsh Basin. *Geol. Mag.* **1989**, *126*, 373–396. [CrossRef]
40. Morton, A.; Hallsworth, C.; Chalton, B. Garnet compositions in Scottish and Norwegian basement terrains: A framework for interpretation of North Sea sandstone provenance. *Mar. Pet. Geol.* **2004**, *21*, 393–410. [CrossRef]
41. Mange, M.A.; Morton, A.C. Geochemistry of heavy minerals. In *Developments in Sedimentology*, 1st ed.; Mange, M.A., Wright, D.T., Eds.; Elsevier: Amsterdam, The Netherlands, 2007; Volume 58, pp. 345–391.
42. Suggate, S.M.; Hall, R. Using detrital garnet compositions to determine provenance: A new compositional database and procedure. *Geol. Soc. Lond. Spec. Publ.* **2014**, *386*, 373–393. [CrossRef]
43. Locock, A.J. An Excel spreadsheet to recast analyses of garnet into end-member components, and a synopsis of the crystal chemistry of natural silicate garnets. *Comput. Geosci.* **2008**, *34*, 1769–1780. [CrossRef]
44. Locock, A.J. An Excel spreadsheet to classify chemical analyses of amphiboles following the IMA 2012 recommendations. *Comput. Geosci.* **2014**, *62*, 1–11. [CrossRef]
45. Hawthorne, F.C.; Oberti, R.; Harlow, G.E.; Maresch, W.V.; Martin, R.F.; Schumacher, J.C.; Welch, M.D. Nomenclature of the amphibole supergroup. *Am. Mineral.* **2012**, *97*, 2031–2048. [CrossRef]
46. Leake, B.E.; Woolley, A.R.; Arps, C.E.; Birch, W.D.; Gilbert, M.C.; Grice, J.D.; Hawthorne, F.C.; Kato, A.; Kisch, H.J.; Krivovichev, V.G.; et al. Nomenclature of amphiboles; report of the Subcommittee on Amphiboles of the International Mineralogical Association Commission on new minerals and mineral names. *Miner. Mag.* **1997**, *61*, 295–310. [CrossRef]
47. GabbroSoft. 2011. Available online: http://www.gabbrosoft.org/ (accessed on 29 May 2019).
48. Morimoto, N. Nomenclature of pyroxenes. *Mineral. Petrol.* **1988**, *39*, 55–76. [CrossRef]
49. Mange, M.A.; Maurer, H. *Heavy Minerals in Colour*, 1st ed.; Chapman: California, CA, USA, 1992; 147p.
50. Andò, S.; Garzanti, E.; Padoan, M.; Limonta, M. Corrosion of heavy minerals during weathering and diagenesis: A catalog for optical analysis. *Sediment. Geol.* **2012**, *280*, 165–178. [CrossRef]
51. Sousa, M.B. Skarns e rochas calco-silicatadas do Complexo xisto-grauváquico do Douro (NE de Portugal)—seu enquadramento litoestratigráfico. *Com. Serv. Geol. Port.* **1981**, *67*, 169–172.
52. Coelho, J.M.D. Os" Skarns" Cálcicos, Pós-Magmáticos, Mineralizados em Scheelite do Distrito Mineiro de Covas, VN de Cerveira, (Norte de Portugal). Ph.D. Thesis, Porto University, Porto, Portugal, 1990; 345p. Available online: http://hdl.handle.net/10216/10227 (accessed on 20 May 2019).
53. Gomes, G.L. Spatial simulation of the W-Sn ore Grades of São Pedro das Águias Skarn Mineral Deposit (Tabuaço, Northern Portugal). Ph.D. Thesis, Porto University, Porto, Portugal, 2016; 102p. Available online: https://run.unl.pt/handle/10362/18446 (accessed on 20 May 2019).
54. Niida, K. Kaersutite, Ti-pargasite, and pargasite from gabbroic rock of the Horoman ultramafic massif, Japan. *J. Jpn. Assoc. Mineral. Petrol. Econ. Geol.* **1977**, *72*, 152–161. [CrossRef]

55. Deer, W.A.; Howie, R.A.; Zussman, J. *An Introduction to the Rock-Forming Minerals*, 2nd ed.; Prentice Hall: Harlow, UK, 1992; 558p.
56. Cascalho, J.; Carvalho, A.M.G. Proveniência dos minerais pesados da plataforma continental portuguesa a norte do paralelo de Espinho. *Gaia* **1993**, *6*, 10–25.
57. Rodrigues, A.; Cascalho, J.; Ribeiro, A.; Dias, J.M.A. *Evidências de actividade vulcânica nas cabeceiras do canhão submarino do Porto. Memórias n. 4*; Museu e Laboratório Mineralógico e Geológico da Faculdade de Ciências da Universidade do Porto: Porto, Portugal, 1995; pp. 305–309.
58. Meinert, L.D. Skarns and skarn deposits. *Geol. Can.* **1992**, *19*, 145–162.
59. Mazurov, M.P.; Grishina, S.N.; Istomin, V.E.; Titov, A.T. Metasomatism and ore formation at contacts of dolerite with saliferous rocks in the sedimentary cover of the southern Siberian platform. *Geol. Ore Dep.* **2007**, *49*, 271–284. [CrossRef]
60. Mazurov, M.P.; Grishina, S.N.; Titov, A.T.; Shikhova, A.V. Evolution of Ore-Forming Metasomatic Processes at Large Skarn Iron Deposits Related to the Traps of the Siberian Platform. *Petrology* **2018**, *26*, 265–279. [CrossRef]
61. Gallien, F.; Abart, R.; Wyhlidal, S. Contact metamorphism and selective metasomatism of the layered Bellerophon Formation in the eastern Monzoni contact aureole, northern Italy. *Mineral. Petrol.* **2007**, *91*, 25–53. [CrossRef]
62. Garzanti, E.; Andò, S.; Vezzoli, G. Settling equivalence of detrital minerals and grain-size dependence of sediment composition. *Earth Plane. Sci. Lett.* **2008**, *273*, 138–151. [CrossRef]
63. Dias, J.M.A.; Nittrouer, C.A. Continental shelf sediments of northern Portugal. *Cont. Shelf Res.* **1984**, *3*, 147–165. [CrossRef]
64. Quaresma, L.S.; Vitorino, J.; Oliveira, A.; da Silva, J. Evidence of sediment resuspension by nonlinear internal waves on the western Portuguese mid-shelf. *Mar. Geol.* **2007**, *246*, 123–143. [CrossRef]
65. Neiheisel, J. Source and distribution of sediments at Brunswick Harbour and vicinity, Georgia. U.S. Army Coastal Engineering. *Cent. Tech. Memo.* **1965**, *12*, 21.
66. Pomerancblum, M. The distribution of heavy minerals and their hydraulic equivalence in sediments of the Mediterranean continental slope of Israel. *J. Sediment. Res.* **1966**, *36*, 162–174. [CrossRef]
67. Doyle, L.J.; Clearly, W.J.; Pilkey, O.H. Mica: Its use in determining shelf depositional regimes. *Mar. Geol.* **1968**, *6*, 381–389. [CrossRef]
68. Doyle, L.J.; Carder, K.L.; Steward, R.G. The hydraulic equivalence of mica. *J. Sediment. Res.* **1983**, *53*, 643–648. [CrossRef]
69. Adegoke, O.S.; Stanley, D.J. Mica and shell as indicators of energy level and depositional regime on the Nigerian shelf. *Mar. Geol.* **1972**, *13*, M61–M66. [CrossRef]
70. Kuenen, P.H. Experimental abrasion: 6. surf action. *Sedimentology* **1964**, *3*, 29–43. [CrossRef]
71. Rodrigues, A.; Magalhães, F.; Dias, J.A. Evolution of the North Portuguese coast in the last 18,000 years. *Quat. Int.* **1991**, *9*, 67–74. [CrossRef]
72. Dias, J.M.A.; Boski, T.; Rodrigues, A.; Magalhães, F. Coast line evolution in Portugal since the Last Glacial Maximum until present—A synthesis. *Mar. Geol.* **2000**, *170*, 177–186. [CrossRef]

MDPI
St. Alban-Anlage 66
4052 Basel
Switzerland
Tel. +41 61 683 77 34
Fax +41 61 302 89 18
www.mdpi.com

Minerals Editorial Office
E-mail: minerals@mdpi.com
www.mdpi.com/journal/minerals

www.ingramcontent.com/pod-product-compliance
Lightning Source LLC
Chambersburg PA
CBHW061800210926
51850CB00006B/608
9783039361304